FOUNDATIONAL *and* APPLIED STATISTICS *for* BIOLOGISTS USING R

FOUNDATIONAL *and* APPLIED STATISTICS *for* BIOLOGISTS USING R

Ken A. Aho
Idaho State University
Pocatello, Idaho, USA

CRC Press
Taylor & Francis Group
Boca Raton London New York

CRC Press is an imprint of the
Taylor & Francis Group, an **Informa** business

A CHAPMAN & HALL BOOK

Cover: View east from the Abiathar Peak in Northeastern Yellowstone National Park. Three mountain goats (*Oreomnos americanus*) are visible at the edge of the summit plateau. Goat weights are approximately normally distributed with mean 90.5 kg and variance 225 kg^2. This distribution is shown in the lower right hand corner of the cover. A random sample from this distribution is shown in the strip at the top. Photo credit: Scott Close. Goat illustration donated by Pearson Scott Foresman (an educational publisher) to Wikimedia Commons, http://commons.wikimedia.org/wiki/ Category:Pearson_Scott_Foresman_publisher; http://all-free-download.com.

CRC Press
Taylor & Francis Group
6000 Broken Sound Parkway NW, Suite 300
Boca Raton, FL 33487-2742

First issued in paperback 2022

© 2013 by Taylor & Francis Group, LLC
CRC Press is an imprint of Taylor & Francis Group, an Informa business

No claim to original U.S. Government works

Version Date: 20150824

ISBN 13: 978-1-03-247741-1 (pbk)
ISBN 13: 978-1-4398-7338-0 (hbk)

DOI: 10.1201/b16126

Library of Congress Cataloging-in-Publication Data

Aho, Ken A.
    Foundational and applied statistics for biologists using R / Ken A. Aho.
        pages cm
    "A CRC title."
    Includes bibliographical references and index.
    ISBN 978-1-4398-7338-0 (hardcover : alk. paper)
    1. Biometry--Textbooks. 2. Statistics--Textbooks. 3. R (Computer program language) I. Title.

QH323.5.A36 2014
570.1'5195--dc23   2013032525

Visit the Taylor & Francis Web site at
http://www.taylorandfrancis.com

and the CRC Press Web site at
http://www.crcpress.com

To JEF and OAA who traveled on this journey

Contents

Preface...xix
Acknowledgments xxi

Section I Foundations

1. Philosophical and Historical Foundations...3
  1.1   Introduction...3
  1.2   Nature of Science 3
  1.3   Scientific Principles..4
        1.3.1   Objectivity..4
        1.3.2   Realism 4
        1.3.3   Communalism..5
  1.4   Scientific Method 5
        1.4.1   A Terse History 6
                1.4.1.1   Experimentation...8
                1.4.1.2   Induction...8
                1.4.1.3   Probability 8
  1.5   Scientific Hypotheses 9
        1.5.1   Falsifiability 9
  1.6   Logic..10
        1.6.1   Induction...10
        1.6.2   Deduction..11
        1.6.3   Induction versus Deduction...11
        1.6.4   Logic and Null Hypothesis Testing 12
                1.6.4.1   Modus Tollens 12
                1.6.4.2   Reductio Ad Absurdum 13
  1.7   Variability and Uncertainty in Investigations 14
  1.8   Science and Statistics..16
  1.9   Statistics and Biology..16
  1.10 Summary..18
  Exercises 18

2. Introduction to Probability...21
  2.1   Introduction: Models for Random Variables.......................................21
        2.1.1   Set Theory Terminology 23
        2.1.2   Philosophical Conceptions of Probability...............................24
  2.2   Classical Probability 26
        2.2.1   Disjoint..28
                2.2.1.1   Boole's Inequality 30
        2.2.2   Independence 31
                2.2.2.1   Bonferroni's Inequality...31
  2.3   Conditional Probability..32

2.4   Odds...34
      2.4.1   Odds Ratio and Relative Risk 34
2.5   Combinatorial Analysis 35
      2.5.1   Multiplication Principle 35
      2.5.2   Permutations...36
      2.5.3   Combinations 37
2.6   Bayes Rule 38
2.7   Summary...43
Exercises 43

3. Probability Density Functions..49
3.1   Introduction...49
      3.1.1   How to Read This Chapter 50
      3.1.2   So, What *Is* Density?..50
      3.1.3   Cumulative Distribution Function.....................................52
3.2   Introductory Examples of pdfs 54
      3.2.1   Discrete pdfs...55
              3.2.1.1   Bernoulli Distribution...55
              3.2.1.2   Binomial Distribution...55
      3.2.2   Continuous pdfs...60
              3.2.2.1   Continuous Uniform Distribution 60
              3.2.2.2   Normal Distribution 61
3.3   Other Important Distributions...65
      3.3.1   Other Discrete pdfs 66
              3.3.1.1   Poisson Distribution..66
              3.3.1.2   Hypergeometric Distribution.................................69
              3.3.1.3   Geometric Distribution...71
              3.3.1.4   Negative Binomial Distribution.............................72
      3.3.2   Other Continuous pdfs 74
              3.3.2.1   Chi-Squared Distribution.......................................74
              3.3.2.2   *t*-Distribution...76
              3.3.2.3   *F*-Distribution..77
              3.3.2.4   Exponential Distribution...79
              3.3.2.5   Beta Distribution 81
              3.3.2.6   Gamma Distribution..82
              3.3.2.7   Weibull Distribution...83
              3.3.2.8   Lognormal Distribution 86
              3.3.2.9   Logistic Distribution...88
3.4   Which pdf to Use?...90
      3.4.1   Empirical cdfs...90
3.5   Reference Tables 91
3.6   Summary...92
Exercises 95

4. Parameters and Statistics 101
4.1   Introduction...101
      4.1.1   How to Read This Chapter 102
4.2   Parameters..103
      4.2.1   Expected Value..103

		4.2.2	Variance	105
		4.2.3	Chebyshev Inequality	106
	4.3	Statistics		106
		4.3.1	Important Considerations	106
		4.3.2	Sampling Error	108
		4.3.3	Gauging Estimator Effectiveness	108
		4.3.4	Types of Estimators	108
		4.3.5	Measures of Location	109
			4.3.5.1 Arithmetic Mean	109
			4.3.5.2 Geometric Mean	110
			4.3.5.3 Harmonic Mean	111
			4.3.5.4 Mode	113
		4.3.6	Robust Measures of Location	113
			4.3.6.1 Median	114
			4.3.6.2 Trimmed Mean	115
			4.3.6.3 Winsorized Mean	115
			4.3.6.4 *M*-Estimators	116
			4.3.6.5 Which Location Estimator to Use?	118
		4.3.7	Measures of Scale	119
			4.3.7.1 Sample Variance	119
			4.3.7.2 Coefficient of Variation	121
		4.3.8	Robust Estimators of Scale	122
			4.3.8.1 Interquartile Range	123
			4.3.8.2 Median Absolute Deviation	123
			4.3.8.3 Which Scale Estimator to Use?	124
		4.3.9	Parameters and Estimators for Distribution Shape	124
			4.3.9.1 Moment Generating Functions	124
			4.3.9.2 Sample Moments and MOM Estimators	126
	4.4	OLS and ML Estimators		127
		4.4.1	Ordinary Least Squares	127
		4.4.2	Maximum Likelihood	128
			4.4.2.1 So, What Is Likelihood?	132
			4.4.2.2 Likelihood versus Probability	132
		4.4.3	MOM versus OLS versus ML Estimation	133
	4.5	Linear Transformations		134
		4.5.1	Transformations and Parameters	135
		4.5.2	Transformations and Statistics	136
	4.6	Bayesian Applications		137
		4.6.1	Priors	138
			4.6.1.1 Noninformative Priors	138
			4.6.1.2 Informative Priors	138
		4.6.2	Conjugacy	138
	4.7	Summary		143
	Exercises			143
5.	Interval Estimation: Sampling Distributions, Resampling Distributions, and Simulation Distributions			149
	5.1	Introduction		149
		5.1.1	How to Read This Chapter	149

5.2   Sampling Distributions 150
       5.2.1   `samp.dist` 151
       5.2.2   Sampling Distribution of $\bar{X}$ 151
               5.2.2.1   Central Limit Theorem 154
       5.2.3   Sampling Distribution of S^2 156
       5.2.4   Sampling Distribution of t^* 156
       5.2.5   Sampling Distribution of F^* 158
5.3   Confidence Intervals 161
       5.3.1   Confidence Interval for μ 161
       5.3.2   Confidence Interval for μ, σ Known 162
               5.3.2.1   Interpreting Confidence Intervals 164
       5.3.3   Confidence Interval for μ, σ Unknown 165
               5.3.3.1   Expressing Precision 165
               5.3.3.2   One-Sided Confidence Intervals 167
       5.3.4   Confidence Interval for σ^2 168
       5.3.5   Confidence Interval for the Population Median 169
       5.3.6   Confidence Intervals and Sample Size 171
       5.3.7   Assumptions and Requirements for Confidence Intervals 172
5.4   Resampling Distributions 172
       5.4.1   Bootstrapping 172
               5.4.1.1   `sample` 173
               5.4.1.2   Bootstrap Confidence Intervals 174
               5.4.1.3   `bootstrap` 178
       5.4.2   Jackknifing 179
               5.4.2.1   `pseudo.v` 181
5.5   Bayesian Applications: Simulation Distributions 182
       5.5.1   Direct Simulation of the Posterior 183
       5.5.2   Indirect Simulation: Markov Chain Monte Carlo Approaches 184
               5.5.2.1   Simple Applications 185
               5.5.2.2   Advanced Applications 188
5.6   Summary 192
Exercises 193

6. Hypothesis Testing 197
6.1   Introduction 197
6.2   Parametric Frequentist Null Hypothesis Testing 197
       6.2.1   Null Hypothesis and Its Motivation 198
       6.2.2   Significance Testing 199
       6.2.3   Models for Null Hypothesis Testing 200
               6.2.3.1   Nonsignificant Results 202
               6.2.3.2   Significant Results 203
       6.2.4   Upper-, Lower-, and Two-Tailed Tests 203
       6.2.5   Inferences for a Single Population Mean 205
               6.2.5.1   One-Sample z-Test 206
               6.2.5.2   One-Sample t-Test 208
       6.2.6   Confidence Intervals and Hypothesis Testing 209
       6.2.7   Inferences for Two Population Means 210
               6.2.7.1   Paired t-Test 211

        6.2.7.2   Pooled Variance *t*-Test 214

        6.2.7.3   Welch's Approximate *t*-Test 217

6.3   Type I and Type II Errors 219

6.4   Power 220

    6.4.1   Sample Adequacy 223

        6.4.1.1   Power in *t*-Tests 224

        6.4.1.2   Effect Size in *t*-Tests 224

6.5   Criticisms of Frequentist Null Hypothesis Testing..................................... 225

    6.5.1   A Final Word 227

6.6   Alternatives to Parametric Null Hypothesis Testing................................... 227

    6.6.1   Permutation Tests.. 227

    6.6.2   Rank-Based Permutation Tests.. 228

        6.6.2.1   Kolmogorov–Smirnov Test... 230

        6.6.2.2   Wilcoxon Sign Rank Test.. 231

        6.6.2.3   Wilcoxon Rank Sum Test.. 233

    6.6.3   Robust Estimator Tests 237

6.7   Alternatives to Null Hypothesis Testing... 237

    6.7.1   Bayesian Approaches 237

    6.7.2   Likelihood-Based Approaches... 240

6.8   Summary... 240

Exercises 241

7. Sampling Design and Experimental Design 247

7.1   Introduction 247

7.2   Some Terminology 247

    7.2.1   Variables 247

        7.2.1.1   Explanatory and Response Variables 248

        7.2.1.2   Categorical, Ordinal, and Quantitative Variables................... 249

        7.2.1.3   Discrete and Continuous Variables 250

        7.2.1.4   Univariate and Multivariate Analysis...................................... 250

        7.2.1.5   Lurking Variables and Confounding Variables 251

7.3   The Question Is: What Is the Question? 253

    7.3.1   Four Types of Questions 253

7.4   Two Important Tenets: Randomization and Replication............................. 254

    7.4.1   Randomization... 254

    7.4.2   Replication 255

7.5   Sampling Design... 256

    7.5.1   Randomized Designs 256

        7.5.1.1   Simple Random Sampling... 256

        7.5.1.2   Stratified Random Sampling... 256

        7.5.1.3   Cluster Sampling 256

    7.5.2   Other Designs... 256

    7.5.3   Comparison of Designs... 257

    7.5.4   Adjustments to Estimators to Accounting for Sampling 258

        7.5.4.1   Finite Population Correction 258

        7.5.4.2   Adjustments for Sampling Design... 259

    7.5.5   Lack of Independence in Samples 262

        7.5.5.1   Time Series Models 262

        7.5.5.2   Psuedoreplication 263

        7.5.6   Other General Sampling Concerns 265
                7.5.6.1   Outliers 265
                7.5.6.2   Measurement Error and Precision 266
                7.5.6.3   Bias 266
                7.5.6.4   Missing Data and Nonresponse Bias 266
                7.5.6.5   Transforming Data 267
                7.5.6.6   Altering Datasets 269
    7.6   Experimental Design 269
        7.6.1   General Approaches 269
                7.6.1.1   Manipulative Experiments versus Observational Studies 269
                7.6.1.2   Randomized versus Nonrandomized Experiments 270
                7.6.1.3   Controls 271
                7.6.1.4   Measuring Appropriate Covariates 272
                7.6.1.5   Prospective and Retrospective Studies 272
        7.6.2   Summary: Inference in the Context of Both Experimental
                and Sampling Design 273
        7.6.3   Classification of Experimental Designs 273
                7.6.3.1   General Linear Models 274
                7.6.3.2   Generalized Linear Models 274
        7.6.4   Regression Designs 275
                7.6.4.1   Simple Linear Regression 275
                7.6.4.2   Multiple Regression 275
        7.6.5   ANOVA Designs 276
                7.6.5.1   ANOVA Terminology 276
                7.6.5.2   Fixed and Random Effects 277
                7.6.5.3   A Compendium of ANOVA Designs 278
        7.6.6   Tabular Designs 286
                7.6.6.1   Three Tabular Designs 287
    7.7   Summary 287
    Exercises 288

Section II   Applications

8. Correlation 295
    8.1   Introduction 295
    8.2   Pearson's Correlation 296
        8.2.1   Association and Independence 297
        8.2.2   Bivariate Normal Distribution 299
        8.2.3   Estimation of ρ 300
        8.2.4   Hypothesis Tests for ρ 302
        8.2.5   Confidence Interval for ρ 306
        8.2.6   Power, Sample Size, and Effect Size 307
    8.3   Robust Correlation 308
        8.3.1   Rank-Based Permutation Approaches 308
                8.3.1.1   Spearman's ρ_s 308
                8.3.1.2   Kendall's τ 312
        8.3.2   Robust Estimator Approaches 315
                8.3.2.1   Winsorized Correlation 315

8.3.2.2 Percentage Bend Criterion, and Sample Biweight
Midvariance 315
8.4 Comparisons of Correlation Procedures 316
8.5 Summary 318
Exercises 318

9. Regression 321
9.1 Introduction 321
9.1.1 How to Read This Chapter 321
9.2 Linear Regression Model 322
9.3 General Linear Models 323
9.3.1 lm 324
9.4 Simple Linear Regression 324
9.4.1 Parameter Estimation 325
9.4.2 Hypothesis Testing 329
9.4.3 An ANOVA Approach 331
9.5 Multiple Regression 333
9.5.1 Parameter Estimation 333
9.5.2 Hypothesis Testing 337
9.5.2.1 Tests Concerning Regression Parameters 337
9.5.2.2 Combined Effect of X on Y 338
9.5.3 An ANOVA Approach 339
9.6 Fitted and Predicted Values 341
9.7 Confidence and Prediction Intervals 343
9.7.1 Confidence Interval for β_k 343
9.7.2 Confidence Intervals for True Fitted Values and Prediction
Intervals for Predicted Values 344
9.8 Coefficient of Determination and Important Variants 346
9.9 Power, Sample Size, and Effect Size 348
9.10 Assumptions and Diagnostics for Linear Regression 348
9.10.1 Independence of Error Terms 350
9.10.2 Normality of Error Terms 352
9.10.3 Constancy of Error Variance 353
9.10.4 The Relationship between X and Y Is Linear 353
9.10.5 Outliers 354
9.10.6 Multicollinearity 355
9.11 Transformation in the Context of Linear Models 359
9.11.1 Optimal Transformation 361
9.11.1.1 Box–Tidwell 361
9.11.1.2 Box–Cox 361
9.12 Fixing the Y-Intercept 364
9.13 Weighted Least Squares 364
9.14 Polynomial Regression 365
9.15 Comparing Model Slopes 368
9.16 Likelihood and General Linear Models 369
9.17 Model Selection 371
9.17.1 Model Selection Approaches 372
9.17.1.1 AIC 372
9.17.1.2 AICc 374

9.17.1.3 BIC 374
9.17.1.4 Mallows' C_p 374
9.17.1.5 PRESS 375
9.17.1.6 Comparison of Selection Criteria 375
9.17.2 Stepwise and All Possible Subset Procedures 377
9.17.2.1 Data Dredging 378
9.18 Robust Regression 378
9.18.1 Bootstrapping Methods 378
9.18.2 Robust Estimators 379
9.19 Model II Regression (*X* Not Fixed) 381
9.19.1 MA Regression 382
9.19.2 SMA Regression 383
9.19.3 RMA regression 384
9.19.4 Assumptions and Additional Comments 385
9.20 Generalized Linear Models 386
9.20.1 Model Estimation 387
9.20.2 glm 387
9.20.3 Binomial GLMs 387
9.20.4 Deviance 389
9.20.5 GLM Inferential Methods 391
9.20.5.1 Wald Test 391
9.20.5.2 Likelihood Ratio Test 391
9.20.5.3 ROC and AUC 392
9.20.5.4 Measures of Explained Variance 394
9.20.5.5 Information-Theoretic Criteria 394
9.20.6 Poisson GLMs 395
9.20.7 Assumptions and Diagnostics 396
9.20.7.1 GLM Residuals 396
9.20.7.2 Model Goodness of Fit 397
9.20.7.3 Dispersion, Overdispersion, and Quasi-Likelihood 399
9.20.7.4 Fitted Probabilities Numerically 0 or 1 400
9.21 Nonlinear Models 400
9.21.1 nls 401
9.21.2 Model Examples 401
9.21.3 Assumptions and Additional Comments 402
9.22 Smoother Approaches to Association and Regression 403
9.22.1 LOWESS 404
9.22.2 Kernel-Based Approaches 404
9.22.3 Splines 405
9.22.4 Generalized Additive Models 408
9.22.4.1 gam 409
9.22.4.2 Assumptions and Additional Comments 410
9.23 Bayesian Approaches to Regression 411
9.24 Summary 413
Exercises 414

10. ANOVA 421
10.1 Introduction 421
10.1.1 How to Read This Chapter 422

10.1.2 Notation...422
10.2 One-Way ANOVA 423
10.2.1 lm Revisited 426
10.3 Inferences for Factor Levels 429
10.3.1 Introduction to Contrasts...429
10.3.2 Orthogonality...431
10.3.3 lm contrast...432
10.3.3.1 Treatment Contrasts..432
10.3.3.2 Helmert Contrasts 433
10.3.3.3 Sum Contrasts..435
10.3.4 Issues with Multiple Comparisons.................................436
10.3.5 Simultaneous Inference Procedures 436
10.3.5.1 Bonferroni Correction.......................................437
10.3.5.2 Least Significant Difference 437
10.3.5.3 Scheffé's Procedure 437
10.3.5.4 Tukey–Kramer Method......................................438
10.3.5.5 Dunnett's Method 439
10.3.5.6 Comparing Simultaneous Inference Procedures....439
10.3.6 General Methods for P-Value Adjustment 441
10.3.7 Depicting Factor-Level Comparisons 442
10.4 ANOVA as a General Linear Model..442
10.5 Random Effects 445
10.5.1 Variance Components 447
10.5.2 Hypothesis Testing 449
10.5.2.1 Likelihood Ratio Test...449
10.5.3 lme and lmer...450
10.6 Power, Sample Size, and Effect Size 453
10.7 ANOVA Diagnostics and Assumptions 454
10.7.1 Independence of Error Terms 454
10.7.2 Normality of Error Terms...456
10.7.3 Constancy of Error Variance 456
10.7.4 Outliers...457
10.7.5 Random and Mixed Effect Models..................................457
10.8 Two-Way Factorial Design 459
10.8.1 Random Effects 463
10.9 Randomized Block Design 466
10.9.1 Random Block Effects 469
10.10 Nested Design 471
10.11 Split-Plot Design..474
10.11.1 aov 475
10.12 Repeated Measures Design 478
10.12.1 Analysis of Contrasts 479
10.12.2 Modern Mixed-Model Approaches.................................482
10.13 ANCOVA..483
10.14 Unbalanced Designs...488
10.14.1 Type II and III Sums of Squares......................................488
10.15 Robust ANOVA 492
10.15.1 Permutation Tests..493
10.15.2 Rank-Based Permutation Tests..493

10.15.3 Robust Estimator Tests..493
10.15.4 One-Way ANOVA..493
    10.15.4.1 Permutation Tests 493
    10.15.4.2 Rank-Based Permutation Tests 493
    10.15.4.3 Robust Estimator Tests 494
10.15.5 Multiway ANOVA 494
    10.15.5.1 Permutation Tests 494
    10.15.5.2 Rank-Based Permutation Tests 494
    10.15.5.3 Robust Estimator Tests 495
10.16 Bayesian Approaches to ANOVA 495
10.17 Summary..496
Exercises 497

11. Tabular Analyses...503
11.1 Introduction...503
    11.1.1 How to Read This Chapter...503
11.2 Probability Distributions for Tabular Analyses 504
11.3 One-Way Formats...506
    11.3.1 Score Test..507
        11.3.1.1 Tests for $\pi = \pi_0$...507
        11.3.1.2 Tests for All $\pi_i = \pi_{i0}$...508
        11.3.1.3 Tests for $\pi_1 = \pi_2 = \cdots = \pi_c$ (Includes Two-Proportion
             Test: $\pi_1 = \pi_2$)..508
    11.3.2 Wald Test..509
        11.3.2.1 Tests for $\pi = \pi_0$...509
        11.3.2.2 Tests for $\pi_1 = \pi_2$ (Two-Proportion Test).....................510
    11.3.3 Likelihood Ratio Test 511
        11.3.3.1 Tests for $\pi = \pi_0$...511
        11.3.3.2 Tests for All $\pi_i = \pi_{i0}$ (Includes Two-Proportion Test) 512
    11.3.4 Comparisons of Methods and Requirements for Valid Inference........512
        11.3.4.1 Tests for $\pi = \pi_0$...512
        11.3.4.2 Tests for All $\pi_i = \pi_{i0}$...513
        11.3.4.3 Tests for $\pi_1 = \pi_2$ (Two-Proportion Test).....................513
11.4 Confidence Intervals for π...513
    11.4.1 Wald Method..513
    11.4.2 Score Method..514
    11.4.3 Agresti–Coull Method 514
    11.4.4 Clopper–Pearson Exact Method 514
    11.4.5 Likelihood Ratio Method..514
    11.4.6 Comparison of Methods and Requirements for Valid Inference..........516
    11.4.7 Inference for the Ratio of Two Binomial Proportions....................516
    11.4.8 Inference for Odds Ratios and Relative Risk 517
        11.4.8.1 Odds Ratio..517
        11.4.8.2 Relative Risk..518
11.5 Contingency Tables..519
11.6 Two-Way Tables..520
    11.6.1 Chi-Squared (Score) Test..521
    11.6.2 Likelihood Ratio Test 521
    11.6.3 Fisher's Exact Test 522

11.6.4 Comparison of Methods and Requirements for Valid Inference..........524
11.7 Ordinal Variables 525
11.8 Power, Sample Size, and Effect Size 526
11.9 Three-Way Tables...527
    11.9.1 Comparison of Methods and Requirements for Valid Inference..........530
11.10 Generalized Linear Models..531
    11.10.1 Binomial GLMs 531
    11.10.2 Log-Linear Models...532
        11.10.2.1 Two-Way Tables 533
        11.10.2.2 Three-Way Tables 533
        11.10.2.3 Four-Way Tables...537
    11.10.3 Comparison of Methods and Requirements for Valid Inference..........537
11.11 Summary...537
Exercises 538

Appendix...541

References 555

Index...575

Preface

Statistical texts and classes within biology curricula generally ignore or fail to instill foundational concepts.[*] This deficiency is not surprising, as these topics are not easy to learn or teach. For example, logic provides a framework for statistical inference, and knowledge of algebra and calculus is required to understand the probability-generating functions used in all popular analyses. Unfortunately, this problem has been exacerbated by advances in statistical software. These tools do not require any knowledge of foundational principles. However, a poor understanding of the theory underlying the algorithms often leads to misapplication of analyses, misunderstanding of results, and invalid inferences.[†]

In this book, I have attempted to bridge the gap between statistical foundations and the myriad statistical applications available to twenty-first century biometrists. I do this by harnessing inherent properties of the **R** computational environment (**R**-core development team 2012) that allow students to actively construct their own sense of what is being learned (*sensu* constructivism, Mayer 1999). Unlike its commercial counterparts, **R** does not lock away its statistical algorithms in black boxes. Instead, one can examine the code of complex procedures step by step to elucidate the process of obtaining the results shown in textbooks. In addition, the outstanding graphical capabilities of **R** allow interactive demonstrations to aid in the instruction of complex statistical concepts. A companion library to this book called *asbio* (Aho 2013) has been crafted specifically for this purpose. A novice-to-command-line interfaces can install and load *asbio*, and can access the central "point and click" book menu by typing just three commands into **R**:

```
install.packages("asbio")
library(asbio)
book.menu()
```

Subsequent access to the menu during new work sessions requires only the last two commands.

All the analyses in this book and all but a handful of its 130 figures were obtained using **R**. The code for the implementation of these procedures is provided either directly in the text or from auxiliary chapter files at the book website (http://www2.cose.isu.edu/~ahoken/book/). An accompanying document "Introduction to **R**" can also be found at this site.

How to Read This Book

Under the premise that strong foundations lead to a deeper understanding of complex applications, I have, in most chapters, juxtaposed sections with introductory and advanced materials. For chapters with such a format (Chapters 3, 4, 5, 9, 10, and 11), I indicate the

---

[*] Among other citations, see Hurlbert (1984), Altman (1991, 1994), Horgan et al. (1991), Yoccoz (1991), Hand (1994), Coste et al. (1995), Finney (1995), Heffner et al. (1996), Feinstein (1997), Bialek and Botstein (2004), Mehta (2004), and Vaux (2012).

[†] See Mehta et al. (2004), VanDongen (2006), Prosser (2010), and Sainani (2011).

location of advanced sections in an initial "how to read this chapter" section. Advanced topics are also specified with a star symbol (★) throughout the book. I have structured the book so that introductory classes can skip "starred" materials with little loss of continuity.

Because of its structure, both introductory and graduate-level biostatistics courses can be taught using this book. To cover materials adequately and decrease frustration, introductory courses should avoid advanced Bayesian topics in Chapters 3, 4, and 5, avoid multifactor ANOVA designs in Chapter 7, cover only simple linear and multiple regression in Chapter 9, cover only one-way ANOVA topics in Chapter 10, and cover only materials up to two-way tables in Chapter 11. A large number of graduate level topics are addressed throughout the book, beginning in Chapter 3. These can be emphasized to varying degrees depending on the interest and needs of students. I assume in all sections of all chapters that students have a familiarity with algebra and general calculus. Linear algebra applications used in Chapters 9 and 10 are described in the book Appendix along with additional mathematical reference materials.

Acknowledgments

I thank Techset Composition for their hard work, and David Grubbs and Ed Curtis from the statistics group at Taylor & Francis. I am especially grateful to David for his unwavering encouragement, diligence, and patience throughout this long process. Useful comments and contributions have been made by a large number of individuals. In alphabetical order, these include: Jim Aho, Kevin Aho, Colden Baxter, Marc Benson, Robert Boik, Terry Bowyer, Dewayne Derryberry, Robert Fisher, Jessica Fultz, Ernest Keeley, John Kie, Al Parker, Teri Peterson, Richard Remington, Yasha Rohwer, and Mike Thomas. Many thanks to Hank Stevens and Andrew Dalby who skillfully reviewed the first seven chapters and the last four (longer) chapters, respectively. A debt of gratitude is also owed to Vern Winston for helping to clear Unix/Linux hurdles for *asbio*, to David Roberts for his expertise in statistics, biology and **R**, and to Kenneth Burnham and David Anderson for tracking down an important quotation from their book on multimodel inference. Finally, I thank the students in BIOL 3316 and BIOL 5599 at Idaho State University, particularly Andrew Caroll, Yevgeniy Ptukhin, and Chan Sang. Of course, despite all this help, errors in the book and gaps in clarity are solely attributable to the author (me). Please address comments and corrections to ahoken@isu.edu.

Section I

Foundations

1

Philosophical and Historical Foundations

The most beautiful thing we can experience is the mysterious. It is the source of all true art and all science. He to whom this emotion is a stranger, who can no longer pause to wonder and stand rapt in awe, is as good as dead: his eyes are closed.

Albert Einstein (1931)

1.1 Introduction

Before beginning a book on statistics for biologists, it will be helpful to take a general view of the concepts that underlie science. It seems reasonable that biologists should be able to understand, to some degree, the general characteristics and goals of science. Further, as scientists using statistics, we should be able to understand the way these goals are addressed by statistical tools. We should bear these ideas in mind as we design our experiments, collect our data, and conduct our analyses.

1.2 Nature of Science

Based on its etymology, *science* (from the Latin *scire* meaning "to know") represents a way of "knowing" about our universe. This conceptualization, however, is still vague. What do we mean by "knowing," and what do we mean by "a way of knowing"; is there more than one way of knowing something? Such philosophical questions go beyond the scope of a book on statistics. Nonetheless, statistics can be considered a practical response to these ideas. For instance, probability, the language of statistics, can be used to quantify both the "believability" of a scientific proposition, and the amount of knowledge/evidence we have concerning a phenomenon.

Because attempts at an all-encompassing definition of science gives rise to still other questions, many authors emphasize only the methodological aspects of science. For example, standard dictionary definitions stress procedures for the accumulation of knowledge. These include "a systematic and formulated knowledge,"[*] "systematized knowledge derived from observation, study, and so on,"[†] and "knowledge covering general truths of the operation of general laws, concerned with the physical world."[‡] Thus, science is concerned with knowledge: a special kind of technical knowledge that may be

---

[*] *The Oxford Pocket American Dictionary of Current English.*
[†] *Webster's New World Dictionary.*
[‡] *Merriam Webster's New Collegiate Dictionary,* 11th edition.

3

intrinsically trustworthy because its methods emphasize precision and objectivity, and useful, because it describes the everyday phenomena of the universe.

Out of necessity, this chapter also explores science methodologically. I briefly address modes of scientific description, the structure and history of the scientific method, the relationship of science to logic, and the correspondence of statistics to the goals and characteristics of empirical science. This approach, however, ignores several fundamental issues. For instance, it does not explain why humans are driven to *do* science. To this end, it has been suggested that science satisfies a basic human need (cf. Tauber 2009). Through its program of discovery, science may provide meaning—a personal context in a societal setting, the biosphere, and the universe.

1.3 Scientific Principles

Science can be loosely subset into the *empirical sciences*, which are concerned with suppositions that are empirically supported (e.g., biology) and the *nonempirical sciences* whose propositions are supported without empirical evidence (e.g., logic and pure mathematics).[*] More than any other enterprise, the field of statistics bridges these branches by quantifying the character of empirical data with mathematical tools.

Scientists have developed specific procedures for observing and considering phenomena, along with unique methods for description (i.e., scientific pictures[†]) that distill technical knowledge into practical forms. Three principles guide these processes: objectivity, realism, and communalism.

1.3.1 Objectivity

The process of science emphasizes *objectivity*. That is, an attempt is made to study and describe phenomena in a way that does not depend on the investigator (Gaukroger 2001). Objectivity has been a primary motivator for the development of the scientific method (Section 1.4) and the statistical tools described in this book, since both seek to provide rigorous and impartial methods for describing data and making inferences (Hempel 1966, p. 11). To facilitate objectivity, a number of procedural steps are often taken by scientists. For example, methods and tools that are not biased (predisposed to certain answers at the expense of other equally valid or more valid answers) should be used for measurements and descriptions. With respect to sampling and experimental design, randomization can be used to select samples representing a phenomenon, and to assign these samples to experimental treatments (see Chapter 7). These steps will tend to decrease investigator bias, and balance out the confounding effect of unmeasured variables.

1.3.2 Realism

Scientists also strive for *realism* (recognizable depictions) in their research. The doctrine of *scientific realism* holds that (1) products of science exist independently of phenomena

---

[*] The empirical sciences can be further subset into the *natural sciences*, including biology, chemistry and physics, and the *social sciences* encompassing anthropocentric topics like sociology, anthropology, and history.

[†] For discussion and comparisons of art and science, see Stent (1972), Hofstadter (1979), and Cech (1991).

```
see.accPrec.tck()
```

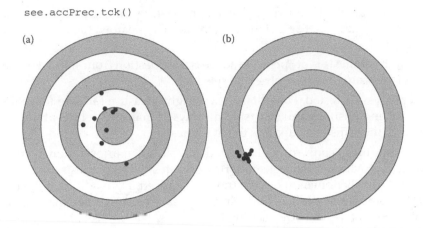

FIGURE 1.1
The concepts of accuracy and precision can be considered interactively using the **R** function `see.accPrec.tck` in **asbio**: (a) high accuracy but low precision; (b) high precision but low accuracy.

being studied, and (2) scientific theories that have been repeatedly corroborated represent truth-directed descriptions of reality (Boyd 1983, Popper 1983, Psilos 1999). Thus, one measure of realism in science is repeatability (i.e., verifiability, *sensu* Hempel 1990). To facilitate repeatability, so that others may substantiate or disprove their results, scientists (1) carefully document their methods of research (including their methods of analysis), and (2) use methods of measurement that are both *accurate* (i.e., give readings close to a "true" reference value) and *precise* (i.e., give very similar measures if repeated for the same object) (see Figure 1.1).

1.3.3 Communalism

The growth of science depends on the refinement, corroboration, and refutation of scientific claims. Because these actions require a community of participants, scientific knowledge can be considered a *communal* process (cf. Shapin and Schaffer 1985). Each scientist is a member of this community, and a contributor to a picture of the universe that has morphed and grown since the dawn of knowledge.

1.4 Scientific Method

While it is helpful to describe general characteristics of scientific description (objectivity, realism, communalism), these are insufficient as a guide for conducting scientific research. Such a framework, however, is provided by the *scientific method*. The process is often presented in general biology texts[*] as four progressive steps:

---

[*] For example, Freeman (2010) and Reece et al. (2011).

1. Observations about the universe and/or preexisting information lead to the formulation of a question. For instance, "Why do male peafowl (peacocks) have elaborate tails?"

2. A testable hypothesis is proposed to address the question from step 1. For instance, "Peacock tails are the result of sexual selection."

3. The hypothesis is tested. For instance, we might measure tail size for males with successful and unsuccessful courtship displays. If males with larger tails are more successful, then this test supports the hypothesis.

4. A decision is made concerning the hypothesis based on information from the test. Regardless of whether the test supports or refutes the hypothesis, more information about the phenomenon under study is acquired. Depending on the decision, additional steps can be taken, including disposal, revision, retesting, and application of the hypothesis to other settings.

This general format has existed for thousands of years. However, perspectives on underlying issues have changed dramatically over this period. These include answers to important questions like "what constitutes evidence?"; "what is the role of experimentation?"; "how do we quantify evidence given that experimental outcomes vary?"; and "how do we weigh evidence from past observations with respect to results from current experiments?"

1.4.1 A Terse History

Aristotle (384–322 BC) is often referred to as the father of the modern scientific method because of his support for a version of *empiricism*, the view that knowledge comes from sensory experience (Pedersen 1993). Aristotle supported the idea that *premises* (propositions prompting a specific conclusion) for "true scientific enquiry" could come from observation. He insisted, however, that genuine scientific knowledge could only come from the application of pure reason, undiluted by observation. Aristotle's empiricism, while limited, was a compromise with the views of his teacher Plato (428–348 BC) who thought that empiricism was dangerous, and that knowledge could only come from rational thought and logic.

While not often considered in histories of science, Aristotle and Plato were preceded by Grecian philosophers such as Thales of Miletus (624–c. 546 BC) and Empodocles of Agrigentum (c. 490–430 BC) who espoused both observation and experimentation. These approaches were, in turn, greatly predated by Egyptian records from at least the 17th century BC describing the basic tenets of modern science (Achinstein 2004).* Still earlier, Babylonian and Mesopotamian methods of astronomy from the dawn of written history (ca. 3100 BC) provide a basis for "all subsequent varieties of scientific astronomy ... if not indeed all subsequent endeavors in the exact sciences" (Aaboe 1974).

---

* Histories of science are often Western-centered. For instance, the Persian *Avicenna* (980–1037 AD), a Muslim physician, scientist, and philosopher outlined a sophisticated and modern version of empiricism that emphasized controlled experimentation as opposed to mere mensurative observation. His contributions, which predate the Renaissance, have been largely overlooked in Western histories of science.

TABLE 1.1

Major Modern (Renaissance–Present) Developments in the Scientific Method

Historical Figure/Event
Galileo Galilei (1564–1642) used work from earlier scientists and his own experiments to support the hypothesis of a Copernican heliocentric solar system. Galileo's empiricism was unique for his time. In the late Renaissance, scientific knowledge was mainly the product of axiomatic deduction from fundamental principles or common notions. As a result, Galileo concealed the role of both experimentation and induction in his conclusions.
Francis Bacon (1561–1626), a contemporary of Galileo, helped to formalize the scientific method and was a strong advocate of induction and empiricism in scientific research. Bacon's penchant for inductive reasoning was demonstrated by the importance he placed on written histories of natural and experimental phenomena. These were to comprise part of his so-called *great instauration* or great restoration of science.
Isaac Newton (1642–1727), originator of many profound ideas in math and physics, demonstrated that mathematics and experimentation could be combined in scientific research. Newton's *Opticks* listed 35 experiments, including a famous test demonstrating that color is the result of objects interacting with light rather than color being generated by the objects themselves. Newton's views of the scientific method were important. He favored an approach in which conclusions were reasoned inductively and deductively from experimental evidence without the use of *a priori* hypotheses. Newton was also opposed to the use of "probable opinion" in research because the aim of science is certain knowledge.
A number of Newton's contemporaries did not share his negative opinion of "probable opinion." For instance, *Christian Huygens* (1629–1695) in his *Treatise on Light* recognized that his "demonstrations did not produce as great a certitude as those of geometry" because premises used by geometers are fixed and based on deductive "incontestable principles." Nonetheless, Huygens claimed a high-degree certainty with his results because his conclusions were supported by a "great number" of his observations.
A large number of mathematicians and scientists, including *Pierre de Fermat* (1601–1665), *Abraham de Moivre* (1667–1754), *Blaise Pascal* (1623–1662), *Gottfried Wilhelm Leibniz* (1646–1716), *Simon Pierre de Laplace* (1749–1827), *Jakob Bernoulli* (1654–1705), *Siméon-Denis Poisson* (1781–1840), *Daniel Bernoulli* (1700–1782) helped develop a mathematical method for quantifying the practical certainty of hypotheses. This approach, called probability, allowed one to view evidence for or against hypotheses as a continuum.[a]
John Maynard Keynes (1883–1946) defined new roles for probability in science by drawing on the mathematical ideas of *Thomas Bayes* (1702–1761). In particular, Keynes claimed that the probability for a hypothesis could be viewed as an investigator's degrees of belief in that hypothesis. Keynes also suggested that prior probabilities could be assigned to a hypothesis representing background knowledge. In this *Bayesian* approach, prior probability could be augmented with current evidence in an inductive manner.
Hans Reichenbach (1891–1953) offered a new perspective on probability. Specifically, he argued that the probability of an event comprised its long-run relative frequency. The *frequentist* view fits well with the use of probability in most modern statistical procedures.
Karl Popper (1902–1994) favored the use of deduction in hypothesis evaluation in his *hypothetico-deductive* approach (Section 1.6.4). Popper's position for deduction (and against induction) was due to *David Hume* (1711–1776) who insisted that the logic of induction is not justifiable because it relies on the "uniformity of nature" (Hume 1740, *qtd in* Selby-Bigge 1996). That is, what happened in the past cannot be assumed to be a perfect predictor for what will happen in the future. However, Popper's view of hypothesis corroboration (repeated nonfalsification of hypotheses) has caused many philosophers of science to claim that Popper's approach was "inductive in the relative sense" (Salmon 1967).

[a] For an overview of the history of statistics and probability, see Owen (1974).

A few figures important to the modern (Renaissance to present) development of the scientific method are described in Table 1.1.[*] The table reveals three trends that are explored in greater detail throughout the remainder of the chapter. First, there has been an increased reliance on experimentation as a method for testing hypotheses. Second, a greater emphasis has been placed on a form of logic called induction. Third, since its invention in the 17th

---

[*] A detailed account of the development of the modern scientific method is beyond the scope of this book. Interested readers are directed to Lloyd (1979), Pedersen (1993), Gower (1997), and Achinstein (2004).

century, probability has been used increasingly, and in increasingly sophisticated ways, to quantify evidence concerning scientific claims.

1.4.1.1 Experimentation

To be taken seriously by his contemporaries, Galileo Galilei (1564–1642) concealed the use of experimentation in his work concerning gravity (Table 1.1). Instead, he attempted to emulate the axiomatic approach of Euclid (fl. 300 BC) in which self-evident truths were deduced from "common notions" (Gower 1997). For instance, instead of summarizing data from his experiments involving balls of different weights rolling down an inclined plane, Galileo presented a thought problem in which two objects, one heavy and one light, were dropped simultaneously from the top of a tower (Galilei 1638). Hypothetically, if the heavier object reaches the ground first, because its greater weight results in greater free falling velocity, then tying the lighter object to the heavier object should slow the heavier object down (and accelerate the lighter object). However, if weight determines velocity, as was first assumed, then the conjoined objects should travel faster than the heavier object alone. Because these arguments are in contradiction, Galileo argued that weight does not influence acceleration due to gravity.

The resistance to experimentation, demonstrated by Galileo, was due to the enduring views of Aristotle and Plato, and the fact that earlier methods of measuring phenomena were imprecise and unrepeatable (Gower 1997). As measurement tools and techniques were improved, and the practical benefits of experiments were demonstrated—for instance, through tests that resulted in effective treatments for disease (Hempel 1966, p. 3)—trust in experimentation increased.

1.4.1.2 Induction

With *inductive reasoning*, we draw conclusions concerning a phenomenon based on accumulated evidence, while acknowledging that the evidence is incomplete (induction is fully explained in Section 1.6.1). The resulting conclusions may be considered "probably correct," but not certain, because they are based on imperfect and incomplete observation. This restriction led many early philosophers to disparage the use of induction in science. However, trust in induction has grown with its demonstrated usefulness, and with improvements in methods for recording, storing, and analyzing data. Modern critics of induction, particularly Karl Popper (1902–1994), have themselves been criticized for refusing to acknowledge the merits of induction, while tacitly embracing its tenets.

1.4.1.3 Probability

Two major conceptions of probability have arisen over time. The first, formalized by Hans Riechenbach (1891–1953), views probability as a limiting frequency. That is, the probability of any event is its relative frequency over an infinite number of trials. This view is consonant with the most commonly used approaches in statistics, including point and interval estimation of parameters (Chapters 4 and 5), and null hypothesis testing (Chapters 6 through 11). The second, attributed to Thomas Bayes (1702–1761), views the probability of an event as the degrees of belief in the truth of that event. Mathematical ideas that underlie this perspective and an increasing trust in induction have resulted in useful methods that create probabilistic statements by synthesizing prior empirical knowledge and current knowledge. The importance of probability to science is revisited in Section 1.8. Probability itself is formally defined and described in Chapter 2.

1.5 Scientific Hypotheses

A *hypothesis* is a statement describing the universe that serves as a starting point for a scientific investigation. All branches of science, including biology, are built on systems of hypotheses. We may be most familiar with hypotheses of causality (e.g., *this* causes *that*) or hypotheses that are comparative (e.g., *this* and *that* are different, or *this* is bigger or smaller than *that*). Hypotheses, however, allow other considerations of phenomena. For instance, they are often used to address association (e.g., *this* increases with *that*), or description, (e.g., *this* is a true characteristic of *that*), or to provide a means for answering, which, in a series of models, is the best at explaining a phenomenon (e.g., *this* is the best model we have for *that*).*

Scientific and nonscientific hypotheses can be distinguished using two criteria. First, a scientific hypothesis will be *testable*. In using the term testable, I mean that an experiment can be designed that provides information concerning the validity of the hypothesis. Of course, the scientific method as described in Section 1.4 explicitly requires testability. Second, many scientists have argued that scientific hypotheses should be falsifiable. *Falsifiability*, popularized by Karl Popper, requires a conceptual outcome that would cause a scientific hypothesis to be rejected. That is, "A theory is potentially a scientific theory if and only if there are possible observations that would falsify (refute) it" (Boyd et al. 1991).

1.5.1 Falsifiability

Popper himself did not advocate discarding a strong hypothesis as soon as negating evidence was encountered, although he has been linked to this idea (cf. Okasha 2002).[†,‡] To quote Popper (1959, p. 32): "once a hypothesis has been proposed and tested, and has proved its mettle, it may not be allowed to drop out without a 'good reason.'" That is, extraordinary falsifications require extraordinary proof (Popper 1959, p. 266).[§] Popper did, however, favor severe testing for scientific hypotheses, and argued that no hypothesis should be immune from scrutiny. That is, "…he who decides one day that scientific statements do not call for any further test, and that they can be regarded as finally verified, retires from the game" (Popper 1959, p. 32).

---

* Science (and statistics) is not solely concerned with hypothesis testing. For example, each year, a large number of descriptive scientific papers are published that do not explicitly test hypotheses (although they may serve as material for hypothesis generation). Still, other scientific studies are concerned with the invention or improvement of tools or methods, for example, measuring instruments and technological advances (although they may be later used to quantify evidence concerning scientific hypotheses).

† See Sober (1999, pp. 46–49) for criticisms and comments on falsification in a biological context.

‡ It can be demonstrated that important advances in science have occurred when the evidence against a hypothesis *did not* lead to its immediate falsification. For instance, astronomers in the 19th century found that the orbit of Uranus was poorly described by Newtonian physics. Instead of rejecting Newton's ideas, however, John Couch Adams and Urbain Leverrier independently predicted that an undiscovered planet was the cause of the orbital discrepancies. Neptune was soon discovered in the exact region forecast by the scientists. On the other hand, consider two other scientific breakthroughs, the theory of relativity and quantum mechanics. These developments demonstrate that Newton's physics gives incorrect results when applied to massively large objects, objects moving at high velocities, or subatomic particles. Blind adherence to Newtonian theories (nonfalsifiability) could have prevented these important discoveries.

§ Strong scientific hypotheses that withstand repeated attempts at falsification are often termed *theories* (Ayala et al. 2008).

Two notable philosophers have presented conceptions of science counter to Popper's severe falsification. They were Imre Lakatos (1922–1974) and Thomas Kuhn (1922–1996). Lakatos (1978) argued that severe falsification was at odds with the possibility of scientific progress. As an alternative, he suggested that scientific theories constituting a "research programme" be classified into two groups. The first would consist of a core set of theories that would be rarely challenged, while the second would be comprised of a protective "auxiliary layer" of theories that could be frequently challenged, falsified, and replaced (Mayo 1996, Quinn and Keough 2002). Kuhn (1963) showed that, in general, falsified hypotheses do not result in the rejection of an overall scientific paradigm, but instead serve to augment and develop it. This state of "normal science" is only occasionally disrupted by a "scientific revolution" in which a paradigm is strongly challenged and overthrown. Kuhn has been described as a critic of scientific rationality (Laudan 1977). However, Kuhn (1970) emphasized that his goal was not to disparage science, but to describe how it worked. That is, science does not progress linearly, but often stumbles and lurches eccentrically (Okasha 2002).

The Popperian, Lakatosian, and Kuhnian conceptions of science each have strengths. Popper presented a clear method for research—propose a hypothesis, and then do everything possible to reject it. If the hypothesis withstands severe testing, then it can be considered a provisional characterization of the universe. Kuhn did not propose a method for scientific research as such, but described how science can proceed nonlinearly (Kuhn 1970). Lakatos proposed taking core theories off the table, something that Popper would have disagreed with, but his views of hypothesis testing were more sophisticated than Popper's (Hillborn and Mangel 1997). In particular, Lakatos proposed that multiple competing hypotheses will always be possible for explaining a phenomenon. As a result, an extant hypothesis is never rejected, but only replaced by a better "research programme." This is in contrast to the Popperian severe tests of individual hypotheses.

1.6 Logic

The approach used by scientists to formulate hypotheses and understand the world is rooted in the precepts of logic. There are two modes of logical reasoning: induction and deduction. Both can be summarized using verbal arguments consisting of *premises* and *conclusions*. *Logic* itself can be defined in the context of a these arguments. Specifically, an inductive or deductive statement is said to be valid or *logically correct* if the premises of the argument support the conclusion (Salmon 1963). Conversely, if the premises do not support the conclusion, then the argument is *fallacious*.

1.6.1 Induction

Formalization of the scientific method by Francis Bacon (Section 1.4.1) and John Stuart Mill (1806–1873) emphasized the importance of *induction*. In this framework, empirical or conceptual premises (e.g., observations, existing theory, and personal insights) lead to a conclusion that is *probably true*, if the premises are true (Russell 1912, Keynes 1921). Here is an example of inductive reasoning:

No bacterial cells that have ever been observed have nuclei.	} Premise
Bacteria do not have nuclei.	} Conclusion

Because of their reliance on uncertain or incomplete premises, inductive arguments have three distinguishing characteristics. First, the conclusion will contain information not present, even implicitly, in the premises. Second, statements in the premises will be more specific compared to more general statements in the conclusion. Third, the conclusion may be false even if the premises are true.

1.6.2 Deduction

Other philosophers of science, notably René Descartes (1596–1650), emphasized the importance of *deduction*. Here, one states a conclusion already contained in the premises (cf. Hempel 1966). An example of deductive reasoning is shown here.

Bacterial cells do not have nuclei.	} Premise 1
Escherichia coli (E. coli) are bacteria.	} Premise 2
E. coli do not have nuclei.	} Conclusion

Deduction has two distinguishing characteristics. First, deductive arguments lead from general premises to a more specific conclusion. For example, in the argument above, the first premise is a statement concerning all bacteria, while the conclusion concerns only one bacterial species. Second, if deductive premises are true, then the conclusions from a logically correct deductive argument *must be true.*

It is clear that mathematical arguments are based on deduction. In a mathematical proof, axioms (which are assumed to be true) are used to derive a conclusion (e.g., a theorem), which, by definition, must also be true.*

1.6.3 Induction versus Deduction

Induction has been criticized as irrational because "believing" in its conclusions means "believing" that the universe never changes. That is, what we learned yesterday is completely applicable today, tomorrow, and so on. Hume (1740) formalized this issue in his *problem of induction*. Many authors, however, have argued that Hume's problem is irrelevant since induction constitutes a fundamental component of human cognition, and that not using it would be crippling (Strawson 1952). Indeed, induction has been increasingly trusted as science has developed.

---

* It is interesting to note that the validity of an argument does not require that that premises are true. For instance, the following represents an argument that is scientifically incorrect (from the perspective of the 21st century), but logically correct:

Flat objects have edges.	} Premise 1
The earth is flat.	} Premise 2
Traveling a long distance in a straight line will cause one to fall off the edge of the earth. } Conclusion	

A fallacious argument can also have both "true" premises and conclusions. Consider the following deductive example:

All mammals are air breathing.	} Premise 1
All cats are air breathing.	} Premise 2
All cats are mammals.	} Conclusion

While the premises and conclusion are both true, the general form of the premises do not support the conclusion. For instance, lizards are air breathing, but are not mammals.

It seems obvious that modern science values and uses both induction and deduction (Salmon 1967, Mentis 1988, Chalmers 1999, Okasha 2002, many others). Deduction is useful for many applications, including mathematics, and null hypothesis testing (see Section 1.6.4). Deductive premises, however, may be the product of induction. Furthermore, the creation of hypotheses from supporting observational evidence is by definition inductive. This association is vital to the establishment of scientific theories and laws (Ayala et al. 2008, p. 11).

EXAMPLE 1.1

Consider a wildlife biologist in Yellowstone National Park. The biologist observes that narrowleaf willow (*Salix angustifolia*) seedling recruitment always seems to decrease as elk (*Cervus canadensis*) populations increase in size. Based on this pattern, she inductively hypothesizes that elk negatively affect willow survival. She further reasons that willow seedlings protected by elk exclosures should have higher recruitment than adjacent groves (controls), which are freely accessible to elk. She gathers appropriate data and tests this hypothesis using mathematical/statistical methods. This would be a test of a deductive argument based on premises from inductive reasoning.

1.6.4 Logic and Null Hypothesis Testing

It is possible to make connections between the precepts of logic and the methods of statistical hypothesis testing described later in this book. In this section, we will examine the logical foundations of an approach called *null hypothesis testing*, that is, testing a hypothesis of "no effect." The null hypothesis testing paradigm is fully explained in Chapter 6.

1.6.4.1 Modus Tollens

Deduction fits well with severe falsification because rejection of a hypothesis based on data is always possible, but empirical confirmation is impossible (Popper 1959). This is because the premises cannot constitute all possible data. To illustrate this, let H be a hypothesis, and let I be an entity that allows evaluation of H. Now, consider the following arguments:

$$
\begin{array}{ll}
\text{If } H \text{ is true, then so is } I & \text{\} Premise 1} \\
\text{But available evidence shows that } I \text{ is not true} & \text{\} Premise 2} \\
H \text{ is not true} & \text{\} Conclusion}
\end{array} \quad (1.1)
$$

$$
\begin{array}{ll}
\text{If } H \text{ is true, then so is } I & \text{\} Premise 1} \\
\text{Available evidence shows that } I \text{ is true} & \text{\} Premise 2} \\
H \text{ is true} & \text{\} Conclusion}
\end{array} \quad (1.2)
$$

In Argument 1.1, we reject hypothesis H using a *logical* form of deduction called *modus tollens*. The statement is deductive because if the premises are true then the conclusion must be true as well. This form of argument is also called *denying the consequent* because the consequence of H is denied, resulting in the refutation of H.

The argument in Argument 1.2 is a form of deduction called *affirming the consequent.** In this case, the conclusion may be false even if the premises are true. Thus, the argument is fallacious. The first premise indicates that *I* is dependent on *H*, not the converse. Thus, the truth of *I* (suggested in the second premise) may not signify the truth of *H*. In addition, the second premise is inconclusive because it consists of incomplete evidence (Hume 1740). Because at least some information concerning *H* is unknown, we cannot prove that the hypothesis *H* is true. At best we can say that we have failed to reject the hypothesis.

Thus, we can only deductively reject or fail to reject a hypothesis whose premises include empirical data. To be precise, one can deductively falsify a hypothesis based on experimental results, but there can be no instance in which one can deductively verify it.

1.6.4.2 *Reductio Ad Absurdum*

Another type of logical argument, important to null hypothesis testing, is *reductio ad absurdum*, which means "reduction to the absurd." Here, one tries to support a hypothesis, *H*, by formulating a hypothesis that represents not-*H* (i.e., the opposite of *H*), and then disproving not-*H*.

> **EXAMPLE 1.2**
>
> One of the most famous examples of *reductio ad absurdum* is a proof, credited to Pythagoras (570–495 BC), which demonstrates that $\sqrt{2}$; is irrational. I will break his arguments into eight steps.
>
> 1. Assume there is a rational number whose square is equal to two.
> 2. This number must be a fraction since there is no integer whose square equals two. Let this fraction be reduced to its lowest common factor. We have
>
> $$(a/b)^2 = 2 \quad \text{and, as a result} \quad a^2 = 2b^2$$
>
> 3. a^2 must be even because the product of two and any other number must be even. In addition, *a* must be even because the square of any even number must be even, while the square of any odd number must be odd.
> 4. Let $a = 2c$ to remind us that *a* is even. We have
>
> $$a = 2c \quad \text{and} \quad a^2 = 4c^2$$

---

* Another logically correct form of deduction is called *affirming the antecedent*. Here is its general form:

If *H*, then *I*	} Premise 1
H	} Premise 2
I	} Conclusion

Like all logically correct deductive arguments, any statement can be selected for *H* and *I*, and the conclusions will be true if the premises are true. In contrast, with affirming the consequence we have

If *H*, then *I*	} Premise 1
I	} Premise 2
H	} Conclusion

As with argument 1.2, the statement above is deductively invalid. This can be demonstrated by substituting statements for *H* and *I* and constructing an argument with true premises and a false conclusion.

If Portland is in Washington state, then it is in the Pacific Northwest	} Premise 1
Portland is in the Pacific Northwest	} Premise 2
Portland is in Washington state	} Conclusion

5. Substituting back into our earlier equation, we have

$$4c^2 = 2b^2 \quad \text{and} \quad 2c^2 = b^2$$

6. From step 3, we see that b must also be even.
7. But if a and b are both divisible by two, then they have not been reduced to their smallest common factor, and this contradicts the assumption in step two.
8. We conclude that it is not possible to have a rational number whose square equals two. That is, we conclude $\sqrt{2}$ is irrational.

This argument follows the form of *reductio ad absurdum*. To prove that $\sqrt{2}$ is irrational, we assume that the elements making up $\sqrt{2}$ are rational. Because we contradict this assumption, we then disprove the hypothesis that $\sqrt{2}$ is rational.[*]

The *asbio* package provides several interactive logic worksheets illustrating important types of logical and fallacious arguments. These can be accessed from the book. menu() or by typing see.logic() in the **R** console after installing and loading *asbio*.

1.7 Variability and Uncertainty in Investigations

The impossibility of proving a deductive argument based on empirical evidence is due to the variability of natural systems, and the fact that all possible data concerning these systems cannot be collected (Section 1.6). If natural phenomena in the universe never varied, then conclusive empirical proof would be possible. In fact, one observation would be sufficient to establish truth/falsehood. However, this does not happen.

In the elk experiment from Example 1.1, there are several potential sources of confounding variability. One possibility is that willow survivorship does not respond linearly to the presence of elk. For example, it may take a large number of elk to cause any sort of decrease in willow seedling recruitment. If this is true, then a researcher must be conscious of this particular relationship to be able to detect the effect of elk. A second confounding outcome will almost certainly occur. This is that replicates in the exclosure and control groves will vary. For instance, it is possible that elk will graze only at some control groves and ignore other control groves to avoid a pack of wolves that move into the area. Another possibility is that some exclosure groves will be unknowingly located on soils that are particularly nutrient poor. As a result, these groves will not respond with increased recruitment even though elk grazing is prevented. In both cases, the negative effect of the elk on the willows will be obscured by sample variability.

All empirical data describing the universe will contain variability that can confound decisions.[†] This extends to even the most venerable scientific "truisms." For instance, it is

---

[*] It is interesting that the followers of Pythagoras, who believed that numbers were sacred, withheld the blasphemous information that $\sqrt{2}$ is irrational.

[†] The natural propensity of experimental outcomes to vary from sample to sample has been given the rather unfortunate name, *sampling error*. The capacity to measure g (or anything else) will also be limited because of *measurement error*. For instance, currently the most precise tool for measuring g is the absolute atom gravimeter (Müller et al. 2008). While these devices are incredibly accurate (they give readings close to the "true" value of g) and precise (i.e., they give very similar repeated measures), they still exhibit a degree of uncertainty $\approx 1.3 \times 10^{-9} \, g$. Sampling error, measurement error, accuracy, and precision are addressed repeatedly in this text.

often taken for granted that the acceleration of objects due to gravity, g, near the earth's surface = $9.8 \text{ m} \cdot \text{s}^{-2}$. However, we now know that g varies with the local density of the earth's surface and the speed of the earth's rotation, which in turn varies with latitude (de Angelis et al. 2009).

It has been suggested that the laws of physics (e.g., laws of electrostatics and thermodynamics) differ from those of biology since they provide tidy deterministic models and often address objects (such as electrons) that are nonvariable (Pastor 2008). However, the empirical behavior of matter at subatomic scales (e.g., electrons) is *not* deterministic. In fact the *Heisenberg uncertainty principle* stipulates that it is impossible, even in theory, to simultaneously know the exact location and momentum of a subatomic particle (Heisenberg 1927). Indeed, the more precisely one attribute is measured, the less precisely the other can be measured. As a result, the field of quantum mechanics defines *stochastic* (nondeterministic) models to describe the workings of subatomic particles.[*] Stochastic processes also underlie much larger presumably deterministic systems, including the orbits of moons and planets in the solar system (Peterson 1993).[†]

Scientific theories that have withstood repeated testing (e.g., the theory of evolution and the theory of relativity) can be trusted as objective, realistic, and corroborated pictures of the universe. However, they should not be treated dogmatically (a behavior called *scientism*). Scientists and others should recognize that the variability implicit in natural phenomena (and measurements of phenomena) means that absolute proof for a hypothesis based on empirical data can never be obtained. Acknowledgment of this state of affairs allows continual refutation and revision of hypotheses, and the continual development and improvement of scientific ideas. Empirical science, of course, is not indifferent to truth. It is more correct, however, to view it as *truth-directed*. That is, it seeks to know, but acknowledges at the outset that it can never know for certain.[‡]

In conclusion, it is important to emphasize two things. First, the refutation of a hypothesis may provide valuable information about the study system by allowing a researcher to eliminate incorrect lines of reasoning, and clarify its true nature. Second, even though empirical evidence cannot be used to prove that a hypothesis is true, it can be used to corroborate a hypothesis. Thus, while nonrefutation is inconclusive, it often provides valuable supporting evidence. The continual accumulation of such evidence, and a lack of negating evidence, may indicate a very strong statement about the natural world, allowing the development of scientific theories and laws (Ayala et al. 2008).

The importance of trustworthy tools to measure the empirical support for hypotheses is obvious. Along with dependable field and laboratory instruments, these include *mathematical models* (descriptions based on mathematical concepts), particularly those associated with statistics.

---

[*] Albert Einstein (1879–1955) found the lack of determinism in quantum mechanics to be unsettling, giving rise to his famous quote: "God does not play dice with the universe."

[†] The orbits of planets are perhaps better described as *chaotic* not *stochastic* (Peterson 1993, Denny and Gaines 2002). While stochastic simply means that a process is nondeterministic, chaotic refers to a process that is deterministic, but very sensitive to initial conditions that determine its outcome. As a practical matter, chaotic processes are often effectively modeled with stochastic tools like probability.

[‡] A view of science as truth-directed is in accordance with *logical positivism* (Okasha 2002) and subsequently, with *scientific realism*. This outlook has been opposed, notably by Kuhn (1963) who insisted that scientific truth is relative to the current scientific paradigm (Laudan 1977). Famous logical positivists included Moritz Schlick (1882–1936), Ernst Mach (1838–1916), and the young Ludwig Wittgenstein (1889–1951). Important scientific realists include Karl Popper (1902–1994) and Richard Boyd (1942–).

1.8 Science and Statistics

It is apparent that empirical scientists are faced with a dilemma. We need to objectively quantify evidence supporting our hypotheses, but we must acknowledge at the outset that the variability of the world makes it impossible to ever prove a hypothesis. A reasonable response to this problem is statistics. Statistical models explicitly acknowledge uncertainty, and through mathematics, allow us to estimate variability and evaluate outcomes probabilistically.

The word *statistics* is conventionally defined as "a branch of mathematics dealing with the collection, analysis, interpretation, and presentation of masses of numerical data."* To empirical scientists, however, this definition seems insufficient. For these individuals statistics allows (1) the expression and testing of research hypotheses within the framework of the scientific method, (2) the use of mathematics to make precise and objective inferences, and (3) the quantification of uncertainty implicit to natural systems. As Ramsey and Schafer (1997) note: "statistics is like grout—the word feels decidedly unpleasant in the mouth, but it describes something essential for holding a mosaic in place."

1.9 Statistics and Biology

The development of modern statistical methods has been largely due to the efforts of biologists or of mathematicians who have dealt almost exclusively with biological systems. Francis Galton (1822–1911) was the first cousin of Darwin, and the inventor of regression analysis (Chapter 9). Karl Pearson (1857–1936) invented the correlation coefficient (Chapter 8), the χ^2 contingency test (Chapter 11), and was one of the first practical users of probability distributions (Chapter 3). Together, Galton and Pearson founded the journal *Biometrika* to define distributional parameters for biological populations with the ambitious goal of quantifying the process of evolution as the result of natural selection (Salsburg 2001). William Sealy Gosset (1876–1937) developed a number of important ideas in mathematical statistics, including the *t*-distribution (Chapter 6), while working as a biometrist for the Guinness brewing company. Ronald A. Fisher (1890–1962) has been hailed by mathematical historians as single-handedly creating the "foundations for modern statistical science" (Hald 1998). His achievements include the invention of maximum likelihood estimation (Chapter 4), the *P*-value (Chapter 6), the null hypothesis (Chapter 6), and the analysis of variance (Chapter 10). Fisher principally analyzed and interpreted biological data during his long career, and also made important scientific contributions as a geneticist and evolutionary biologist. Indeed, he has been described as "the greatest of Darwin's successors" (Dawkins 1995). Jerome Cornfield (1912–1979) was not only a developer of ideas associated with Bayesian inference (Chapter 6) and experimental design (Chapter 7) but also made important contributions to epidemiology, including the identification of carcinogens. A.W.F. Edwards (1935–), whose books have clarified the importance of likelihood in inferential procedures, has worked principally as a geneticist and evolutionary biologist. The list goes on and on (Table 1.2).

---

* *Merriam Webster's New Collegiate Dictionary*, 11th edition.

TABLE 1.2

An Incomplete Hagiography of Important 20th- and 21st-Century Statisticians with Strong Biological Connections

Name	Statistical Contributions	Biological Connections/Contributions
Karl Pearson (1857–1936)	With *Francis Galton* (1822–1911) invented the correlation coefficient. Played a key role in the early development of regression analysis. Invented the chi-squared test, and the method of moments for deriving estimators.	Founded the journals *Biometrika* and *Annals of Human Genetics*. Involved in epidemiology and medicine. Evaluated evidence for the theory of evolution with species probability distributions.
R.A. Fisher (1890–1962)	Enormously important figure in the development of mathematical and applied statistics. Contributions include the *F*-distribution, ANOVA, maximum likelihood estimation, and many other approaches.	A leading figure in evolutionary ecology, quantitative population genetics, and population ecology. Biostatistician at the Rothamsted agricultural experimental station.
Jerzy Neyman (1894–1981)	With K. Pearson's son *Egon Pearson* (1895–1980), and battling R.A. Fisher, developed a framework for null hypothesis testing, and the concepts of type I and type II error.	Founded biometric laboratories in Poland. Established methods for testing medicines still used today by the United States Food and Drug Administration.
W.S. Gosset (1876–1937)	Proposed and developed the *t*-distribution. To assuage fears of his employer (Guinness) that he would give away brewery secrets, Gosset published under the pseudonym "student."	For many years studied methods to maximize barley yield.
A.N. Kolmogorov (1903–1987)	Made large contributions to the understanding of stochastic systems, particularly Markov processes. Created applied statistical innovations, including the Kolmogorov–Smirnov test.	Generalized and expanded the Lotka–Volterra model of predator–prey interactions, and postulated models for the structure of the brain.
G.W. Snedecor (1881–1974)	With *W.G. Cochran* (1909–1980) wrote the most frequently cited statistical text in existence: "Statistical Methods." The *F*-distribution is often called Snedecor's *F*-distribution to acknowledge the fact that he handcrafted the first *F* cumulative distribution tables.	Deeply involved in agricultural research. Published influential biometric texts including: "Statistical Methods Applied to Experiments in Agriculture and Biology."
Gertrude Cox (1900–1978)	With *David Box* (1924–) developed maximum likelihood methods for optimal variable transformations in linear models. With W.G. Cochran published an important book on experimental design.	Editor of the *Biometrics Bulletin* and of *Biometrics*, and founder of the International Biometric Society.
J. Cornfield (1912–1979)	Contributed to applications in Bayesian inference. Formalized the importance of observational studies. Played a role in the early development of linear programming.	Strong research interests in human health, particularly the effects of smoking. Developed important epidemiologic experimental approaches.
John Nelder (1924–2010)	With *Peter McCullagh*, developed the unifying concept of generalized linear models (GLMs). Proponent for the inferential use of likelihood.	Strong interests in ornithology and evolutionary biology. Wrote a book on computational biology. Served as head of biostatistics at the Rothamsted experimental station.
A.W.F. Edwards (1935–)	Strong proponent of the inferential use of likelihood. Lucid statistical writer.	Geneticist and evolutionary biologist. Developed methods for quantitative phylogenetic analysis.

Note: Statistical terms in the table are explained in Chapters 2 through 11.

The need for biologists to be able to understand and apply statistical and mathematical models has never been greater. This is true for two reasons. First, biologists are faced with the tremendous responsibility of being able to accurately, precisely, and objectively describe the world. This includes the complex global problems we currently face, for example, global climate change, epidemics, human aging and disease, alteration of nutrient cycles, increased resource demands from overpopulation and cultural development, habitat fragmentation, loss of biodiversity, acidification of rain, and ozone thinning. Second, developments in statistical theory and computer technology have provided a bewildering new array of analytical tools for biologists. If handled correctly and intelligently, these tools provide an unprecedented medium to describe our world, and address our problems. Conversely, if these tools are handled inappropriately, they will produce ambiguous or incorrect results, and give rise to distrust of scientific claims.

1.10 Summary

This chapter provides a motivation for the development of the remainder of the book, and a philosophical and historical context for both science and statistics. Scientists seek to provide truth-directed descriptions of the world. To accomplish this, scientific descriptions are based on three principles: objectivity, realism, and communalism. The scientific method is built on hypotheses, that is, statements describing the universe, which serve as starting points for investigations. Scientific hypotheses have two characteristics: testability and (arguably) falsifiability. Testability is explicit to the scientific method, while falsifiability requires the possibility of refutation. The approaches used by scientists to understand the natural world are based on precepts of logic. Two modes of logic are possible: induction and deduction. The conclusions from an inductive argument contain information not present, even implicitly, in the premises. Conversely, a deductive conclusion is contained entirely in the premises. All natural systems vary, including those studied by biologists. Thus, information about those systems will be incomplete. Because of this, absolute proof of the truth of a hypothesis concerning the natural world is impossible. In the face of these issues, biologists require a way to describe the trends and variability in the systems they study, and to quantify evidence concerning their hypotheses (despite this variability). A rational solution to this problem is the discipline of statistics.

EXERCISES

1. Distinguish "empirical science" and "nonempirical science."
2. Define the terms "objectivity," "scientific realism," and "communalism."
3. Describe two characteristics of a scientific hypothesis. Given these characteristics, list three hypotheses that could fall under the heading "scientific hypothesis" and three that could fall under the heading "nonscientific hypothesis." Justify your reasoning.
4. Describe similarities and differences in the views on falsifiability of Popper, Kuhn, and Lakatos.
5. How does the first premise of scientific realism given in this chapter provide a connection to falsificationism?

6. Define the terms "fallacious" and "logically correct."

7. Distinguish inductive and deductive reasoning. Give an example of each in the context of a biological investigation.

8. Create an example of a logically correct argument with false premises, and a fallacious argument with true premises.

9. Give examples of *modus tollens* (*denying the consequent*) and *affirming the consequent* arguments. Explain why the latter is logically incorrect.

10. Give an example of *reduction ad absurdum*.

11. Go through the interactive logic worksheet provided by installing and loading the *asbio* package and typing see.logic(). Come up with your own syllogistic arguments illustrating *modus ponens, post hoc ergo propter hoc*, and confusing cause and effect. Indicate whether these arguments are logical or fallacious.

12. According to the author, why should science be described as "truth-directed"?

13. Why does the field of statistics fit well with the goals and realities of empirical science?

14. Choose an individual from Table 1.2 and create an expanded biography, or prepare a brief summary for another biostatistician.

2

Introduction to Probability

We cannot understand the scientific method without a previous investigation of the different kinds of probability.

Bertrand Russell (1948)

Misunderstanding of probability may be the greatest of all impediments to scientific literacy.

Stephen Jay Gould

2.1 Introduction: Models for Random Variables

In mathematics, a *variable* is an entity whose value or meaning can change, causing the evaluation of an expression to change. For instance, the venerated function for obtaining the instantaneous velocity of a falling body is

$$f(t) = gt,$$

where $g = 9.8$ m $\cdot$ s^{-2} and t = time in seconds. Both t and $f(t)$ are variables. Any value greater than or equal to zero can be assigned to t, and $f(t)$ will vary directly with t. In contrast, g is a *constant* with a single fixed value.

In applied statistics, we generally view a variable as a phenomenon whose outcomes can vary from measurement to measurement, for instance, the height of students in a biometry class or the surface area of individual leaves on a quaking aspen tree. A very important type of variable, whose outcomes *cannot* be known preceding a measurement or trial, is called a *random variable*. The behavior of random variables must be defined using probability. Consider a fair coin flip. Outcomes from individual tosses (i.e., heads and tails) will vary, and toss outcomes will be unknown preceding a toss. Nonetheless, we generally assume that heads will occur for 50% of tosses (probability = 0.5). Given appropriate methods for acquiring data, student height and leaf area, mentioned above, can also be regarded as random variables. Both will vary from outcome to outcome, and the results from these outcomes will generally be impossible to predict in advance of measurements. However, these measurements will commonly demonstrate probabilistic behavior.

Deterministic models use only nonrandom variables, and given the same inputs *will always produce the same result*. The equation for free falling velocity is deterministic because a particular input for *t* will always produce the same result, *f(t)*.[*] Deterministic models are often extremely useful, particularly as heuristic representations.[†] However, they will almost never perfectly characterize observations of real-world phenomena. For instance, the measurements of the velocity of a falling body will be inexact because the gravitational constant *g* will vary with altitude, latitude, air resistance, and because measuring devices for *g* have inherent error (Chapter 1).

Probabilistic models incorporate random variables. As a result, given the same user inputs they *need not produce the same result*. Instead of returning the *exact* outcome from a deterministic abstraction, probabilistic models can be used to find the most *probable* outcome in the context of data. A probabilistic version of the free falling velocity equation would be

$$f(t) = gt + \varepsilon.$$

In this case, ε represents a random variable that can take on any real number in its support. Conventionally, the average value of ε will be 0, and values near 0 will occur with the highest likelihood. Note that, in the formulation above, *t* remains nonrandom. However, *f(t)* now contains both deterministic and random qualities because it is the sum of *gt* and ε. Thus, probabilistic models consider both random *and* deterministic processes.

EXAMPLE 2.1

Biologists have long marveled at the properties of spider silk. This substance is as strong, per unit weight, as steel, and more energy absorbing than Kevlar, the material in bullet proof vests. The silk from Darwin's bark spider (*Caerostris darwini*)—whose webs span 25 m streams—is the strongest known biological material (Price 2010).

The strength of spider silk is largely due to the amino acid characteristics of its structural proteins, which are frozen into pleated crystalline sheets (Denny and Gaines 2002). This, however, does not explain the exceptional capacity of spider silk to flex and absorb energy without breaking, *à la* Kevlar. The mechanism underlying this characteristic was elucidated in a series of experiments by Gosline et al. (1984).

It is well known that most materials will tend to expand when heated, and will shrink when cooled. However, several substances, including rubber polymers, have *negative thermal expansion*. That is, they will shrink when heated. When Gosline et al. applied heat to spider silk, the strands of web contracted. This indicated the presence of rubber-protein chains connecting the crystalline sheets. The crystals were moved around by reconformation of the chains, but remained otherwise unchanged. Thus, spider silk is energy absorbent because it is partially composed of rubber-like polymers.

Figure 2.1 shows the results of one of the Gosline et al. (1984) experiments. The line in the figure and the variability around the line are both important parts of a probabilistic model called linear regression (Chapter 9). The line itself indicates a deterministic negative association between strand length and temperature: length decreases linearly with temperature (Figure 2.1a). However, there is also a degree of randomness demonstrated by scatter around the line. There even appears to be a pattern to this variability, as

---

[*] The line between deterministic and random processes is often blurry. A coin flip can be considered a random variable. However, if one knew the height of a flip above the ground, the angular velocity initially applied to the coin, the effects of air resistance, and other important mechanistic information, one should be able to correctly decide whether a flip will result in a head *a priori* (Denny and Gaines 2002). Thus, the result of random variable may not be *unknowable*, but merely *unknown* because other variables determining the result are too difficult to measure.

[†] Interactive examples of famous deterministic biological models can be accessed from the *Biology* tab in the `book.menu()`.

Fig.2.1()##see Ch. 2 code

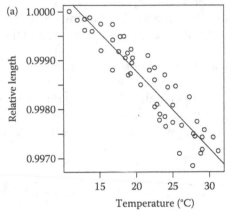
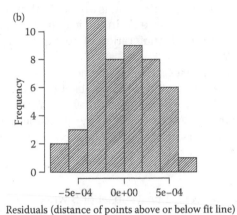

FIGURE 2.1

A linear regression (probabilistic model) of spider web strand length as a function of temperature. (a) The data and a fitted regression line. (b) The distribution of residuals (distances above or below the fit line), represented with a histogram. A deterministic model could be expressed as strand length $= \beta_0 - \beta_1$ (temp), where β_0 and β_1 represent the slope and Y-intercept of the fit line, respectively. Conversely, a probabilistic (statistical) model would be of strand length $= \beta_0 - \beta_1$ (temp) $+ \varepsilon$, where ε describes the natural variability in the system, and $\sum_{i=1}^{n} \varepsilon_i = 0$.

described by a histogram (Figure 2.1b; see Section 3.1). In particular, while most points are near the line, a few are relatively distant. The vertical distances of the observed points from the fit line are called residuals (see Chapter 9). The symmetric, bell-shaped distribution of the residuals allows inferences regarding the "true" slope of the fitted line.

Assuredly, deterministic physical laws set general boundaries on the behavior of the universe. We know, for instance, that energy can neither be created nor destroyed, and that all systems move to higher states of entropy. Within this framework, however, chaotic or even random elements appear omnipresent. For example, at the inception of the big bang, the uncertainty principle (Chapter 1) may have produced randomness in the distribution of particles resulting in the current nonuniform configuration of celestial bodies (Hawking 1988). As a species living within an apparent deterministic/random duality, humans have evolved to handle probabilistic constraints. Indeed, natural selection itself requires variability and randomness as grist for its mill. Nonetheless, it is remarkable that we now address these properties both consciously and pragmatically. We ask: "where should I plant my crops to insure the highest probability of seed germination?" or even "which is the most likely scenario for global climate in the next 100 years?"

2.1.1 Set Theory Terminology

Probability is often introduced using notation and terminology from a branch of mathematics called *set theory*. A *set* is simply a collection of distinct objects. Sets are generally specified using capital letters, for example, A and B, while objects comprising sets are listed inside curly brackets. For instance, $A = \{♀, ♂\}$ is a set containing possible gender outcomes from sexual reproduction. *Elements* are the individual objects comprising a set, and are indicated with the term ∈. For instance, in the example above, it is true that ♂ ∈ *A*. That is, a female offspring is an element of *A*. If an object *a* is not an element of a set *A*, then this is denoted as *a* ∉ *A*. In the idiom of set theory the process of obtaining an observation is an *experiment*, an iteration of an experiment is a *trial*, and an observed result of a trial is an *outcome*.

In many cases, all of the elements of set *A* will also be contained in a second set *B*. In this case, we can say that *A* is a subset of *B*, or *A* ⊂ *B*. We denote that *A* is *not* a subset of *B* as *A* ⊄ *B*. The *sample space* or *universal set*, termed *S*, will contain all possible outcomes of an experiment. We can define any plausible *event*, for example, *A* or *B*, as a particular subset of the universal set. Consider an experiment involving two coin tosses. The universal set is

$$S = \{HH, HT, TH, TT\}.$$

Let the *event A* be a subset of *S* whose outcomes contain at least one tail. Then we have

$$A = \{HT, TH, TT\} \subset S.$$

A special set called the *empty set* is denoted as ∅. The empty set will contain no elements with respect to any defined events of interest. For instance, let *B* denote the implausible outcome: two coin tosses without a head or a tail, then

$$B = \emptyset.$$

The probability of the event *A* can be designated as *P(A)*. The philosophical meaning, and the methods used for calculating *P(A)*, will concern the remainder of this chapter.

2.1.2 Philosophical Conceptions of Probability

Although the mathematical precepts of probability are well established (and will be described shortly), there are a number of divergent philosophical interpretations of probability. The two most common are probabilities as *frequencies* and *degrees of belief* (cf. Wasserman 2004).[*]

In the *frequentist interpretation of probability P(A)* describes the limiting relative frequency of *A*. That is, *P(A)* is the proportion of *A* outcomes as the total number of trials in an experiment goes to infinity. For example, it can be demonstrated that the proportion of heads from a series of fair coin flips will approach the constant 0.5 as the number of trials grows large, that is, *P(Head)* = 0.5 (Figure 2.2).

A slightly more complex example involves throws of a six-sided die. Here, we have six possible outcomes. Given a fair die, we would assume that the proportion of each of these outcomes will approach 1/6 as the number of throws grows large, that is, the probability of any outcome from the die is expected to be 0.167 (Figure 2.3).

Because it is impossible to observe an infinite number of trials, a frequentist interpretation of *P(A)* is an idealization. Within the frequentist paradigm, we assume that there is a true value for *P(A)*, although it may be unknown. In addition, we assume no prior knowledge with respect to the outcome of a random trial. This is in contrast to the degrees of belief approach.

---

[*] A large number of other interpretations for probability are also possible. See Gillies (2000, 2002) and Carnap (1963) for more information.

```
anm.coin.tck()
```

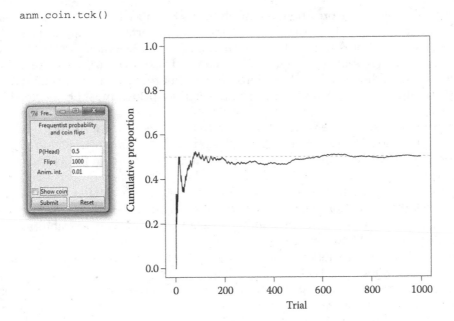

FIGURE 2.2

The frequentist conception of probability illustrated with coins flips. As observations (fair coin flips) accumulate, the proportion of heads approaches 0.5.

```
anm.die.tck()
```

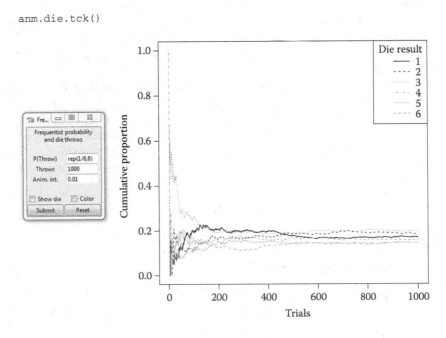

FIGURE 2.3

Convergence in probability for a fair six-sided die.

The *degrees of belief* interpretation of probability is tied to what are known as *Bayesian* statistical methods. For a Bayesian, $P(A)$ is a measure of certainty; a quantification of an investigator's belief that A is true. This interpretation assumes that a number of explanations for a phenomenon may be valid, and that a fixed value for $P(A)$ is neither necessary nor desirable. In addition, a Bayesian approach does not require that only current data be used to describe phenomena. Instead prior information (priors) must be specified to augment sample data.

There are many cases where probabilities only make sense from the degrees of belief perspective. Consider the statement: "An explosion on the moon documented by Gervase of Canterbury in 1178 AD, was due to a meteorite impact, resulting in the 14-mile (in diameter) lunar crater Giordano Bruno" (*q.v.* Hartung 1976). This *single* outcome is "possible," the lunar feature in question was probably formed around the specified time, but it does not describe an outcome from a potential distribution of outcomes, and does not allow limiting frequency interpretation for probability. Instead, an associated probability statement requires the degrees of belief approach.

There are also many situations where use of priors makes sense. Consider a simple example involving genetics. The gene for hemophilia is carried on the X chromosome. As a result, a woman will either express the disease (if she carries the disease allele on both of her X chromosomes) or be a carrier (if she carries the allele on a single X chromosome). Males on the other hand will always express hemophilia (if they carry the disease allele) because they have only one X chromosome. Consider now a woman named Mary whose mother is a carrier, and whose father did not have hemophilia. Because Mary received an X chromosome from each parent, she has a 50% chance (= probability 0.5) of carrying the gene. Suppose, now that Mary has two sons, each of whom is disease free. Surely, this new information should now be used to update the priors, and reduce the estimated probability that Mary is a carrier. Bayesian methods allow such calculations.[*]

Bayesian priors can be subjective (Gillies 2002). Because objectivity is a primary goal of scientific description (Chapter 1) Bayesian methods have been criticized as being unscientific. Bayesians, however, contend that specification of priors allows an efficient and inductive approach to science, since prior information is used to supplement current sample data (Gotelli and Ellison 2004).

Differences in frequentist and Bayesian perspectives are most important in the context of inferential procedures, for example, parameter estimation and hypothesis testing (Wasserman 2004). They are irrelevant to the mathematical precepts of probability itself. Extensive comparisons of Bayesian, likelihood, and frequentist inferential approaches are provided in Chapters 4 through 6, 9, and 10. Most of the inferential procedures in this book use a frequentist interpretation of probability. Bayesian approaches and alternatives, however, are described in relevant situations.

2.2 Classical Probability

As we familiarize ourselves with the behavior of a random variable, we may become aware of probabilistic patterns. Consider the system described in Example 2.1. Although

---

[*] This exact problem is readdressed in Example 6.19.

the association between spider web strand length and temperature is not deterministic, higher temperatures *tend* to produce longer (stretchier) strands.[*]

The work of geneticist Gregor Mendel (1822–1884) provides another good example of this process. Mendel observed two general color phenotypes, purple and white, for domesticated pea plants (*Pisum sativum*). Further, he found that a cross of purple and white homozygotes produced approximately 75 purple and 25 white individuals in 100 offspring. Thus, he reasoned that the probability of a purple-flowered individual given from this cross would be 75/100, while white-flowered individuals would have a probability of 25/100.

We note that this approach for defining probability tacitly (and reasonably) assumes a frequentist viewpoint. In particular, assume that an experiment contains N possible outcomes, each of which is equally likely. Let $N(A)$ be the number of times that outcomes result in the event A. Then, it follows that[†]

$$P(A) = N(A)/N. \tag{?.1}$$

Given the definition in Equation 2.1, several general rules can be established. First, since $P(A)$ represents a proportion, then $P(A)$, must always be a number between 0 and 1:

$$0 \leq P(A) \leq 1. \tag{2.2}$$

We refer to the outcome when A *does not occur* as A *complement*, or A'. It follows from Equations 2.1 and 2.2 that

$$P(A') = 1 - P(A). \tag{2.3}$$

The sample space, S, comprises all possible outcomes. As a result it has a probability of one:

$$P(S) = 1. \tag{2.4}$$

The probability of the empty set will always be 0. That is

$$P(\emptyset) = P(S') = 0. \tag{2.5}$$

Venn diagrams, introduced by John Venn (1834–1923), provide a tool for visualizing sets and their relationships in sample space. These plots will generally be delimited by a rectangle representing the universal set, S. The rectangle, in turn, will almost always contain one or more geometric shapes representing subsets of S. These will usually be depicted with circles, although other shapes are possible.[‡] If we assume that the rectangle representing S has unit area, then we can represent the probability of an event A with a shape whose proportional area in S will be equivalent to $P(A)$. This approach provides a graphical bridge between set theory and probability. Figure 2.4 shows the estimated sample space for the

---

[*] It is interesting that R.A. Fisher (1936) insisted that Mendel's data were too deterministic and were likely fudged.

[†] In biological experiments, the number of possible trials, N, will frequently be unknown or, in the frequentist ideal, will be infinite. In this case, if we assume that a true value for $P(A)$ exists (*à la* the frequentist perspective), it will only be possible to estimate $P(A)$ using a finite number of samples, n (see Chapter 11).

[‡] Complex shapes will often be necessary to illustrate the relationships of a large number of events (see Ruskey and Weston 2005).

Venn.tck()

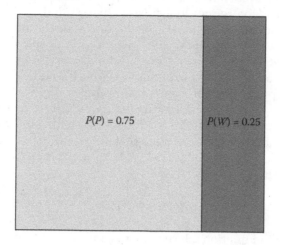

FIGURE 2.4
Venn diagram for flower color outcomes for a cross of purple and white homozygous individuals with complete dominance of the purple allele.

color of flowers for a single pea offspring given complete dominance of the purple allele, and random mating of homozygous dominant and homozygous recessive parents.[*]

EXAMPLE 2.2

Consider a sample space that compares blood types in the United States. Based on large sample sizes, it is estimated that the probability of encountering an individual with blood type O, $P(O)$, equals 0.44, while the probability of encountering individual with blood type B, $P(B)$, equals 0.1 (Stanford School of Medicine 2013). A Venn diagram of unit area can contain the probabilities of all possible blood types (i.e., A, B, AB, and O) for a single individual. The probabilities for O and B are shown in Figure 2.5.

2.2.1 Disjoint

If two events *cannot* co-occur simultaneously (i.e., they have no common elements), then we call them *mutually exclusive* or *disjoint*. Because one cannot get a head *and* a tail on a single coin flip, then head and tail outcomes are disjoint. Flower colors for an individual pea plant and blood types for a single U.S. citizen are also disjoint. That is, an individual cannot have two distinct blood types, and, given complete dominance, a flower cannot simultaneously be two different colors.

If, for two outcomes A and B, we wanted to know the probability of the event A *or* B, we would express this as $P(A \cup B)$. The term $\cup$ is called *union*.

If A and B are disjoint, then

$$P(A \cup B) = P(A) + P(B).$$ (2.6)

---

[*] The figure was created using the function Venn in *asbio*, which makes approximate Venn probability diagrams for two events. More sophisticated Venn diagrams (including those that can represent the interactions of more than two events) are possible with functions from the packages *venn*, and *gplots*.

Venn.tck()

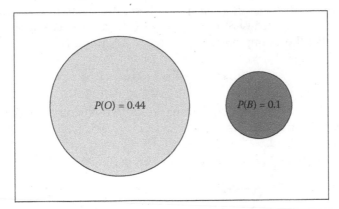

FIGURE 2.5
Expression of the probability of blood types O and B in the United States.

Thus, the probability of getting a head *or a* tail in a single coin toss, is $P(H \cup T) = 1.0$, while the probability of randomly encountering a U.S. citizen with an O *or* B blood type is $P(O \cup B) = 0.44 + 0.1 = 0.54$.

We can also imagine situations where events are *not* disjoint. Consider an herbivore that feeds on plant A with a probability of 0.3, plant B with a probability of 0.3, and plants A *and* B with a probability of 0.09 (Figure 2.6). We denote the probability of A *and* B as $P(A \cap B)$. The term $\cap$ is called *intersect*. If $P(A \cap B) > 0$ (i.e., if the probability of A and B both occurring is greater than zero), then A and B will overlap within a Venn diagram, indicating that they are not disjoint (Figure 2.5). Conversely, if A and B are mutually exclusive then $P(A \cap B) = 0$, because the event $A \cap B$ describes the empty set, $\varnothing$.

If A and B are *not* mutually exclusive, we can still calculate the union of A and B; however, the computation is now more complex:

$$P(A \cup B) = P(A) + P(B) - P(A \cap B). \tag{2.7}$$

Venn.tck()

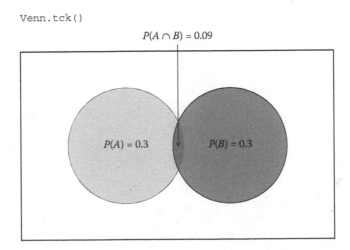

FIGURE 2.6
Venn diagram for the herbivore problem posed. Because $P(A \cap B) > 0$, A and B are *not* mutually exclusive.

This formulation is necessary if $P(A \cap B) > 0$, because if $P(A \cap B)$ is not subtracted when calculating the union of A and B, then it will be added to the probability of the union of A and B twice, causing $P(A \cup B)$ to be overestimated. This is known as the *inclusion-exclusion principle*. For the example shown in Figure 2.6, we have

$$P(A \cup B) = 0.3 + 0.3 - 0.09 = 0.51.$$

We can extend Equation 2.7 to include more than two outcomes. To calculate the probability of the union of three outcomes, A, B, and C, we have

$$P(A \cup B \cup C) = P(A) + P(B) + P(C) - P(A \cap B) - P(B \cap C) - P(A \cap C)$$
$$+ P(A \cap B \cap C). \tag{2.8}$$

To find the correct probability, we must again subtract the sample space containing the intersections. We can denote the union of a set of sets $A_1, A_2, \ldots, A_k$ as $\bigcup_{i=1}^{k} A_i$, and the probability of this union as $P\left(\bigcup_{i=1}^{k} A_i\right)$.

2.2.1.1 Boole's Inequality

From the definitions in Equations 2.6 through 2.8, it is possible to specify an upper bound to the probability of any union by using a rule called *Boole's inequality.*[*] This rule states that if $A_1, A_2, \ldots, A_k$ are events, then

$$P\left(\bigcup_{i=1}^{k} A_i\right) \le \sum_{i=1}^{k} P(A_i). \tag{2.9}$$

This must be true because $P\left(\bigcup_{i=1}^{k} A_i\right) = \sum_{i=1}^{k} P(A_i)$ if all events in A are disjoint, and $P\left(\bigcup_{i=1}^{k} A_i\right) < \sum_{i=1}^{k} P(A_i)$ if any events in A are not disjoint.

EXAMPLE 2.3

If $P(A) = 0.4$ and $P(B) = 0.3$, what is the maximum value for $P(A \cup B)$? The answer is

$$0.4 + 0.3 = 0.7.$$

The *difference* of A and B is defined to be the sample space in A, but *not in $A \cap B$*. This can be expressed as $A \cap B'$. The probability of this event is

$$P(A \cap B') = P(A) - P(A \cap B). \tag{2.10}$$

For example, in Figure 2.6 we have

$$P(A \cap B') = 0.3 - 0.09 = 0.21.$$

---

[*] Named for George Boole (1815–1864), an English mathematician and logician.

2.2.2 Independence

If, when A occurs it does not affect the probability of B, then we say that A and B are *independent*. An example would be a head on one coin flip (event A), and a head on a second consecutive coin flip (event B). If A occurs, this *will not* affect the probability of B occurring.

If A and B are independent, then

$$P(A \cap B) = P(A)P(B). \tag{2.11}$$

Because coin tosses are independent, the probability of two consecutive heads in a two coin toss sample space is $P(H \cap H) = 0.5 \times 0.5 = 0.25$. Note that in this sample space we assume that $P(T \cap H)$ is distinct from $P(H \cap T)$. That is, order matters, but only in distinguishing heads from tails. We cannot distinguish one head from another head, or one tail from another tail. As a result, there are four outcomes in a two coin toss universe: $S = \{HH, HT, TH, TT\}$, and we have $P(H \cap H) = P(T \cap T) = P(H \cap T) = P(T \cap H) = 0.25$. Methods for enumeration of events in S will be discussed in greater detail in Section 2.5.

If two nonzero probability events A and B are disjoint, then they cannot also be independent. This is mathematically true because $P(A \cap B) = 0$, but $P(A)P(B) > 0$. The definition of independence given Equation 2.11 also suggests that independent events should be exceedingly rare. This is because, while there are an infinite number of ways that the union of A and B can satisfy $P(A \cap B) > 0$, only one of these satisfies $P(A \cap B) = P(A)P(B)$.

2.2.2.1 Bonferroni's Inequality

An important extension of Boole's inequality called *Bonferroni's inequality*[*] can be used to find the lower bounds for the probability of the intersection of a finite number of potentially dependent events. The upper bound to the probability of an intersection occurs under independence. The Bonferroni inequality is

$$P\left(\bigcap_{i=1}^{k} A_i\right) \geq 1 - \sum_{i=1}^{k} P(A_i')$$

$$\geq \sum_{i=1}^{k} P(A_i) + (1 - k), \tag{2.12}$$

where $\bigcap_{i=1}^{k} A_i$ denotes the intersection of k sets: $A_1, A_2, \ldots, A_k$. As inductive support, consider the case when $k = 2$. We have

$$1 \geq P(A_1 \cup A_2),$$

$$1 \geq P(A_1) + P(A_2) - P(A_1 \cap A_2),$$

$$P(A_1 \cap A_2) \geq P(A_1) + P(A_2) - 1,$$

$$P(A_1 \cap A_2) \geq P(A_1) + P(A_2) + (1 - k).$$

---

[*] Named for the Italian mathematician Carlo Emilio Bonferroni (1892–1960).

EXAMPLE 2.4

Assume that $P(A_1) = P(A_2) = P(A_3) = 0.8$. Under independence, we have $P(A_1 \cap A_2 \cap A_3) = 0.8^3 = 0.512$. However, we have not established independence. The lower bound for the probability of this intersection, allowing for dependence, is $(0.8 + 0.8 + 0.8) + (1 - 3) = 0.4$.

EXAMPLE 2.5 SIMPSON'S DIVERSITY INDEX AND PROBABILITY

Alpha-diversity considers both species richness (the number of species) and species evenness (homogeneity in species abundance) for a sample. A widely used measure of alpha-diversity, Simpson's index (Simpson 1949), is calculated as

$$D_0 = 1/\sum_{i=1}^{s} p_i^2 \quad \text{or as} \tag{2.13}$$

$$D_1 = 1 - \sum_{i=1}^{s} p_i^2, \tag{2.14}$$

where p_i is the proportional abundance of the ith species at a site $i = \{1,2,\ldots,s\}$. The functions require different interpretations. D_1 gives the probability of reaching into a plot twice and pulling out two different species, while D_0 is the inverse of the probability of reaching into a plot twice and pulling out two individuals of the same species. D_1 is more frequently used since it is more straightforward to interpret.

We can think of p_i as the probability of randomly encountering/sampling the ith species. Let A be the event, we encounter/sample two individuals and both are from species 1. If encounters with individuals occur independently, then $P(A) = P(sp_1 \cap sp_1) = p_1^2$. By the same logic, if B is the event that two individuals are encountered, and both individuals are from species 2, then $P(B) = P(sp_2 \cap sp_2) = p_2^2$. A and B can be considered two disjoint events from a universe representing outcomes where two species are sampled. Thus, $P(A \cup B) = P(A) + P(B)$. It follows from this that the probability of encountering/sampling the same species twice for *any* of the total number of recorded species is simply the union of their squared proportional abundances. One minus this probability is the probability of encountering two different species in a two sample universe. This is the recommended index, D_1.

2.3 Conditional Probability

It is not difficult to think of a situation where events are *not* independent. For instance, let A be the outcome that a student (let us call him Joe) takes an ornithology test on Monday, and let B be the outcome that Joe passes if he takes the same test on Tuesday. Clearly, B is not independent of A. Joe will do better the test on Tuesday if he took the same test on Monday. If events A and B are *not* independent, then we will need to use *conditional probability* when calculating $P(A \cap B)$.

We denote the probability of B (Joe passes a test on Tuesday) given A (he takes the same test on Monday) as $P(B|A)$. This is pronounced: the probability of B *given* A. It is important to note that $P(A|B) \neq P(B|A)$. For instance, consider the conditionality of skin spots and measles. While $P(\text{spots}|\text{measles})=1$, it is also true that $P(\text{measles}|\text{spots}) \neq 1$.

We find $P(B|A)$ and $P(A|B)$ using

$$P(B|A) = \frac{P(A \cap B)}{P(A)},$$

$$P(A|B) = \frac{P(A \cap B)}{P(B)}. \qquad (2.15)$$

For instance, let $A = 0.4$, $B = 0.5$, and $P(A \cap B) = 0.3$, we have

$$P(B|A) = \frac{0.3}{0.4} = 0.75.$$

We see that $P(B|A)$ will be undefined if $P(A) = 0$, while $P(A|B)$ will be undefined if $P(B) = 0$. From Equation 2.15, we have the following important derivation:

$$P(A \cap B) = P(B|A)P(A). \qquad (2.16)$$

Note that this does not invalidate the representation of $P(A \cap B)$ in Equation 2.11. This is because if A and B are independent, then

$$P(B|A) = P(B) \quad \text{and} \quad P(A|B) = P(A). \qquad (2.17)$$

Thus, if A and B are independent, then

$$P(A \cap B) = P(B)P(A).$$

The complement of $P(A|B)$ is $P(A'|B)$. Because the probability of the union of an event and its complement will be 1, it follows that

$$P(A'|B) = 1 - P(A|B). \qquad (2.18)$$

EXAMPLE 2.6 AN ILLUSTRATION OF THE IMPORTANCE OF INDEPENDENCE

Sudden infant death syndrome (SIDS) is the unexpected sudden death of a child under the age of one in which an autopsy does not show an explainable cause of death. SIDS cases have been reduced as parents have increasingly heeded pediatrician recommendations to place babies on their backs when sleeping. However, the factors that underlie SIDS remain poorly understood.

Until the early 2000s, parents in England were often convicted of murder when more than one SIDS case occurred in a family. This was largely because one particular "expert" witness for the prosecution claimed that there was only a one in 73 million chance that two children could die from SIDS in the same family. The rationale was based on the observation that the rate of SIDS in nonsmoking families was 1/8500. Thus, the probability of two deaths would be

$$\frac{1}{8500} \times \frac{1}{8500} = \frac{1}{72,250,000}.$$

However, the reasoning behind this conclusion was flawed. The prosecution had no basis for assuming that SIDS deaths in the same family were independent events. In fact, reevaluation of the data indicated that a baby was approximately 10 times more likely to die from SIDS if a sibling was also a SIDS victim (Hill 2004). In light of this information, the British government decided to review the cases of 258 parents convicted of murdering their children, and 5000 other cases where children were taken away from their parents because of an earlier SIDS death.

2.4 Odds

Closely related to concept of probability are *odds*. It is helpful to define this term using the context of gambling from which it is derived. Specifically, the odds in favor of the outcome A is the amount a gambler would be willing to pay if A *does not* occur compared to a one unit compensation if A *does* occur (cf. Edwards et al. 1963). The odds of A, notated $\Omega(A)$, can be calculated as

$$\Omega(A) = \frac{P(A)}{1 - P(A)} = \frac{P(A)}{P(A')}. \tag{2.19}$$

Thus

$$P(A) = \Omega(A)P(A'), \tag{2.20}$$

$$P(A) = \frac{\Omega(A)}{1 + \Omega(A)}. \tag{2.21}$$

If $P(A) < 0.5$, then $\Omega(A) < 1$, and as $P(A)$ increases from 0.5 to 1, then $\Omega(A)$ increases from 1 to ∞. For example, if $P(A) = 0.6$, then the odds of A are $\Omega(A) = (0.6/0.4) = 1.5$. This can be interpreted as A is 1.5 times more likely to occur than to not occur.

EXAMPLE 2.7

Cooper et al. (1993) studied the effectiveness of zidovudine (AZT) for the treatment of asymptomatic HIV+ patients. This study was of interest because while the effectiveness of AZT for preventing further loss of T-helper cells for patients with full-blown AIDS had been demonstrated in earlier studies, its effectiveness in cases with less severe disease development was unknown. Cooper et al. found that the disease progressed in 28% of asymptomatic HIV+ patients receiving a placebo, and in 16% of the patients receiving AZT. Based on these results, the odds of the disease progressing for placebo patients was 0.28/0.72 = 0.39, while the odds of the disease progressing for AZT patients was 0.16/0.84 = 0.19.

2.4.1 Odds Ratio and Relative Risk

The ratio of the *odds* for two outcomes is their *odds ratio*.

Using the information from Example 2.7, we see that the odds of the HIV progressing in the placebo patients compared to the AZT patients is 0.39/0.19 = 2.04. Thus, the odds for the disease progressing in the placebo group are approximately twice as high as for the AZT group.

The ratio of the *probability* of two events has been given the name *relative risk*. For instance, in Example 2.7, the relative risk of HIV progressing in placebo patients, compared to AZT patients is 0.28/0.16 = 1.75. Thus, the probability that the disease will progress in placebo patients is almost twice that of AZT patients.

2.5 Combinatorial Analysis

Counting methods are fundamentally tied to probability since they allow the determination of the number of outcomes in the sample space. Although the enumeration of points in S will occasionally be straightforward, in many cases it will be extremely difficult, and will require mathematical approaches. We will examine several elementary approaches in this section.

Three characteristics will affect the number of outcomes that occur in sample space. These are (1) whether outcomes are obtained with replacement or without replacement (2) whether or not one considers individual objects to be distinguishable, and (3) whether or not the order of distinguishable objects matters. When sampling *without replacement* objects are chosen one at a time, and are not replaced before the next choice. Conversely, when sampling *with replacement*, these objects would be replaced, and could conceivably be chosen again. To explain the importance of order and distinguishability, consider the side-by-side seating arrangement of three individuals on an advisory panel, two females and one male. There are three possible configurations: $\{♀♂♂, ♂♂♀, ♂♀♂\}$. In this case, order matters, but only when distinguishing males from females. This is because the females are *indistinguishable*. However, assume that the females are *distinguishable* as $♂_1♂_2$. Now, we have six possible arrangements: $\{♀♂_1♂_2, ♂_1♂_2♀, ♂_1♀♂_2, ♀♂_2♂_1, ♂_2♂_1♀, ♂_2♀♂_1\}$. As the number of distinct objects for sampling increases, the number of possible distinct outcomes in sample space also increases.

2.5.1 Multiplication Principle

A large number of counting methods are based on a concept called the *multiplication principle*, summarized in Theorems 2.1 and 2.2.

Theorem 2.1

If one operation can be performed n_1 ways and a second operation can be performed n_2 ways, then there are $n_1 n_2$ ways to perform both operations at the same time.

∎

The multiplication principle can be extended to more than two operations. Specifically, if the ith of r successive operations can be performed in n_i ways, then the total number of ways to carry out all operations is given by

$$N(S) = \prod_{i=1}^{r} n_i = n_1, n_2, \ldots, n_r, \tag{2.22}$$

where $N(S)$ is the total number of possible outcomes in sample space. This idea can be summarized (if n_i's are equal) with the following theorem.

Theorem 2.2

If there are n outcomes for each of r trials in an experiment, then there are n^r possible outcomes in the sample space.

                 ■

Both Theorems 2.1 and 2.2 can be easily demonstrated using biological examples.

EXAMPLE 2.8

A medical researcher is curious how survivorship from lymphoma can be enhanced by dosage combinations of two different drugs. She plans on randomly assigning one of three different dosages of drug 1 and one of two different dosages of drug 2 to subjects. Thus, she will have $3 \times 2 = 6$ distinct treatments in her experiment.

EXAMPLE 2.9

The litter size for domestic dogs (*Canis familiaris*) generally ranges between 6 and 10 pups (HSUS 2013). How many different sorts of litters (combinations of male and female pups) are possible for a litter size of six? How about a litter size of 10?

We have two possible outcomes for each pup (male or female). Thus, for a litter size of 6, we have 2^6 possible outcomes. For instance, one possibility is three male pups followed by three female pups. For a litter size of 10, we have 2^{10} possible outcomes.

```
2^6
[1]  64
2^10
[1]  1024
```

Note that here we assume that the event ♀♀♀♂♂♂ is different from ♂♂♂♀♀♀ (these represent distinct ways to get three male and three female offspring), but that individual male and female outcomes are indistinguishable from each other.

The multiplication principle can also be used in situations when outcomes *are* distinguishable. Consider a group of five distinct females. If five selections from this group are made *with replacement* (i.e., a single individual can be chosen twice or more), and *females are distinguishable*, that is, the selection ♀$_1$♀$_2$ is considered different from ♀$_2$♀$_1$, then there are $5^5 = 3125$ possible distinct outcomes.

The theorems above assume sampling *with replacement*. A version of the multiplication principle can also be used when sampling *without replacement*. The resulting outcomes will be either permutations or combinations.

2.5.2 Permutations

The number of *permutations* enumerates the ways we can arrange distinguishable objects when sampling without replacement and order matters. The number of permutations of any n distinguishable objects is $n!$ where

$$n! = n \cdot (n-1) \cdot (n-2), \ldots, \cdot 1. \tag{2.23}$$

We note that

$$0! = 1! = 1. \tag{2.24}$$

Thus, the number of different arrangements of five distinguishable female offspring is $5! = 120$ ($♀_1♀_2♀_3♀_4♀_5$ and $♀_3♀_2♀_4♀_5♀_1$ are two such arrangements).

```
factorial(5)
[1] 120
```

The number of permutations (*nPerm*) of the n distinguishable objects taken r at a time is

$$nPerm = n!/(n-r)! \tag{2.25}$$

For instance, if we choose two females from the group of five females without replacement (the same individual cannot be chosen twice), and permuted each selection, then this describes the number of permutations when choosing two individuals at a time. We have

$$5!/(5-2)! = 120/6 = 20.$$

```
factorial(5)/factorial(3)
[1] 20
```

The answer can also be obtained by simply finding 5×4. This is because the number of permutations can also be defined as a factorial multiplication up to (but not including) $n - r$:

$$nPerm = n \cdot n - 1 \cdot, \ldots, \cdot n - r. \tag{2.26}$$

2.5.3 Combinations

If we are sampling without or with replacement, but order does not matter, then we are interested in the number of *combinations*. The number of combinations of n distinct objects taken r at a time without replacement is

$$\binom{n}{r} = \frac{n!}{r!(n-r)!}. \tag{2.27}$$

The term $\binom{n}{r}$, is called the *binomial coefficient*, and for the given indices can be pronounced: "*n* choose *r*." It is important to several discrete probability density functions introduced in Chapter 3. The binomial coefficient can also be considered the number of combinations of n indistinguishable objects of which r are of one type and $n - r$ are the other, when sampling with replacement, and order matters.

We can see that there are 10 ways to have puppy litter of size 5 with 2 females. That is

$$\binom{5}{2} = \frac{5!}{2!(5-2)!} = \frac{120}{2(6)} = 10.$$

The function choose provides the binomial coefficient.

```
choose(5,2)
[1] 10
```

2.6 Bayes Rule

Bayes rule inverts the formula for conditional probability. That is, it allows the computation of $P(A|B)$ from $P(B|A)$, and provides the basis for all Bayesian statistical methods.[*] The usefulness of Bayes rule to biologists is evident in the current explosion of papers in biological journals that use Bayesian approaches (for review papers, see Guisan and Zimmermann 2000, Sullivan and Joyce 2005, and McNamara et al. 2006).

The structure of Bayes rule allows one to use prior beliefs or preexisting evidence (i.e., *priors*) about a parameter or hypothesis, θ, in combination with sample data, conditioned by θ, to create a distribution that inverts conditionality. As noted earlier in this chapter, the use of Bayesian priors is a controversial issue in statistics that has been fiercely debated for centuries (Section 2.1.2, Efron 1986). For now, we only note that a prior distribution is considered "proper" if it does not depend on the current data, and sums (or integrates, in the case of continuous priors) to one. Bayes rule has the form

$$P(\theta \mid \text{data}) = \frac{P(\text{data} \mid \theta)P(\theta)}{P(\text{data})}. \tag{2.28}$$

In Equation 2.28, we solve for the *posterior distribution*, $P(\theta|\text{data})$. This will be an inversion of the conditionality that is actually observed, $P(\text{data}|\theta)$. The quantity $P(\text{data}|\theta)$ is often referred to as *data distribution* (Gelman et al. 2003) or the *likelihood function* (see Chapter 4). The probabilities for the *prior distribution* (weights for θ) are denoted $P(\theta)$. The denominator for Bayes rule, $P(\text{data})$, is called the *total probability*. Its form will depend on the form of θ. In the case that outcomes in θ are discrete or categorical, we have

$$P(\text{data}) = \sum_{k=1}^{c} P(\text{data} \mid \theta_k)P(\theta_k),$$

where c is the number of priors. The priors represent the so-called Bayesian *states of nature*. The notation in the summation indicates that one should sum across all values of θ.

The form of $P(\text{data})$ follows from the fact that for any event A, $P(A)$ can be described as

$$P(A) = P(A \cap B) + P(A \cap B'). \tag{2.29}$$

That is, the probability of A is the probability of the intersection of A and another event B, plus the probability of the intersection A and the complement of B. Given this, $P(\text{data})$ can be described as

$$P(\text{data}) = P(\text{data} \cap \theta) + P(\text{data} \cap \theta').$$

Let r represent the number of states of nature not included in θ, that is, $r = c - 1$. We have

$$P(\text{data}) = P(\text{data} \cap \theta) + P(\text{data} \cap \theta'_1) + \cdots + P(\text{data} \cap \theta'_r).$$

---

[*] Bayes rule is generally attributed to Thomas Bayes, an 18th-century Presbyterian minister and part-time mathematician (Chapter 1). Bayes' ideas are now utilized in many cutting-edge statistical techniques. Bayes, however, never published a mathematical paper during his lifetime (his famous theorem was included in a manuscript published 2 years after his death). Bayes and Pierre Simon Laplace (Chapter 1) are often given independent credit as the first to invert conditional probability.

From the definition of conditional probability in Equation 2.16, we have

$$P(\text{data}) = P(\text{data} \mid \theta)P(\theta) + P(\text{data} \mid \theta_1')P(\theta_1') + \cdots + P(\text{data} \mid \theta_r')P(\theta_r'),$$

giving the correct form of the denominator in Equation 2.28.

It is left as a simple exercise to show that Equation 2.28 can be derived from simple algebraic manipulation of the equations of conventional conditional probability (i.e., Equations 2.15 and 2.16).

From Equation 2.18, it follows that

$$P(\theta' \mid \text{data}) = \frac{P(\text{data} \mid \theta')P(\theta')}{P(\text{data})} = 1 - P(\theta \mid \text{data}).$$

Thus, the posterior odds of $\theta \mid$data are

$$\frac{P(\text{data} \mid \theta)P(\theta)}{P(\text{data})} \bigg/ \frac{P(\text{data} \mid \theta')P(\theta')}{P(\text{data})} = \frac{P(\text{data} \mid \theta)P(\theta)}{P(\text{data})} \times \frac{P(\text{data})}{P(\text{data} \mid \theta')P(\theta')}$$

$$= \frac{P(\text{data} \mid \theta)P(0)}{P(\text{data} \mid \theta')P(\theta')} = \frac{P(\text{data} \mid \theta)}{P(\text{data} \mid \theta')}\Omega(\theta). \qquad (2.30)$$

The posterior requires the degrees of belief interpretation because it concerns a conditionality that is not (or cannot be) directly observed. For Bayesian inferences, priors will represent (1) previous knowledge, unrelated and independent of the current experiment; (2) personal weights (beliefs) concerning mechanisms underlying the experiment; or (3) noninformative weights representing a lack of knowledge concerning the experiment (see Section 4.6).

In this chapter, we will use Bayes rule in a very simple way: to calculate posterior probabilities, based on categorical priors, considered one prior at a time. In the biological sciences, Bayes rule is generally utilized in much more sophisticated ways. For example, it can be used to derive quantitative posterior distributions for parameters describing allele frequencies and phylogenetic characters given prior and likelihood distribution functions for those parameters.[*] Advanced Bayesian topics are addressed in later chapters. These include informative and noninformative priors (Section 4.6), simulation distributions (Section 5.5), hypothesis testing (Section 6.7), and analytical models (Chapters 9 and 10).

EXAMPLE 2.10 AN INTRODUCTORY ILLUSTRATION

From previous research, it is believed that three varieties of a species of oak (*Quercus* sp.) occur at equal frequencies on a landscape. Thus each variety has an equal probability, 1/3, of being randomly selected. Assume that variety one will always die when infected by a particular fungus, variety two will die with probability 0.5, and variety three will be unaffected by the infection. A tree is randomly chosen from the landscape, which has died from the fungal infection. What is the probability that it came from variety one?

---

[*] Even more complex are models with multiparameter priors and sampling distributions, and *Bayesian hierarchical models* in which the dependence of the priors on other parameters is acknowledged and modeled, and the dependence of these parameters on still other parameters is modeled, and so on. See Chapters 5, 9, and 10.

Without carefully thinking about the problem, one might claim that the answer is 1/3, the probability of randomly encountering variety one. However, this is incorrect.

We have the following form to our problem:

$$P(\text{variety}_1 \mid \text{death}) = \frac{P(\text{death} \mid \text{variety}_1)P(\text{variety}_1)}{\sum_{k=1}^{3} P(\text{death} \mid \text{variety}_k)P(\text{variety}_k)}.$$

Thus, we have

$$P(\text{variety}_1 \mid \text{death}) = \frac{(1)(1/3)}{(1)(1/3) + (1/2)(1/3) + 0} = \frac{1/3}{3/6} = 6/9 = 2/3.$$

We know from Equations 2.19 and 2.20 that the odds of $\text{variety}_1 \mid$ death are

$$\Omega(\text{variety}_1 \mid \text{death}) = P(\text{variety}_1 \mid \text{death})/P(\text{variety}_1' \mid \text{death}) = (2/3) / [1 - (2/3)] = 2.$$

That is, it is twice as likely that a dead tree will come from variety one than from one of the other two varieties.

The entire posterior distribution will consist of three outcomes:

$$P(\text{variety}_1 \mid \text{death}) = 2/3,$$

$$P(\text{variety}_2 \mid \text{death}) = \frac{(1/2)(1/3)}{3/6} = \frac{1/6}{3/6} = \frac{6}{18} = \frac{1}{3},$$

$$P(\text{variety}_3 \mid \text{death}) = 0.$$

We note that the prior probabilities (1/3, 1/3, 1/3) sum to one because they are legitimate priors, and the posterior probabilities also sum to one (2/3, 1/3, 0) because of the standardization provided by the denominator in Bayes rule. The function `Bayes.disc` performs Bayesian analyses given discrete data and priors (Figure 2.7). The function `Bayes.disc.tck` provides an interactive GUI for `Bayes.disc`.

EXAMPLE 2.11 BAYES RULE AND CYSTIC FIBROSIS DIAGNOSIS

Cystic fibrosis, a deadly lung disorder, is one of the most common autosomal recessive diseases. This is true in part because there are more than 1500 mutational forms of the cystic fibrosis transmembrane regulatory (CFTR) gene (Ratbi et al. 2007). With current diagnostic methods, the true-positive rate (i.e., sensitivity) for detecting CFTR mutations is approximately 0.99; thus the false-negative rate is 0.01. The true-negative rate (i.e., specificity) of this test is generally assumed to be 100%, but it ignores human error. As a result, we will let this probability be 0.998, and let the false-negative probability be 0.002. These data are summarized in Table 2.1.

True-positive rate = $P(\text{POS} \mid \text{CF}+) = 0.99$
False-positive rate = $P(\text{POS} \mid \text{CF}-) = 0.002$
True-negative rate = $P(\text{NEG} \mid \text{CF}-) = 0.998$
False-negative rate = $P(\text{NEG} \mid \text{CF}+) = 0.01$

A question of great interest is "what is the probability that a person is a carrier of some mutational variety of CFTR but will not know it because their test was negative." That

```
Bayes.disc.tck()
```

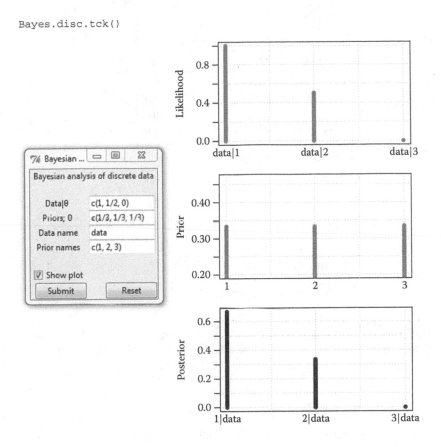

FIGURE 2.7
Likelihood, prior, and posterior probability distributions for the oak infection problem.

is, "what is $P(CF+|NEG)$?" We do not know this, although we do have an estimate for the probability of the conditional outcome, $P(NEG|CF+) = 0.01$.

There will be two possible states of nature resulting in the priors: $P(CF+)$ and $P(CF-)$. As a result, we set up our particular problem as follows:

$$P(CF+ | NEG) = \frac{P(NEG | CF+)P(CF+)}{\sum_{k=1}^{2} P(NEG | CF+_k)P(CF+_k)}.$$

Our priors will concern probabilities for the presence/absence of CFTR mutations in the population at large. These probabilities will in turn depend on ethnicity (Table 2.2).

TABLE 2.1

Detection of CFTR Mutations

Diagnostic Test Result	CF±	
	CF+	CF−
POS	0.99	0.002
NEG	0.01	0.998

TABLE 2.2

Carrier Risk of CF for Five Ethnic Groups

Ethnicity	Carrier Probability
Non-Hispanic Caucasian	0.040
Hispanic American	0.017
African American	0.016
Ashkenazi Jewish	0.042
Asian American	0.011

To find $P(CF+|NEG)$ for Hispanic Americans, we have

$$= \frac{0.01 \cdot 0.017}{\left[P(NEG|CF+)P(CF+)\right] + \left[P(NEG|CF-)P(CF-)\right]}$$

$$= \frac{0.01 \times 0.017}{0.01 \times 0.017 + 0.998 \times 0.983}$$

$$= \frac{0.00017}{0.981204} = 0.0001732565.$$

We weight our observed data with prior information, allowing more concise inferential statements about Hispanic Americans and cystic fibrosis. Again, the function `Bayes.disc.tck` can be used summarize our results (Figure 2.8).

```
                CF+                 CF-
Likelihood    0.0100000000        0.9980000
Prior         0.0170000000        0.9830000
Posterior     0.0001732565        0.9998267
```

Our likelihoods consist of $P(NEG|CF+)$ and $P(NEG|CF-)$, that is, the false-negative rate and true-negative rate from the experimental trial. The posterior distribution also consists of two outcomes $P(CF+|NEG)$ and $P(CF-|NEG)$. Again, we are particularly interested in $P(CF+|NEG)$. If our priors are correct, then 0.0173% ($\approx$ 173/1,000,000) of the Hispanic carriers that pass the test will be carriers.

```
Bayes.disc.tck()
```

FIGURE 2.8
Specification for the cystic fibrosis problem.

2.7 Summary

This chapter introduces the topic of probability. Probability is useful for modeling random variables (those whose outcomes cannot be known prior to measurement).

- Two general interpretations for probability are possible. In the frequentist paradigm, the probability of A, $P(A)$, is simply the limiting relative frequency of A. Conversely, from a Bayesian perspective, $P(A)$ is measure of an investigator's belief that A is true. Differences in frequentist and Bayesian conceptualizations do not affect the mathematical underpinnings of probability.

- Numerically, a probability is always a proportion: a number in [0,1]. When discussing events, for example, A and B, it is important to be able to determine whether these are disjoint, and/or independent. These characteristics will affect the computation of the probability of A *or* B, that is, $P(A \cup B)$, and the probability or A *and* B, that is, $P(A \cap B)$. If two events are not disjoint, then we must account for $P(A \cap B)$ to calculate $P(A \cup B)$. Conversely, if two events are not independent, we must have knowledge of the conditional probability $P(A|B)$ or $P(B|A)$ (they are not inverses) to compute $P(A \cap B)$.

- Methods of combinatorial analysis are important for determining the number of events and/or distinct outcomes in the sample space. Permutations are ways we can arrange distinguishable objects when sampling without replacement and order matters. Combinations are ways we can sample with or without replacement and order does not matter.

- The chapter concludes with a description of simple Bayesian approaches for analysis of discrete priors and single posterior outcomes.

EXERCISES

1. Answer the following questions contrasting deterministic and probabilistic models:
   a. Create a mathematical model that defines the circumference of a circle, C, as a linear function of its diameter, D. Is this a deterministic model? Why?
   b. Studies have shown that species of plants often demonstrate the −3/2 self-thinning rule. That is, after achieving a certain density of plants 2 log(individuals) will tend to be lost from a site for each 3 log(grams of biomass) gained. Is this a deterministic model? Why?
   c. The model in (b) does not work with perfect consistency for any species, prompting the model: log(biomass) = −3/2 log(density) + ε, where ε is a random variable. Is this a deterministic model? Why?

2. Distinguish frequentist and Bayesian conceptions of probability.

3. Simulate 1000 coin tosses of a fair coin using the defaults of `anm.coin.tck()`. Does the proportion of heads generally approach 0.5 after 5 coin flips? After 100 flips? After 1000 flips?

4. Simulate rolls from a loaded die using `anm.die.tck()`. Does "loading" the dice affect convergence in probability? That is, do probabilities for outcomes still converge to particular proportions given a large number of flips? Explain.

5. Let the sample space S represent two coin tosses using a fair coin. Let A be outcomes in S with at least one head, and let B be an outcome with two heads. Which of the following are true?

   a. $S \subset A$

   b. head $\cap$ head $\in A$

   c. head $\cap$ tail $\notin A$

   d. $S \supset B$

   e. $A \not\subset B$

6. Which of the following are true?

   a. If $P(A \cup B) = P(A)$, then $B \subset A$.

   b. If $P(A \cap B) = P(B)$, then $B \subset A$.

7. Create Venn diagrams for the following scenarios. Use the *asbio* function Venn.tck to maximize figure accuracy.

   a. $P(A) = 0.65$, $P(B) = 0.35$, $P(A \cap B) = 0.0$

   b. $P(A) = 0.4$, $P(B) = 0.5$, $P(A \cap B) = 0.4$; is $A \subset B$?

   c. $P(A) = 0.7$, $P(B) = 0.5$, $P(A \cap B) = 0.35$

   d. $P(A) = 0.7$, $P(B) = 0.5$, $P(A \cap B) = 0$; was there a problem executing this figure? Why? Be specific.

8. Address the following questions.

   a. Draw a Venn diagram representing $P(A) = 0.5$, $P(B) = 0.5$, $P(A \cap B) = 0.25$. Use the *asbio* function Venn.tck to maximize figure accuracy.

   b. Are A and B disjoint?

   c. Are A and B independent?

   d. Let $P(A) = 0.5$, $P(B) = 0.5$, and let $P(A \cap B) \neq 0$. Can $P(A \cap B)$ conceivably equal a probability other than 0.25? If so draw some. How many other legitimate possibilities are there? What does this say about the rarity of independence?

9. A scientist is concerned with the occurrence of big sage brush *Artemisia tridentata* in a sagebrush-steppe ecosystem. She knows that the probability of finding *A. tridentata* at habitat $A = 0.2$, the probability of finding it at habitat $B = 0.4$, and the probability of finding it at both habitat A and $B = 0.08$.

   a. Create a Venn diagram for probabilities associated with this problem (use the *asbio* function Venn.tck to maximize figure accuracy).

   b. Are A and B mutually exclusive? Why?

   c. Are A and B independent? Why?

   d. What is $P(A \cup B)$?

   e. What is $P(A \cap A')$?

   f. What is $P(A' \cap B')$?

   g. What is $P(A' \cup B')$?

   h. What is $P(A \cup B')$?

   i. What is $P(A' \cup B)$?

   j. What is $P(A|B)$?

   k. What is $P(B|A)$?

10. Assume that we expand our study in Question 9 to include a third habitat, C. Assume further that $P(C) = 0.3$, $P(A \cap C) = 0.05$, $P(B \cap C) = 0.03$, and $P(A \cap B \cap C) = 0.01$. Find $P(A \cup B \cup C)$.

11. Use Venn.tck() to represent a situation where $A \subset B$ (any legitimate probabilities can be used). Without using math, intuit the answers for $P(A|B)$, and $P(B|A)$. Now check your answers using the formula for conditional probability.

12. Let $P(A) = 0.3$, $P(B) = 0.7$, and $P(C) = 0.05$.

    a. According to Boole's inequality, what is the largest possible value for $P(A \cup B \cup C)$?

    b. According to Bonferroni's inequality, what are the largest and smallest possible values for $P(A \cap B \cap C)$?

    c. Prove that the two variations shown for Equation 2.12 are equivalent.

13. What are all the possible outcomes (in head and tails) for a toss using five coins. (see Theorem 2.2)?

14. How many ways can a 20-question test be answered when each question has four possible multiple choice answers?

15. "Rapid HIV tests" provide results for the presence/absence of HIV antibodies in a matter of minutes. One such test, which uses oral fluids, has a probability of 0.004 of producing a false-positive result (Branson 2005). Given that 500 people, who are antibody free, are tested using this method, how many will receive a false positive result? Assume that test results for different individuals are independent.

16. Using the definitions of $r!$ and $n!$ (Equation 2.23) and $\binom{n}{r}$ (Equation 2.27), prove that

$$\binom{n}{r} = \binom{n-1}{r-1} + \binom{n-1}{r}.$$

17. DNA contains the nucleotide bases (G, A, T, C). How many different ways can the bases be rearranged in a 4 base pair strand?

18. A restriction endonuclease enzyme is used to cut the DNA in Question 17 resulting in two pieces of DNA, each 2 bases long.

    a. How many arrangements are possible if order is important?

    b. How many arrangements are possible if order is not important?

19. A fair 12-sided die (a dodecahedron) is thrown. What is the probability of

    a. Getting a 9 on one roll and at least a 10 on a second roll?

    b. Getting a 6 or a 7?

    c. The sum of two tosses being less than 6?

20. While varying with season, the litter size for brown spiny field mice (*Mus plantythrix*) on dry land sites in India is approximately four (Raj 1994).

    a. List all the possible outcomes in terms of male and female offspring for this litter size.

b. How many outcomes are possible? How did you mathematically verify this?

c. What is the probability of each outcome?

d. Are each of the outcomes mutually exclusive? Why? What will this allow you to do?

e. Given your answer in (c), are each of the outcomes independent? Why?

f. What is the probability of getting three males in a litter of three?

g. What is the probability of getting three females in a litter of four?

h. What is the probability of getting at least one female in a litter of four?

i. What is the probability of getting no females?

21. For litters of spiny field mice (*Mus plantythrix*) of size 4 (see Question 20), let $P(A)$ be the probability of getting exactly 2 females, $P(B)$ be the probability of getting at least 1 female, and $P(C)$ be the probability of getting no females.

a. What is $P(B|A)$?

b. What is $P(B'|A)$?

c. What is $P(A|C)$?

d. Are A and B independent? What about A and C? What about B and C? Defend your answers.

22. If $P(A) = 0.3$, find the odds of A, $\Omega(A)$. Also, find $\Omega(A')$. Finally, compute the odds ratio: $\Omega(A)/\Omega(A')$. Interpret your results.

23. Distinguish the definitions for permutations and combinations. How do you calculate each?

24. We have a deck of 12 cards and 3 cards are repeatedly drawn from the deck and then replaced. How many possible hands are there?

25. In Question 24, how many hands are possible when sampling without replacement? Assume order matters.

26. In Question 25, how many hands are possible when sampling without replacement if order does not matter?

27. Prove that Bayes rule can be derived from the conventional formula for conditional probability, that is, Equations 2.15 and 2.16.

28. Prove Equation 2.29.

29. In Example 2.10, what is $P(\text{variety}_1|\text{death} \cup \text{variety}_2|\text{death} \cup \text{variety}_3|\text{death})$?

30. Recent reports indicate that 57% of college students are women. In 2008, the American Heart Association reports that 23.1% of men and 18.3% of women are smokers. If we can assume that college attendance and smoking are independent (i.e., 23.1% of college men and 18.3% of college women are smokers), then

a. What is the probability that a college student observed smoking is a man?

b. What is Ω (smoking|man)?

31. Rao et al. (1998) reported on the utility of using computerized tomography (CT) as a diagnostic test for patients with clinically suspected appendicitis. Traditional clinical methods of diagnosis, and diagnosis using the aid of CT were used on 100 patients. Whether patients actually had appendicitis was determined later by examining the appendix following an appendectomy.

Here are the clinical results.

	Presence of Appendicitis	
Clinical Diagnosis	APP+	APP–
Definite appendicitis (DA)	18	5
Probably or equivocally appendicitis (PA)	28	33
Probably not, or definitely no appendicitis (NA)	7	9
Total	53	47

Here are the CT-aided results.

	Presence of Appendicitis	
CT-Aided Diagnosis	APP+	APP–
Definite appendicitis (DA)	50	1
Probably or equivocally appendicitis (PA)	2	0
Probably not, or definitely no appendicitis (NA)	1	46
Total	53	47

The 1996 probability of appendicitis among U.S. citizens (APP+) was approximately 0.00108 (Ott and Longnecker 2004).

a. Determine $P(DA|APP+)$, $P(DA|APP-)$, $P(NA|APP+)$, and $P(NA|APP-)$ for clinical diagnosis method and the CT method (this does not require Bayes theorem).

b. Find the probability that a patient *did* have appendicitis (APP+) given that the determination was definite appendicitis (DA). Calculate this for both the clinical diagnosis method and the CT method.

c. Find the probability that a patient *did not* have appendicitis (APP–) given that the determination was DA for the CT method. Note: this may be higher than you might think.

d. What is the sensitivity and specificity of diagnoses with and without CT? Ignore PA data.

e. Given results in (c) and (d) find the posterior odds of APP–|DA compared to APP+|DA.

f. Regardless of the problems with the CT procedure, Rao et al. (1998) thought it would likely reduce health care costs. Does this seem reasonable? Base your answer on your results from (d) and (e).

3

Probability Density Functions

> The theory of probabilities is at bottom nothing but common sense reduced to calculus.
>
> Laplace, *Théorie analytique des probabilités*, 1820

3.1 Introduction

In Chapter 2, we learned how to calculate probabilities for events that comprised unions, intersections, and conditional and reversed conditional associations of variable outcomes. In this chapter, we will discover that it is often possible (and extremely useful) to express the probability distributions of random variables with mathematical functions.

The examples and exercises in Chapter 2 dealt only with *categorical* random variables that lack numerically meaningful outcomes. As an illustration, in Example 2.2, we examined probabilities associated with the blood types A, B, O, and AB. Instead of labeling blood-type outcomes with their actual names, we could have also used the numbers 1–4. However, despite this system, the categories would remain quantitatively uninterpretable. Blood type 4 would not suddenly be four times larger (in any sense) than blood type 1. Outcomes from *quantitative* variables, the focus of this chapter, will have an exact numeric meaning. That is, for a quantitative variable X, the outcome $X = 4$ will always be exactly four times as large as the outcome $X = 1$, with respect to the baseline $X = 0$.

Quantitative random variables will either be discrete or continuous. *Discrete random variables* have a limited number of distinct outcomes. Consider a variable representing the counts of mountain goats at a particular location. We can count 0 mountain goats, 1 mountain goat, 2 mountain goats, and so on. The variable outcomes will be limited to *natural numbers* (nonnegative integers). We cannot, for example, count 2.37 goats. *Continuous random variables* have an infinite number of distinct outcomes in any interval. Consider the weights of individual goats in a population. Our capacity to differentiate goat weights will depend on the precision of our scale. Although such a weighing instrument does not exist, we can imagine one that is able to distinguish goat weights for infinitely small differences, allowing a continuous range of outcomes.

A *probability distribution* assigns probabilities to outcomes from a random variable. A mathematical expression that defines the probability distribution for a quantitative random variable is called a *probability density function* or *pdf*. A pdf is used to determine density (defined below) for particular outcomes. Density, in turn, can be used directly or indirectly (depending on whether a random variable is discrete or continuous) to compute probability.

Pdfs describing discrete random variables are often called *probability mass functions* or *pmfs*, while variable outcomes, x_i, are called *mass points*. However, because it will be clear in this text whether discrete or continuous random variables are being used, the term pdf is used to describe the probability-generating functions of *both* types of variables (cf. Bain and Engelhart 1992). This approach corresponds to the naming conventions for probability generating algorithms in **R**.

3.1.1 How to Read This Chapter

This chapter contains both introductory and advanced materials. For most introductory classes, coverage of Section 3.1, Section 3.2 (which introduces the binomial and normal distributions), and Section 3.4 will be sufficient. Descriptions of additional pdfs in Section 3.3 will be more appropriate for advanced students. Introductory classes, however, will be helped by (1) reading materials concerning the χ^2, t-, F-distributions in Section 3.3 as these topics are reintroduced in Chapter 5, and (2) reviewing the exponential and geometric distributions as these pdfs are discussed in Sections 4.3.9 and 4.5, respectively. Topics in Section 3.3 can also be temporarily skimmed or skipped by advanced classes without loss of continuity, and examined in association with later advanced topics including Bayesian approaches and generalized linear models starting in Chapter 4.

3.1.2 So, What *Is* Density?

A pdf is a function, denoted $f(x)$, that varies with quantitative random variable outcomes, x. The random variable itself is denoted by X. The output generated by a pdf is called *density*.

For a discrete random variable, the density of an outcome is simply its relative frequency given all possible trials. That is, discrete density corresponds to a frequentist conception of probability. As a result, a discrete pdf has the general form

$$f(x) = P(X = x),\tag{3.1}$$

where $x \in X = \{x_1, x_2, ...\}$.

A graphical tool called a *histogram* plots the frequency or the relative frequency of quantitative outcomes within specified ranges (i.e., *bins*). Consider a random variable X in which the outcome 1 occurs 2 times, the outcome 2 occurs 5 times, and the outcome 3 occurs 3 times. In a histogram, we might plot the outcome frequencies as a function of three bins ($0 < X \le 1$, $1 < X \le 2$, and $2 < X \le 3$; Figure 3.1). We can then make the sum of areas of the bins equal one. In this case, because exactly one outcome is allocated to each bin, the bin heights will give the relative frequencies of the outcomes (right axis). Assuming that the observed frequencies perfectly describe true patterns in X, the bin heights will be equivalent to both discrete density and probability.*

As with discrete density, continuous density is equivalent to the height of a histogram describing the frequency of outcomes, standardized so that the total area of the bins equals one. However, because there are infinite numbers of distinct outcomes in any continuous interval, precise depiction of the variable requires an infinite number of bins. Because of

---

* While histograms serve as useful first approximations of distributions, they will represent random variables with imprecision for three reasons. First, the number of bins will dramatically affect depiction of the distribution. Second, outcomes from discrete random variables can only occur at distinct values, making bins unnecessary and misleading. Third, continuous random variables will require the impossibility of an infinite number of bins for precise representation. Other, more exact graphical approaches for pdf depiction are used in the remainder of this chapter.

```
hist(c(rep(1, 2), rep(2, 5), rep(3, 3)), breaks = c(0, 1, 2, 3))
axis(4, at = 0:5, labels = seq(0, 0.5, 0.1))
mtext("Density", 4, line = 2.7, cex = 1.2)
```

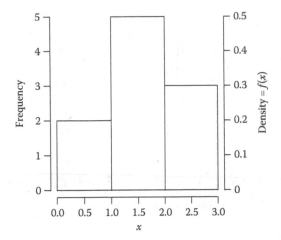

FIGURE 3.1
Histogram for a simple random variable with a bin for each of the three possible outcomes. Frequencies (counts) are shown on the left-hand axis and relative frequencies are shown on the right-hand axis. Because (1) the variable outcomes are discrete, (2) there is a bin for each outcome, and (3) we assume that the frequencies in the bins reflect the true proportions from the random variable, the relative frequencies will be equivalent to both probability and density.

this representation (an infinite number of infinitely small bins) the top of the histogram appears smooth.* Snapshots of this process are shown in Figure 3.2.

The height of the pdf curve provides insight into patterns of relative frequencies for outcomes in a continuous distribution. However, unlike a discrete distribution, it will not represent probability at $X = x$. Indeed, as we will see, $P(X = x)$ will always equal zero for a continuous pdf.

We can summarize the preceding descriptions of discrete and continuous distributions with two statements:

1. Both discrete and continuous pdfs calculate a quantity called density. The "height" of any pdf "curve" at an outcome x will equal the density of x, given as $f(x)$.

2. Density is equivalent to probability for a discrete pdf, but not for a continuous pdf.

Whether a pdf defines a discrete or continuous distribution, it will be *valid* only if two conditions are met:

1. $f(x) \geq 0$ for all real x
2. $\sum_x f(x) = 1$ for a discrete pdf, or $\int_{-\infty}^{\infty} f(x)dx = 1$ for a continuous pdf

---

* This describes the method for computing a *Riemann sum*. Bernhard Riemann (1826–1866) was a student of Carl Friedrich Gauss (1777–1855) at the University of Göttingen. Riemann made large contributions to the fields of math and physics. For instance, his concept of space and geometry was utilized by Einstein in his theory of general relativity (Stewart 2003). In a Riemann sum, the abscissa (X-axis) is divided into intervals representing the bottom of rectangles where the top of the rectangles end at the curve of interest. The sum of the areas of the rectangles can be used to approximate an integral (the area under the curve).

```
see.pdf.conc.tck()
```

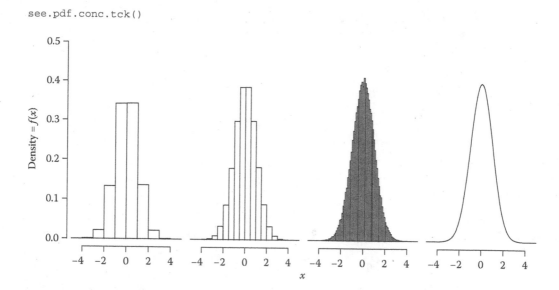

FIGURE 3.2
Conceptualization of a continuous pdf as the height of a histogram with infinitely small bins.

That is, (1) the densities for a particular outcome must each be greater than or equal to 0, and (2) if a random variable is discrete, the sum of densities for all possible outcomes must equal 1, and, if a random variable is continuous, the total area under the density curve, that is, the definite integral from $-\infty$ to ∞, must equal 1.

3.1.3 Cumulative Distribution Function

A *cumulative distribution function (cdf)* for a random variable X is denoted $F(x)$. Cdf output is obtained by summing or integrating the pdf, $f(x)$, between $-\infty$ and an outcome x, depending on whether X is discrete or continuous, respectively. Thus, a cdf gives the *lower tail probability* $P(X \le x)$ for both discrete and continuous random variables.

For a discrete random variable, the cdf represents the probability of the union of contiguous but disjoint events less than or equal to x (Chapter 2):

$$F(x) = P(X \le x) = \sum_{x_i \le x} f(x_i). \tag{3.2}$$

For instance, for the discrete random variable shown in Figure 3.1 we have

$$F(2) = P(X \le 2) = f(1) + f(2) = 0.2 + 0.5 = 0.7.$$

For a continuous random variable, the cdf is a function whose relation to its pdf, $f(x)$, can be expressed as

$$F(x) = P(X \le x) = \int_{-\infty}^{x} f(t)dt. \tag{3.3}$$

For instance, the continuous random variable shown in Figure 3.2 is symmetric and centered at zero. Because the area under the pdf curve must equal 1, we know that

$$F(0) = P(X \le 0) = 0.5.$$

From Equation 3.3, we see that integration is theoretically necessary to calculate the probability for a continuous random variable. Closed functional forms for some continuous cdfs, however, allow one to bypass this step. The probability $P(X = x)$ is uninformative in a continuous distribution because the results of integration at a point will always be 0. Specifically, the upper and lower endpoints for definite integration will be the same. Thus, the solution of the integral will be the area of a line, and the area of a line is zero.

Instead, we calculate continuous probability for a *range* of outcomes. Let X be a continuous random variable with pdf $f(x)$, and let $\{a, b\} \in X$, where $a < b$. To obtain $P(a \le X \le b)$, we find $\int_a^b f(x)dx$.

Because the probability of any exact continuous outcome $(X = x)$ will always be zero, this also means that $P(a < X < b) = P(a \le X < b) = P(a < X \le b) = P(a \le X \le b)$. That is, the probabilities associated with "less than" and "less than or equal to" inequalities will be equal, and the probabilities associated with "greater than" and "greater than or equal to" inequalities will also be equal. This will *not* be true for a discrete pdf because $P(X = a)$ and $P(X = b)$ can each be greater than zero.

The *inverse cdf or quantile function* is denoted as F^{-1} and gives x for a specified lower tailed probability. That is, for a random variable X, the function provides the outcome x associated with the probability $P(X \le x)$. For example, if $P(X \le 3) = 0.4$, then $F(3) = 0.4$, and $F^{-1}(0.4) = 3$.

Two other biologically important extensions of cdfs are *survivorship functions* and *hazard functions*. If a random variable T has pdf $f(t)$, giving densities for survival time (i.e., time until death), then the survivorship function, denoted $R(t)$, can be calculated as $1 - F(t)$, where $F(t)$ is the cdf of T. Because $F(t)$ will give the probability of survivorship *up to* time t, the survivorship function represents the probability of survival beyond time t. The *hazard function*, denoted $h(t)$, is the ratio of the pdf and the survivorship function. That is, $h(t) = f(t)/R(t)$ (Section 3.3.2.4).

EXAMPLE 3.1 TABULAR EXPRESSION OF A DISCRETE PDF

Imagine that you are an alpine ecologist studying the demographics of mountain goats (*Oreomnos americanus*). You observe a certain ridge top for a long period of time and decide that the probabilities of goat counts in this area can be described by Table 3.1. The table here serves as a pdf. Nothing would be gained by summarizing this variable mathematically.

Because the densities of all outcomes are greater than or equal to 0, and the sum of all the densities equals 1, this appears to be a valid pdf. The pdf and cdf are graphically represented in Figure 3.3.

TABLE 3.1

Probabilities of Mountain Goat Counts

x = Goat count	0	1	2	3	4
$f(x)$ = Probability of a particular goat count	0.5	0.3	0.1	0.05	0.05

```
##pdf
x <- c(0, 1, 2, 3, 4)
f.x <- c(0.5, 0.3, 0.1, 0.05, 0.05)
plot(x, f.x, xlab = "x", ylab = "f(x)", pch 16)
segments(x, 0, x, f.x)

##cdf
x <- c(0, 1, 2, 3, 4)
F.x <- c(0.5, 0.8, 0.9, 0.95, 1.0)
plot(x, F.x, xlab = "x", ylab = "F(x)", pch = 16)
segments(x, F.x, x + 1, F.x)
points(1:4, F.x[1:4])
```

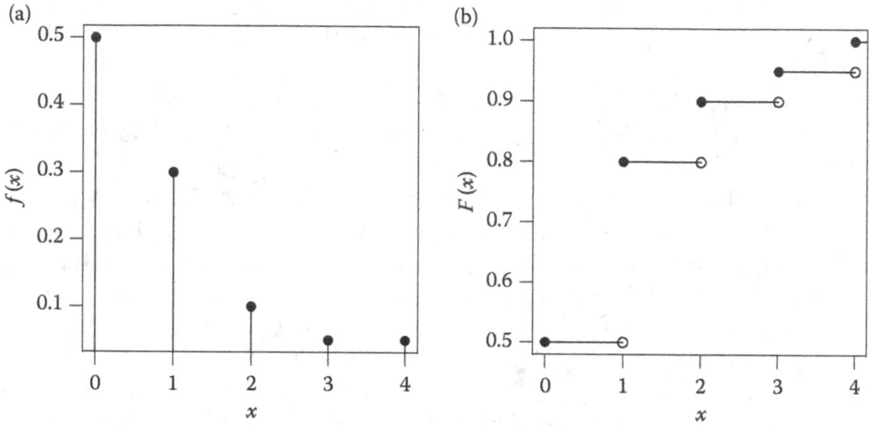

FIGURE 3.3

The pdf (a) and cdf (b) from the goat sighting distribution in Example 3.1. Filled dots indicate that the point is included in the density and/or cumulative distribution functions. Open dots indicate the point is not included in the cdf.

Discrete density is equal to probability. Thus, the probability of seeing two goats is $f(2) = P(X = 2) = 0.1$ (Figure 3.3a). The cdf (Figure 3.3b) provides the cumulative probabilities for outcomes in the pdf. These are the sums of the lengths of the lines in Figure 3.3a, for all $x_i \leq x$. For example, the probability of seeing two or fewer goats is 0.9. This is because $F(2) = P(X \leq 2) = 0.1 + 0.3 + 0.5 = 0.9$. The distribution of goat counts is only defined for the set {0, 1, 2, 3, 4}. As a result, for any outcome $X \geq 4$, the cdf will always equal one because it will always include the universal set (Figure 3.3b).

3.2 Introductory Examples of pdfs

Conventional statistical methods use a handful of pdfs with well-understood mathematical properties. Procedures built on these algorithms are called *parametric* because they use pdfs with known parametric forms. A *parameter* can be defined as a fixed numeric characteristic of a pdf. Infinitesimally changing the parameters for any pdf will result in an infinite number of distinct distributions.

3.2.1 Discrete pdfs

Discrete pdfs describe the probabilistic form of discrete random variables (see Table 3.1). Consequently, these functions are often used to define the probability for the number of successes in a single trial (either a 0 or 1) or counts (e.g., population numbers for a species on a landscape).

3.2.1.1 Bernoulli Distribution

The *Bernoulli distribution* is arguably the simplest useful pdf. It defines the probability of success for a single random binary trial.[*] A *binary trial* has only two possible outcomes, for example, presence/absence, life/death, male/female, and head/tail. Either outcome from any of these examples could be defined as a "success" depending on the focus of the investigator. The Bernoulli pdf has the form

$$f(x) = \pi^x(1-\pi)^{1-x},$$

(3.4)

where
1. x defines the number of successes for a single binary trial, $x \in \{0, 1\}$.
2. $0 < \pi < 1$.

The parameter π in Equation 3.4 defines the probability of a binary success. Note that $P(X = 0) = 1 - \pi$, and $P(X = 1) = \pi$. Thus, the Bernoulli cdf is simply

$$F(x) = \begin{cases} 1-\pi & x = 0 \\ 1 & x = 1 \end{cases}.$$

(3.5)

> **EXAMPLE 3.2**
>
> The probability of a live male birth in the United States in 2002 was approximately 0.5117 (Davis 2007). Thus, counting a male birth as a success, and assuming the independence of gender outcomes, the probability that a child born in 2002 was female is
>
> $$P(X = 0) = f(0) = 0.5117^0(1 - 0.5117)^1 = 0.4883.$$

3.2.1.2 Binomial Distribution

The *binomial distribution* defines the probability of a particular number of successes given n independent and identically distributed Bernoulli trials. Its pdf is

$$f(x) = \binom{n}{x}\pi^x(1-\pi)^{n-x},$$

(3.6)

---

[*] The Bernoulli distribution was named for Jakob Bernoulli (1655–1705), an early advocate and developer of probability density functions for use in scientific experiments.

where
1. The individual Bernoulli trials defining the distribution are independent, with an unchanging probability of success.
2. x is the number of successful trials, $x \in \{0, 1, 2,\ldots,n\}$.
3. $n > 0$.
4. $0 < \pi < 1$.

The binomial distribution has two parameters: the number of trials, n, and the probability of success for a single trial, π. The outcome from a binomial variable will always be a nonnegative integer with an upper bound at n. The latter characteristic feature distinguishes this pdf from the Poisson distribution (Section 3.3.1.1), which does not have an upper bound to its number of successes.

The binomial pdf follows directly from the rules of probability described in Chapter 2. Let X be a sample space containing two independent Bernoulli trials. Now, let the event A be a success occurring with a fixed probability π, and let B be a failure with a fixed probability $1 - \pi$. Because outcomes are independent, the probability of two successes is $P(A \cap A) = P(A)P(A) = \pi^2$, the probability of two failures is $P(B \cap B) = P(B)P(B) = (1 - \pi)^2$, and the probability of one success and one failure is $2P(A \cap B) = 2P(A)P(B) = 2\pi(1 - \pi)$. Note that we multiply $P(A \cap B)$ by 2 because there are two ways of getting one success and one failure in the sample space: $A \cap B$ and $B \cap A$. The number two comes from the quantity $\binom{2}{1} = 2$, which gives the total number of combinations for 1 success and $2 - 1 = 1$ failure (Chapter 2).

Extending this conceptualization to n trials, the probability of x successes is

$$P\left(\bigcap_{i=1}^{x} A_i\right) = \prod_{i=1}^{x} P(A_i) = \pi^x$$

and the probability of $n - x$ failures is

$$P\left(\bigcap_{i=1}^{n-x} B_i\right) = \prod_{i=1}^{n-x} P(B_i) = (1 - \pi)^{n-x}.$$

Thus, the product $\pi^x(1 - \pi)^{n-x}$ gives the probability for one of the combinations of x successes and $n - x$ failures. The quantity $\binom{n}{x}$ gives the total number of combinations for x successes and $n - x$ failures. Finally, the product $\binom{n}{x}\pi^x(1 - \pi)^{n-x}$ gives the probability of x successes in n trials.

The binomial distribution and all other pdfs are generally indicated using an abbreviation for the name of the distribution followed by the parameter values in parentheses. If a random variable X has a binomial distribution we commonly denote this as $X \sim \mathrm{BIN}(n, \pi)$. The tilde (~) here means "follows." We note that the Bernoulli pdf is a special case of the binomial distribution. Specifically, because $\binom{1}{1} = \binom{1}{0} = 1$, it is equivalent to $\mathrm{BIN}(1, \pi)$.

```
see.bincdf.tck ()
```

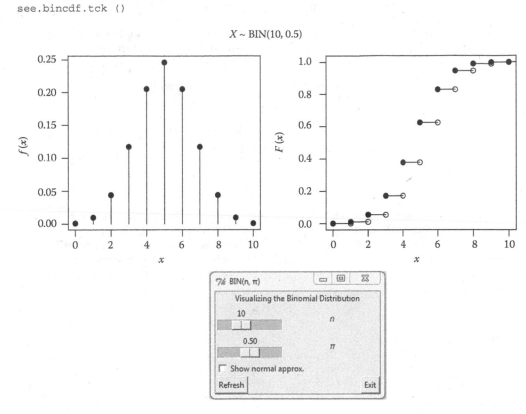

FIGURE 3.4
The pdf and cdf for $X \sim$ BIN(10, 0.5). Characteristics of the functions are revealed by moving sliders, which in turn control parameter values. Filled dots indicate that the point is included in the density and/or cumulative distribution functions. Open dots indicate the point is not included in the cdf.

The effect of parameter values on the binomial pdf and cdf can be interactively demonstrated using the function `see.bincdf.tck` (Figure 3.4). Note that the distribution is symmetric whenever $\pi = 0.5$.[*]

The binomial distribution is most frequently used by biologists to define the probability for an observed number of binary successes when randomly sampling from a population *with* replacement.

EXAMPLE 3.3

Native cutthroat trout (*Salmo clarki*) in Yellowstone Lake (Yellowstone National Park, Wyoming, USA) are currently being threatened by nonnative lake trout (*Salvelinus namaycush*). Juvenile cutthroats are highly vulnerable to the larger piscivorous lake trout. In fact, one mature lake trout can eat 41 juvenile cutthroats per year (Ruzycki et al. 2003). In 1994, the year that the first confirmed lake trout was caught in Yellowstone Lake, a gill net survey found that approximately 68% of fish in the lake were cutthroats (Kaeding et al. 1995). Suppose that recent catch and release data from anglers showed

---

[*] Interactive GUIs for all the pdfs in this chapter can be obtained by typing `see.pdfdriver.tck()`.

that of 50 fish caught in the lake only 25 were cutthroats.[*] What is the probability of this outcome given 1994 fish proportions?

Assuming that all fish species have an equal probability of being caught with the sampling method being used, and individual catches are independent, we have

$$f(25) = \binom{50}{25}0.68^{25}0.32^{25} = \frac{50\,!}{25\,!(50-25)!}0.68^{25}0.32^{25}$$

$$= 1.26 \cdot 10^{14} \times 6.50 \cdot 10^{-5} \times 4.25 \cdot 10^{-13} = 0.0035.$$

The binomial pdf is provided by the function dbinom.

```
dbinom(25, 50, 0.68)
[1] 0.00349346
```

This small probability indicates that is extremely unlikely that 25 cutthroats would be caught out of 50 total fish if the parameter value $\pi = 0.68$ is still correct. In fact, we would expect this outcome to occur only 0.3% of the time. Given the recent catch and release data, the current proportion of cutthroats in the lake is probably much lower than 0.68.

It is always helpful to create a visual representation of problems associated with pdfs (Figure 3.5).

```
shade.bin.tck()
```

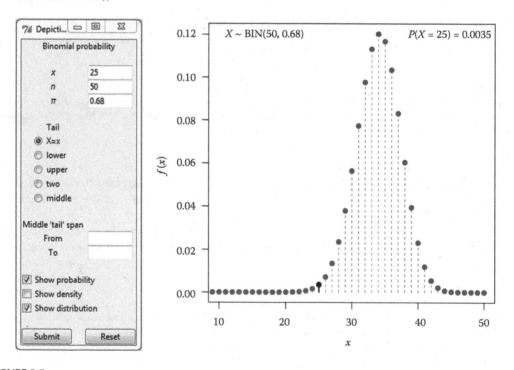

FIGURE 3.5
Depiction of $P(X = 25)$ for $X \sim$ BIN(50, 0.68) in the cutthroat trout example. The height of the short black line at $X = 25$ is both the density $f(25)$ and the probability $P(X = 25)$.

---

[*] The conceptual projection P(cutthroat) = 0.5 is extremely optimistic (cf. Koel et al. 2005).

We note that the probability of catching 25 cutthroat trout over 50 trials given $\pi = 0.68$ is the same as catching 25 noncutthroats over 50 trials, if $\pi = 1 - 0.68$. This characteristic is due to the principles of the binomial coefficient and a rule called the *binomial theorem* (see Exercise 6 at the end of the chapter).

```
dbinom(25, 50, 0.32)
[1] 0.00349346
```

We might also want to know $P(X \le 25)$, the probability of catching 25 or fewer cutthroats, that is, 0 fish, or 1 fish, up to and including 25 fish. For this calculation, we would use the binomial cdf. That is, we would find

$$P(X \le 25) = \sum_{x_i \le 25} f(x_i) = \sum_{x_i \le 25} \binom{50}{r_i} 0.68^{x_i} 0.32^{50 - x_i}.$$

Calculating this probability by hand would be arduous. The densities requiring summation are shown as vertical solid black lines in Figure 3.6. Luckily, the binomial cdf is provided in **R** by the function pbinom.

```
pbinom(25, 50, 0.68)
[1] 0.006124875
```

One should be aware that while pbinom(25, 50, 0.68) gives $P(X \le 25)$, as in our example above, the code pbinom(25, 50, 0.68, lower.tail = FALSE) *does not* give $P(X \ge 25)$. To get this *upper tail probability* I specify: pbinom(24, 50, 0.68, lower.tail = FALSE). This issue will not be a concern for continuous cdfs because, in this case, $P(X = x) \equiv 0$.

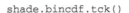

```
shade.bincdf.tck()
```

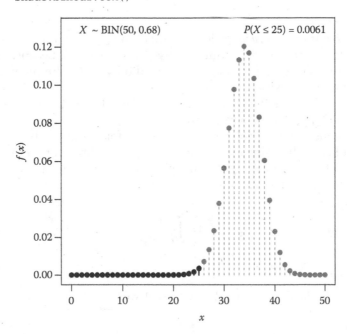

FIGURE 3.6
The lower tailed probability, $P(X \le 25)$ for $X \sim \text{BIN}(50, 0.68)$.

3.2.2 Continuous pdfs

Continuous pdfs define density for all possible real outcomes in their *support* (the region for which a random variable X is defined). Continuous variables exist for all branches and hierarchical levels of biology. They include, but are not limited to, measures describing subcellular functions (e.g., enzymatic rates, electrical membrane potentials, diffusion rates, and molecular weights), individual organisms (e.g., blood pressure, turgor pressure, photosynthetic rate, respiration rate, temperature, height, weight, girth, volume, migration distance, and travel velocities), populations (e.g., competition indices, self-thinning rates, inbreeding coefficients, and foraging optimality measures), communities (e.g., community dissimilarity, cost/benefits of interspecific interactions, and resistance and resilience to disturbance), ecosystems (e.g., habitat patch shape and nutrient fluxes), and the biosphere (e.g., global temperature, global CO_2 fluxes, and ozone density).

3.2.2.1 Continuous Uniform Distribution

The simplest continuous distribution is the *continuous uniform distribution*. It is often used as a naïve model to represent processes in which all possible continuous outcomes have the same likelihood. The pdf is simply

$$f(x) = \frac{1}{b-a}, \tag{3.7}$$

where
   1. Outcomes, x, are continuous.
   2. $a \le x \le b$.

The pdf has two parameters: a is the lower limit of the support (minimum possible value of X), while b is the upper limit (maximum possible value of X). Because the density will be equal for all possible outcomes, and will depend only on the limits of the distribution, x is not required in the density function.

If a random variable X has the pdf in Equation 3.7, we denote this as $X \sim \text{UNIF}(a, b)$. The distribution UNIF(2.5, 3) is shown in Figure 3.7.

For any continuous pdf, the probability $P(a \le X \le x)$ for $(x \ge a)$ can be calculated by integrating the pdf of X, while using a and x as the respective lower and upper limits of integration. When a is the true lower limit of the pdf, then the result is the cdf.

Thus, the continuous uniform cdf is

$$F(x) = \int_a^x \frac{1}{b-a} dt. \tag{3.8}$$

EXAMPLE 3.4

Let $X \sim \text{UNIF}(2.5, 3.0)$, then to find $P(2.5 \le X \le 2.8)$, we have

$$\int_{2.5}^{2.8} \frac{1}{b-a} dt = \int_{2.5}^{2.8} \frac{1}{3-2.5} dt = \int_{2.5}^{2.8} 2\, dt = 2t \big|_{2.5}^{2.8} = 2(2.8) - 2(2.5) = 0.6.$$

```
see.unifcdf.tck()
```

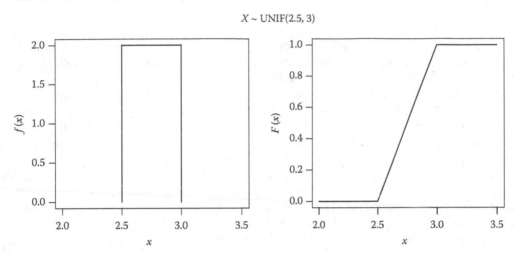

FIGURE 3.7
Pdf and cdf for $X \sim UNIF(2.5, 3.0)$ depicting Example 3.4. As with other valid pdfs, we see that regardless of the way we manipulate the parameters the area under the pdf curve always equals one.

This agrees with the cdf depiction in Figure 3.7. We can check our calculus using **R**.

```
integrand <- function(x) 2
integrate(Vectorize(integrand), 2.5, 2.8)
0.6 with absolute error < 6.7e-15
```

Note that we use the Vectorize function here since the integrand is simply a constant and not a function of x.

Figure 3.7 clearly demonstrates that the continuous pdfs *do not* directly provide probability. The total area under the pdf curve in Figure 3.7 equals one because it is a rectangle whose consecutive sides have length 2 and 0.5, and $2 \times 0.5 = 1$. However, because $b - a = 0.5$, the density for any outcome is 2, and we know that a probability must be in the interval [0, 1].

3.2.2.2 Normal Distribution

The most commonly used continuous pdf in statistics is the *normal distribution*.[*] It is used to represent processes in which the most likely outcome is the average. The distribution is symmetric around the average. Thus, outcomes an identically increasing distance above and below the average, are identically and increasingly unlikely. The normal pdf is

$$f(x) = \frac{1}{\sigma\sqrt{2\pi}} e^{-\frac{1}{2}\left(\frac{x-\mu}{\sigma}\right)^2}, \tag{3.9}$$

where
1. Outcomes, x, are continuous.
2. $x \in \mathbb{R}$.

---

[*] C.F. Gauss made a large number of contributions to mathematics, and the normal (Gaussian) distribution is attributed to him. The normal distribution was described previous to Gauss by Abraham de Moivre (1752–1833) in 1805 (although Gauss claimed that he had been using it since 1795). Salsburg (2001) noted that the normal distribution was likely discovered by Daniel Bernoulli (1700–1782), predating either Gauss or de Moivre.

3. $\mu \in \mathbb{R}$.

4. $\sigma > 0$.

The notation $x \in \mathbb{R}$ indicates that support for the pdf encompasses all real numbers. While sharing symbology with the Bernoulli probability of success, the term π in Equation 3.9 represents the irrational constant 3.14159. . . .

The normal distribution has two parameters, μ and σ, representing the *mean* and the *standard deviation* of the pdf, respectively. Squaring the standard deviation gives the distribution *variance*, σ^2. Essentially, all random variables will have derivable means, variances, and standard deviations (see Tables 3.4 and 3.5).[*] Only rarely, however, will they be explicitly present in the pdf. In general usage, the mean of a pdf is denoted $E(X)$, while the variance and standard deviation of a pdf are denoted $\mathrm{Var}(X)$ and $\mathrm{SD}(X)$, respectively. That is, if X is a normal random variable, then $E(X) = \mu$, $\mathrm{Var}(X) = \sigma^2$, and $\mathrm{SD}(X) = \sigma$ (Chapter 4).

The normal pdf is bell shaped and symmetric, while the cdf (derived by integrating the pdf) is sigmoidal (Figure 3.8). By manipulating the sliders from `see.normcdf.tck`, it is clear that the parameter μ controls the location of the peak in density or *mode*. Thus, μ is called a *location parameter*. Altering σ produces noticeable changes in density of the pdf in the vicinity of the mean. Because σ and σ^2 change local density concentrations by stretching/shrinking the pdf, they are called *scale parameters*.

If a random variable X has the pdf in Equation 3.9, we denote this as $X \sim N(\mu, \sigma^2)$. Figure 3.9 shows the distribution $N(0, 1)$.

In a normal probability distribution, the interval $\pm 1\sigma$ from μ contains approximately 68% of the distributional area, the interval $\pm 2\sigma$ from μ contains approximately 95% of the area of the distribution, and the interval $\pm 3\sigma$ from μ mean contains 99.7% of the area (Figure 3.9). This is known as the *empirical rule*.

`see.normcdf.tck()`

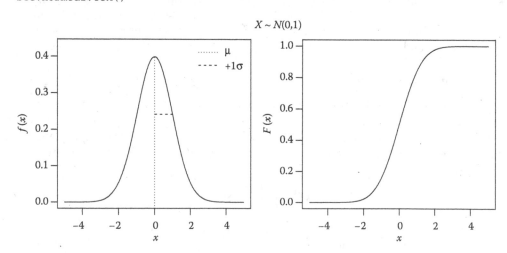

FIGURE 3.8
The standard normal pdf and cdf, that is, $X \sim N(0, 1)$.

---

[*] Exceptions include the Cauchy distribution, which is a legitimate pdf without derivable values for $E(X)$ or $\mathrm{Var}(X)$.

Fig.3.9()##See Ch.3 code

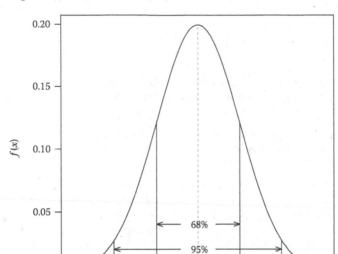

FIGURE 3.9
A normal distribution, showing locations ±1σ, ±2σ, and ±3σ from the mean.

3.2.2.2.1 *Standard Normal Distribution*

The *standard normal distribution* or *Z-distribution* can be denoted as $N(0, 1)$. Thus, it is a normal distribution with a mean of zero, and a variance of one (Figure 3.8). We can standardize any normal distribution to be standard normal using

$$Z = \frac{X - \mu}{\sigma}, \tag{3.10}$$

where μ is the mean and σ is the standard deviation from a normal random variable X. Once standardized, outcomes in X are expressed as standard deviations away from its mean, μ. Any outcome, x, from any normal distribution becomes a standard normal outcome (*z-score*) by applying

$$z = \frac{x - \mu}{\sigma}. \tag{3.11}$$

EXAMPLE 3.5

In a summary for the center for disease control, McDowell (2005) reported that the height of non-Hispanic white women in the United States over 20 years of age was approximately normally distributed with a mean of 162.9 cm and a standard deviation of 7.71 cm. Given this, what proportion of these women are ≤153 cm tall?

Using Equation 3.11, we find that

$$z = \frac{153 - 162.9}{7.71} = -1.284047.$$

This *z*-score specifies that a height of 153 cm is 1.284 standard deviations below the mean height of 162.9 cm.

We can calculate the proportion of women less than or equal to 153 cm tall by solving

$$\int_{-\infty}^{153} \frac{1}{7.71\sqrt{2\pi}} e^{-\frac{1}{2}\left(\frac{x-162.9}{7.71}\right)^2} dx.$$

However, this might be a slightly hairy procedure by hand. In fact, because the integrand does not have a closed form, it will require techniques such as an asymptotic series for integration. Luckily, we can use numerical calculus routines built into **R**.

```
integrand <- function(x){1/sqrt(2 * pi) * exp(-1/2 * x^2)}
integrate(integrand, lower=-Inf, upper=-1.284047)
0.09956276 with absolute error<2.4e-07
```

To get the lower tailed probability, we integrate the Z-distribution using the limits -1.284047 and $-\infty$. Even though this probability includes nonexistent heights less than 0 cm, it will not differ appreciably from $P(0 \leq X \leq 153)$.

As a result of our calculations, we can say that approximately 10% of white non-Hispanic women are less than or equal to 153 cm tall (Figure 3.10). That is

$$P(X \leq 153) = P[Z \leq (153 - 162.9)/7.71] = P(Z \leq -1.284047) = 0.09956276.$$

Because the function dnorm represents (by default) the standard normal pdf, we could have also typed

```
integrate(dnorm,-Inf, -1.284047)
0.09956276 with absolute error<2.4e-07
```

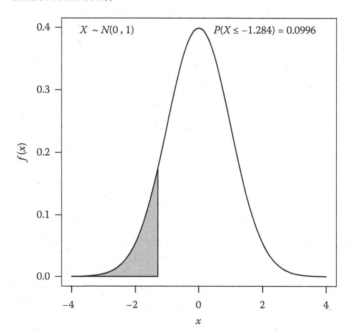

FIGURE 3.10

The distribution of U.S. white non-Hispanic women's heights expressed as Z-quantiles (i.e., standard deviations away from the mean height). The probability $P(X \leq -1.284)$ is overlaid.

We can bypass integration functions entirely by using the **R** function pnorm. By default, pnorm is the cdf for the standard normal distribution.

```
pnorm(-1.284047)
[1] 0.09956276
```

Finally, by specifying the outcome whose lower tailed probability we are interested in (153), along with the mean (162.9) and standard deviation (7.71) of the original normal distribution, we can bypass even the z-standardization step.

```
pnorm(153, mean = 162.9, sd = 7.71)
[1] 0.0995628
```

From definitions in Section 3.1, we know that the definite integral of a normal pdf from $-\infty$ to ∞ (i.e., all possible outcomes of X, encompassing the entire density curve) equals one. That is

$$\int_{-\infty}^{\infty} \frac{1}{\sigma\sqrt{2\pi}} e^{-\frac{1}{2}\left(\frac{x-\mu}{\sigma}\right)^2} dx = 1. \tag{3.12}$$

```
pnorm(Inf)
[1] 1
```

Because of this, we also know that a lower tail probability is the complement of an upper tailed probability. For instance,

$$P(X > -1.284) = 1 - P(Z < -1.284) = P(Z < 1.284) = 0.90043.$$

```
pnorm(1.284)
[1] 0.900429
```

It is often desirable to define an outcome that serves as the upper bound to a particular proportion of the population. For instance, we might be interested in the height that demarks the lower seventy-fifth percent of all heights in the population. This would be the outcome x associated with $P(X \le x) = 0.75$. For this application, we use the normal inverse cdf (quantile function) provided by the **R** function qnorm.

```
qnorm(0.75, mean = 162.9, sd = 7.71)
[1] 168.1003
```

75% of non-Hispanic white women are less than or equal to 168.1 cm tall.

★ 3.3 Other Important Distributions

While many conventional statistical procedures rely on the binomial and normal pdfs, there are many other distributions important to biologists.[*] The reader should be aware that the examples shown in the next several sections represent only general approaches.

---

[*] Most of the pdfs described in this chapter, including those in this section, belong to an important group of distributions called the exponential family. The exponential family of distributions has crucial applications in Bayesian analyses and sophisticated linear models (see Appendix, Chapters 4, 5, 9 through 11).

They do not encompass the incredible range of biological phenomena that have been (and can be) represented by pdfs.

3.3.1 Other Discrete pdfs

3.3.1.1 Poisson Distribution

The *Poisson distribution*[*] is functionally similar to the binomial distribution. That is, the distribution gives the probability for a defined number of successes, x, given a series of independent trials. In fact, the Poisson pdf will be equivalent to the binomial distribution as the number of trials goes to infinity, while the expected number of successes remains fixed, and will approximate the binomial distribution if n is large and π is small. Unlike the binomial distribution, however, the Poisson distribution has no upper limit to its support. There is no explicit maximum value for x. This is because the Poisson distribution considers successes in the context of a fixed success rate instead of a fixed sample size.

The Poisson pdf has the form

$$f(x) = \frac{e^{-\lambda}\lambda^x}{x!}. \tag{3.13}$$

where
1. The number of observed successes in disjoint intervals, x, will be independent, $x \in \{0,1,2,3,\dots\}$.
2. The probability of observing two or more successes in a very small interval will approach zero.
3. Events (successes) occur at a constant (random) rate, λ, over set intervals of space or time.
4. $\lambda > 0$.

The lone parameter λ describes the rate of successful outcomes (e.g., the number of organisms encountered per unit time).

If a random variable X has the pdf in Equation 3.13, then this is denoted as $X \sim \text{POI}(\lambda)$. The pdf and cdf for POI(5.8) are shown in Figure 3.11.

The variance and the mean of a Poisson random variable both $= \lambda$. Because of this, as the Poisson mean increases, a larger proportion of the distribution occurs in the tails, and the distribution becomes squashed (Figure 3.11). This property is shared by a number of other distributions that derive the mean and the variance from the same parameter, including the exponential, lognormal, and χ^2 distributions.

The pdf and cdf of several discrete distributions, including the binomial and Poisson, can approximate the normal distribution. For instance, if $X \sim \text{BIN}(n, \pi)$, and n is large, then X will approximately follow $N[n\pi, n\pi(1 - \pi)]$. Similarly, if $X \sim \text{POI}(\lambda)$, then this distribution

---

[*] The Poisson distribution was named after the French mathematician, geometer, and physicist Simeon-Denis Poisson (1781–1840).

```
see.poiscdf.tck()
```

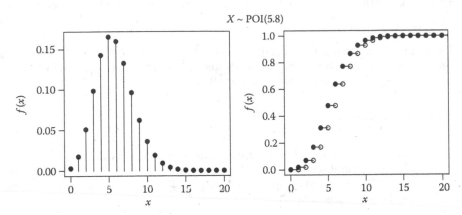

FIGURE 3.11
The pdf and cdf for $X \sim \text{POI}(5.8)$. Note that this is the expected Poisson distribution for the middle *Larrea tridentata* age class in Example 3.6. Open dots indicate that the point is not included in the cdf.

will be approximately normal, $N(\lambda, \lambda)$, for large values of λ. For a visual demonstration, type `see.bincdf.tck()` and/or `see.poiscdf.tck()`, and click on the "show normal approximation" widget.[*]

Because of their assumed properties (see above), Poisson distributions are most often used by biologists to represent spatial or temporal randomness of counts of organisms in space or time (Schabenberger and Gotway 2005, Shumway and Stoffer 2000).

EXAMPLE 3.6

Desert shrubs have conventionally been assumed to have regular (uniform) distributions in space (Greig-Smith and Chadwick 1965; review in Barbour 1973 who referred to this convention as "desert dogma"). Numerous studies have attributed these spatial patterns to inherently strongly competitive interactions between individuals (e.g., Beals 1968). An extensive investigation of *Larrea tridentata* (creosote bush) distributions in the Mojave and Sonoran deserts, however, found that this convention was probably an oversimplification (Phillips and MacMahon 1981). In particular, the authors found that young creosote bush seedlings were often clumped in space, intermediate-aged individuals were often randomly distributed, and only mature individuals were regularly distributed. The following explanations were given: (1) young *Larrea* plants will be clumped because of limited seed dispersal, clonal reproduction, and patchy resource availability, (2) intermediate-age plants will appear randomly distributed because of self-thinning in dense seedling patches as a result of intraspecific competition, and (3) mature plants with extensive root systems will be even more competitive resulting in regular distributions of *Larrea* individuals.

At their Gila Bend study site, Phillips and MacMahon (1981) took counts of *L. tridentata* in 25 randomly established 10 m × 10 m plots. Individual plants were categorized into

---

[*] The binomial approximation will be quite good if $n\pi$ and $n(1 - \pi) \geq 20$. The approximation will be helped by a continuity correction in the range $5 < n\pi < 20$ (see Ott and Longnecker 2004). It will be unsatisfactory if $n\pi$ and $n(1 - \pi) < 5$, or if $\pi \approx 0$, because then the binomial distribution will be strongly positively skewed (Chapter 4). It will also be unsatisfactory if $\pi \approx 1$, because then the binomial distribution will be strongly negatively skewed.

life stage classes based on areal cover. These were: class 1 (10^2–10^3 cm^2), class 2 (10^3–10^4 cm^2), and class 3 (10^4–10^5 cm^2). The data are contained in the dataframe `larrea` in *asbio*.

```
data(larrea)
```

The arithmetic mean (the sum of counts divided by the number of counts) serves as an estimator for the parameter λ (Chapter 4). Because each of the three columns in `larrea` represents an age class, the estimates for λ are then

```
colMeans(larrea)
class1 class2 class3
 7.84   5.80   4.04
```

Because random spatial distributions are Poisson distributed, we can test the hypothesis of random configurations by creating frequency histograms of observed *Larrea* data in age classes 1, 2, and 3, and compare them to the expected frequencies of POI(7.84), POI(5.8), and POI(4.04) (Figure 3.12).

```
Fig.3.12() ## See Ch. 3 code
```

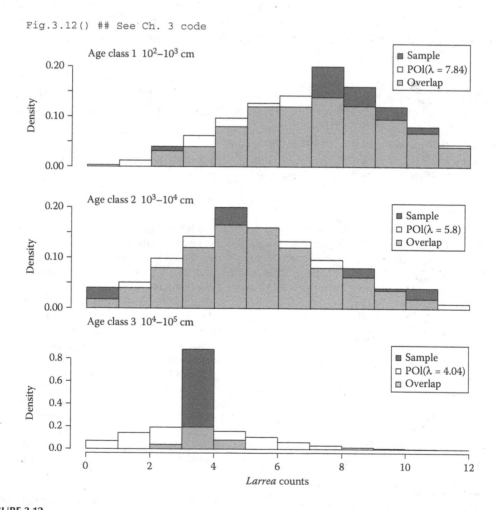

FIGURE 3.12
Density histograms for observed counts of *Larrea tridentata* versus expected counts assuming a Poisson distribution (i.e., *Larrea* individuals randomly distributed in space). Densities are shown for three different age (size) classes.

The histograms in Figure 3.12 represent the density distribution of observed and expected counts. Distributions of individuals in class 1 and 2 are well described by Poisson distributions, because the densities of observed and expected counts largely overlap. This indicates that the distribution of individuals in these age classes is more or less random. Conversely, there is little overlap between observed and Poisson expected counts in the largest age class. In fact these counts are highly uniform, not random, indicating a regular distribution.

3.3.1.2 Hypergeometric Distribution

Recall that the binomial distribution defines the probability for x independent successes when sampling over n binary trials *with* replacement. That is, x would define the number of fish caught from catch and release fishing (Example 3.3). Conversely, the *hypergeometric distribution* defines this probability when sampling *without* replacement. That is, x would define the number of fish caught as a harvest season progresses. As a result, hypergeometric trials are not independent because they result in a contracted sample space that affects the probability of the next outcome.

The hypergeometric pdf has the form

$$f(x) = \binom{M}{x}\binom{N-M}{n-x} \bigg/ \binom{N}{n},$$ (3.14)

where
1. x is the number of successful trials, $x \in \{0,1,2,\ldots,n\}$.
2. $N \in \{1,2,3,\ldots\}$.
3. $n \in \{0,1,2,\ldots,N\}$.
4. $M \in \{0,1,2,\ldots,N\}$.

The pdf requires three parameters: n is the number of trials, M is the number of items in the group of interest (i.e., the number of possible successful outcomes), and N stipulates the total number of selectable items (i.e., possible successes+possible failures).

If a random variable X has the pdf in Equation 3.14, we denote this as $X \sim \text{HYP}(n, M, N)$. The distribution HYP(9, 9, 30) is shown in Figure 3.13.

The hypergeometric pdf works by finding the relative frequency for x in a shrinking sample space. It does this by first finding the number of combinations for x successes given a population of M total successful outcomes $\binom{M}{x}$. This quantity is then multiplied by the number of combinations for $n - x$ failures from a population of $N - M$ unsuccessful outcomes $\binom{N-M}{n-x}$. This product is divided by the combinations for samples of size n from a population of size N, $\binom{N}{n}$.

The hypergeometric distribution is frequently used by biologist to model population counts when organisms are sampled without replacement. For instance, Templeton et al. (2011) captured eastern collared lizards (*Crotaphytus collaris*) on mountains in Missouri, marked them, and then released them. The method allowed mark/recapture estimation of population sizes, and ensured that the same lizard was not repeatedly captured and

```
see.hypercdf.tck()
```

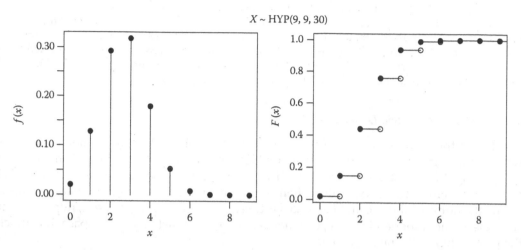

FIGURE 3.13

The pdf and cdf for HYP(9, 9, 30). That is, given 9 samples obtained without replacement from a population with 9 possible successes, and $30 - 9 = 21$ possible failures, what are the probabilities for x successful trials? Open dots indicate that the point is not included in the cdf.

analyzed. The authors used sample data to estimate hypergeometric parameters for the study system.

EXAMPLE 3.7

The Greater Yellowstone Area (GYA) is a contiguous set of natural ecosystems located in the North-Central Rocky Mountains that comprise an area of 80,000 km². Garrott et al. (2011) compiled data collected from 18 GYA mountain goat herds over 35 years. Using this information, Flesch and Garrott (2011) found that the long-term kid:adult ratio of the herds was approximately $M/(N - M) = 0.27$, while the average population size was 1927 individuals. Recently, 88 goats were randomly sampled from the GYA without replacement (sampled goats were kept in a holding pen by a survey team) and 32 were found to be kids. Assuming the (unlikely) independence of kid and adult captures, what is the probability of this outcome?

We have, $n = 88 + 32 = 120$ (the number of recent captures) and $N = 1927$ (total population size). To define M (the number of kids) we have

$$\frac{M}{N - M} = 0.27; \quad \frac{M}{1927 - M} = 0.27; \quad M(1 + 0.27) = 0.27(1927); \quad M = \frac{0.27(1927)}{1.27} = 410 \text{ kids.}$$

As a result, we have

$$P(32 \text{ kids} \mid M, N, n) = f(32) = \binom{410}{32}\binom{1517}{88} \Big/ \binom{1927}{120}$$

$$= \left\{ \frac{410\,!}{32\,!(410 - 32)\,!} \cdot \frac{1517\,!}{88\,!(1517 - 88)!} \right\} \Big/ \frac{1927\,!}{120\,!(1927 - 120)!}$$

$$= 1.5501 \times 10^{192} / 5.210883 \times 10^{193} = 0.02974689.$$

In **R**, we have

```
x <- 32; M <- 410; NminusM <- 1517; n <-120
dhyper(x, M, NminusM, n)
[1] 0.02974689
```

This improbable but real outcome occurred because the recent samples were taken in a nonrepresentative region where goat populations were growing rapidly (the Northeast corner of Yellowstone National Park), resulting in an unexpectedly high kid:adult ratio.

3.3.1.3 Geometric Distribution

The *geometric distribution* provides the probability that x independent Bernoulli failures occur prior to obtaining the first success. The pdf can be expressed as

$$f(x) = \pi(1-\pi)^x, \tag{3.15}$$

where
1. The Bernoulli trials defining the distribution are independent with an unchanging probability of success, π.
2. x is the number of unsuccessful trials, preceding the first success, $x \in \{0,1,2,\ldots,n\}$.
3. $0 < \pi < 1$.

The only required parameter, π, defines the probability of a single Bernoulli success.

If a random variable X has the pdf in Equation 3.15, then we denote this as $X \sim \text{GEO}(\pi)$. The distribution GEO(0.17) is shown in Figure 3.14.

The geometric pdf works mathematically because the quantity $\pi(1-\pi)^x$ gives the probability of the intersection of one success following x failures. There is no need for multiplication by the binomial coefficient (*à la* the binomial or hypergeometric pdf) because there is only one way to have one success following a series of identical failures.

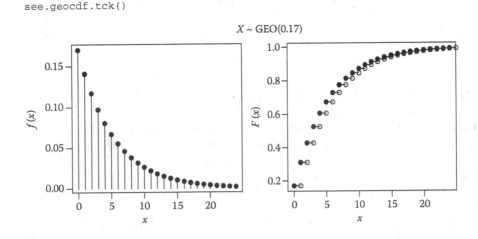

FIGURE 3.14
The pdf and cdf for GEO(0.17) representing the stopping distances of white crowned sparrows in Example 3.8. Open dots indicate that the point is not included in the cdf.

The geometric pdf has been used to model the extinction probability of biological popu-lations (Lebreton et al. 2007), the distributions of genetic polymorphisms (Vinson et al. 2005), and other applications (see below).

EXAMPLE 3.8

The geometric distribution has been used to model species dispersal when the *stopping probability* (the probability that an organism will stay at a habitat) is constant for a series of habitats.* Specifically, if x is the observed number of habitats that a population dis-perses into ($x = 0$ for the home habitat), and π is the stopping probability, then $f(x)$ gives the proportion of the population that will disperse into x habitats before stopping.

The stopping probability for a population of male white crowned sparrows (*Zonotrichia luecophyris*) was estimated to be $\pi = 0.17$ (Baker and Mewalt 1978; see Figure 3.14). Given this, what fraction of the population will disperse into two habitats before stopping?

$$f(x) = 0.17 \times 0.83^2 = 0.117113.$$

This is confirmed in **R**:

```
dgeom(2,0.17)
[1] 0.117113
```

3.3.1.4 Negative Binomial Distribution

The *negative binomial distribution* gives the probability that x independent Bernoulli failures will occur prior to obtaining the rth success. Its pdf has the form

$$f(x) = \binom{x+r-1}{r-1} \pi^r (1-\pi)^x, \tag{3.16}$$

where
1. The Bernoulli trials defining the distribution are independent with an unchang-ing probability of success, π.
2. x is the number of unsuccessful trials, preceding the rth success, $x \in \{0, 1, 2, \ldots, n\}$.
3. $r \in \{0, 1, 2, \ldots\}$.
4. $0 < \pi < 1$.

There are two parameters: r is the number of successes and π is the probability of an individual Bernoulli success.

If a random variable X has the pdf in Equation 3.16, we denote this as $X \sim NB(r, \pi)$. The geometric distribution is a special case of the negative binomial distribution, since it equals $NB(1, \pi)$. The distribution $NB(3, 0.76)$ is shown in Figure 3.15.

The mathematical mechanism in Equation 3.16 is similar to the binomial pdf. The quan-tity $\pi^r (1 - \pi)^x$ gives the probability of the intersection of all successes and all failures for one of the possible combinations of r successes following x failures. The quantity $\binom{x+r-1}{r-1}$

---

* See Miller and Carroll (1989) for a criticism of this approach.

```
see.nbincdf.tck()
```

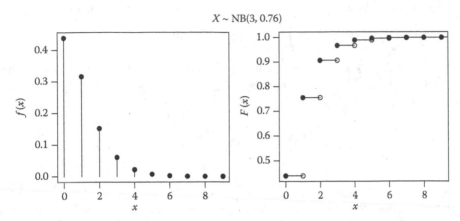

FIGURE 3.15

The pdf and cdf for NB(3, 0.76) defining probabilities for the number of nesting failures preceding three successfully hatched glaucous-winged gull eggs in Example 3.9. Open dots indicate that the point is not included in the cdf.

gives the total number of combinations for getting x failures preceding r successes. The product of these expressions gives the probability that the rth success will occur after x failures.

Among other applications (see below), biologists have used the negative binomial distribution to detect aggregation and variation in census surveys (Shaw and Dobson 1995), and to model genetic predispositions for animal fertility (Tempelman and Gianola 1999) and mortality (Varona and Sorenson 2009).

EXAMPLE 3.9

The successful hatching probability for glaucous-winged gull (*Larus glaucescens*) eggs in Glacier Bay Alaska is about 0.76 (Zador and Piatt 2007; see Figure 3.15). Assuming the independence of hatches, what is the probability that a gull nest site will fail once before successfully hatching three eggs?

$$f(2) = \binom{1+3-1}{3-1} 0.76^3 0.24^1 = 3 \times 0.421875 \times 0.24 = 0.316.$$

In **R**, we have

```
dnbinom(1,3,0.76)
[1] 0.3160627
```

For noninteger values for r (see applications below), the following expression can be used for the negative binomial distribution:

$$f(x) = \Gamma(x+r)/[\Gamma(r)x!]\pi^r(1-\pi)^x, \tag{3.17}$$

where $x \in \{0, 1, 2,...\}$, $r > 0$, $0 < \pi < 1$, and $\Gamma(x)$ represents the *gamma function* applied to x. The gamma function is defined as $\Gamma(x) = (x - 1)!$ for positive nonzero integers, while the complete gamma function is $\Gamma(x) = \int_0^1 \left[\ln\left(\frac{1}{t}\right) \right]^{x-1} dt$, for all $x > 0$.

An extension of Equation 3.17 is often used by ecologists and epidemiologists to describe the distribution of organisms in time or space (e.g., Lindén and Mäntyniemi 2011, many others). This is

$$f(x) = \frac{\Gamma(x + k)}{\Gamma(k)x!} \left(\frac{k}{k - m} \right)^k \left(1 - \frac{k}{k - m} \right)^x, \tag{3.18}$$

for $x \in \{0, 1, 2,...\}$, $k > 0$, and $m > 0$.

Here, k represents a dispersion parameter (not required to be an integer) and m is the mean number of counts. Equation 3.18 is often used as a distributional framework of organism counts that are *overdispersed*. That is, the variance of the counts is greater than their mean. Overdispersion indicates that organisms are clumped, potentially due to positive interactions among individuals, or clustered resources. This will be reflected by small values of k. Conversely, when counts are randomly distributed, k will be large. In these cases, the variance will equal the mean, and the Poisson distribution (Section 3.3.1.1) will provide the correct representation of the count distribution. Equation 3.18 will be equivalent to the Poisson pdf when $k = \infty$. As a result, the negative binomial pdf encompasses the Poisson pdf, and can be used to represent both clumped *and* random configurations (see Exercise 13 at the end of the chapter).

3.3.2 Other Continuous pdfs

Aside from the normal distribution (Section 3.2.2.2), a number of other continuous pdfs are frequently used by biologists. Several of these can be derived from the normal distribution, while several others can be used to mimic normal distributions (Figure 3.16).[*]

3.3.2.1 Chi-Squared Distribution

The χ^2 distribution (pronounced "kī-squared"), along with the F, and t, distributions (described below), are defined only by their *degrees of freedom*: the number of independent pieces of information that exist concerning an estimable parameter.[†] Degrees of freedom parameters are denoted here (and elsewhere) using the Greek letter ν (pronounced "new"). All three pdfs are employed extensively in null hypothesis testing (Chapter 6), and many examples of their uses are given in later chapters.

The χ^2 *distribution* is frequently used for testing the null hypothesis that observed and expected frequencies are equal (Chapters 6, 7, 9 through 11). It also describes the sampling

---

[*] Conventional parameterization of these pdfs requires reuse of symbols, resulting in potential confusion. It should be emphasized, however, that in general parameters are not exchangeable across distributions. For instance, for the gamma and Weibull distributions described in this section, $X \sim GAM(\kappa, 2)$, does not require that $X \sim WEI(\beta, 2)$, despite the fact that the symbol θ is used to represent the second parameter in both distributions. Note, however an exception occurs in the case $\kappa = \beta = 1$ (see Example 3.15).

[†] The term "degrees of freedom" is used for these distributional parameters because test statistics, which have these distributions under true null hypotheses, will have these degrees of freedom (see Chapter 6).

```
Fig.3.16()##see Ch. 3 code
```

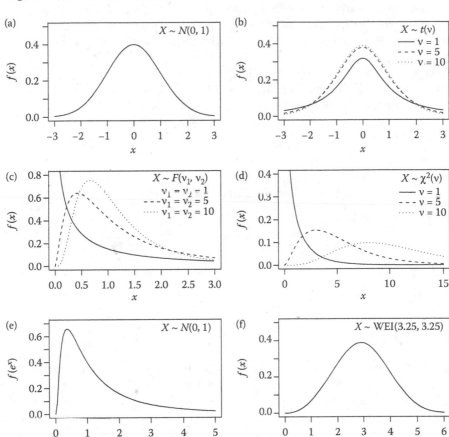

FIGURE 3.16
Continuous distributions that are derived from, or which can be similar to, normal distributions. The (a) Z-(standard normal), (b) t-distribution, (c) F-distribution, (d) χ^2 distribution, (e) lognormal distribution, and (f) Weibull distribution. Distributions (a) through (e) are all derived from the normal distribution. The properties of the t-, F-, and χ^2 distributions change dramatically as their degrees of freedom change. Weibull (and gamma) parameters can be manipulated to create distributions similar to a normal distribution.

distribution (Chapter 5) of the sample variance (Chapter 4), taken from a normal parent distribution, after linear transformation (Chapters 4 and 5). The pdf has the following form:

$$f(x) = \frac{1}{2^{v/2}\Gamma(v/2)} x^{(v/2)-1} e^{-x/2},$$ (3.19)

where
1. Outcomes, x, are continuous and independent.
2. $x \geq 0$.
3. $v \geq 0$.
4. $\Gamma(.)$ is the gamma function, described in Section 3.3.1.4.

The χ^2 distribution results from the summing of independent, squared, standard normal distributions. Specifically, if $Z \sim N(0, 1)$, and $X = \Sigma_{i=1}^{\nu} Z_i^2$, then X will follow a χ^2 distribution with ν degrees of freedom, denoted $X \sim \chi^2(\nu)$. A squared standard normal distribution will follow $\chi^2(1)$. As a result, outcomes from χ^2 distributions must be greater than or equal to 0.

Increasing the degrees of freedom will cause the mode of the χ^2 distribution to shift to the right (Figure 3.16). This is because, in a χ^2 distribution, the degrees of freedom are a location parameter, equivalent to the mean (Chapter 4). The χ^2 pdf and cdf can be viewed interactively by typing `see.chicdf.tck()`.

EXAMPLE 3.10

If $X \sim \chi^2(10)$, what is $P(X \geq 7)$? See Figure 3.17.
We can calculate this probability by finding $1 - P(X < 7)$,

```
1 - pchisq(7, 10)
[1] 0.725445
```

or by obtaining the upper tail directly.

```
pchisq(7, 10, lower.tail = FALSE)
[1] 0.725445
```

3.3.2.2 t-Distribution

The *t-distribution* represents a standard normal random variable divided by the square root of the quotient of a chi-square random variable and its degrees of freedom. That is, if $Z \sim N(0, 1)$, $V \sim \chi^2(\nu)$, and $T \sim Z/\sqrt{V/\nu}$, then $T \sim t(\nu)$. The t pdf is

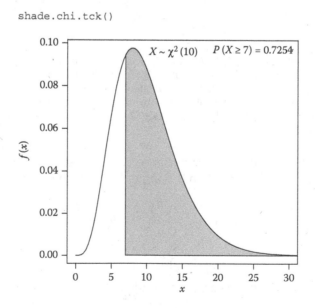

FIGURE 3.17
The pdf for the random variable $X \sim \chi^2(10)$. Shading for the probability $P(X \geq 7)$ is overlaid.

$$f(x) = \frac{\Gamma\left(\dfrac{v+1}{2}\right)}{\Gamma\left(\dfrac{v}{2}\right)} \frac{1}{\sqrt{v\pi}} \left(1 + \frac{x^2}{v}\right)^{-\frac{v+1}{2}}, \tag{3.20}$$

where
1. Outcomes, x, are continuous.
2. $x \in \mathbb{R}$.
3. $v > 0$.
4. $\Gamma(.)$ is the gamma function.

Like the standard normal distribution, the t-distribution is symmetric about zero. In fact, the t-distribution asymptotically converges to the standard normal distribution as $v \to \infty$. For smaller values of v, the t-distribution is platykurtic (flatter) compared to the Z-distribution (Figure 3.16). This can be demonstrated by typing see.tcdf.tck() and overlaying the standard normal distribution.

The t-distribution has been used to directly model biological variables like seed dispersal distances (Clarke et al. 1999), but is used most often as a pivotal quantity in confidence interval estimation (Chapter 5), and as a null distribution in tests for comparing the mean of one population to a certain value, or for comparing two hypothesized populations (Chapter 6).

EXAMPLE 3.11

If $X \sim t(5)$, what is $P(-2 \geq X \geq 2)$? See Figure 3.18.
We can calculate this *two-tailed probability* by finding $P(X \leq -2) + 1 - P(X \leq 2)$, by finding $P(X \leq -2) + P(X \geq 2)$, or (since the distribution is symmetric) by obtaining $2P(X \leq -2)$ or $2P(X \geq |-2|)$.

```
2* pt(-2, 5)
[1] 0.1019395
```

3.3.2.3 F-Distribution

The *F-distribution* represents the distribution of the ratio of two independent χ^2 random variables, each divided by their degrees of freedom. That is, if $V_1 \sim \chi^2(v_1)$, $V_2 \sim \chi^2(v_2)$, and $X = \dfrac{V_1/v_1}{V_2/v_2}$, then $X \sim F(v_1, v_2)$. Because of this derivation, the F-distribution requires specification of two separate degrees of freedom, v_1 and v_2. These are called the *numerator degrees of freedom* and *denominator degrees of freedom*, respectively. The F pdf has the form

$$f(x) = \frac{\Gamma\left(\dfrac{v_1 + v_2}{2}\right)}{\Gamma\left(\dfrac{v_1}{2}\right)\Gamma\left(\dfrac{v_2}{2}\right)} \left(\frac{v_1}{v_2}\right)^{\frac{v_1}{2}} x^{\left(\frac{v_1}{2}-1\right)} \left(1 + \frac{v_1}{v_2}x\right)^{-\frac{v_1+v_2}{2}}, \tag{3.21}$$

```
shade.t.tck()
```

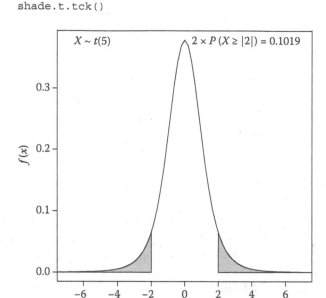

FIGURE 3.18

The pdf for the random variable $X \sim t(5)$. Shading for the probability $P(-2 > X > 2)$ is overlaid.

where
1. Outcomes, x, are continuous.
2. $x > 0$.
3. $v_1 > 0$.
4. $v_2 > 0$.
5. $\Gamma(.)$ is the gamma function.

The distribution can be viewed interactively by typing `see.Fcdf.tck()`.

The F-distribution is utilized by a number of important statistical procedures, including ANOVA (analysis of variance), which test the null hypothesis that two or more normal populations with equal variances have the same mean (Chapter 10). While the name of the F-distribution honors R.A. Fisher, it was George W. Snedecor who made the distribution widely available for use by researchers by painstakingly integrating the F pdf for many combinations of degrees of freedom without a computer (Fisher 1973). As a result, the F-distribution is often called Snedecor's F-distribution or the Fisher–Snedecor distribution.

EXAMPLE 3.12

If $X \sim F(15, 12)$, what is $P(1 \le X \le 3)$? See Figure 3.19.

We can calculate this "middle" probability by obtaining $P(X \le 3) - P(X \le 1)$.

```
pf(3, 15, 12) - pf(1, 15, 12)
[1] 0.4772689
```

```
shade.F.tck()
```

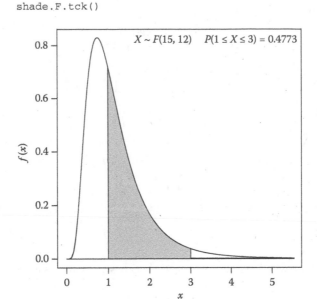

FIGURE 3.19
The pdf for the random variable $X \sim F(15,12)$. Shading for the probability $P(1 \leq X \leq 3)$ is overlaid.

3.3.2.4 Exponential Distribution

The *exponential distribution* is a continuous conflation of the geometric distribution and Poisson distributions. It defines the density of waiting times (failures) before the next Poisson (random) outcome. The exponential pdf is

$$f(x) = \theta e^{-x\theta}, \tag{3.22}$$

where
1. Outcomes, x, are continuous.
2. $x \geq 0$.
3. $\theta > 0$.

θ is called a *rate parameter* because it is a function of the distribution's location and a scale parameter. Its value defines the exponential hazard function (see below).

The exponential cdf has a *closed form*. That is, integration is not required for the calculation of probability. Its form is

$$P(X \leq x) = F(x) = 1 - e^{-x\theta}. \tag{3.23}$$

If a random variable X has the pdf in Equation 3.22, we denote this as $X \sim EXP(\theta)$. The pdf and cdf for $EXP(1.1)$ are shown in Figure 3.20.

The exponential pdf has been used in many biological applications, including species area curves (Gleason 1922, Miller and Wiegert 1989), and competition and dispersal models

```
see.expcdf.tck()
```

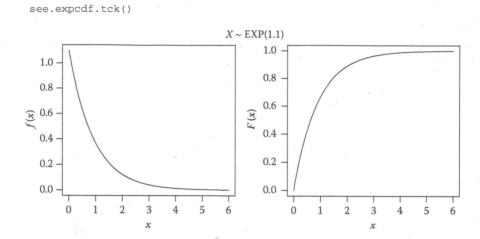

FIGURE 3.20
The pdf and cdf for EXP(1.1) which describes the survivorship of song sparrows in Example 3.13.

(McCarthy 1997).[*] Given the definition for the exponential cdf in Equation 3.23, the exponential survivorship function is $e^{-x\theta}$, resulting in the hazard function $h(x) = \theta e^{-\theta x}/e^{-\theta x} = \theta$. Because the exponential hazard function is fixed at the parameter θ it is said to have a *no-memory property*. This makes the probability of exponential survivorship invariant across age classes. That is, a 90-year-old organism with an exponential hazard function will have the same probability of dying as a 10-year-old organism.[†] The geometric pdf is the only other memory-less distribution.

EXAMPLE 3.13 EXPONENTIAL SURVIVORSHIP

Bird species often have type II survivorship curves. That is, log(number of survivors) will decrease linearly over time because mortality rates will be more or less equal in each age class (Ricklefs and Miller 2000). As a result, bird survivorship is often effectively modeled with memory-less exponential pdfs. Smith (2006) followed a group of 115 male song sparrows (*Melospiza melodia*) from time of hatching to death 6 years later on Mandarte Island, British Columbia. A cohort life table summarizing this data is shown in Table 3.2.

As an estimate of θ, I find the mean bird age (Pereira and Daily 2006). I do this by multiplying the proportions of birds each age class by their age, and then taking the sum (Chapter 4). I assume the average age in each age class is midway between the specified age classes.

```
n.x <- c(115, 25, 19, 12, 2, 1, 0)
theta <- sum(n.x/sum(n.x) * seq(0.5, 6.5, 1))# weighted mean for age
theta
[1] 1.143678
```

The resulting exponential survivorship curve is shown in Figure 3.21. We see that the function provides an effective representation of song sparrow survivorship.

---

[*] Alternative formulations of the exponential pdf are frequently used. I use this parameterization because it is the one used by the *R base* library. In this form, the expected value of the exponential distribution is $1/\theta$ (Example 3.15).

[†] This is often expressed as $P[X > a + t | X > a] = P[X > t]$, where a is a constant and t represents time; that is, the fact that a has already occurred has no bearing on what occurs at time t.

TABLE 3.2

Cohort Life Table for Male Song Sparrows from Mandarte Island, BC

Age in Years (x)	Number of Birds Alive (n_x)	Survivorship (l_x)
0	115	1.0
1	25	0.217
2	19	0.165
3	12	0.104
4	2	0.017
5	1	0.009
6	0	0.0

```
Fig.3.21() ## See Ch. 3 code
```

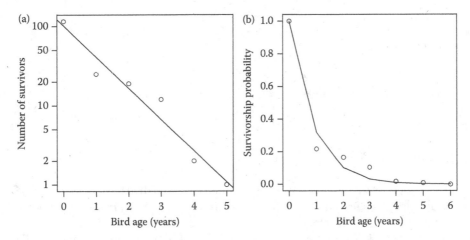

FIGURE 3.21
Survivorship of male song sparrows on Mandarte Island, BC. (a) Demonstration of type II survivorship and (b) an exponential survivorship curve.

A large number of other parametric approaches are also possible for estimating survivorship, including the Weibull distribution, described in Section 3.3.2.7 (see Lynch and Fagan 2009 for a review). One simple nonparametric (nonpdf) alternative is *Kaplan–Meier survivorship* (Kaplan and Meier 1958, Pollock et al. 1989), which explicitly accounts for *censored data* (i.e., experimental units lost before the end of a study). Type ?km for more information.

3.3.2.5 Beta Distribution

The *beta distribution* is the result of complex transformation of a random F variable. In particular, if $Y \sim F(v_1, v_2)$, and $X = \dfrac{(v_1/v_2)Y}{1+(v_1/v_2)Y}$, then $X \sim \text{BETA}(v_1/2, v_2/2)$. The pdf has the form

$$f(x) = \frac{\Gamma(\alpha+\beta)}{\Gamma(\alpha)\Gamma(\beta)} x^{\alpha-1}(1-x)^{\beta-1}, \tag{3.24}$$

```
see.betacdf.tck()
```

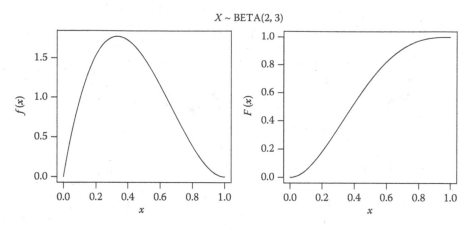

FIGURE 3.22
The pdf and cdf for BETA(2, 3).

where
    1. Outcomes, x, are continuous.
    2. $0 < x < 1$.
    3. $\alpha > 0$.
    4. $\beta < 0$.
    5. $\Gamma(.)$ is the gamma function.

In the beta pdf, α and β are so-called *shape parameters*, which affect the *shape* of a distribution without relocating it (which a location parameter does) or stretching or shrinking it (which a scale parameter does).

If a random variable X has the pdf in Equation 3.24, we denote this as $X \sim$ BETA(α, β). The pdf and cdf for BETA(2, 3) are shown in Figure 3.22.

Because *beta distribution* is only defined between 0 and 1, it has been used to model biological variables defined as percentages or proportions, for example, the cover of vegetation on a landscape (Chen et al. 2006). It is also frequently used in Bayesian analyses. This is because the distribution BETA(1, 1) is equivalent to UNIF(0, 1) allowing it to be used as a so-called noninformative prior distribution (see Examples 4.18 and 4.19).

3.3.2.6 Gamma Distribution

The *gamma distribution* is named after the gamma function described in Section 3.3.1.4. It has the form

$$f(x) = \frac{1}{\theta^\kappa \Gamma(\kappa)} x^{k-1} e^{-x/\theta},$$
(3.25)

where
    1. Outcomes, x, are continuous.
    2. $x > 0$.
    3. $\theta > 0$.

```
see.gamcdf.tck()
```

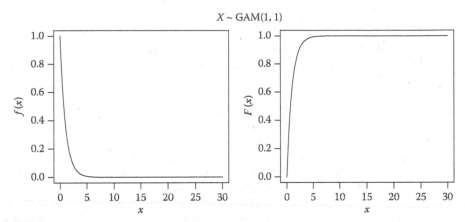

FIGURE 3.23
The pdf and cdf for GAM(1, 1).

    4. $\kappa > 0$.
    5. $\Gamma(.)$ is the gamma function.

In the gamma pdf, κ is a shape parameter, and θ is a scale parameter.

If a random variable X has the pdf in Equation 3.25, we denote this as $X \sim$ GAM(κ, θ). The distribution GAM(1, 1) is shown in Figure 3.23.

The gamma distribution is most frequently used for representing phenomena with highly right-skewed probability distributions. For instance, it has been used to model abundances of marine invertebrates, which are strongly concave, and poorly fit by lognormal and other pdfs (Schmidt and Garbutt 1985). The distribution, however, is extremely flexible and can be used to mimic other pdfs, including the normal distribution. When $\kappa = 1$ the gamma distribution is equivalent to EXP($1/\theta$). For example, GAM($\kappa = 1$, $\theta = 2$) is equivalent to EXP($\theta = 1/2$). When $\kappa = \nu/2$ and $\theta = 2$, the gamma distribution is equivalent to a χ^2 distribution with ν degrees of freedom. In Bayesian applications, the gamma distribution is a so-called conjugate prior distribution for a number of likelihood functions including the exponential pdf (for the parameter θ) and the Poisson pdf (for the parameter λ), see Chapter 4, particularly Example 4.20.

3.3.2.7 Weibull Distribution

The *Weibull distribution*[*] has the following form:

$$f(x) = \frac{\theta}{\beta^\theta} x^{\theta-1} e^{-(x/\beta)^\theta},$$ (3.26)

where
    1. Outcomes, x, are continuous.
    2. $x > 0$.

---

[*] The Weibull distribution is named after Waloddi Weibull (1887–1979), a Swedish engineer, scientist, and mathematician who first described it in 1951.

3. $\beta > 0$.
4. $\theta > 0$.

In the Weibull pdf, θ is a shape parameter, and β is a scale parameter. The distribution is occasionally specified with a third parameter that defines location. Like the exponential distribution the Weibull cdf has a closed form:

$$F(x) = 1 - e^{-(x/\beta)^{\theta}}. \tag{3.27}$$

If a random variable X has the pdf in Equation 3.26, then we denote this as $X \sim \text{WEI}(\beta, \theta)$. The pdf and cdf for WEI(13.3, 13.3) are shown in Figure 3.24.

Like the gamma pdf, the Weibull distribution can approximate a large number of other distributions, including normal distributions (Figure 3.16). The distribution WEI($\beta = 1, \theta$) is equivalent to EXP($1/\theta$). Thus, assuming that θ represents the same number, WEI($\beta = 1, \theta$) = GAM($\kappa = 1, \theta$) = EXP($1/\theta$).

The Weibull distribution is probably most frequently used by biologists for survivorship analyses. If pdf outcomes specify lifespans, then the Weibull distribution provides a distribution in which the hazard function is proportional to the power of $\theta + 1$. Thus, $\theta < 1$ indicates that the likelihood of mortality decreases over time (i.e., type III survivorship demonstrated in marine invertebrates and annual plants), a value of $\theta = 1$ indicates that likelihood of mortality is constant, and will be defined by β (type II survivorship, demonstrated in the exponential hazard function), and a value of $\theta > 1$ indicates that the likelihood of mortality increases over time (type I survivorship, demonstrated by humans and large mammals).

Among its other uses, the *Rayleigh distribution*, WEI(β, 2), defines the distance from one organism to its nearest neighbor when spatial patterns of a biological population are Poisson (randomly) distributed (cf. Garret and Bowden 2002). Other applications include

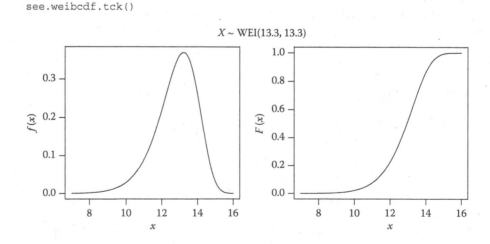

FIGURE 3.24
The pdf and cdf for WEI(13.3, 13.3) representing DBH in cm for gray alder trees in Eastern Finland in Example 3.14.

modeling extreme events such as annual maximum one-day rainfalls (Sharma and Singh 2010), and river discharges (Clarke 2002), and representing negatively skewed distributions (see below).

EXAMPLE 3.14

The Weibull distribution has been used to model DBH (diameter at breast height) in mature forest stands because it can represent unimodal distributions with strong negative skew (Merganic and Sterba 2006). This property occurs in trees because physiological factors will tend to impose an approximate upper limit to tree diameter, and most mature trees will often be near this limit. Kärki et al. (2000) found that the distribution of DBH (in centimeters) of gray alder trees (*Alnus incerta*) in eastern Finland was approximately WEI(13.3, 13.3) (Figure 3.24). Given this model, what proportion of alders is expected to be greater than or equal to 14 cm in diameter?

For the upper tailed probability, $P(X \geq 14)$, we have

$$P(X \geq 14) = 1 - \left[1 - e^{-(14/13.3)^{13.3}} \right] = 1 - (1 - 0.1383143) = 0.1383143.$$

```
pweibull(14, shape = 13.3, scale = 13.3, lower.tail = FALSE)
[1] 0.1383143
```

We might also be interested in the quantile that demarks a particular proportion of the alder distribution. For instance, we might ask: "what is the DBH that demarks the smallest 35% of the alders?" To answer this question, we would use the inverse Weibull cdf.

```
qweibull(0.35, 13.3, 13.3)
[1] 12.48396
```

Thus, 35% of trees have a DBH less than or equal to 12.5 cm.

EXAMPLE 3.15 EXPONENTIAL, GAMMA, AND WEIBULL DISTRIBUTIONS

For insects in the genus *Aphytis*, the time in hours between encounters with predators is exponentially distributed with $\theta = 3.2$ (Heimpel et al. 1997). What is the probability of consecutive encounters occurring within 15 min (i.e., 0.25 h or less)?

The exponential cdf gives

$$P(X \leq x) = F(x) = 1 - e^{-(0.25)(3.2)} = 1 - 0.449329 = 0.550671.$$

In **R**, we have

```
pexp(0.25, 3.2)
[1] 0.550671
```

We can use the gamma distribution or Weibull distribution to get the same answer.

```
pgamma(0.25, shape = 1, scale = 1/3.2)
[1] 0.550671
pweibull(0.25, shape = 1, scale = 1/3.2)
[1] 0.550671
```

3.3.2.8 Lognormal Distribution

If a random variable X has a *lognormal distribution*, and $Y = \log(X)$, then Y will have a normal distribution. Correspondingly, if Y is normally distributed then e^Y will be lognormally distributed (Figure 3.16). The lognormal pdf has the following form:

$$f(x) = \frac{1}{x\sigma\sqrt{2\pi}}e^{-\frac{1}{2}\left(\frac{\ln(x)-\mu}{\sigma}\right)^2}, \tag{3.28}$$

where
  1. Outcomes, x, are continuous.
  2. $x > 0$.
  3. $\mu \in \mathbb{R}$.
  4. $\sigma > 0$.

Like the normal distribution, μ serves as a location parameter, while σ is a scale parameter.
  If a random variable X has the pdf in Equation 3.28, then this is denoted as $X \sim \text{LOGN}(\mu, \sigma^2)$. The pdf and cdf for $\text{LOGN}(0, 1)$ are shown in Figure 3.25.
  Many variables in biology have lognormal distributions. That is, they cannot be less than zero, are right-skewed (Chapter 4) and are normally distributed after log-transformation (Chapter 7). Documented examples include distributions of fruit sizes, flower sizes, and population sizes for a wide variety of organisms, including diatoms, plants, fish, insects, and birds (Limpert et al. 2001).

EXAMPLE 3.16 PRESTON'S LOGNORMAL? DISTRIBUTION

A major issue in ecology is the effective description of species abundance in an ecosystem (i.e., how many species are rare, how many are common, and how many are intermediate in ubiquity). A large number of models have been proposed (see Magurran

```
see.lnormcdf.tck()
```

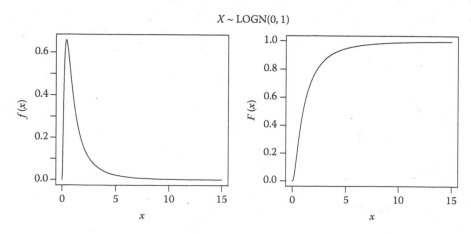

FIGURE 3.25
The pdf and cdf for LOGN(0, 1).

(1988) for a review of methods, and Hubbell (2001), for an overview of his own methodology). Preston (1948) proposed that after a $\log_2$ transformation, species abundances, grouped in bins representing a doubling of abundance, would be normally distributed.

The lognormal pdf is normal after log transformation. Thus, Preston assumed an underlying lognormal distribution for species abundances *before* transformation. As a result, Preston's lognormal distribution (described below) is *not* a valid pdf, and is emphatically *not* a lognormal pdf.

We will apply Preston's idea to tropical tree counts data made on Barro Colorado Island in Gatun Lake in central Panama (Condit et al. 2002). The researchers recorded 225 tree species in 50 1-hectare contiguous plots. We can use Preston's (1948) method to create a frequency distribution by placing abundances into bins (octaves) based on $\log_2$ species abundance and plotting these versus number of species. We fit a Gaussian curve using the overall model:

$$n = n_0 \times e^{-(aR)^2}, \tag{3.29}$$

where n_0 is the number of species contained in the modal octave, n is the number of species contained in an octave, R represents octaves from the modal octave, and a is an unknown parameter (Preston 1948). The relationship between Pearson's lognormal distribution and the normal pdf is easily demonstrated. It should be visually apparent that Equation 3.29 is an adaptation of Equation 3.30 given below:

$$f(x) = e^{-(x)^2}. \tag{3.30}$$

This is the formula for a Gaussian distribution, which underlies the normal pdf. If Equation 3.30 was scaled so that it had unit area beneath its curve, then it would be equivalent to the standard normal pdf. For an interactive demonstrative of the derivation of the Gaussian distribution from a simple exponential power function, type see. exppower.tck().

We will estimate a with a method called nonlinear least squares (see Section 9.21, Magurran 1988). For the Barro Colorado Island example, we obtain the estimate $n = 39e^{-(-0.271R)^2}$.

A graphical expression of the Barro Colorado Island analysis is shown in Figure 3.26.

```
data(BCI.count)
BCI.preston <- apply(BCI.count, 2, sum)
Preston.dist(BCI.preston)
$Equation
[1] "n = 39 * exp(-0.271 * R)^2"
$Est.no.of.spp
[1] 254.8617
$Pct.sampling.completed
[1] 0.8828319
```

We can see that Preston's method effectively models common species (those with $\log_2$ abundance >4). However, it does not fit the data well for rarer species. In fact, no observations are available for the extreme left side of the curve. Preston called a line placed at the 0th octave the *veil line*. He argued that species with abundances below the veil line have not been detected due to inadequate sampling.

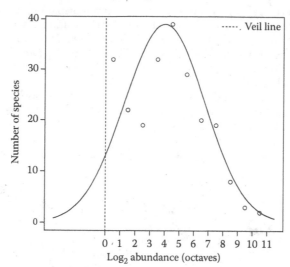

```
Fig.3.26()#See Ch. 3 code
```

FIGURE 3.26

Preston's lognormal distribution applied to tree species counts on Barro Colorado Island in central Panama. Points indicate species counts within octaves.

The area beneath a Preston's lognormal curve does not equal one; instead it is equivalent to the total number of species predicted for the sampled area, $n_{species}$:

$$n_{species} = \int_{-\infty}^{\infty} n dR = n_0 \sqrt{\pi}/a. \tag{3.31}$$

For our example $n_{species} = 255$, indicating that Condit et al. (2002) only recorded 88% of the possible species in the area in their survey. More recently developed methods for estimating species richness use resampling approaches and are discussed in Section 5.4.

3.3.2.9 Logistic Distribution

The *logistic distribution* is very similar in shape to the normal distribution although it is more sharply peaked at is its mode. The pdf has the form

$$f(x) = \frac{e^{-(x-\beta)/\theta}}{\theta \left[1 + e^{-(x-\beta)/\theta}\right]^2}, \tag{3.32}$$

where
1. Outcomes, x, are continuous.
2. $x \in \mathbb{R}$.
3. $\beta \in \mathbb{R}$.
4. $\theta > 0$.

Here, θ is a scale parameter, and β is a location parameter.

If a random variable X has the pdf in Equation 3.32, we denote this as $X \sim \text{LOGIS}(\beta, \theta)$. The pdf for $X \sim \text{LOGIS}(0, 1)$ is shown in Figure 3.27.

Solving for x in the logistic cdf results in the so-called *logit function* used in *logistic generalized linear models* (GLMs; Chapter 9). Logistic GLMs are widely used for population survivorship and species presence/absence model (e.g., Aho et al. 2011, many others).

EXAMPLE 3.17 LOGISTIC DISTRIBUTION AND LOGIT FUNCTION

Like the exponential and Weibull distributions, the logistic cdf has a closed form:

$$P(X \le x) = F(x) = \frac{1}{1 + e^{-(x-\beta)/\theta}}$$

$$= \frac{1}{1 + 1/e^{(x-\beta)/\theta}} = \frac{1}{\left(e^{(x-\beta)/\theta} + 1\right)/e^{(x-\beta)/\theta}} = \frac{e^{(x-\beta)/\theta}}{1 + e^{(x-\beta)/\theta}}. \tag{3.33}$$

If we let $\beta = 0$ and $\theta = 1$, we have a sigmoidal cdf centered at 0 (Figure 3.27). That is,

$$P(X \le x) = F(x) = \frac{1}{1 + e^{-x}}. \tag{3.34}$$

If we solve for x in Equation 3.34, we have inverse cdf of the logistic distribution. That is, we have a function that provides quantiles for the logistic cdf given a particular probability. Letting $\pi = F(x)$, we have

$$\pi = \frac{1}{1 + e^{-x}}; \quad \frac{1}{\pi} = 1 + e^{-x}; \quad \frac{1-\pi}{\pi} = e^{-x}; \quad -\ln\left(\frac{1-\pi}{\pi}\right) = x;$$

$$\ln\left(\frac{\pi}{1-\pi}\right) = x. \quad \text{Rules of logarithms: } \ln(1/b) = -\ln(b)$$

This is known as the logit function, and is equivalent to the natural log of the odds of π (Equation 2.19). Similarly, finding the inverse cdf for the normal distribution gives the *probit function*.

```
see.logiscdf.tck()
```

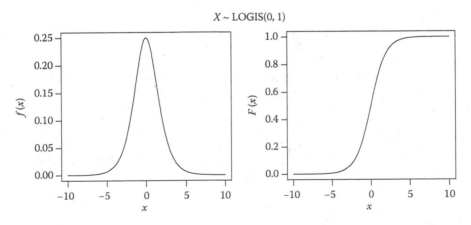

FIGURE 3.27
The pdf and cdf for LOGIS(0, 1).

3.4 Which pdf to Use?

A major concern for biologists is deciding which pdf best represents the phenomena underlying their data. A myriad of hypothesis testing procedures to specifically address the question: "how likely are the data given a particular null distribution?" are introduced in later chapters.

Pdf choice will largely be determined by the type of data one has, or the statistical procedure one wishes to use. For instance, if data are discrete, one is generally limited to discrete pdfs (although a number of transformations have been suggested to "convert" discrete biological data to continuous outcomes (e.g., Greig-Smith 1983), or to account for jagged ordinal categories (Johnson et al. 2004)). The type of scientific question will also determine the correct use of pdfs. For example, if the phenomenon of interest is the number of independent binary successes when sampling with replacement, then the binomial distribution will provide the proper probabilistic model.

If a studied phenomenon is continuous, then normal distributions are desirable for three reasons. First, normal distributions are symmetric. That is, given that X is centered at 0, $P(X < -x)$ will be equivalent to $P(X > x)$. This facilitates interpretation and probability calculations for lower and upper tailed hypothesis tests (Chapter 6). Second, normal distributions are simply very effective for describing many real biological variables (e.g., height, weight, and girth). Third, the sampling distributions of many statistics (Chapter 4) become normally distributed given large sample sizes, regardless of the distributional characteristics of the sampled parent distribution (Chapter 5). In this text, we often (but not always) use distributions that are normal or that are derived from normal distributions.

3.4.1 Empirical cdfs

If it is unclear which distribution will do the best job of representing a particular phenomenon then one can derive a cdf from observed data. This is called an *empirical cdf* or *ecdf*.

Let $y_1, y_2, \ldots, y_n$ be a ordered set of data $y_1 < y_2 < \cdots < y_n$, and let $i = \{1, 2, \ldots, n\}$, then the ecdf is

$$Fn(x) = \begin{cases} 0 & x < y_1 \\ i/n & y_i \leq x < y_{i+1} \\ 1 & y_n \leq x \end{cases} \tag{3.35}$$

The term $Fn(x)$ denotes that the empirical cdf is based on data (of size n). One can often compare an ecdf to cdfs of various distributions to help decide on the appropriateness of a distributional model.

EXAMPLE 3.18

The dataset `catsM` from the library *boot* contains body weight and heart weight data for 97 male cats used in digitalis experiments (Fisher 1947).

```
library(boot)
data(catsM)
```

The second column in `catsM` contains body weights (in kilograms). We will use the function ecdf to create an ecdf for this data.

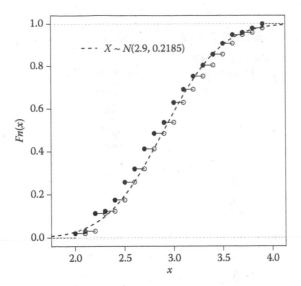

FIGURE 3.28
The empirical cdf for Example 3.18. A normal cdf, $X \sim N(2.9, 0.2185)$, is overlaid. Open dots indicate that the point is not included in the ecdf.

```
cats.ecdf < - ecdf(catsM[,2])
```

Using the ecdf, we can ask what is the empirical probability of observing a cat under 3 kg?

```
length(catsM[,2][catsM[,2] < 3])/97
[1] 0.5360825
```

Here is a plot of the ecdf (Figure 3.28).

```
plot(cats.ecdf, main = "")
curve(pnorm(x,mean(catsM[,2]), sd(catsM[,2])),
from = 1.5,to = 4.2,add = T,col = 2,lwd = 2)
legend("topleft",inset = 0.1, bty = "n", legend = "X ~ N(2.9, 0.2185)",
lwd = 2,col = 2)
```

We can see that a normal distribution represents these data fairly well (Figure 3.28). As a result, we get approximately the same answer for $P(X < 3)$ using an appropriate normal distribution.

```
pnorm(3,mean(catsM[,2]),sd(catsM[,2]))
[1] 0.5846917
```

3.5 Reference Tables

I conclude this chapter with useful reference information concerning pdfs. Table 3.3 lists the distributional functions found in the **R** *base* package. Tables 3.4 and 3.5 summarize important characteristics of all the discrete and continuous distributions discussed in this chapter.

TABLE 3.3

Probability Distributions Available in the **R** base Package[a]

Distribution	Continuous (C) or Discrete (D)	Function Name[b]	Parameter Arguments
Beta	C	beta	α = shape1, β = shape2
Binomial	D	binom	n = size, π = prob
Chi-squared	C	chisq	v = df
Exponential	C	exp	θ = rate
F	C	f	v_1 = df1, v_2 = df2
Gamma	C	gamma	κ = shape, θ = scale
Geometric	D	geom	π = prob
Hypergeometric	D	hyper	M = m, $M - N$ = k, n = n
Lognormal	C	lnorm	μ = meanlog, σ = sdlog
Logistic	C	logis	β = location, θ = scale
Negative binomial	D	nbinom	r = size, π = prob
Normal	C	norm	μ = mean, σ = sd
Poisson	D	pois	λ = lambda
Student's t	C	t	v = df
Uniform	C	unif	a = min, b = max
Weibull	C	weibull	β = shape, θ = scale

[a] Many other useful distributions are available in supplemental **R** libraries. For instance, see package *SuppDists*.

[b] Placing a d in front of the function name (e.g., dnorm) provides the density for a distribution given an outcome, x. Placing a p in front of the function (e.g., pnorm) provides the cumulative distribution function (i.e., probabilities at or below a particular outcome), Placing a q in front of the function (e.g., qnorm) provides the inverse cdf or quantile function, that is, the outcome (quantile), given a particular probability from 0–1. Placing an r in front of the function (e.g., rnorm) allows random number generation.

3.6 Summary

- A pdf allows derivation of the probabilistic density for the outcome of a random variable.

- For both discrete and continuous random variables, density at $X = x$ can be thought of as the height of a pdf curve (or discrete pdf line). Density will be equivalent to probability $P(X = x)$ for a discrete random variable. However, calculation of probability for continuous random variables requires integration of the pdf.

- The normal distribution is the most frequently used pdf in conventional statistical analyses. It is bell shaped and symmetric. A standard normal, or Z-distribution will be centered at zero and have unit variance.

- A large number of other well-understood distributions are also used frequently by biologists, including discrete pdfs (binomial, negative binomial, Poisson, geometric, and hypergeometric distributions), and continuous pdfs (t-, χ^2, t, F-, exponential, Weibull, gamma, beta, lognormal, and logistic distributions).

TABLE 3.4

Discrete Probability Distributions Commonly Used by Biologists

Distribution Name	pdf	Mean, Var, Mode, MGF
Binomial $X \sim \text{BIN}(n,\pi)$ Gives probability for x successes over n trials. $n = 1$ gives a Bernoulli pdf. $0 < \pi < 1$ $x \in \{0,1,2,\ldots,n\}$	$\binom{n}{x}\pi^x(1-\pi)^{n-x}$ where π is the probability of success, and n = number of trials	$E(X) = n\pi$ $\text{Var}(X) = n(1-\pi)\pi$ $\text{Mode}(X) = (n+1)\pi$ $\text{MGF} = [\pi e^t + (1-\pi)]^n$
Hypergeometric $X \sim \text{HYP}(n, M, N)$ Gives the probability of selecting x individuals belonging to group 1 without replacement given n total selections $n \in \{1,2,\ldots,N\}$ $M \in \{0,1,2,\ldots,N\}$ $x \in \{0,1,2,\ldots,n\}$	$\binom{M}{x}\binom{N-M}{n-x}\bigg/\binom{N}{n}$ where M is the number of elements in the group of interest (i.e., group 1), N = the total number of elements in groups 1 and 2, and n = the number of elements that are selected in the experiment	$E(X) = nM/N$ $\text{Var}(X) = n\dfrac{M}{N}\left(1-\dfrac{M}{N}\right)\dfrac{N-n}{N-1}$ $\text{Mode}(X) = \dfrac{(n+1)(m+1)}{N+2}$
Geometric $X \sim \text{GEO}(\pi)$ Gives the probability of the first success following x failures $0 < \pi < 1$ $x \in \{0,1,2,\ldots\}$	$\pi(1-\pi)^x$ where π is the probability of success	$E(X) = 1/\pi$ $\text{Var}(X) = (1-\pi)/\pi^2$ $\text{Mode}(X) = 0$ $\text{MGF} = \dfrac{\pi e^t}{1 - [(1-\pi)e^t]}$
Negative binomial $X \sim \text{NB}(r,p)$ Gives the probability of obtaining the rth success after x failures $0 < \pi < 1$ $r \in \{0,1,2,\ldots\}$ $x \in \{0,1,2,\ldots\}$	$\binom{x+r-1}{r-1}\pi^r(1-\pi)^x$ where π is the probability of success	$E(X) = r/(1-\pi)\pi^2$ $\text{Var}(X) = r(1-\pi)/\pi$ $\text{MGF} = \pi^r[1 - (1-\pi)e^t]^{-r}$
Poisson $X \sim \text{POI}(\lambda)$ The probability that an event occurs x times in a finite observation space $\lambda > 0$ $x \in \{0,1,2,\ldots\}$	$\dfrac{e^{-\lambda}\lambda^x}{x!}$ where λ indicates the rate of success	$E(X) = \lambda$ $\text{Var}(X) = \lambda$ $\text{Mode}(X) = \lambda$ $\text{MGF} = e^{\lambda(e^t-1)}$

Note: The pdfs give both probability and density at $X = x$. The mean, variance, mode, and moment generating functions are listed when these exist for the pdf. Moment generating functions (MGFs) are explained in Chapter 4.

TABLE 3.5

Some Continuous Probability Distributions Commonly Used by Biologists

	pdf	Mean, Var, Mode, MGF
Beta $X \sim \text{BETA}(\alpha,\beta)$ $\alpha > 0$ $\beta > 0$ $0 < x < 1$	$\dfrac{\Gamma(\alpha+\beta)}{\Gamma(\alpha)\Gamma(\beta)} x^{\alpha-1}(1-x)^{\beta-1}$	$E(X) = \dfrac{\alpha}{\alpha+\beta}$ $\text{Var}(X) = \dfrac{\alpha\beta}{(\alpha+\beta)^2(\alpha+\beta+1)}$ $\text{Mode}(X) = \dfrac{\alpha-1}{\alpha+\beta-2}$
Chi-square $X \sim \chi^2$ $(\nu)\nu \geq 0$ $x \geq 0$	$\dfrac{1}{2^{\nu/2}\Gamma(\nu/2)} x^{(\nu/2)-1}e^{-x/2}$	$E(X) = \nu$ $\text{Var}(X) = 2\nu$ $\text{Mode}(X) = \nu - 2 \text{ for } \nu \geq 2$ $\text{MGF} = \left(\dfrac{1}{1-2^t}\right)^{\nu/2}$
Exponential $X \sim \text{EXP}(\theta)$ $\theta > 0$ $x \geq 0$	$\theta e^{-x\theta}$	$E(X) = 1/\theta$ $\text{Var}(X) = 1/\theta^2$ $\text{Mode}(X) = 0$ $\text{MGF} = \dfrac{\theta}{\theta - t}$
Gamma $X \sim \text{GAM}(\kappa,\theta)$ $\theta > 0$ $\kappa > 0$ $x > 0$	$\dfrac{1}{\theta^\kappa \Gamma(\kappa)} x^{k-1}e^{-x/\theta}$	$E(X) = \kappa\theta$ $\text{Var}(X) = k\theta^2$ $\text{Mode}(X) = \dfrac{\theta-1}{\kappa} \text{ for } \theta \geq 1$ $\text{MGF} = \left(\dfrac{1}{1-\theta t}\right)^k$
Logistic $X \sim \text{LOGIS}(\beta,\theta)$ $\theta > 0$ $\beta \in \mathbb{R}$ $x \in \mathbb{R}$	$\dfrac{e^{-(x-\beta)/\theta}}{\theta\left[1+e^{-(x-\beta)/\theta}\right]^2}$	$E(X) = \beta$ $\text{Var}(X) = \dfrac{1}{3}\pi^2\theta^2$ $\text{Mode}(X) = \beta$
Lognormal $X \sim \text{LOGN}(\mu,\sigma^2)$ $\sigma > 0$ $\mu \in \mathbb{R}$ $x > 0$	$\dfrac{1}{x\sigma\sqrt{2\pi}} e^{-\frac{1}{2}\left(\frac{\ln(x)-\mu}{\sigma}\right)^2}$	$E(X) = \mu + \dfrac{1}{2}\sigma^2$ $\text{Var}(X) = e^{(2\mu+\sigma^2)}(e^{\sigma^2}-1)$ $\text{Mode}(X) = \mu - \sigma^2$
Normal $X \sim N(\mu,\sigma^2)$ $\sigma > 0$ $\mu \in \mathbb{R}$ $x \in \mathbb{R}$	$f(x) = \dfrac{1}{\sigma\sqrt{2\pi}} e^{-\frac{1}{2}\left(\frac{x-\mu}{\sigma}\right)^2}$	$E(X) = \mu$ $\text{Var}(X) = \sigma^2$ $\text{Mode}(X) = \mu$ $\text{MGF} = e^{\mu t + \sigma^2 t^2/2} c$
Snedecor's F $X \sim F(\nu_1,\nu_2)$ $\nu_1 > 0$ $\nu_2 > 0$ $x > 0$	$\dfrac{\Gamma\left(\dfrac{\nu_1+\nu_2}{2}\right)}{\Gamma\left(\dfrac{\nu_1}{2}\right)\Gamma\left(\dfrac{\nu_2}{2}\right)}\left(\dfrac{\nu_1}{\nu_2}\right)^{\frac{\nu_1}{2}} x^{\left(\frac{\nu_1}{2}-1\right)}\left(1+\dfrac{\nu_1}{\nu_2}x\right)^{-\frac{\nu_1+\nu_2}{2}}$	$E(X) = \dfrac{\nu_2}{\nu_2 - 2}$ $\text{Var}(X) = \dfrac{2\nu_2^2(\nu_1+\nu_2-2)}{\nu_1(\nu_2-2)^2(\nu_2-4)}$
Studentized-t $X \sim t(\nu)$ $\nu > 0$ $x \in \mathbb{R}$	$\dfrac{\Gamma\left(\dfrac{\nu+1}{2}\right)}{\Gamma\left(\dfrac{\nu}{2}\right)} \dfrac{1}{\sqrt{\nu\pi}}\left(1+\dfrac{x^2}{\nu}\right)^{-\frac{\nu+1}{2}}$	$E(X) = 0$ $\text{Var}(X) = \dfrac{\nu}{\nu-2}$ $\text{Mode}(X) = 0$

TABLE 3.5 (continued)

Some Continuous Probability Distributions Commonly Used by Biologists

	pdf	Mean, Var, Mode, MGF
Uniform $X \sim$ UNIF(a,b) $a \le x \le b$	$\dfrac{1}{b-a}$	$E(X) = \dfrac{b+a}{2}$
		$\text{Var}(X) = \dfrac{(b-a)^2}{12}$
		$\text{MGF} = \dfrac{e^{bt} - e^{at}}{t(b-a)}$
Weibull $X \sim$ WEI(β,θ) $\beta > 0$ $\theta > 0$ $x >$	$\dfrac{\theta}{\beta^\theta} x^{\theta-1} e^{-(x/\beta)^\theta}$	$E(X) = \beta\Gamma\left(1 + \dfrac{1}{\theta}\right)$
		$\text{Var}(X) = \beta^2\left[\Gamma\left(1 + \dfrac{2}{\theta}\right) - \Gamma^2\left(1 + \dfrac{1}{\theta}\right)\right]$

Note: The pdfs give density at $X = x$. The area under the pdf curve provides probability. The mean, variance, mode, and moment generating functions (Chapter 4) are listed when these exist.

EXERCISES

1. Let X be a discrete random variable whose pdf is described in the table given here:

x	−1	0	1
$f(x)$	1/8	6/8	1/8

   Find the following:
   a. $P(X = 0)$
   b. $P(X < 10)$
   c. $P(0 > X > 0)$
   d. $P(X \le 1)$
   e. $F(1)$
   f. $F^{-1}(7/8)$
   g. $R(x)$, that is, the survivorship function
   h. $f(x)/R(x)$, that is, the hazard function

2. Assume that leaf biomass (in grams) from the plant *Salix arctica* follows the pdf, $f(x) = 2(x + 1)^{-3}$ for $x > 0$.

   Find the probability of a leaf being between 0 and 3 g, that is, $P(0 < X < 3)$. Solve by hand using calculus. Verify your result using **R**.

3. Let X be a continuous random variable with the pdf $f(x) = 3x^2$ if $0 < x < 1$. Find the following by hand, using calculus if necessary. Verify your results using **R**.
   a. $P(X = 0.5)$, explain your answer
   b. $P(X < 1)$, explain your answer
   c. $P(X \le 1)$, explain your answer
   d. $P(X > 1)$, explain your answer

   e. $P(X < 0.7)$

   f. $P(0.1 < X < 0.6)$

4. Let X be a continuous random variable with the pdf $f(x) = e^{-x}$, for $x > 0$. Find the following by hand, using calculus if necessary. Verify your results using **R**.

   a. $P(X < 0)$, explain your answer

   b. $P(X > 0)$, explain your answer

   c. $P(X < 0.5)$

   d. $P(X \leq 0.5)$, explain your answer

   e. $P(0.1 < X < 1)$

5. You are working on a mark–recapture study of boreal toads, *Bufo boreas*. You predict that there is a 40% chance of capturing a marked toad for each of 30 traps that you establish.

   a.   What is the probability that *exactly* 20 marked toads will be found in the 30 traps? Calculate this by hand, and confirm your calculation using **R**.

   b.   What is the probability that *up to* 20 marked toads will be found? Use **R**.

   c.   How do these values change if your estimated probability of recapture is 30%, not 40%?

6. The *binomial theorem* establishes a general formula for the expansion of $(a + b)^n$. The theorem states if n is a positive integer, and if a and b are any two numbers then

$$(a + b)^n = \binom{n}{0}a^n + \binom{n}{1}a^{n-1}b + \cdots + \binom{n}{x}a^{n-x}b^x + \cdots + \binom{n}{n}b^n.$$

There are important shared characteristics of the binomial expansion and the binomial pdf. In particular, for the $(x + 1)$th term in the expansion we have $\binom{n}{x}a^{n-x}b^x$.

A number of interesting patterns are evident in *Pascal's triangle* (which represents the coefficients for binomial expansions). For instance, the sum of the indicated components at a higher level of Pascal's triangle equals the indicated component of the lower level given here.

$$\binom{0}{0} \qquad = \qquad 1$$

$$\binom{1}{0}\binom{1}{1} \qquad = \qquad 1 \ \ 1$$

$$\binom{2}{0}\binom{2}{1}\binom{2}{2} \qquad = \qquad 1 \ \ 2 \ \ 1$$

$$\binom{3}{0}\boxed{\binom{3}{1}}\boxed{\binom{3}{2}}\binom{3}{3} \qquad = \qquad 1 \ \ \boxed{3} \ \ \boxed{3} \ \ 1$$

$$\binom{4}{0}\binom{4}{1}\boxed{\binom{4}{2}}\binom{4}{3}\binom{4}{4} \ \ = \ \ 1 \ \ 4 \ \ \boxed{6} \ \ 4 \ \ 1$$

To demonstrate the generality of this, and other Pascal triangle patterns, prove

a. $\dbinom{n}{0} = \dbinom{n}{n}$

b $\dbinom{n}{1} = \dbinom{n}{n-1}$

c. $\dbinom{n}{x} + \dbinom{n}{x+1} = \dbinom{n+1}{x+1}$

    d. What are the implications of the symmetric character of Pascal's triangle [demonstrated in (a) and (b) above] for the binomial pdf?

7. A herd of mountain goats is believed to contain 30 males and 20 females. If, for a single random capture, a male or female is equally easy to capture, what is the probability of capturing 5 females and 1 male in six attempts? Use **R** to calculate probabilities

    a. With replacement

    b. Without replacement

8. What is the probability that a predator will capture its third prey item after six independent attempts given that its probability of capturing prey per attempt is 0.3?

9. Find the following probabilities using pnorm in **R**. Also, draw the Z-distribution curve, and shade the correct area under the curve. To create graphs (and check results) one can use the function shade.norm.

    a. $P(Z < 2.57)$

    b. $P(Z > 1.20)$

    c. $P(Z < -1.04)$

    d. $P(-1.96 > Z > 1.96)$

    e. $P(-1.96 < Z < -0.61)$

10. Stanford–Binet IQ test scores are normally distributed with a mean μ, of 100, and a standard deviation σ, of 16. Let Stanford–Binet IQ be described by the random variable X. Use pnorm and qnorm to find

    a. $P(X = 50)$, justify your answer

    b. $P(X > 100)$, justify your answer

    c. $P(X \geq 100)$, justify your answer

    d. $P(X < 70)$

    e. $P(X \geq 123)$

    f. $P(83 < X < 120)$, depict this using shade.norm

    g. $P(83 > X > 120)$, depict this using shade.norm

    h. $P(83 > X < 120)$, depict this using shade.norm

    i. 20% of test scores are less than what IQ?

    j. 70% of test scores are more than what IQ?

11. With respect to the negative binomial distribution:

    a. Mathematically show that $NB(1, \pi) = GEO(\pi) = BIN(0, \pi)$.

b. Graphically demonstrate (a) for $\pi = 0.7$ with see.nbin.tck(), see.geo.tck(), and see.nbin.tck().

c. Mathematically show that $\Gamma(1) = \Gamma(2) = 1$.

★ 12. Are outcomes from the hypergeometric distribution independent? Why or why not?

★ 13. Bliss and R.A. Fisher (1953) examined female European red mite counts (Panonychus ulmi) on McIntosh apple trees [*Malus domestica* (McIntosh)]. Counts of the mites on 150 leaves are shown here.

Mites per leaf	0	1	2	3	4	5	6	7	8
Leaves observed	70	38	17	10	9	3	2	1	0

The investigators tested the usefulness of the negative binomial pdf given as Equation 3.18 for describing this count distribution. The function ML.k provides maximum likelihood estimates (Chapter 4) for negative binomial dispersion parameter, m, and mean, k.

a. Run the ML.k function by specifying: ML.k(Leaf.obs, Mites.per.leaf).

b. Interpret the estimates for m and k.

c. Get the probabilities of the mite counts by using the ML estimates for the negative binomial parameters in dnbinom, that is,

```
p <- dnbinom(Mites.per.leaf, size = k, mu = m)
```

d. Multiply the probabilities from (c) by the total number of leaves to get the mite frequency distribution, under a negative binomial distribution, that is,

```
round(p * 150, 1)
```

e. Repeat (c) and (d) with a Poisson distribution, using the mean leaf count as the estimate for λ, that is, lambda.hat <- mean(Leaf.obs), using pp <- dpois(Mites.per.leaf, lambda.hat) to obtain the Poisson probabilities, and using round(pp* 150, 1) to obtain expected Poisson frequencies.

f. Which expected frequency distribution (negative binomial or Poisson) agrees with the observed frequency distribution better? Why?

g. Support your conclusions in (f) by creating an ecdf, and overlaying the Poisson and negative binomial cdfs.

★ 14. Assume $X \sim t(4)$.

a. What does the expression $X \sim t(4)$ *mean*?

b. Use **R** to find the probability *densities* for the following values of X: $x = -2, 3$, and 4.

★ 15. Assume $X \sim t(7)$, and use **R** to answer the questions.

a. Find $F^{-1}(0.4)$.

b. Find the value of x associated with the upper tail probability 0.62.

c. Find the probability *density* associated with $x = -0.2631669$.

d. Find $P(X < -0.2631669)$.

★ 16. For the following questions, find the correct probabilities using `pnorm`, `pt`, `pF`, or `pchisq`; also draw associated distribution curves, and shade the correct area under the curve using the functions `shade.norm`, `shade.t`, `shade.F`, and `shade.chi`.

    a. Let $X \sim t(7)$, find $P(X < 0)$, explain your answer.

    b. Let $X \sim t(10)$, find $P(X < 2)$.

    c. Let $X \sim \chi^2(7)$, find $P(X < 7)$, also find this probability using the gamma distribution.

    d. Let $X \sim \chi^2(5)$, draw a figure with 0.025 probability in the upper and lower tails.

    e. Let $X \sim F(10,4)$, find $P(1 < X < 2)$.

    f. Let $X \sim F(10,4)$, draw a figure with 0.025 probability in the upper and lower tails.

★ 17. Complete the following. Show all code.

    a. Create an **R** object, `X`, with 10000 random observations from $N(0, 1)$ using `rnorm`.

    b. Square `X`, call this object `Y`.

    c. Create an ecdf of `Y`. Now find the empirical probability $P(Y < 1)$.

    d. Assume $X \sim \chi^2(1)$ and calculate $P(X < 1)$. Comment on your results.

★ 18. Assume that the number of sightings for bald eagles (*Haliaeetus leucocephalus*) at a site over a 4-week period is Poisson distributed (temporally random) with a mean $\lambda = 10$. What is the probability of having 15 sightings over a 4-week period?

★ 19. In your own words describe how the normal, χ^2, t-, and F-distributions are related to each other.

★ 20. Mathematically and graphically demonstrate that BETA(1, 1) = UNIF(0, 1).

★ 21. Graphically demonstrate, using *asbio* GUI functions, that WEI($\beta = 1, \theta$) = GAM($\kappa = 1, \theta$) = EXP($1/\theta$), if θ represents the same number.

★ 22. If the probability of extinction of an endangered population is estimated to be 0.1 every year, and the probabilities of extinction/survival are independent from year to year, then what is the expected time for P(extinction) = 0.9? Calculate "by hand" and confirm your answer is correct using **R**.

★ 23. Derive the exponential inverse cdf from its closed form cdf.

★ 24. Assume $X \sim$ EXP(2), and find the following by hand using the closed form exponential cdf. Confirm your results using **R**.

    a. $P(X = 0.5)$

    b. $P(X > 1)$

    c. $P(X < -1)$

    d. $P(0.2 > X > 0.7)$

    e. $P(0.1 < X < 0.6)$

★ 25. Redo Question 24 using the gamma and Weibull cdfs.

★ 26. Shown here is a cohort life table for the mountain lion (*Puma concolor*) in the Pacific Northwest. Data are from Lambert et al. (2006).

Age (Years)	l_x	Age (Years)	l_x
0	0.185	8.0	0.027
1.5	0.123	9.0	0.022
3.0	0.048	10.0	0.021
4.0	0.043	11.0	0.020
5.0	0.038	12.0	0.016
6.0	0.035	13.0	0.000
7.0	0.030		

    a. Create an exponential survivorship curve.

    b. Create a Kaplan–Meier survivorship curve using the function km from *asbio*.

    c. Comment on your results.

★   27. Times between sightings of bald eagles at a site are thought to be exponentially distributed with $\theta = 4$ h (i.e., rate $= 0.25$ h). What is the probability of seeing a bald eagle within 15 min, 1 h, and 2 h of arriving at the site? Solve using the exponential, gamma, and Weibull distributions.

★   28. The following questions concern Weibull survivorship analyses:

    a. Derive the survivorship function and the hazard function for the Weibull distribution.

    b. Mathematically show that the Weibull hazard function is constant (and defined by β) if $\theta = 1$.

    c. Demonstrate, using graphs, that the hazard functions decreases with age (specified as x) if $\theta < 1$, and increases with age if $\theta > 1$.

★   29. The datasets bryceveg in *labdsv* and dune in library *vegan* contain plant community data. Fit Preston's lognormal distribution to these data. Comment on your results.

4

Parameters and Statistics

To understand God's thoughts we must study statistics, for these are the measures of his purpose.

Florence Nightingale

4.1 Introduction

When we make an *inference,* we draw a conclusion based on evidence. Suppose you are a botanist in charge of a rare plant inventory. From extensive field work, you find that an endemic plant is absent in samples from a forest where it had been previously reported. As a result, you conclude that the species is in decline, and will require protective management.

The tendency to arrive at particular conclusions has helped *Homo sapiens* to survive and thrive, even if our inferences are not always correct. For example, the perception of danger will often trigger a *fight or flight* response in hominids. In this process, the autonomic nervous system signals to the body to release adrenaline, increase heart rate, and otherwise prepare for fighting or fleeing. Interestingly, recent evidence suggests that the human fight or flight response errors on the side of caution (Rakison 2009).[*] That is, we may be hard-wired to see the world in a slightly paranoid way. There are at least two possible reasons for this. First, natural selection may have favored nervous ancestral organisms, hyper-alert to hazards, such as predators, over excessively calm organisms that were easy prey. Second, characteristics influenced (or uninfluenced[†]) by natural selection will not always be the "perfect response to the environment." Instead, they will be the result of a long series of evolutionary compromises, based on a collection of phenotypes and genotypes arising from stochastic processes. Clearly, these constraints are unlikely to give rise to a species whose inferences are flawless.

Other simple examples of "human" predispositions for invalid inference include *cognitive illusions* that tend to fool the intellect and senses (Piattelli-Palmarini 1994).[‡] Consider Figure 4.1. Note that although the dots at the line intersections are white (the **R**-code for

---

[*] See discussion of Type I and Type II hypothesis testing error, Chapter 6.

[†] For example, genetic drift.

[‡] Still other examples include the human inability to resist "appeals to emotion" (e.g., flattery, fear, and ridicule), confirmation bias (in which one searches for evidence to confirm predilections), and heuristic processes that tend to ignore available information (Gigerenzer and Gaissmaier 2011).

```
illusions(3)
```

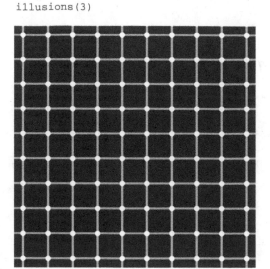

FIGURE 4.1
Are the dots black or white? The figure is based on the code from Yihui Xie's package *animation*. (Adapted from Xie, Y. 2012. Animation: A gallery of animations in statistics and utilities to create animations. R-package version 2.1. http://cran.r-project.org/web/packages/animation/index.html.)

the figure tells us so[*]), it is difficult to visually determine whether they are black, white, or gray.

Logical/mathematical tools such as statistics can serve as "thinking aids" to help us consider the workings of complex phenomena that may fool our intuition (cf. Kokko 2007). In this way, they may help us avoid, or at least account for our propensities to inferentially err.

Statistics is largely concerned with two types of *inferential procedures*. These are (1) the estimation of unknown parameters describing a population, based on sample data (evidence) and (2) hypothesis testing, in which data summaries concerning unknown parameters are compared to hypothetical parameter values. Parameter estimation is addressed in this chapter and in Chapter 5, and while hypothesis testing is described in Chapter 6.

4.1.1 How to Read This Chapter

By in large, materials in this chapter are appropriate for introductory statistics students although a review of elementary calculus may be necessary (see Appendix). Readers should be aware, however, that a portion of Section 4.3.6 is devoted to robust *M*-estimators, Section 4.3.9.1 introduces moment generating functions (MGFs), and Section 4.6 concerns the derivation of Bayesian posterior distributions using conjugate priors. These topics may be overwhelming or at least confusing to less advanced students. This material can

---

```
* par(bg = "black")
  plot.new()
  x = seq(0, 1, length = 10)
  y = seq(0, 1, length = 10)
  abline(v = x, h = y, col = "gray", lwd = 6)
  points(rep(x, each = 10), rep(y, 10), col = "white", cex = 3, pch = 20)
```

be skipped without loss of continuity, provided that extensions of these topics are to be overlooked in the remaining chapters.

The proof of $E(S^2)$ in Section 4.3.7, and derivations of ordinary least squares (OLS) and maximum likelihood (ML) estimators of μ in Sections 4.4.1 and 4.4.2 may also be challenging to some students, and require consultation with the Appendix. These procedures add important mathematical rigor, but may be ignored if they are hampering conceptual understanding of these topics.

4.2 Parameters

To a statistician, a *population* is the set of all possible outcomes from a random variable. Note that this definition differs from the biological definition of a population: a group of individuals of the same species living in the same area at the same time. From a conventional (non-Bayesian) perspective, a *parameter* is a fixed numeric characteristic describing an entire population; for instance, the maximum tree height from an entire population of tree heights. Recall that specific parameters were required to define pdfs in Chapter 3.

4.2.1 Expected Value

The *expected value* of a random variable X is denoted $E(X)$ and represents the arithmetic mean (Section 4.3.5) of a statistical population. $E(X)$ can be calculated directly from its pdf. For any discrete pdf, $f(x)$, defining a random variable X, the expected value is

$$E(X) = \sum_x x f(x). \tag{4.1}$$

That is, $E(X)$ is the sum of the products of all possible numeric outcomes and their corresponding densities.

For any continuous pdf, $f(x)$, defining a random variable X, the expected value is

$$E(X) = \int_{-\infty}^{\infty} x f(x) \, dx. \tag{4.2}$$

That is, to find $E(X)$, we compute the definite integral of a function resulting from the multiplication the pdf and x at the closed bounds for the support of X.

Expectations for transformations of X can be made using the approach demonstrated in Equations 4.1 and 4.2. For instance, the expectation $E(X^2)$ for a continuous pdf is simply

$$E(X^2) = \int_{-\infty}^{\infty} x^2 f(x) \, dx.$$

EXAMPLE 4.1

Consider the discrete probability distribution for goat counts from Table 3.1. Using Equation 4.1 we see that

$$E(X) = \sum xf(x) = 0(0.5) + 1(0.3) + 2(0.1) + 3(0.05) + 4(0.05) = 0.85,$$

$$E(X^2) = \sum x^2 f(x) = 0(0.5) + 1(0.3) + 4(0.1) + 9(0.05) + 16(0.05) = 1.95.$$

EXAMPLE 4.2

Next, consider the continuous uniform distribution. If $X \sim UNIF(a,b)$, then $f(x) = 1/(b-a)$ (Table 3.5). Because the maximum and minimum values for this distribution occur at the parameters b and a, respectively, we have

$$E(X) = \int_{-\infty}^{\infty} xf(x)\,dx = \int_a^b x\left(\frac{1}{b-a}\right)dx$$

$$= \frac{1}{b-a}\int_a^b x\,dx$$

$$= \frac{1}{b-a}\left(\frac{b^2}{2} - \frac{a^2}{2}\right)$$

$$= \frac{1}{b-a}\left(\frac{b^2 - a^2}{2}\right)$$

$$= \frac{(b-a)(b+a)}{2(b-a)} = \frac{b+a}{2}.$$

$$E(X^2) = \int_a^b x^2\left(\frac{1}{b-a}\right)dx$$

$$= \frac{b^3 - a^3}{3(b-a)}$$

$$= \frac{(b-a)(a^2 + ab + b^2)}{3(b-a)} \quad \text{(rule for factoring cubic polynomials)}$$

$$= \frac{a^2 + ab + b^2}{3}$$

4.2.2 Variance

Another important parameter is the *variance* of a random variable X, denoted $\text{Var}(X)$. It quantifies the amount of dispersion or "spread" in a distribution. This is defined to be

$$\text{Var}(X) = E[X - E(X)]^2$$
$$= E(X^2) - 2E(X)^2 + E(X)^2$$
$$= E(X^2) - E(X)^2. \tag{4.3}$$

The standard deviation of X is the positive square root of the variance of X. That is

$$\text{SD}(X) = \sqrt{\text{Var}(X)}. \tag{4.4}$$

EXAMPLE 4.3
For the goat count example in Table 3.1, we have

$$\text{Var}(X) = E(X^2) - E(X)^2 = 1.95 - 0.85^2 = 1.2275,$$

$$\text{SD}(X) = \sqrt{1.2275} = 1.1079.$$

EXAMPLE 4.4
For the continuous uniform distribution, we have

$$\text{Var}(X) = E(X^2) - E(X)^2 = \frac{a^2 + ab + b^2}{3} - \frac{(b+a)^2}{4}$$

$$= \frac{4(a^2 + ab + b^2)}{12} + \frac{3(-b^2 - 2ab - a^2)}{12}$$

$$= \frac{a^2 - 2ab + b^2}{12} = \frac{(b-a)^2}{12},$$

$$\text{SD}(X) = \sqrt{\frac{(b-a)^2}{12}} = \frac{b-a}{\sqrt{12}}.$$

$E(X)$ and $\text{Var}(X)$ can be calculated for any pdf in Tables 3.4 and 3.5 using these methods. For instance, if $X \sim N(\mu, \sigma^2)$, then

$$E(X) = \mu, \tag{4.5}$$

$$\text{Var}(X) = E(X^2) - \mu^2 = \sigma^2. \tag{4.6}$$

4.2.3 Chebyshev Inequality

Recall from Chapter 2 that approximately 68%, 95%, and 99.7% of a normal distribution is within ±1, ±2, and ±3 standard deviations of the mean, respectively. An important generalization of this concept, called the *Chebyshev*[*] *inequality*, is given as

$$P\{|X - E(X)| \geq c[SD(X)]\} \leq \frac{1}{c^2}, \tag{4.7}$$

where X is a random variable with a calculable mean and standard deviation, $E(X)$ is the expectation of X, $SD(X)$ is the standard deviation of X, and c is a constant. By letting $c = 2$, we have

$$P\left[|X - E(X)| \geq 2SD(X)\right] \leq 0.25.$$

Thus, at most 25% of the distribution of X will be *more than* 2 standard from $E(X)$, and at least 75% of outcomes will be *within* two standard deviations of the population mean. We know that this is true for normal distributions because, in this case, $1 - P(|X - \mu| \geq 2\sigma) \approx 0.95$. In fact, the Chebyshev inequality holds for all possible discrete and continuous distributions!

4.3 Statistics

In general, it will be impossible to observe all possible outcomes from a phenomenon (if we could, then inferential statistics would be unnecessary). Instead, we use estimating algorithms, called *estimators*, to approximate unknown population parameters. Such estimates are called *statistics*. With statistics, we attempt to make inference to an entire population using only samples from the population.

4.3.1 Important Considerations

Before going further, four points should be emphasized. First, if a sample comprises an entire population, then statistical summaries will serve a descriptive instead of an inferential role. This is because statistics now represent fixed parameter values. Second, statistical summaries are only necessary when they make the result of an experiment or data from a series of samples easier to interpret or visualize (Vaux 2012). In fact, if the sample size is extremely small, it will often be more informative to present all the data points in a table or graph. Third, estimators generally assume that observations used to calculate estimates are independently obtained from the population. If this is not true, then statistics can only be used to make inference to the sample and not the underlying population (see Chapter 7). Fourth, numerical summaries of samples are often inadequate for describing a population, or for that matter, a sample (Chaterjee and Firat 2007). Thus, at all stages of analysis, statistics should be accompanied with graphical representations of data. As John Tukey noted: "the picture-examining eye is the best finder we have of the wholly unanticipated."

---

[*] Named for the Russian statistician Pafnuty Lvovich Chebyshev (1821–1894).

Such visual approaches include histograms, discussed in Chapter 3, *stemplots* which show the numerical distribution of individual data points (Figure 4.2a), *bar charts* whose bar lengths denote quantities associated with categories (Figure 4.2b), *scatterplots*, which depict the association of two quantitative variables using paired observations to define Cartesian coordinates (Figure 4.2c), and *lineplots*, which connect scatterplot points to demonstrate or

```
Fig4.2()## see Ch. 4 code
```

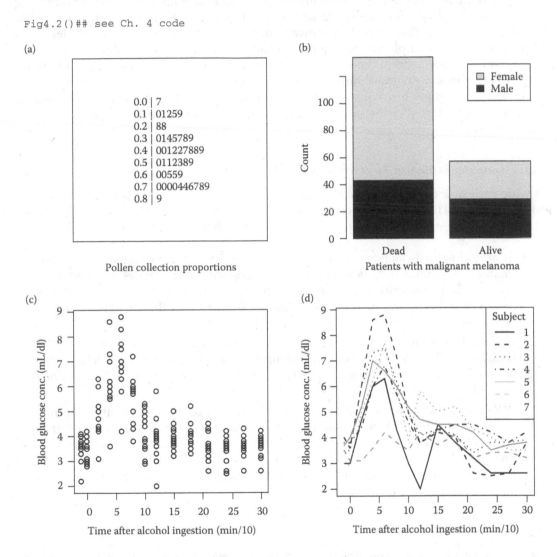

FIGURE 4.2
Four types of graphical data summaries. (a) A stemplot is the quantitative counterpart of a histogram. "Leaves" constitute the final (rightmost) digits in a quantitative observation while "branches" constitute all other values. The plot here shows pollen proportions removed from glacier lilies in Colorado by bees. (Adapted from Harder, L. D., and Thomson, J. D. 1989. *American Naturalist* 133: 323–344.) With respect to the two largest stems, there was one observation of 0.89, four observations of 0.7, two observations of 0.74, and one observation each of 0.76, 0.77, 0.78, and 0.79. (b) A stacked barplot describing the survival of patients with malignant melanoma. (Adapted from Andersen, P.K. et al. 1993. *Statistical Models Based on Counting Processes*. Springer-Verlag, New York, NY.) (c) A scatterplot showing blood glucose levels for experimental subjects after ingestion of alcohol. (Adapted from Hand, D. J. and Crowder, M. J. 1996. *Practical Longitudinal Data Analysis*, Chapman and Hall, London.) (d) A lineplot distinguishing the seven subjects in (c).

clarify association (Figure 4.2d). Many additional graphical tools and techniques are introduced in later chapters, and in the introduction to **R** appendix.

4.3.2 Sampling Error

An important and somewhat confusing term associated with estimators is sampling error. *Sampling error* refers to the variability in statistical estimates due to incomplete repeated sampling. The name is somewhat misleading because this "error" describes natural sample-to-sample variability as a result of the fact that variables vary (hence the name) and the fact that the entire population cannot be sampled at once. It is unrelated to error in sampling technique by the investigator. Increases in Var(X) will always increase the sampling error of statistical estimates taken from X, as will smaller sample sizes. Chapter 5 describes the sampling distributions of statistics whose variances define sampling error.

4.3.3 Gauging Estimator Effectiveness

A wide variety of estimators have been developed to summarize data and make inference to populations. This has prompted the establishment of criteria for evaluating estimator efficacy. The three most frequently used are bias, consistency, and efficiency.

- If an estimator is *unbiased*, then repeated estimates of the parameter by the estimator will demonstrate neither predispositions for overestimates nor underestimates. Mathematically, this means that the expected value of the estimator (the mean of its sampling distribution, Chapter 5) will equal the parameter, that is, E(statistic) = parameter.
- If the estimator is *consistent*, then as the sample size increases, the accuracy of the estimator (Chapter 1) increases. If a sample contains the entire population, then an estimator with *Fisher consistency* (Fisher 1922) will always provide an estimate equal to the parameter.[*]
- If an estimator is *efficient*, then it will provide a more precise (less variable) estimate of the parameter than other estimators for a particular sample size. For instance, if data are taken randomly from a normally distributed parent population, then the arithmetic mean (Section 4.3.5.1) will provide a more efficient estimate for the population mean, μ, than the sample median (Section 4.3.6.1) despite the fact that both are unbiased for μ.

4.3.4 Types of Estimators

Two sorts of estimators can be distinguished. *Point estimators* provide estimates for parameters with a single unknown value, and are the focus of this chapter. *Interval estimators*, described in Chapter 5, estimate the bounds of an interval that is expected, preceding sampling, to contain a parameter for a specified proportion of estimates.

---

[*] The distinction between the terms unbiased and consistent may seem unclear. However, not all consistent estimators will be unbiased. Consider any continuous distribution. The sample minimum and sample maximum will be both Fisher consistent *and* biased for the population minimum and maximum, respectively. This is because a finite sample will almost always fail to include the population extremes, while a sample of the entire population will always contain both values.

Three types of point estimators can be distinguished. These are measures of location, scale, and distributional shape. Measures of *location* estimate the so-called *typical* or *central values* from a sample. These include the sample arithmetic mean and the sample median. Measures of *scale* quantify data variability, spread, or dispersion. They include the sample variance and sample interquartile range (IQR). Finally, there are measures that quantify the *shape* of the data distribution. These include the sample skewness and sample kurtosis.

4.3.5 Measures of Location

4.3.5.1 Arithmetic Mean

The most frequently used location estimator is the *arithmetic mean*. It is commonly denoted $\bar{X}$ (pronounced x-bar) and is calculated by summing measurements for all n observations taken on a variable, and dividing by n:

$$\bar{X} = \frac{\sum_{i=1}^{n} X_i}{n}. \tag{4.8}$$

$\bar{X}$ is ubiquitous to empirical scientific studies for three reasons. First, its sampling distribution (Chapter 5) will be asymptotically normal (regardless of the parent distribution) facilitating inferential tests. Second, the arithmetic mean will be unbiased and consistent for $E(X)$ for any distribution (for proof, see Chapter 5). Third, $\bar{X}$ will be a maximum efficiency estimator for $E(X)$ when the parent distribution underlying the data is normal.

In later chapters, the sample arithmetic mean will simply be referred to as the sample mean.

EXAMPLE 4.5

African elephants (*Loxodonta africana*) often thrive in the protected confines of national parks. However, their overall population numbers have dwindled from 5–10 million at the turn of the 19th century to around 500,000 individuals today. These decreases are largely due to poaching in response to the ivory trade (see Thornton and Currey 1991). Annual counts of African elephants in Addo National Park South Africa from 1994 to 2003 (Gough and Kerley 2006) are

```
eleph <- c(220, 232, 249, 261, 284, 315, 324, 326, 377, 388)
```

To find the arithmetic mean, we take the sum of the observations, and then divide by the number of observations.

```
sum(eleph)/10
[1] 297.6
```

The function mean provides arithmetic means.

```
mean(eleph)
[1] 297.6
```

This simple example can be used to demonstrate the potential subtleties of statistical inference. We note, first of all, that the entire park was surveyed each year, and that (in theory) every elephant was counted. Thus, the sample comprises the entire population, making the sample mean equivalent to the population mean. Note, however, that in the unlikely event that the years 1993–2004 were a random sample of years, the sample arithmetic mean would merely allow inference to the true mean count for this population of years. Second, the arithmetic mean, a numerical summary, captures the typical value, but does not reflect that elephant numbers have been increasing in the park. A graphical summary would be better suited for communicating this trend. For instance, try

```
barplot(eleph, names = seq(1994, 2003, 1), ylab = "Elephant counts")
```

Just as a realization from the random variable X is denoted with x, we denote parameter *estimates*, the realizations of *estimators*, with lower case letters. For instance, the arithmetic mean is an *estimator* for the parameter $E(X)$, and is written as $\bar{X}$. Conversely, if we collected data, calculated the arithmetic mean, and found that the *estimate* equaled 3, we would write this as $\bar{x} = 3$. This dichotomy emphasizes that an estimator can be considered a random variable with a population consisting of all possible statistical estimates (Chapter 5).

4.3.5.2 Geometric Mean

The arithmetic mean is not the best location measure for all types of data. For instance, $\bar{X}$ will provide biased (high) estimates for data resulting from a multiplicative process. An important example is the growth rate of a biological population.

The growth rate at time t is conventionally denoted λ_t, and calculated as N_{t+1}/N_t, where N_t is the number of individuals at time t. Consider censuses of a biological population taken at times 0, 1, 2, and 3. To find the long-term growth rate, N_3/N_0, we note that

$$\frac{N_3}{N_0} = \frac{N_1}{N_0} \times \frac{N_2}{N_1} \times \frac{N_3}{N_2}. \quad \text{However,} \quad \frac{N_3}{N_0} \neq \frac{N_1}{N_0} + \frac{N_2}{N_1} + \frac{N_3}{N_2}.$$

Thus, growth rate is the result of a multiplicative, not an additive process.

If X is a random variable with a lognormal distribution, and $Y = \ln(X)$, then the transformation $e^{E(Y)}$ provides the *population geometric mean* of X. That is

$$G(X) = exp\{E[\ln(X)]\}. \tag{4.9}$$

We estimate $G(X)$ with the *sample geometric mean*:

$$GM = \sqrt[n]{\prod_{i=1}^{n} X_i}. \tag{4.10}$$

GM will never exceed $\bar{X}$ and will be much smaller than $\bar{X}$ given a large degree of variability among observations.[*]

EXAMPLE 4.6

Exotic species are a topic of great concern for ecosystem ecologists. Mountain goats (*Oreomnos americanus*) in Yellowstone National Park are non-native since they are the

---

[*] This is due to a mathematical property called *Jensen's inequality*, which stipulates that $\sqrt[n]{x_1 x_2, \ldots, x_n} \leq \frac{x_1 + x_2 + \cdots + x_n}{n}$.

descendants of transplants by the Montana Department of Fish and Wildlife to surrounding mountain ranges in the 1940s and 1950s (Aho 2006). The size of the Yellowstone mountain goat population has been monitored using aerial surveys since 2003 with the following results: $N_{2003} = 120$, $N_{2004} = 161$, $N_{2005} = 140$ (White 2003, Davis 2007).

We can calculate two annual growth rates; between years 2003 and 2004, and between years 2004 and 2005:

$$\lambda_{2003} = \frac{N_{2004}}{N_{2003}} = \frac{161}{120} = 1.342 \quad \text{and} \quad \lambda_{2004} = \frac{N_{2005}}{N_{2004}} = \frac{140}{161} = 0.8696.$$

The arithmetic mean of these rates is 1.1058.

```
mean(c(1.342,0.8696))
[1] 1.1058
```

The geometric mean is $\sqrt{1.342 \times 0.8696} = 1.080279$. The function G.mean from *asbio* can be used to calculate the geometric mean.

```
G.mean(c(1.342,0.8696))
[1] 1.080279
```

The arithmetic mean rate results in an overestimate of the population size in 2005:

$$1.1058 \times 132.696 = 146.7352 \text{ goats.}$$

The geometric mean of the rates, however, gives the correct population size in 2005:

$$1.080279 \times 129.6335 = 140.0404 \text{ goats.}$$

4.3.5.3 Harmonic Mean

The arithmetic mean is also an inappropriate location estimator when a phenomenon of interest is based on the average of a set of reciprocals. A common example is the time for travel, given varying velocities, but a common distance. The equation for velocity is

$$v = d/t,$$

where d is distance and t is time. Solving for time, we have $t = v^{-1}d$. If d is a known constant, we need only to multiply this number by the arithmetic mean of the observed reciprocal velocities to get total travel time. However, for this average to be in the original units, we must take reciprocal of the arithmetic mean of the reciprocal velocities. This results in the *harmonic mean*.

The *population harmonic mean* can be expressed as the function of arithmetic and geometric means. In particular, let N be the total number of possible outcomes from X, then

$$H(X) = \frac{[G(X)]^N}{\bar{X} \left(\frac{\prod_{i=1}^{N} X_i}{X_1}, \frac{\prod_{i=1}^{N} X_i}{X_2}, \ldots, \frac{\prod_{i=1}^{N} X_i}{X_N} \right)}, \tag{4.11}$$

where $\bar{X}(.)$ indicates take the arithmetic mean. The *sample harmonic mean* is calculated as

$$\mathrm{HM} = \left(\frac{1}{n} \sum_{i=1}^{n} \frac{1}{X_i} \right)^{-1}.$$ (4.12)

That is, *HM* is the reciprocal of the arithmetic mean of a set of reciprocals.

EXAMPLE 4.7

Consider a cheetah (*Acinonyx jubatus*) that moves 2 km at 20 km/h and then moves the same distance again at 30 km/h. In this case, the appropriate measure for the average velocity of the cheetah is the harmonic mean of the velocities:

$$\frac{1}{0.5(1/20\,\mathrm{km/h} + 1/30\,\mathrm{km/h})}$$

$$= \frac{1}{0.5(3/60\,\mathrm{km/h} + 2/60\,\mathrm{km/h})} = \frac{120\,\mathrm{km/h}}{5} = 24\,\mathrm{km/h}.$$

The average velocity is *not* the arithmetic mean, 25 km/h.
The function H.mean can be used to calculate the harmonic mean.

```
H.mean(c(30,20))
[1] 24
```

The total travel time for the cheetah (in hours) is the total distance traveled (4 km) divided by the harmonic mean, that is

```
4/24
[1] 0.1666667
```

Note that if one wanted to find the total distance traveled for a set of velocities given a common time, then we have $d = v/t$. In this case, we would divide the arithmetic mean for the nonreciprocal (observed) velocities by the known value t. Because the velocities are not reciprocals, obtaining the reciprocal of their average is unnecessary, allowing the straightforward use of the arithmetic mean.

The harmonic mean is also frequently used in the context of population genetics. R.A. Fisher (see Chapter 1) and evolutionary geneticist Sewall Wright (1889–1988) coined the term *idealized population* to describe a biological population with (1) random reproduction, (2) equal numbers of males and females, and (3) nonoverlapping generations. This framework leads to the *effective population size*, defined as the number of breeding individuals in an idealized population that would be required to obtain the same amount of allelic variability as the population under consideration (Wright 1938). Departure from the idealized population will make the effective population size smaller than a census of the individuals in a population.

The harmonic mean of population sizes from nonoverlapping generations can be used to estimate effective population size. This is because a *population bottleneck* (an extremely small population) will strongly decrease the magnitude of the harmonic mean compared

with the arithmetic mean* just as it strongly decreases the size of the gene pool in generations following the bottleneck.[†]

4.3.5.4 Mode

Yet another measure of typical location is the mode. The *population mode* is the location
of peak density in a distribution. We estimate the population mode with the *sample mode*,
which is simply the most frequently observed outcome in a sample. The function Mode
from *asbio* can be used to calculate sample modes.[‡]

```
Mode(finch)
[1] 280
```

4.3.6 Robust Measures of Location

In the preceding pages, we have already seen that $\bar{X}$ is a poor location estimator for data
resulting from multiplicative processes, or for measures expressed with respect to a second common measure. $\bar{X}$ also works poorly in the context of strongly asymmetric pdfs.
This is because the measure will be pulled toward extreme values in the long tail of the
distribution, causing it to poorly represent the majority of the data. Surprisingly, this is
because $E(X)$ will also be pulled in the direction of the longer tail. For a demonstration,
type shade.chi(3, nu = 3,tail = "upper"). Note that only 39% of the distribution is to
the right of the true mean, 3. Thus, $\bar{X}$ is a poor estimate of the typical value in a skewed
distribution because the parameter it estimates, $E(X)$, no longer clearly represents the typical value.

Outliers are unusual observations in a sample that may be the result of measurement
error. Such outcomes may also strongly "pull on" $\bar{X}$ and decrease its inferential effectiveness. In the language of robust statistics, the *breakdown point* of an estimator is defined to be
the percentage of arbitrarily large outliers that are required to produce an arbitrarily large
estimate (Venebles and Ripley 2002). Because $\bar{X}$ can be made arbitrarily large by a single
outlier, its breakdown point is said to be 0%.

A number of *robust* estimators of location have been developed that are resistant to the
effects of outliers, although they will have less efficiency than $\bar{X}$ if the underlying parent
population is normal. These include the sample median, the trimmed and Winsorized
mean, and measures known as *M*-estimators.

---

* In a situation where a set of nonidentical numbers is subject to a *mean-preserving spread* (i.e., elements in the set
  are spread apart, leaving the arithmetic mean unchanged), and values remain greater than zero, the harmonic
  mean will always be less than the geometric mean, and the geometric mean will always be less than the arithmetic mean.
† Our current organism of interest, the cheetah, is famous for having a small effective population size because
  of bottleneck that occurred during the late Pleistocene (≈10,000 YBP). This has led to a number of unexpected
  results. For instance, because they have similar underlying genetics, skin grafts between unrelated cheetahs are
  never rejected (Menotti-Raymond and O'Brien 1993). Small effective population sizes can have negative consequences for biological populations. This is because such populations tend to have fewer evolutionary options
  in the context of natural selection, and are more susceptible to extinction.
‡ The function mode in the *R base* package sets the storage mode of an object and is thus unrelated to the sample
  mode. The function Mode from *asbio* calculates sample mode(s).

4.3.6.1 Median

The *population median* is the 50th *percentile* of its pdf. That is, 50% of a distribution will lie below and above its median.* The *sample median* is the middle value from a set of *n ordered responses* (data arranged from low to high). If *n* is odd, then the median is the $(n + 1)/2$th response from a set of ordered data. If *n* is even, then the median is the average of the two middle ordered values.

> **EXAMPLE 4.8**
>
> Consider counts of medium ground finches (*Geospiza fortis*) at Daphne Major Island in the Galapagos from 1976 to 1984 (Gibbs and Grant 1987):
>
> ```
> finch <- c(1220, 400, 380, 298, 280, 200, 297, 280, 1250)
> s <- sort(finch); s[(length(s)+ 1)/2]
> [1] 298
> ```
>
> The function median can be used to calculate the sample median.
>
> ```
> median(finch)
> [1] 298
> ```

The mean and the median can be contrasted by the fact that the median is the central value in an ordered distribution, while the mean is the "center of gravity" for the distribution. Note that to "balance" the distribution in Figure 4.3, the mean is pulled in the direction of the outliers.

The sample median has an asymptotic relative efficiency of only 64% compared to the given normality (Wilcox 2005), meaning, for large values of *n*, its sampling distribution (Chapter 5) has 64% more variability than sampling distribution of $\bar{X}$. However, it has

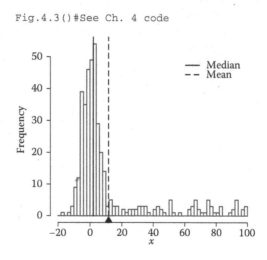

Fig.4.3()#See Ch. 4 code

FIGURE 4.3

A right-skewed distribution with the mean and median overlaid. The pyramid under the mean indicates that this point is the center of gravity, that is, the balance point for the distribution.

---

* Question: what are the mean, median, and mode of the standard normal distribution? Answer: Since the standard normal distribution is unimodal, symmetric and centered at 0 then, mean = median = mode = 0.

a breakdown point of 50%. That is, it can tolerate up to 50% gross errors in sample data before it can be made arbitrarily large.

4.3.6.2 Trimmed Mean

Trimmed means and Winsorized means (discussed in Section 4.3.6.2) are robust estimators of location that should be used carefully. This is because they overtly alter raw data, by eliminating or uniformly transforming extreme observations. Such an approach may hinder valid inferences, often with dangerous results. For instance, detection of the Antarctic ozone hole was delayed for several years because a sensor, the total ozone mapping spectrometer, ignored low O_3 levels as outliers (Kandel 1991, p. 110). Trimmed and Winsorized means are primarily described here because of their important role in robust analogs of conventional statistical procedures (see Wilcox 2005, and examples in Chapters 8 through 10).

The *trimmed mean* is simply the sample mean calculated for the central $1 - 2\kappa$ part of an ordered sample, where κ is a proportion of observations, from 0 to 0.5, trimmed from each tail. When $\kappa = 0$, the trimmed mean is equivalent to the arithmetic mean, and when $\kappa = 0.5$, it is equivalent to the median. We can find trimmed means in **R** using mean(x, trim = κ). Trimmed datasets can be created using the function trim.me from *asbio*.

> **EXAMPLE 4.9**
>
> In the early 1990s, managers for an oil refinery northeast of San Francisco agreed with local air quality regulators to reduce carbon monoxide emissions. Baselines for these reductions were to be based on CO measurements (in ppm) made by refinery personnel, and by independent measurements from the BAAQMD (Bay Area Air Quality Management District) for roughly the same time period. It was apparent that the measurements from both sources contained outliers, prompting the use of robust location estimators. The dataset is available in *asbio* under the name refinery.
>
> ```
> data(refinery)
> tapply(refinery[,1], refinery[,2], function(x) mean(x, trim =.2))
> BAAQMD refinery
> 18.21429 58.94737
> ```
>
> The refinery trimmed mean with $\kappa = 0.2$ was more than three times the BAAQMD trimmed mean. These results prompted regulators to question the validity of refinery measurements given the economic incentive to overestimate baseline carbon monoxide emissions.

4.3.6.3 Winsorized Mean

With *Winsorization*, we replace responses that are not in the central $1-2\kappa$ part of an ordered sample with the minimum and maximum responses of this central part of the sample.[*] For instance, consider 20% Winsorization ($\kappa = 0.2$) of a dataset with 10 observations. Two observations from each tail of the ordered data will be Winsorized.

```
x <- c(0.001, 0.002, 1, 2, 2.2, 4, 5, 6, 15, 17)
```

---

[*] Named for engineer and biostatistician Charles P. Winsor (1895–1951).

Applying the *asbio* function `win` (whose default is $\kappa = 0.2$), we have

```
win(x)
[1]  1.0 1.0 1.0 2.0 2.2 4.0 5.0 6.0 6.0 6.0
```

Winsorized statistics, including the sample mean, are calculated using Winsorized data. Thus, our Winsorized sample mean is

```
mean(win(x))
[1]  3.42
```

The optimal amount of Winsorization and trimming will vary with the proportion of outliers in the data. Several studies have indicated that optimal trimming usually varies from $\kappa = 0.0$ to 0.25 (Hill and Dixon 1982, Wu 2002). For general applications in large datasets where outliers are a concern, Wilcox (2005) recommended using $\kappa = 0.2$. However, neither trimming nor Winsorization should be used without careful consideration of the consequences of discarding extreme observations. A common response to outliers is to analyze the data both with and without the outlying points to see whether the resulting conclusions differ. If they do, then one would report both results, reconsider the statistical approach being used, or reevaluate the underlying research/statistical assumptions. Methods of handling outliers in the context of particular analytical models are described in Chapters 6 through 10.

4.3.6.4 M-Estimators

A family of statistics called *M-estimators* (for maximum-likelihood type estimators, see Section 4.4.2) have been developed that include robust estimators of location. *M*-estimators of a location can be conceptualized as weighted arithmetic means, where observations further from the center of an ordered sample receive less weight. The arithmetic mean, median, trimmed mean, and Winsorized mean can all be considered special cases of *M*-estimators (Venebles and Ripley 2002, Maronna et al. 2006). Other robust *M*-estimators include Huber's ψ (pronounced si), Hampel's ψ, and Tukey's bisquare function (Venebles and Ripley 2002, Huber 2004).

Huber's method results in the weighting function

$$W_c(x) = \min\left\{1, \frac{c}{|x - \hat{\mu}|}\right\},$$

(4.13)

where $\hat{\mu}$ (pronounced mew-hat) is a location estimator and c is a *bend criterion*, supplied by the user, that defines the central region of an ordered sample and the degree of weighting. The value $c = 1.345$ gives 95% asymptotic efficiency of the arithmetic mean assuming normality. The value $c = 1.28$ gives 90% asymptotic efficiency but has less sensitivity to outliers in a contaminated dataset. The later value is the default for *M*-estimator functions in *asbio*. The function `see.M` reveals that if $|x - \hat{\mu}|$ is greater than c, then the observation x will be down-weighted (Figure 4.4).

It is important to note that both Equation 4.13 and Figure 4.4 are merely conceptualizations. This is because weights depend on $|x - \hat{\mu}|$, but $\hat{\mu}$ depends on the weights. Because of this framework, a solution for $\hat{\mu}$ is the result of an *iteratively weighted least squares* process in which weights are updated during a series computational iterations. This procedure is described below:

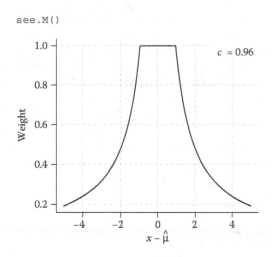

FIGURE 4.4

Weighting function for the Huber *M*-estimator for different values of the bend criterion, *c*. At $c = 0$, the Huber *M*-estimator is equivalent to the sample median; at $c = \infty$, it is equivalent to the arithmetic mean. Thus, *c* determines the scaled distance at which maximum weighting (weight = 1) stops.

1. Calculate *A*:

$$A = \sum_{i=1}^{n} \psi\left(\frac{X_i - \hat{\mu}_k}{\text{MAD}}\right), \tag{4.14}$$

where $\hat{\mu}_k$ is the sample median, MAD is the median absolute deviation (see Section 4.3.8.2), and $\psi(.)$ is a minmax weighting function. For Huber's ψ, we have

$$\psi(x) = max[-c, \min(c, x)]. \tag{4.15}$$

2. Calculate *B*:

$$B = \sum_{i=1}^{n} \psi'\left(\frac{X_i - \hat{\mu}_k}{\text{MAD}}\right), \tag{4.16}$$

where $\psi' = \begin{cases} 1, & \text{if } -c \leq x \leq c \\ 0, & \text{otherwise} \end{cases}$.

3. Calculate $\hat{\mu}_{k+1}$:

$$\hat{\mu}_{k+1} = \hat{\mu}_k + \frac{\text{MAD} \times A}{B}. \tag{4.17}$$

4. Finally, we use an iterative least squares process to solve for the *M*-estimate, $\hat{\mu}$. In particular, if $\left|\hat{\mu}_{k+1} - \hat{\mu}_k\right|$ is less than some tolerance,* we let $\hat{\mu} = \hat{\mu}_{k+1}$. Otherwise, we substitute $\hat{\mu}_{k+1}$ for $\hat{\mu}_k$ in Equation 4.14 and proceed again with steps 1–4.

---

* Conventionally given to be 1×10^{-7}.

Steps 1–4 represent an application of the so-called *Newton–Raphson method* (see Ypma 1995). This general approach is used in a myriad of computationally intensive mathematical and statistical applications, including numerical calculus, the identification of Bayesian posterior distribution modes (Chapter 5), generalized linear model derivation (Chapter 9), and mixed model parameter estimation (Chapter 10).

EXAMPLE 4.10

Fires from 1988 constituted the largest conflagration in the history of Yellowstone National Park. The dataframe `fire` lists burned areas, in hectares^2, for 10 Yellowstone stream catchments (Robinson et al. 1994). We will use Huber's M-estimator, with $c = 1.28$, to obtain a typical value for catchment burn area.

```
data(fire)
```

The function `huber.NR` from *asbio* displays the Raphson–Newton iterations for a Huber M-estimate. The function `huber.mu` provides final (converged) estimates.

```
huber.NR(fire$fire, c=1.28, iter=10)
[1]  986.075 1222.038 1213.260 1213.260 1213.260 1213.260 1213.260
[8] 1213.260 1213.260 1213.260 1213.260
```

After two iterations, the answer converges to 1213.260.

As with the example above, M-estimators generally converge after only a few iterations. As a result, M-estimators based on a single iteration have desirable asymptotic properties. In particular, given large samples, a *one-step M-estimator* performs similarly to a fully iterated M-estimator, and is much easier to calculate. The first element from the output above (986.075) is the sample median. The second element (1222.038) is the Huber one-step estimate.

For a rigorous mathematical/theoretical treatment of M-estimators, including a formal motivation for Huber's ψ, see Huber (2004). For gentler introductions, see Wilcox (2005) or Maronna et al. (2006). For further discussion of M-estimator functions in **R**, see Venables and Ripley (2002).

4.3.6.5 Which Location Estimator to Use?

I have introduced a variety of location measures in Sections 4.3.5 and 4.3.6; however, it will rarely be necessary to calculate more than one location estimate for a sample. An important skill is to present the results of analyses in an uncluttered manner, and this certainly applies to summary statistics.

The arithmetic mean is the most frequently used measure of location for four reasons. First, it is unbiased for $E(X)$ for any distribution in which $E(X)$ exists. Second, the weak law of large numbers (Chapter 5) guarantees that $\bar{X}$ is consistent for $E(X)$. Third, under normality, $\bar{X}$ will have maximum efficiency for $E(X)$. Fourth, the central limit theorem (Chapter 5) ensures that the sampling distribution of $\bar{X}$ will become normal as sample sizes become large, simplifying inferential procedures for $E(X)$.

The geometric mean and the harmonic mean have specialized uses in biology. The geometric mean should be used as a location measure for multiplicative processes such as population growth rates. The harmonic mean should be used to find the average of set of reciprocals that are commonly expressed as nonreciprocals (e.g., velocities allowing calculation of travel time given a common distance) or as an estimator of effective population size.

Robust approaches may provide more useful location estimates, given distributional asymmetry and/or outliers. Because of its simplicity and high breakdown point, the most commonly used robust estimator is the sample median. Trimmed and Winsorized means may have strong (and dangerous) effects on inference and are rarely used by biologists, although see Nair et al. (2008) and Xia and Lo (2008) for applications in human health and physiology. *M*-estimators should be appealing to biologists because of their outlier resistance and high efficiency (e.g., Silvapulle 2001). For instance, Niselman et al. (1998) found that *M*-estimators outperformed the arithmetic mean in bioequivalence assays comparing drug concentration and plasma concentration. However, these measures are rarely implemented because of their complexity, the unavailability of necessary software, and the lack of associated null hypothesis testing procedures.

4.3.7 Measures of Scale

Measures of location cannot quantify the variability or dispersion implicit in measures of natural phenomena. This absence gives rise to arguably the most important class of estimators in statistics: measures of *scale* that quantify data variability and dispersion. As with measures of location, there are many ways to consider scale [see Lax (1975, 1985) for a comparison of 150 algorithms]. This section addresses only the most important and frequently used scale estimators.

4.3.7.1 Sample Variance

The most commonly used scale estimator is the *sample variance*, denoted as S^2. It is an unbiased and consistent estimator of the true variance, $\text{Var}(X)$, for any parent distribution. It has the form

$$S^2 = \frac{\sum_{i=1}^{n}(X_i - \bar{X})^2}{n-1} = \frac{\sum_{i=1}^{n}X_i^2 - n\bar{X}^2}{n-1}. \tag{4.18}$$

Observed differences from mean, $X_i - \bar{X}$, are called *deviations*, while the numerators in both versions of Equation 4.18 are the *sum of squares* (i.e., the sum of the squared deviations). To obtain the sample variance, we divide the sum of squares by the number of observations minus one. The quantity $n - 1$ constitutes the *degrees of freedom* for the sample variance. It describes how many independent units of information are available to estimate the true variance. In this case, if we know the values of $n - 1$ of the observations, and the sample mean, we will always be able to figure out the value of the last observation. Thus, we only have $n - 1$ independent pieces of information concerning $\text{Var}(X)$. If all possible outcomes from X are observed, then the population variance, $\text{Var}(X)$, can be obtained (not merely estimated) by dividing the sum of squares by n, not $n - 1$.

★ In the proof below, we can see why we need to divide by $n - 1$ in Equation 4.18 in order for S^2 to be unbiased for σ^2.

If X is a random variable with $E(X) = \mu$, and $\text{Var}(X) = \sigma^2$, then

$$E(S^2) = E\left[\frac{1}{n-1}\sum_{i=1}^{n}(X_i - \bar{X})^2\right] \quad \text{From (4.18)}$$

$$= E\left[\frac{1}{n-1}\sum_{i=1}^{n}(X_i^2 - 2X_i\bar{X} + \bar{X}^2)\right] \quad \text{Algebra}$$

$$= E\left[\frac{1}{n-1}\left(\sum_{i=1}^{n}(X_i^2) - \sum_{i=1}^{n}(2X_i\bar{X}) + \sum_{i=1}^{n}\bar{X}^2\right)\right] \quad \text{Distribute summation}$$

$$= E\left[\frac{1}{n-1}\left(\sum_{i=1}^{n}(X_i^2) - 2n\bar{X}^2 + n\bar{X}^2\right)\right] \quad \text{Summation rules}$$

$$= E\left[\frac{1}{n-1}\left(\sum_{i=1}^{n}(X_i^2) - n\bar{X}^2\right)\right]$$

$$= \frac{1}{n-1}\left(\sum_{i=1}^{n}\left(E(X_i^2)\right) - nE\left(\bar{X}^2\right)\right) \quad \text{Distribute expectations}$$

$$= \frac{1}{n-1}\left(\sum_{i=1}^{n}\left(\sigma^2 + \mu^2\right) - nE\left(\bar{X}^2\right)\right)$$

$$= \frac{1}{n-1}\left(\left(n\sigma^2 + n\mu^2\right) - n\left(\frac{\sigma^2}{n} + \mu^2\right)\right) \quad \begin{array}{l}\text{Mean and variance for the sampling}\\ \text{distribution of } \bar{X}, \text{ see Chapter 6}\end{array}$$

$$= \frac{1}{n-1}\left(n\sigma^2 + n\mu^2 - \sigma^2 - n\mu^2\right)$$

$$= \frac{1}{n-1}\left(n\sigma^2 - \sigma^2\right)$$

$$= \frac{\sigma^2(n-1)}{n-1} = \sigma^2.$$

Thus, the denominator $n-1$ is required to obtain unbiased estimates for σ^2. Dividing by n would result in a biased (low) estimator. Intuitively, this occurs because deviations from the sample mean will be smaller than deviations from μ because we calculate $\bar{X}$ in a way that minimizes deviations (Section 4.4.1). This discrepancy is addressed by dividing by $n-1$ instead of n.

The *sample standard deviation*, denoted as S, is the positive square root of the sample variance:

$$S = \sqrt{S^2}. \tag{4.19}$$

While S^2 is unbiased for $\text{Var}(X)$, S is biased (low) for $\text{SD}(X)$. However, the bias asymptotically goes to zero with increasing sample size because S is consistent for $\text{SD}(X)$.[*] S allows increased interpretability, compared to S^2, because (1) its units will be in the units of the original observations (units for S^2 will be the original units, squared), and (2) the underlying normality of the parent population (the one we sample from) allows inferences based on the cutoffs of the empirical rule (Section 3.2.2.2).

EXAMPLE 4.11

Soil nutrient characteristics may be particularly important in controlling alpine ecosystem productivity (Aho et al. 1998). Below are soil nitrate and phosphorous measures for 10 alpine sites on Abiathar Peak (3331 m) in Yellowstone National Park (Table 4.1).

To calculate the sample variance for soil nitrate, we calculate the sum of squares of the NO_3 data and divide by the degrees of freedom.

```
NO3 <- c(6, 3, 12, 9, 2, 3, 7, 2, 6, 8)
sum((NO3 - mean(NO3))^2)/9
[1] 11.06667
```

The functions `var` and `sd` can be used to calculate the variance and standard deviation in **R**.

```
var(NO3)
[1] 11.06667
sd(NO3)
[1] 3.32666
```

4.3.7.2 Coefficient of Variation

Another commonly used scale measure is the coefficient of variation. The *population coefficient of variation* is

$$\frac{\text{SD}(X)}{E(X)}, \tag{4.20}$$

while the *sample coefficient of variation* is calculated as[†]

$$CV = \frac{S}{\overline{X}}. \tag{4.21}$$

---

[*] S will tend to underestimate σ for small sample sizes (Gurland and Tripathi 1971). To account for this an estimate, s, can be multiplied by the correction factor C_n:

$$C_n = \frac{[(n-1)/2]^{0.5}\Gamma[(n-1)/2]}{\Gamma(n/2)}.$$

The correction factor is generally ignored by researchers and will be irrelevant for $n > 30$, since $C_{30} = 1.00866$, and decreases further for larger numbers.

[†] Equation 4.21 will be biased low for $\text{SD}(X)/E(X)$ given small sample sizes (Sokal and Rohlf 1998). An estimator corrected for this bias is $CV(1 + 1/n)$.

TABLE 4.1

Alpine Soils Data from Abiathar Peak, Yellowstone National Park

Site	NO$_3$ (lb/acre)	P (ppm)
1	6	20
2	3	19
3	12	21
4	9	21
5	2	23
6	3	14
7	7	15
8	2	14
9	6	18
10	8	19

Source: Aho, K. 2006. *Alpine Ecology and Subalpine Cliff Ecology in the Northern Rocky Mountains.* PhD dissertation, Montana State University, 458 pp.

The coefficient of variation scales the sample standard deviation using the arithmetic mean. Unlike the standard deviation, the coefficient of variation is expressed without units of measurement (since both the standard deviation and the mean will have the same units). As a result, it is useful for comparing the variability of populations independent of units. For example, Tilman (1996) calculated variability in primary productivity for sections of Minnesota prairie over 4 years using CV. He found that CV decreased as prairie species richness increased, indicating that prairie ecosystem stability was enhanced by diversity.*

EXAMPLE 4.12

Although the nitrate and phosphorous data in Example 4.9 are measured using different units (Table 4.1), we can use the coefficients of variation to compare their variability.

```
P <- c(20, 19, 21, 21, 23, 14, 15, 14, 18, 19)
sd(P)/mean(P)
[1] 0.1703284
sd(NO3)/mean(NO3)
[1] 0.5735621
```

Nitrate is much more variable than phosphorous at the 10 sites.

4.3.8 Robust Estimators of Scale

Because they are calculated from squared deviations, S, S^2, and CV are even more strongly affected by outliers than the sample mean, although like the mean, each has a breakdown point of 0%. As a result, a number of robust alternatives have been developed. These include the interquartile range (IQR) and the median absolute deviation (MAD). When robust measures of location are used, associated robust measures of scale (e.g., IQR with the sample median, or MAD with M-estimators) should also be reported.

---

* For a statistical discussion of this result, see Tilman (1996), Doak et al. (1998) and the response to Doak et al. by Tilman et al. (1998).

4.3.8.1 Interquartile Range

A quantile function [i.e., inverse cdf (Chapter 3)] obtains the value that a random variable will be less than or equal to, given a particular probability. The quantile at probability p is equivalent to the pth *percentile*. For instance, the value that halves a distribution is called both the 0.5 quantile and the 50th percentile (*and* the population median).

```
qnorm(0.5)
[1] 0
```

Quantiles include *deciles* (the distribution divided into 10ths) and *quartiles* (the distribution divided into fourths). The median is the *2nd quartile*.

For any probability distribution, the region between the *1st quartile and 3rd quartile* contains the middle 50% of a distribution. To find the *sample interquartile range (IQR)*, we find the medians of the ordered data, which lie below and above the median. These are Q_1 and Q_3, which are estimators for the first and third population quantiles. Q_1 is subtracted from Q_3 to obtain the sample interquartile range:

$$IQR = Q_3 - Q_1. \tag{4.22}$$

```
quantile(NO3,.75)-quantile(NO3,0.25)
[1] 4.75
```

The interquartile range can be calculated using the function IQR.[*]

```
IQR(NO3)
[1] 4.75
```

Analysts often consider observations greater than $1.5 \times$ IQR from the median to be outliers.[†]

4.3.8.2 Median Absolute Deviation

Another robust estimator of scale is *MAD*. Recall that this measure is used to scale Huber *M*-estimators of location (Section 4.3.6.4). It is obtained by finding the median of the absolute value of the differences of observations and the sample median:

$$MAD = c \times \text{median } |X_i - \text{median}(X)|. \tag{4.23}$$

Generally, we let $c = 1.4826$ since this allows MAD to be consistent for the population standard deviation.

The function mad from library *MASS* calculates MAD.

```
median(abs(NO3-median(NO3)))*1.4826
[1] 4.4478
library(MASS)
mad(NO3)
[1] 4.4478
```

---

[*] The function IQR uses a different method to calculate the interquartile range than the one described here. It relies on the function quantile to calculate 1st and 3rd quartile. The function quantile produces quantiles corresponding to specified probabilities.

[†] Sample quantiles based on ordered data, including Q_1, Q_2, and Q_3, are often termed *order statistics*.

4.3.8.3 *Which Scale Estimator to Use?*

The sample variance is the most frequently used estimator of scale because of its desirable mathematical qualities. S^2 is unbiased and consistent for any $\text{Var}(X)$, and maximally efficient for σ^2 under normality (Graybill 1976). While biased for small sample sizes, the sample standard deviation provides increased interpretability compared to S^2 because it allows inferences concerning the empirical rule, and will be in the units of the original observations.

While robust, the MAD and IQR are not efficient given normality. In fact, MAD has an asymptotic relative efficiency of 37% compared to the sample standard deviation, given normality. MAD, however, is more robust to outliers than IQR over a wide range of conditions (Venables and Ripley 2002).

4.3.9 Parameters and Estimators for Distribution Shape

In Sections 4.3.5 through 4.3.8, we learned about estimators for location and scale. We can also describe distributions with respect to their shape. *Skewness* describes the symmetry of a distribution, and *kurtosis* describes its peakedness. The population skewness and kurtosis are denoted as γ_1 and γ_2, respectively, and are based on distributional expected values:

$$\gamma_1 = \frac{E[(X - E(X))^3]}{[SD(X)]^3}, \tag{4.24}$$

$$\gamma_2 = \frac{E[(X - E(X))^4]}{[\text{Var}(X)]^2}. \tag{4.25}$$

For the symmetric distributions, skewness will equal zero, that is, $\gamma_1 = 0$, and the population mean and median will be equal. Conversely, if $\gamma_1 \neq 0$, the mean will be drawn toward the long tail. That is, the mean will be to the right of the median in a distribution with a longer right tail (see Figure 4.3). A distribution with a "long" right-hand tail, and a squashed left-hand tail will be *positively skewed*, resulting in $\gamma_1 > 0$. Conversely, a distribution with a "long" left-hand tail and a squashed right-hand tail will be *negatively skewed*, resulting in $\gamma_1 < 0$.

If a distribution is normal, it will be *mesokurtic*, and γ_2 will equal 3. As a result, the parameter $\gamma_{2\_excess} = \gamma_2 - 3$ is generally used to define the kurtosis relative to the normal distribution. Thus, normal distributions will have $\gamma_{2\_excess} = 0$, strongly peaked (*leptokurtic*) distributions will have $\gamma_{2\_excess} > 0$, and flat (*platykurtic*) distributions will have $\gamma_{2\_excess} < 0$.

★   ### 4.3.9.1 *Moment Generating Functions*

γ_1 and γ_2 are most easily derived using a special type of expectation called a *moment generating function* or *MGF*. The moment generating function for a random variable X is

$$M_X(t) = E(e^{tx}). \tag{4.26}$$

Given this definition, the MGF for a discrete random variable is

$$M_X(t) = \sum_x e^{tx} f(x), \tag{4.27}$$

while for a continuous random variable, we have

$$M_X(t) = \int_{-\infty}^{\infty} e^{tx} f(x) \, dx. \tag{4.28}$$

When they exist, an MGF will be specific to a particular pdf. That is, the normal MGF has a specific form, distinct from the binomial MGF (Tables 3.4 and 3.5). The term t is included in the MGFs of all pdfs (see Tables 3.4 and 3.5). In the process of moment generation, one finds derivatives of the MGF with respect to t, and then takes t equal to zero in the differential.

Letting $t = 0$ in the first derivative of $M_X(t)$, we have the *first moment*, which will be equivalent to $E(X)$. Letting $t = 0$ in the second derivative of $M_X(t)$ gives the *second moment*, which is equivalent to $E(X^2)$, and so on. In this way, MGFs allow relatively straightforward calculation of higher-order expectations [e.g., $E(X^3)$, $E(X^4)$], and derivation of pdfs.

EXAMPLE 4.13

The MGF for the exponential distribution (see Section 3.3.2.4) is $\theta/\theta - t$ (Table 3.5). Thus,

$$\frac{d}{dt} \frac{\theta}{\theta - t} = \frac{\theta}{(\theta - t)^2}.$$

Letting $t = 0$, we have

$$\frac{\theta}{(\theta - 0)^2} = \frac{1}{\theta}.$$

This is the first moment, and will be equivalent to $E(X)$ (Table 3.5). Under the same approach, and using the definitions of expectations and variances, we find

$$E(X^2) = \frac{2}{\theta^2}, \quad E(X^3) = \frac{6}{\theta^3}, \quad \text{Var}(X) = \frac{2}{\theta^2} - \left(\frac{1}{\theta}\right)^2 = \frac{1}{\theta^2}, \quad \text{and} \quad \text{SD}(X) = \sqrt{\frac{1}{\theta^2}} = \frac{1}{\theta}.$$

Expanding Equation 4.24, we have

$$\gamma_1 = \frac{E(X^3) - 3E(X^2)E(X) + 3E(X)E(X)^2 - E(X)^3}{[\text{SD}(X)]^3}$$

Substituting, we have

$$\gamma_1 = \frac{6\theta^{-3} - 3(2\theta^{-2})\theta^{-1} + 3(\theta^{-1})\theta^{-2} - \theta^{-3}}{\theta^{-3}} = \frac{6\theta^{-3} - 6\theta^{-3} + 3\theta^{-3} - \theta^{-3}}{\theta^{-3}} = \frac{2\theta^{-3}}{\theta^{-3}} = 2.$$

Thus, the skewness of an exponential distribution equals 2.

4.3.9.2 Sample Moments and MOM Estimators

We can estimate γ_1 and γ_2 with the estimators G_1 and G_2, which are based on *sample moments*. Sample moments, which can be used to obtain so-called *method of moments (MOM)* parameter estimators, are related to MGFs described in the previous section (which allow derivation of parameters), although explicit connections between the two are not given here. The interested reader is directed to Hansen (2002). The *ith central sample moment* is given by

$$m_j = \frac{1}{n} \sum_{i=1}^{n} (X_i - \bar{X})^j. \tag{4.29}$$

We define the first central sample moment to be the average deviation of all observations from the arithmetic mean. The result will be zero, because by definition the sum of the differences of observations from the arithmetic mean will always be zero. The second central sample moment will be the biased (low) version of the sample variance discussed in Section 4.3.7. We calculate G_1 and G_2 using

$$G_1 = \frac{m_3}{m_2^{3/2}}, \tag{4.30}$$

$$G_2 = \frac{m_4}{m_2^2}. \tag{4.31}$$

By adjusting Equation 4.31 so that normal distributions will have a kurtosis of 0, we have

$$G_{2\_excess} = \frac{m_4}{m_2^2} - 3. \tag{4.32}$$

The functions for `skew` and `kurt` in *asbio* provide estimates for skewness and kurtosis using the procedures described above. To use the MOM estimators (i.e., Equations 4.30 and 4.31), specify `method = "moments"`. To the use the excess method for calculating kurtosis (i.e., Equation 4.32), specify `method = "excess"`. To use the unbiased method for calculating skewness and kurtosis (see footnote),* use `method = "unbiased"`.

EXAMPLE 4.14

Here are the sample skewness and kurtosis for a random sample of 1,000,000 from the exponential distribution using sample moments estimators.

```
set.seed(1)# This will create the random distribution used here.
x <- rexp(1000000)
skew(x, "moments")
```

---

* The estimators of γ_1 and γ_2 shown in Equations 4.30 and 4.31 are biased low, although this will be irrelevant for large sample sizes. Unbiased estimators are

$$G_{1\_unbiased} = \frac{n}{(n-1)(n-2)} \sum_{i=1}^{n} \left(\frac{X_i - \bar{X}}{S} \right)^3 ; G_{2\_unbiased} = \frac{n(n+1)}{(n-1)(n-2)(n-3)} \sum_{i=1}^{n} \left(\frac{X_i - \bar{X}}{S} \right)^4 - \frac{3(n-1)^2}{(n-2)(n-3)}$$

```
[1] 1.999791
kurt(x,"moments")
[1] 9.030814
```

These summaries indicate that the exponential distribution is positively skewed and sharply peaked relative to a normal distribution (see Figure 3.20).

4.4 OLS and ML Estimators

The three most commonly used approaches for deriving estimators are *MOM* (described in Section 4.3.9.2), *ordinary least squares* (OLS), and *maximum likelihood* (ML). The latter two procedures are described in this section and use different approaches to optimize efficiency.

4.4.1 Ordinary Least Squares

OLS estimators minimize a function that represents the difference of a parameter and its estimator, given data. Let the sample data have the form $x_i = f_i(\theta) + \varepsilon_i$, where $f_i(\theta)$ is a linear function for x_i (i.e., it is the deterministic "fit" part of the statistical model, see Section 2.1) and ε_is are random *error terms*, with $E(\varepsilon_i) = 0$. With OLS, we identify an estimator of the parameter θ in $f_i(\theta)$ for which Q (the sum of the squared error terms) is minimized:

$$Q = \sum_{i=1}^{n} \left(x_i - f_i(\theta)\right)^2. \tag{4.33}$$

We can demonstrate that $\overline{X}$ is the OLS estimator for $E(X)$ using *optimization* from calculus.

For notational compactness, we let $E(X) = \mu$. We first take the derivative of Equation 4.33 with respect to the parameter μ:

$$Q = \sum_{i=1}^{n} (x_i - \mu)^2$$

Chain rule; ignore summation.

$$\frac{dQ}{d\mu} = 2\sum_{i=1}^{n} (x_i - \mu)(-1).$$

We then take $dQ/d\mu$ equal to zero, replace μ with its estimator, $\hat{\mu}$, and solve for $\hat{\mu}$.

$$2\sum_{i=1}^{n} (x_i - \hat{\mu})(-1) = 0$$

$$\sum_{i=1}^{n} (x_i - \hat{\mu}) = 0 \quad \text{Divide both sides by} -2$$

$$\sum_{i=1}^{n} x_i - n\hat{\mu} = 0 \quad \text{Summation rules}$$

$$\hat{\mu} = \frac{\sum\limits_{i=1}^{n} x_i}{n}$$

Thus, the arithmetic mean is the OLS estimator for $E(X)$ for any distribution.[*] That is, if $\bar{X}$ is used to estimate $E(X)$, then the sum of squared errors will be smaller than for any other possible estimator of $E(X)$.

EXAMPLE 4.15

Let us visualize how the least squares derivation works. The dataframe `trag` contains heights (in cm) for 20 salsify plants (*Tragapogon dubius*), an invasive ubiquitous to disturbed areas in sagebrush steppe, randomly sampled from the Barton Road Long Term Experimental Research site near Pocatello, Idaho. We will find an estimate for $E(X)$ that minimizes the sum of squared residuals for these data under Equation 4.33 (Figure 4.5).

```
data(trag)
```

Q is minimized at 61.1 cm, which is the arithmetic mean.

```
mean(trag)
[1] 61.1
```

4.4.2 Maximum Likelihood

With maximum likelihood estimation (MLE), we find estimate, $\hat{\theta}$, that maximizes the *likelihood function*

$$f(\text{data} \mid \theta) = \mathcal{L}(\theta \mid \text{data}) = \prod_{i=1}^{n} f(x_i \mid \theta), \qquad (4.34)$$

where $f(\text{data} \mid \theta)$ is a probability density function applied to observations in a sample of size n, and θ is a parameter we want to estimate that is required by the pdf. In ML estimation, observations (x_is) are assumed to be independent and identically distributed. The conditionality in $f(\text{data} \mid \theta)$ and $\mathcal{L}(\text{data} \mid \theta)$ is switched in Equation 4.34 to show that in likelihood estimation, the data are fixed while parameter values are varied.

---

[*] We use the *second derivative test* to assure ourselves that this optimization is a minimum. That is, we take the second derivative of Q with respect to μ, substitute for $\hat{\mu}$ for μ, and check the sign of the result. If the sign is positive then the optimization represents a minimum value the function Q. If the sign is negative then the optimization represents a maximum value for Q.

$$\frac{\delta Q^2}{\delta \mu^2} = \frac{\delta Q}{\delta \mu} - 2\sum_{i=1}^{n} (x_i) + n\mu = n.$$

Since n is always positive, the sample mean minimizes Q with respect to the parameter μ.

```
anm.ls(trag)
```

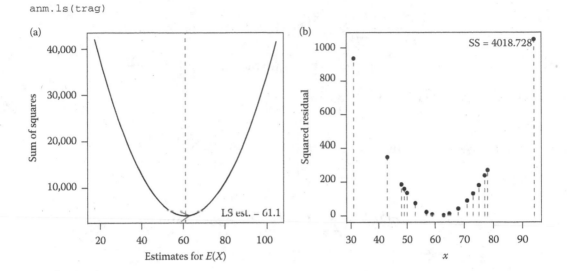

FIGURE 4.5

Sums of squares estimation for $E(X)$ for data in Example 4.15. The sum of the squared residuals is minimized at a value equivalent to the sample mean, $\bar{X}$. (a) The sum of square function, that is, residuals as a function of $E(X)$ estimates. (b) The squared residuals for the optimized least squares function. The sum of the heights of the lines is equivalent to the minimized sum of squares.

It is more computationally straightforward to find the log-likelihood since this allows us to get rid of both the product sign in Equation 4.34, and the exponential notation associated with a number of important distributions (e.g., the normal distribution). This approach is valid because log-transformations are monotonic and will thus preserve the likelihood function maxima and minima.

$$\ell(\theta \mid \text{data}) = \ln\left[\prod_{i=1}^{n} f(x_i \mid \theta)\right] = \sum_{i=1}^{n} \ln f(x_i \mid \theta) \tag{4.35}$$

If we assume a normal distribution and we want to find an ML estimate of μ, we would let $\mu = \theta$ in Equation 4.35, then hold the other required parameter for the normal pdf, σ^2, constant. As our ML estimate, we use the outcome that maximizes the normal log-likelihood function.

★ The derivation below shows that $\bar{X}$ is the ML estimator for μ (assuming normality). The normal likelihood function for μ is

$$f(\text{data} \mid \mu, \sigma^2) = \mathcal{L}(\mu \mid \sigma^2, \text{data}) = \prod_{i=1}^{n} \frac{1}{\sigma\sqrt{2\pi}} \exp-\left[\frac{(x_i - \mu)^2}{2\sigma^2}\right] \tag{4.36}$$

Manipulating Equation 4.36, we have

$$\mathcal{L}(\mu \mid \sigma^2, \text{data}) = \left(\frac{1}{\sigma\sqrt{2\pi}}\right)^n \prod_{i=1}^{n} \exp\left[-\frac{(x_i - \mu)^2}{2\sigma^2}\right] \quad \text{Rule for products}: \prod_{i=1}^{n} x = x^n.$$

$$= \frac{1}{\left(\sigma\sqrt{2\pi}\right)^n} \exp\left(\sum_{i=1}^{n} \frac{(x_i - \pi)^2}{2\sigma^2}\right) \quad \text{Rule for multiplying numbers}$$

with exponents when bases are the same: $\prod e^{y_i} = e^{\sum y_i}$.

Taking the natural log of our result, we have

$$\ell = \ln\left[\mathcal{L}(\mu \mid \sigma^2, \text{data})\right] = \ln\left[\frac{1}{\left(\sigma\sqrt{2\pi}\right)^n} \exp\left(-\sum_{i=1}^{n} \frac{(x_i - \mu)^2}{2\sigma^2}\right)\right]$$

$$= -n\ln\sigma(2\pi)^{1/2} - \sum_{i=1}^{n} \frac{(x_i - \mu)^2}{2\sigma^2} \quad \text{Rules for logarithms}$$

$$= -n\left(\ln\sigma + \frac{1}{2}\ln 2\pi\right) - \frac{1}{2\sigma^2} \sum_{i=1}^{n} (x_i - \mu)^2 \quad \text{Algebra, rules for logarithms}$$

This is the normal log-likelihood function.
We maximize this function to find the maximum likelihood estimator for μ

$$\frac{d\ell}{d\mu}\left[-n\left(\ln(\sigma) + \frac{1}{2}\ln(2\pi)\right) - \frac{1}{2\sigma^2} \sum_{i=1}^{n} (x_i - \mu)^2\right]$$

$$= 0 + \frac{d\ell}{d\mu}\left[-\frac{1}{2\sigma^2} \sum_{i=1}^{n} (x_i - \mu)^2\right] \quad \text{Zeroed terms did not include } \mu.$$

$$= (-1)\left(-\frac{2}{2\sigma^2}\right) \sum_{i=1}^{n} (x_i - \mu) \quad \text{Chain rule}$$

$$= \frac{1}{\sigma^2} \sum_{i=1}^{n} (x_i - \mu)$$

We now substitute $\hat{\mu}$ for μ, take the function equal to 0, and solve for $\hat{\mu}$.

$$\frac{1}{\sigma^2} \sum_{i=1}^{n} (x_i - \hat{\mu}) = 0$$

$$\sum_{i=1}^{n} (x_i - \hat{\mu}) = 0$$

$$\sum_{i=1}^{n} x_i - n\hat{\mu} = 0 \quad \text{Rules of summations}$$

$$\frac{\sum_{i=1}^{n} x_i}{n} = \hat{\mu}.$$

Thus, the maximum likelihood estimator for the mean of a normal distribution is the arithmetic mean.

EXAMPLE 4.16

We will visually demonstrate ML estimation of μ using the *Tragapogon* data from Example 4.14 (Figure 4.6).

We see that the log-likelihood function is also maximized at the value of the arithmetic mean, 61.1 cm. Graphical demonstrations using the Poisson and binomial distributions are left as exercises at the end of the chapter.

The normal log-likelihood function for $E(X)$ will still be maximized at $\bar{X}$ given nonnormal parent distributions. However, an incorrect identification of the underlying distribution will generally hamper inferential procedures (particularly for small sample sizes). This is evident in Figure 4.7. The data in the figure constitute a random sample of size 20 from an exponential parental distribution.

```
anm.loglik(trag, dist = "norm", parameter = "mu")
```

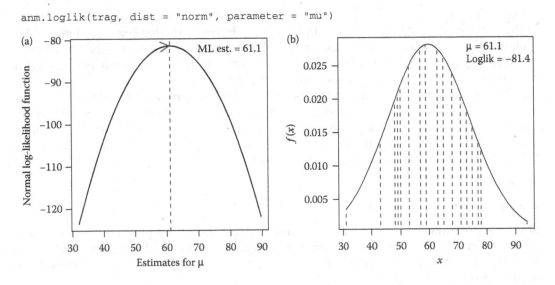

FIGURE 4.6
Normal maximum likelihood estimation of μ, for Example 4.16. (a) Shows the normal log-likelihood function is maximized at 61.1, which is equivalent to the arithmetic mean. (b) Shows the densities which resulted in maximizing the log likelihood function. That is, it shows the normal distribution (with expected value, $\mu = 61.1$) that fit the observed data best.

```
set.seed(1)
x <- rexp(20)
anm.loglik(x, dist = "norm",parameter = "mu")
anm.loglik(x, dist = "exp")
```

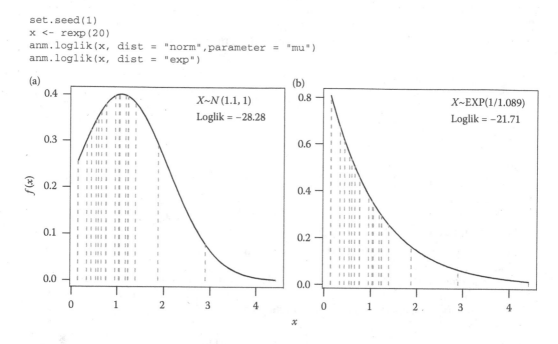

FIGURE 4.7

(a) Normal maximum likelihood estimation of μ, and (b) exponential maximum likelihood estimation for 1/θ, for a sample taken from an exponential distribution. The log-likelihood functions are maximized at the values of 1.089 and 1/1.089, respectively. 1.089 is the value of the arithmetic mean.

Estimates for μ (the expectation of the normal pdf) and θ (the inverse of the expectation of the exponential pdf) are both maximized at $\bar{X}$. However, the log-likelihood associated with the exponential pdf is much higher since the observations occur at higher densities in the optimal exponential distribution, compared to the optimal normal distribution (Figure 4.7). Assumptions of normality are clearly inappropriate here.

It is also interesting to examine the effect of sample size on log-likelihood functions. As we decrease the sample size, less information is contained in the likelihood function, and the function takes on a platykurtic appearance as a result of decreased estimator precision. Compare the normal log-likelihood functions for the full *Tragapogon* dataset in Example 4.12, and a subset of that data (Figure 4.8).

4.4.2.1 So, What Is Likelihood?

From the notation in Equation 4.34, likelihood, $\mathcal{L}(\theta|\text{data})$, is simply the product of data densities, given a specified value for the parameter θ, which defines the underlying assumed distribution. Maximum likelihood occurs when a ML estimate is substituted for θ (Figures 4.6 through 4.8).

4.4.2.2 Likelihood versus Probability

Many authors, particularly Bayesians, present likelihood as a special type of probability (cf. Gelman et al. 2003). This perspective is useful in the case of discrete likelihood (see

```
trag.sub <- sample(trag, 10, replace = F); p <- seq(35, 80,.1)
anm.loglik(trag, parameter = "mu", dist = "norm", poss = p,
ylim = c(-300, -60), xlim = c(0, 100))
anm.loglik(trag.sub, parameter "mu", dist = "norm", poss = p,
ylim = c(-300, -60), xlim = c(0, 100))
```

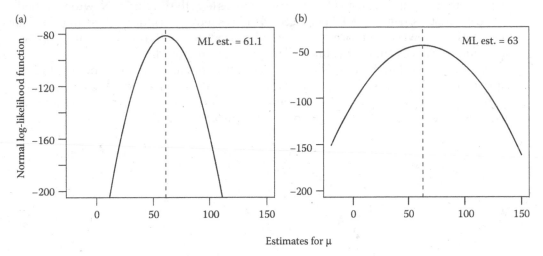

Estimates for μ

FIGURE 4.8
(a) and (b) demonstrate what happens to log-likelihood functions as sample size decreases. A sample size of 20 was used in (a), while in (b), this was reduced to 10. The log-likelihood function appears squashed in (b) because it contains less information. Because the data vary in (a) and (b), the units in the Y-axis are not useful for comparative purposes. Unlike a valid pdf, the area under a likelihood function is not required to equal 1 or, for that matter, any other number.

Equation 2.28). This is because discrete density (i.e., mass) is equivalent to probability and as a result, under independence:

$$\mathcal{L}(\theta \mid data) = \prod_{i=1}^{n} f(x_i \mid \theta) = \bigcap_{i=1}^{n} P(x_i \mid \theta).$$

One should remember, however, that when calculating probability, data vary with respect to fixed parameter values, while in likelihood estimation, the data are fixed and parameter values are varied. Continuous likelihood is definitively *not equal to* probability. Consider the fact that densities from continuous pdfs may exceed 1 (see Example 3.4). Continuous likelihoods (based on the product of densities) may therefore also be greater than 1. As a consequence of this probabilistic impossibility, and to assist generality, we should associate both discrete and continuous likelihood with density, not probability. That is, we should define $\mathcal{L}(\theta|data)$ as $f(data \mid \theta) = \prod_{i=1}^{n} f(x_i \mid \theta)$.

4.4.3 MOM versus OLS versus ML Estimation

MOM, OLS, and ML are the most frequently used methods for deriving point estimators. MOM point estimators are generally easy to calculate; however, they may be inefficient, biased (particularly for small sample sizes), and, in some situations, produce impossible results (see below). OLS estimators are generally unbiased, consistent, and have high efficiency. For ML estimation, we must specify a probability distribution for a statistical model or algorithm that uses MLEs. Given these conditions, ML estimators are asymptotically unbiased as $n \rightarrow \infty$, generally have high efficiency (minimum variability compared to other estimators), are consistent, and have asymptotically normal sampling distributions (see Chapter 5). Likelihood equations, however, may be intractable in many situations, making parameter estimation impossible without computers using iterative procedures (e.g., the Raphson–Newton method). Unlike ML estimators, neither MOM nor OLS estimators have distributional assumptions for point estimation. However, for formal inferential procedures, both normality and *homoscedasticity* (equal variances among treatments) are generally required.

For commonly used parameters, ML and OLS estimators are generally identical under assumptions of normality. The exception is the parameter σ^2, for which the ML estimator is both the population variance and the second sample moment (see Section 4.3.7), and will be biased low (see Exercise 21 at the end of the chapter). Specifically

$$\hat{\sigma}^2_{ML} = \frac{\sum_{i=1}^{n}\left(X_i - \bar{X}\right)^2}{n}$$

$$= (n-1)S^2/n, \qquad (4.37)$$

where S^2 is the variance estimator described in Equation 4.18.

The most frequently used estimation framework in applied statistics is the *general linear model* (Chapters 7 through 10). If these models are balanced (equal replication for each treatment) and if normality assumptions are met, then ML and OLS estimators of model parameters will be identical. For several other major classes of statistical models, however, OLS estimation is often inappropriate. *GLMs*, discussed in Chapters 9 through 11, use only maximum likelihood for parameter estimation, while *nonlinear models* introduced in Chapter 9 use an iteratively weighted algorithm for estimation. Population variance estimation with the so-called *random* and *mixed effect models* (Chapter 10) will be hampered by both MOM estimation (which may produce impossible outcomes <0) and ML estimates (which will be biased). As a result, these methods generally use an approach called *restricted maximum likelihood* or *REML.*[*]

4.5 Linear Transformations

It is often desirable or necessary for an investigator to mathematically alter (*transform*) a random variable or a set of data. For instance, a geometric random variable describes the number of independent failures that occur before obtaining the first success (Section 3.3.1.3). However, we may be interested in the number of *trials* required to obtain the first success (thus, including the first success). This would be equivalent to adding 1 to each outcome, *x*, from *X*.

Thus, if $X \sim GEO(\pi)$, then

$$f(x) = \pi(1-\pi)^x \quad x \in \{0,1,2,\dots\}.$$

---

[*] Other methods for parameter estimation include generalized least squares (Chapter 10) of which weighted least squares (Chapter 9) is an example, and permutation methods (Chapter 5).

Now, let $X' = X + 1$, which requires that $X = X' - 1$. Then, we have

$$f(x') = \pi(1-\pi)^{x'-1} \quad x' \in \{1,2,3,\ldots\}.$$

The distribution $f(x')$ is also called the geometric pdf by some authors, although it should not be confused with the form of the geometric pdf described in Chapter 3.

The above is an example of a *linear transformation* because

$$X' = a + bX, \tag{4.38}$$

where a and b are constants, and X' is a transformation of the original variable X.

Linear transformations include simple addition, subtraction of constants (or other variables), and multiplication of variables by constants. Conversely, in linear transformation, variables *cannot* be used as exponents and two variables cannot be multiplied or divided by each other.

4.5.1 Transformations and Parameters

Linear transformations of random variables result in straightforward changes to parameters. In particular, if X is a random variable, a and b are constants, and $Z = a + bX$, then

$$E(Z) = a + bE(X), \tag{4.39}$$

$$Var(Z) = b^2 Var(X). \tag{4.40}$$

Conversely, if $Z = a - bX$, then

$$E(Z) = a - bE(X), \tag{4.41}$$

$$Var(Z) = b^2 Var(X). \tag{4.42}$$

Thus, addition or subtraction of a constant affects the mean, but not the variance. Furthermore, multiplication by a constant results in new mean that will be a product of the constant and the old mean, and a new variance that will be product of the old variance and the squared constant.

We can also form linear combinations of random variables. For instance, if X and Y are random variables and $Z = X + Y$, then

$$E(Z) = E(X) + E(Y), \tag{4.43}$$

$$Var(Z) = Var(X) + Var(Y) + 2Cov(X,Y). \tag{4.44}$$

Conversely, if $Z = X - Y$, then

$$E(Z) = E(X) - E(Y), \tag{4.45}$$

$$Var(Z) = Var(X) + Var(Y) - 2Cov(X,Y). \tag{4.46}$$

Note that in Equation 4.46, the variances are still added even though the variables are being subtracted. $Cov(X,Y)$ is the *covariance* of X and Y. It describes the way two variables tend to covary (i.e., increase or decrease) when they occur together. If X increases when Y increases, then $\text{Cov}(X,Y) > 0$. If X increases when Y decreases (and vice versa), then $\text{Cov}(X,Y) < 0$. When X and Y are independent (Chapter 2), we can ignore the covariance term because $\text{Cov}(X,Y)$ will be equal to 0. Covariance is explained in greater detail in Chapter 8.

Linear combinations of normal random variables will also be normally distributed. For instance, if X and Y are independent random variables with $X \sim N(\mu_X, \sigma_X^2), Y \sim N(\mu_Y, \sigma_Y^2)$, and $Z = X + Y$, then

$$Z \sim N(\mu_X + \mu_Y, \sigma_X^2 + \sigma_Y^2). \tag{4.47}$$

Conversely, if $Z = X - Y$, then

$$Z \sim N(\mu_X - \mu_Y, \sigma_X^2 + \sigma_Y^2). \tag{4.48}$$

Other distributions often have similar properties. For instance, if X and Y are independent chi-squared random variables with $X \sim \chi^2(v_X)$ and $Y \sim \chi^2(v_Y)$, then $X + Y \sim \chi^2(v_X + v_Y)$. This can be proven using the chi-squared MGF.[*]

4.5.2 Transformations and Statistics

The rules for linear transformation of random variables can be extended to samples (nonlinear data transformations to meet statistical model assumptions are described in Chapters 7 and 9). In Table 4.2, assume x is a sample of size n and let c be any constant. In parallel to parameters, adding or subtracting c from each observation in the sample results in a new sample mean, $\bar{x} \pm c$, (although this transformation does not affect the sample variance), while multiplying by c affects both estimates of location and scale.

EXAMPLE 4.17

A temperature measured in the United States using degrees Fahrenheit must be expressed in degrees Celsius to be understood by the rest of the world. Over 20 years, the mean maximum June temperature in a desert location is 95°F with a variance of 5°F². What is the mean and variance in degrees Celsius?

Since $C = \dfrac{5}{9}(F - 32)$, converting from Fahrenheit to Celsius will involve a linear transformation of the individual sample observations. Our new sample mean and variance are

$$\bar{x}_C = \frac{5}{9}(95 - 32) = 35°C, \quad s_C^2 = \left(\frac{5}{9}\right)^2 \cdot 5 = 1.54.$$

---

[*] Let $X \sim \chi^2(v_X)$, $Y \sim \chi^2(v_Y)$, and $Z = X + Y$. By definition, we have

$$M_Z(t) = E(e^{tz}) = M_{X+Y}(t) = E(e^{t(x+y)}) = E(e^{tx}e^{ty}) = E(e^{tx})E(e^{ty}) = M_X(t)M_Y(t).$$

Substituting the chi-square MGF, $\left(\dfrac{1}{1-2^t}\right)^{v/2}$, we have:

$$M_Z(t) = \left(\frac{1}{1-2^t}\right)^{v_X/2}\left(\frac{1}{1-2^t}\right)^{v_Y/2} = \left(\frac{1}{1-2^t}\right)^{(v_X+v_Y)/2}, \text{ which is a chi-squared MGF with VX + VY degrees of freedom.}$$

TABLE 4.2

Sample Mean, Median, Standard Deviation, and Variance after the Data
Are Linearly Transformed by Adding or Subtracting a Constant c or
Multiplying by a Constant c

	Estimate Given Original (Raw) Data and Transformations		
	Original Data: x	x ± c	x × c
Location Estimate			
Arithmetic mean	$\bar{x}$	$\bar{x} \pm c$	$\bar{x} \times c$
Median	Median(x)	Median(x) $\pm c$	Median(x) $\times c$
Scale Estimate			
Standard deviation	s	s	$s \times c$
Variance	s^2	s^2	$s^2 \times c^2$

4.6 Bayesian Applications

For a Bayesian, parameters of distributions describing random variables are not fixed.
Instead, they are themselves random variables with derivable parameters. This has
important implications for inference since, instead of point or interval estimates for a
parameter, an entire distribution must be obtained. This population of parameters will be
contained in the posterior distribution (see Section 2.6). That is, it will be defined by the
term $P(\theta|\text{data})$ in

$$P(\theta \mid \text{data}) = \frac{P(\text{data} \mid \theta)P(\theta)}{P(\text{data})}. \tag{2.28}$$

The denominator in Equation 2.28 is simply a normalizing constant that makes the integral of the posterior distribution equal to 1. If we drop the denominator (whose derivation
is often the most difficult aspect of Bayesian analyses), then Bayes rule can be simplified to

$$P(\theta|\text{data}) \propto P(\text{data}|\theta)P(\theta), \tag{4.49}$$

where $\propto$ means "proportional to." The right side of Equation 4.49 is the *unnormalized posterior distribution* (Gelman et al. 2003). As we already know from Section 2.6, the term
$P(\text{data}|\theta)$ is the *likelihood function* and $P(\theta)$ is the *prior probability*.

In applications from Section 2.6, priors and likelihoods consisted of a small number of
categorical outcomes. Thus, to create posterior distributions, we simply plugged in the-
tas, representing categories, one at a time into Bayes theorem. However, it is also possible to explicitly define the priors and likelihood as quantitative random variables. In this
case, because discrete density is equivalent to probability, while continuous density is not,
Equation 4.49 can be more generally expressed (Albert 2007) as

$$f(\theta|\text{data}) \propto f(\text{data}|\theta)f(\theta), \tag{4.50}$$

where $f(\theta|\text{data})$ is the unnormalized posterior density distribution, $f(\text{data}|\theta)$ is the likeli-
hood function, that is, $f(\text{data}|\theta) = \mathcal{L}(\theta|\text{data})$, and $f(\theta)$ is the prior density function.

4.6.1 Priors

Bayesian methods entail specification of priors with the goal of creating updated knowledge by combining new data (the result from an investigator's research) with prior knowledge (e.g., the results from previous research). The requirement of priors is by far the most contentious aspect of Bayesian statistics. This is because there may be little known about the phenomenon under investigation leading to priors that are contrived out of convenience or that are merely the subjective opinions of the investigator.

From a mathematical perspective, proper prior distributions are independent of the current data, and like a valid pdf, will integrate to one (Section 3.1). Priors that are mathematically improper may also be used in Bayesian analyses. In this case, however, the posterior distribution should be carefully checked to ensure that it has a finite integral (see Gelman et al. 2003, Chapter 2). There are two classes of priors, noninformative and informative.

4.6.1.1 Noninformative Priors

When an investigator has no knowledge of the distribution of a parameter of interest and/ or does not wish to cloud inferences with potentially biased information, then *noninformative priors* are generally used. Many approaches have been proposed. The *principle of insufficient reason* (ascribed to Pierre-Simon Laplace, see Chapter 1) and an often-cited paper by Edwards et al. (1963) are frequently used to justify the use of the uniform distribution UNIF(0, 1), which is equivalent to BETA(1, 1), as a noninformative prior. Similar noninformative priors include BETA(0, 0) and BETA(0.5, 0.5). Another approach is *Jefferys' invariance principle*, which gives the prior density $f(\theta) \propto [J(\theta)]^{1/2}$, where $J(\theta)$ is the so-called Fisher's information for θ (see Gelman et al. 2003, p. 62). It should be emphasized that while noninformative priors may appear innocuous, they may still have profound and unexpected effects on the posterior distribution (see Exercise 31 at the end of the chapter, and Gelman 2009).

4.6.1.2 Informative Priors

Informative priors are used when an investigator wishes to impose particular distributional assumptions on the distribution of the parameter of interest. As with noninformative priors, informative priors can be either conjugate or nonconjugate (see below).

4.6.2 Conjugacy

In Bayesian probability, if the posterior distribution follows the same parametric form as the prior, then the prior and posterior can be referred to as *conjugate* (Gelman et al. 2003). Important qualities will hold if the likelihood function, allowing conjugacy, belongs to a common family of distributions called the *exponential family*, which includes normal, log-normal, chi-squared, exponential, gamma, Poisson, and in certain cases, the binomial family of distributions (see Appendix). Such likelihood functions will always have a conjugate prior. That is, multiplication of the likelihood functions by said priors will always produce a conjugate posterior distribution. Further, the conjugate prior will often also be a member of the exponential family. Conjugacy is mathematically convenient for three reasons. First, the posterior distribution will have a known parametric form, facilitating the expression of important summaries such as the expected value for the distribution of a parameter (a very Bayesian idea). Second, these distributions often provide a good approximation to reality. Third, they can serve as building blocks in more complicated models; for example, mixtures of prior conjugate distributions (Gelman et al. 2003). Unfortunately, however, the conjugate family may poorly represent prior knowledge, resulting in imprecise or invalid inferences. As a result, it may often be necessary to use less convenient but more realistic nonconjugate prior distributions. Indeed, for complicated models, conjugacy may not even be possible (Gelman et al. 2003). In the case of nonconjugacy, description of the posterior distribution will often require iterative simulation procedures. This topic is addressed in Section 5.5.

EXAMPLE 4.18 BINOMIAL LIKELIHOOD AND CONJUGATE BETA PRIOR

Consider a situation where the likelihood function can be characterized with a binomial pdf:

$$f(\text{data} \mid \theta) = \binom{n}{x} \theta^x (1 - \theta)^{n-x}.$$

We can drop the binomial coefficient in the likelihood function because it will be held constant as we vary θ. Without the binomial coefficient, the binomial pdf will still be proportional to the full pdf (see Section 6.5). Thus, the likelihood function can be expressed as

$$f(\text{data} \mid \theta) = \mathcal{L}(\theta \mid \text{data}) \propto \theta^x (1 - \theta)^{n-x}.$$

Assume that priors for θ can be represented by a beta prior distribution; $\theta \sim \text{BETA}(\alpha, \beta)$. We have

$$f(\theta) = \frac{\Gamma(\alpha + \beta)}{\Gamma(\alpha)\Gamma(\beta)} \theta^{\alpha-1} (1 - \theta)^{\beta-1}.$$

The parameters for the prior distribution and posterior distributions are called *hyperparameters*. The beta prior distribution is indexed by two hyperparameters, α and β. We can drop the term $\dfrac{\Gamma(\alpha + \beta)}{\Gamma(\alpha)\Gamma(\beta)}$ in the prior distribution because it does not involve θ, and since α and β are held constant as we vary θ. Doing this, we have

$$f(\theta) \propto \theta^{\alpha-1} (1 - \theta)^{\beta-1}.$$

Substituting into Equation 4.50, we have

$$f(\theta \mid \text{data}) \propto \theta^x (1 - \theta)^{n-x} \theta^{\alpha-1} (1 - \theta)^{\beta-1}.$$

We add the exponents with the same base and get

$$f(\theta \mid \text{data}) \propto \theta^{\alpha-1+x} (1 - \theta)^{\beta-1+n-x}.$$

Because of conjugacy, the posterior has the same parametric form as the prior. In particular, $f(\theta \mid \text{data})$ is the likelihood function to yet another beta distribution, $\text{BETA}(\alpha + x,$

$\beta + n - x$). The expected value of this distribution is now easily found. From Table 3.5, we have

$$E(\theta \mid \text{data}) = \frac{\alpha + x}{\alpha + x + \beta + n - x} = \frac{\alpha + x}{\alpha + \beta + n}.$$

EXAMPLE 4.19 PROBABILITY OF A U.S. MALE BIRTH: BINOMIAL LIKELIHOOD, AND A NONINFORMATIVE, CONJUGATE, BETA(1, 1) PRIOR

Davis et al. (2007) reported that the probability of a live male birth in United States in 2002 was approximately 0.5117. This was a decline from the reported 1970 level of 0.5134 and was attributed to a larger proportion of male fetal deaths compared to females in 2002. Assume that we perform random samples of 2002 U.S. birth records of size 5, 50, 100, and 1000 (Table 4.3).

We can use the binomial likelihood function to model the number of male births, X, that is

$$f(\text{data} \mid \theta) \propto \theta^x (1 - \theta)^{n-x},$$

where n is the total number of births, x is the observed number of male births, and θ is the probability of a single male birth (Table 3.5).

To find the likelihoods, we supply the required parameters for the binomial pdf from estimates using sample data. We then evaluate $f(\text{data} \mid \theta)$ for a range of θ values given in support of the prior.

To find the posterior distribution, we multiply the likelihood by the priors, θ. For simplicity, we will use BETA(1, 1) as a noninformative prior distribution.

As a result, we have

$$f(\theta \mid \text{data}) = \theta^x (1 - \theta)^{n-x} \frac{\Gamma(2)}{\Gamma(1)\Gamma(1)} \theta^0 (1 - \theta)^0$$

$$\propto \theta^x (1 - \theta)^{n-x} \theta^0 (1 - \theta)^0$$

$$\propto \theta^x (1 - \theta)^{n-x}.$$

We know that the beta prior is conjugate to the binomial likelihood function and will result in a beta posterior distribution (Example 4.18). In fact, we see that $\theta \mid \text{data} \sim \text{BETA}(\alpha = x + 1,\ \beta = n - x + 1)$. Because the expected value for any beta

TABLE 4.3

Hypothetical Sampling of 2002 U.S. Birth Records

Sample Size	Number of Male Births	Number of Female Births	Proportion of Male Births	$E(\theta)$ from Posterior pdf
5	3	2	0.6	0.571429
50	26	24	0.52	0.519231
100	51	49	0.51	0.509804
1000	512	488	0.512	0.511976

distribution is $\alpha/(\alpha + \beta)$, then for our largest sampling effort, we have $E(\theta|\text{data}) = (512 + 1)/[512 + 1 + (1000-512 + 1)] = 513/1002 = 0.511976$.

We note that this value is slightly smaller than the proportion of males in the sample, 0.512. This is because the expected value of the posterior pdf is a compromise between the expectation of the uniform prior distribution, $E(X) = 0.5$ and the sample proportion of males. As the sample size increases, the prior distribution will have less influence on the posterior distribution and the likelihood function will have more influence (Table 4.3, Figure 4.9). That is, fewer reasonable values for θ are possible as sample size increases. In general, the posterior density distribution will be more leptokurtic (acutely peaked) than either the prior or likelihood distribution. This is because it will contain more information than either single distribution.

Fig.4.9()#See Ch. 4 code.

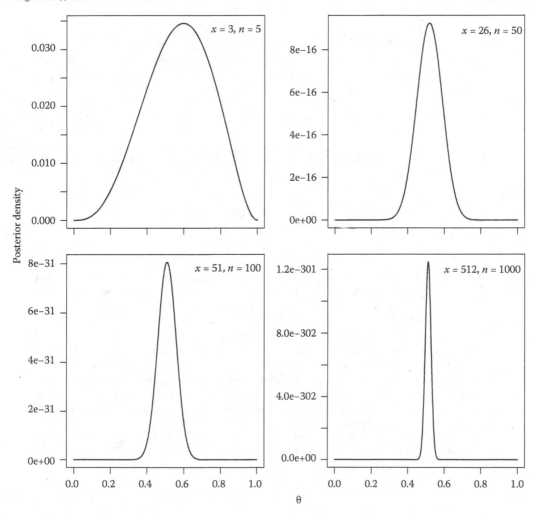

FIGURE 4.9
Unnormalized posterior density functions for θ, given five different sample sizes; that is, five different values for x (number of male births) and corresponding values of n (total births).

EXAMPLE 4.20　MODELING ASTHMA MORTALITY: POISSON LIKELIHOOD AND AN INFORMATIVE AND CONJUGATE GAMMA PRIOR

Asthma is exacerbated by ozone, respirable particulates, and nitrogen dioxide associated with heavy traffic and allergens including cockroach, mouse, and rat droppings. As a result, urban areas tend to have higher mortality from asthma than rural settings (Byrd and Joad 2006). In 2009, the crude asthma mortality rate in Bronx County, New York, was found to be 3.8 persons per 100,000 people (Centers for Disease Control, accessed 1/28/2013). We will model the true long-term asthma mortality rate of Bronx County using a Bayesian approach. Note that we will use λ here to denote the parameter of interest, not the conventional Bayesian θ to avoid confusion with hyperparameters below.

A Poisson likelihood function is often used for epidemiological rate data because the model assumes random exchangeability of intervals of disease exposure (Chapter 3, Gelman et al. 2003). Thus, we can use

$$\text{data}|\lambda \sim \text{POI}(\lambda).$$

The gamma distribution is the conjugate of the Poisson distribution. To find sensible hyperparameter values for this prior, we consider the overall long-term (1999–2009) U.S. asthma mortality rate, which is approximately 1.2 deaths per 100,000 individuals per year. For our prior density distribution, we will use

$$\lambda \sim \text{GAMMA}(2,0.6).$$

The expected value for this distribution is 1.2 (Table 3.5). The resulting posterior distribution will be another gamma distribution. In particular

$$\lambda \,|\, \text{data} \sim \text{GAMMA}\left(\kappa + \sum_{i=1}^{n} x_i, \frac{\theta}{(n\theta + 1)} \right),$$

where λ is the mortality rate we are interested in, x is the number of observed asthma deaths, that is, 3.8 (per 100,000), n is the number of observations (1 in our case), and κ and θ are the prior hyperparameters. Because we used $\kappa = 2$, $\theta = 0.6$, we have $\lambda|\text{data} \sim \text{GAMMA}(5.8, 0.375)$.

This distribution has the expected rate of 2.17 deaths per 100,000 individuals and constitutes a substantial shrinkage from the observed rate of 3.8 deaths per 100,000 individuals. This shift toward the prior occurred because (1) the sample size is small ($n = 1$) causing the priors to overwhelm the likelihood function and (2) information from nonurban areas was used to create the prior densities that poorly represent the conditions in Bronx County. This outcome illustrates the inferential danger of coupling misspecified priors with small samples.

Assume that we gather 19 more years of data from Bronx County and find that the rate, 3.8 deaths per 100,000 individuals, remains invariant. That is, 76 people die from asthma over 20 years. We have

$$\lambda|\text{data} \sim \text{GAMMA}(2 + 76, 0.6/(20(0.6) + 1)).$$

Because we have dramatically increased the sample size (from 1 to 20), the priors have much less influence on the posterior and the expected mortality rate is 3.6 deaths per 100,000 people (close to what is specified in the likelihood function).

4.7 Summary

- Important distributional parameters including $E(X)$ and $Var(X)$ can be mathematically derived using their respective pdfs.

- Parameter values are rarely known; so, parameter estimates from sample data are used in their place. Effective parameter estimators will have three desirable characteristics. They will (1) be unbiased for the parameter that they estimate, (2) converge to the parameter (be increasingly accurate) as the sample size grows large, and (3) have high efficiency (precision) in estimating the parameter. Three different classes of point estimators were discussed in this chapter: (1) measures of location, (2) measures of scale, and (3) measures of shape.

- The most common location estimator, $\bar{X}$, is an unbiased, consistent, and maximum efficiency estimator for $E(X)$; however, its efficiency will decrease as sample contamination (i.e., outliers) increases. Robust location measures resistant to contamination include the sample median and a family of statistics called M-estimators. The most common scale estimator, S^2, is an unbiased, consistent, and high-efficiency estimator for $Var(X)$; however, it is also poorly resistant to outliers. Robust estimators of scale include the IQR and the MAD. Estimators for distributional shape include G_1 (skewness), which measures symmetry, and G_2 (kurtosis), which measures peakedness.

- Three different general approaches are generally used to derive estimators: method of moments (MOM), ordinary least squares (OLS), and maximum likelihood (ML). MOM estimators can be considered empirical representations of moment generating functions (MGFs). OLS estimators minimize a function that represents the sum of squared deviations of the data from the estimator. ML estimators maximize an assumed likelihood function that varies potential parameter estimates while holding the data constant. Likelihood is the product of the densities of data points when using a particular parameter estimate. OLS and ML estimators are generally identical if a normal distribution is assumed for the population represented by the sample data.

- Particular rules apply for deriving parameters, given the linear transformations of random variables. Distributions can also be mathematically combined resulting in completely different but identifiable distributions. Important examples include Bayesian posterior distributions resulting from multiplication of the likelihood function and a conjugate prior density distribution.

EXERCISES

1. Distinguish the words *parameter, estimator,* and *estimate.*
2. Define the terms *expected value* and *true variance.*
3. Using the rules for expectations in the Appendix, prove the derivation in Equation 4.3.
4. Let X be a discrete random variable with the pdf $f(x) = x/8$ if $x = 1$, 2, or 5, and $f(x) = 0$; otherwise, find
   a. $E(X)$
   b. $Var(X)$
   c. $E(2X + 3)$

5. Let X be a continuous random variable with the pdf $f(x) = 3x^2$ if $0 < x < 1$; using calculus, find

    a. $E(X)$

    b. $Var(X)$

    c. $E(2X + 3)$

6. Using the equation for the exponential cdf, show that the median of the exponential distribution is $-\ln(0.5)/\theta$.

7. Using the definition for the expected value of a discrete distribution, prove that the Bernoulli expected value is π (i.e., the probability for success).

8. Let X be a discrete random variable whose pdf is described in the table below:

X	-1	0	1
$f(x)$	1/8	6/8	1/8

   Use the Chebychev inequality to find the probability bound for X if $c = 2$, that is, find $P\{|X - E(X)| \geq c[SD(X)]\}$. Hint: Let c be any constant, then $P(|X-E(X)| \geq c) = P(-c \geq X - E(X) \leq c)$.

9. For the sample $x = \{2, 3, 1, 5, 4, 5.5, 90, 0.001, 95, 4, 5, 10\}$, calculate the arithmetic mean, mode, trimmed mean ($\kappa = 0.2$), and Winsorized mean ($\kappa = 0.2$). Accomplish this by hand (without using the existing R functions for these particular statistics); then check your results using any existing R functions including those in *asbio*.

10. For the data in Question 9, compare the sample mean, median, and Huber's M-estimator of location using $c = 0$, $c = 1.28$, and $c = \infty$. Use R for calculations. Comment on your results.

11. Medium ground finch (*Geospiza fortis*) population numbers in the Galapagos vary enormously from year to year since their reproduction is strongly tied to precipitation and the region is strongly affected by El Niño and La Niña climatic oscillations (Gibbs and Grant 1987). Here are G. *fortis* counts on Daphne Major Island from 1976 to 1984:

```
finch <- c(1220, 400, 380, 298, 280, 200, 297, 280, 1250)
```

    a. Calculate the arithmetic, geometric, and harmonic means using any R function for calculations.

    b. Justify the use of the harmonic mean.

12. Describe what is happening for each step in the function huber.NR in *asbio* (use the R appendix to help you).

13. For the data in Question 9, calculate the sample standard deviation, variance, Winsorized variance, coefficient of variation, interquartile range, and median absolute deviation. Accomplish this by hand (without using existing R functions for these particular statistics); then check your results using any existing R functions, including those in *asbio*.

14. The following questions relate to appropriateness of statistical measures:

    a. Why might we want to use a robust M-estimator instead of the mean as a measure of location?

    b. When would we want to use the geometric or harmonic mean?

c. Why might we want to use the interquartile range instead of the variance as a measure of dispersion?

d. What are some nice features about the standard deviation?

e. Draw three distributions: one that is positively skewed, one that is negatively skewed, and one that is symmetric. Show (approximately) where the mean and the median would be for these distributions.

15. Write **R** functions for $\bar{X}$ and S^2 without using the existing functions `mean`, `sd`, or `var`.

16. Prove that the two versions of Equation 4.18 are equivalent.

17. For the data in Question 9, calculate the sample skewness and kurtosis using both MOM and unbiased procedures. Accomplish this by hand (without using existing **R** functions for these particular statistics), then check your results using any existing **R** functions including those in *asbio*. Interpret your results.

18. Derive the kurtosis of the exponential distribution using the exponential MGF. Hint:

$$(X - E(X))^4 = X^4 - 4X^3E(X) + 6X^2E(X)^2 - 4XE(X)^3 + E(X)^4.$$

19. Let

$$X_1 \sim \chi^2(v_1)$$

$$X_2 \sim \chi^2(v_2),$$

$$\vdots$$

$$X_n \sim \chi^2(v_n)$$

$$Y = X_1 + X_2 + \cdots + X_n.$$

Use the chi-squared MGF to find the distribution of Y. Assume that all X_i are independent.

20. (difficult) Using the MGF definition, derive the exponential MGF.

21. Assuming X and Y are independent, use MGFs in Tables 3.4 and 3.5 to answer the following questions:

a. Let X and $Y \sim \text{GAMMA}(\kappa,\theta)$. Find the distribution of $X + Y$.

b. Let $X \sim \text{BIN}(\pi,n_x)$, $Y \sim \text{BIN}(\pi,n_y)$. Find the distribution of $X + Y$.

22. Define likelihood in your own words using information from this chapter.

23. In your own words, distinguish the characteristics of MOM, ML, and OLS estimators.

24. We have the following data: 10.3, 12.4, 10.5, 13.1, 16.0, 13.3, 13.1, 14.7, 13.9, 14.6, 16.1, 16.2, 14.8, 14.3, 16.0, 12.1, 17.1, 14.5, 16.0, 17.6.

a. Assume these data are from a normal distribution and use the function `anm.loglik.tck` to plot the log-likelihood function for μ. How does the ML estimate for μ compare to the sample arithmetic mean?

  b. Calculate the normal log-likelihood by hand for $\mu = 13$ and $\mu = 14$. Fix σ at its ML estimate.

25. Repeat Questions 24a after randomly deleting 10 observations. What happened to the log-likelihood function? Why did this happen?

26. Using the data from Question 24, use the function `anm.loglik.tck` and plot the log-likelihood function for σ^2. What is the ML estimate for σ^2? Is the ML estimate the same as the sample variance? Comment, given the fact that the estimator, S^2, is unbiased for σ^2.

27. We have the following count data: 3, 2, 1, 0, 0, 4, 2, 5, 4, 5, 6, 4, 3, 2, 1.

  a. Assume the data are from a Poisson distribution and use the function `anm.loglik.tck` to plot the log-likelihood function for λ. How does the ML estimate for λ compare to the sample arithmetic mean?

  b. Calculate the Poisson log-likelihood by hand for $\lambda = 3$ and $\lambda = 3.4$.

28. We have the following binary data: 1, 0, 1, 0, 0, 1, 0, 0, 0, 0, 0 (where 1 indicates success).

  a. Use the function `anm.loglik.tck` to plot the binomial log-likelihood function for π. How does the ML estimate for π compare to the sample arithmetic mean?

  b. Calculate the binomial log-likelihood by hand for $\pi = 0.23$ and $\pi = 0.45$.

29. Let $X \sim N(2, 4)$ and let $Y \sim N(3, 2)$. Define the distribution of

  a. $3X + 7$

  b. $\dfrac{Y}{10}$

  c. $3(X - Y)$

30. Bob and Jimmy Joe are middle-distance runners. The 1600 m (metric mile) times for Bob can be represented by a random variable, B, which is normally distributed with a mean μ of 260 s and a variance σ^2 of 20 s; that is, $B \sim N(260, 20)$. The 1600 m times for Jimmy Joe can be represented by a random variable, J, which is normally distributed with a mean μ of 265 s and a variance σ^2 of 17 s; that is, $J \sim N(265, 17)$.

  a. Give the joint distribution of $J - B$.

  b. Use **R** to find the probability that Jimmy Joe beats Bob in a 1600 m race? That is, find $P(J - B < 0)$.

  c. Why *does* $P(J - B < 0)$ give the probability that Jimmy Joe would beat Bob?

★ 31. Jimmy Joe claims he has magical abilities and can determine from a collection of five cards {A♠, K♠, Q♠, J♠, 10♠}, which individual card you are holding. To test his abilities, you decide to use a Bayesian approach. You create a vector containing the following probabilities, θ, for success:

```
theta <- c(0.0, 0.1, 0.2, 0.3, 0.4, 0.5, 0.6, 0.7, 0.8, 0.9, 1.0)
```

You decide to use the following priors, $f(\theta)$, knowing that Jimmy will randomly choose the right card $1/5 = 20\%$ of the time.

```
f.theta <- c(0.005, 0.005, 0.95, 0.005, 0.005, 0.005, 0.005, 0.005,
0.005, 0.005, 0.005)
```

The likelihood function, $f(\text{data}|\theta)$, for a single outcome will be binomial. That is

```
f.data.given.theta <- theta^x * (1 - theta)^nmx
```

where x is the number of successes and nmx is the number of failures.

Jimmy repeats the trick 10 times. He is right 5 times and wrong 5 times.

a. Compute the unnormalized posterior densities.
b. Make a plot of the unnormalized posterior densities as a function of θ; specify type = "h" in the plot function.
c. What is the most likely outcome in the posterior density?
d. Redo the experiment with noninformative priors, f.theta <- rep(1/11, 11). Does this change the results?

32. Assume $\theta \sim \text{BETA}(\alpha,\beta)$ and data$|\theta \sim \text{BETA}(\gamma + 1, \eta + 1)$ for some arbitrary $\alpha, \beta, \gamma, \eta$.

a. Find the unnormalized posterior distribution, $f(\theta|\text{data})$.
b. Plot the three distributions $f(\theta)$, $f(\text{data}|\theta)$, and $f(\theta|\text{data})$ on a single graph. Use the entire space for which the distributions are defined, [0,1], as the abscissa. When plotting, divide the outcomes from functions by their respective sums to facilitate comparisons. Comment on the results.

5

Interval Estimation: Sampling Distributions, Resampling Distributions, and Simulation Distributions

... Whenever a large sample of chaotic elements are taken in hand and marshaled in the order of their magnitude, an unsuspected and most beautiful form of regularity proves to have been latent all along.

W. S. Gosset (Student)

... Confidence is the first requisite to great undertakings.

Samuel Johnson (1709–1784)

5.1 Introduction

From a frequentist perspective, there are two types of parameter estimation: point and interval. With the former, we compute an estimate for a parameter with a single fixed but unknown value (see Chapter 4). With the latter, we estimate the bounds of an interval that preceding sampling contains a parameter at a given probability. This chapter concerns interval estimation, and resampling and simulation methods that allow interval estimation.

EXAMPLE 5.1

To investigate the influence of sexual activity on animal lifespan, Partridge and Farqaur (1981) applied five mating partner configurations to male fruit flies (*Drosophila melano-gaster*). These were (1) one virgin female per day, (2) eight virgin females per day, (3) one newly inseminated female per day, (4) eight newly inseminated females per day, and (5) no added females. In Figure 5.1, the bar heights are treatment means, while the overlaid intervals have bounds that with repeated sampling will contain the true mean 95% of the time. Interestingly, we see that sexual activity appears to increase male fruit fly lifespan.

5.1.1 How to Read This Chapter

Sections 5.1 through 5.3 describe topics appropriate for an introductory class in biostatistics. Specifically, Section 5.2 presents a wide variety of sampling distributions, while Section 5.3 introduces conventional methods for confidence interval derivation. If advanced pdf materials in Chapter 3 have been skipped, materials in Section 5.2 should be conjoined with descriptions of the χ^2, t-, and F-distributions in Sections 3.3.2.1, 3.3.2.2, and 3.3.2.3, respectively. Derivations of sampling distributions in Section 5.2 add mathematical rigor but may be skipped if they are hampering the conceptual understanding of topics.

```
data(fly.sex)
with(fly.sex, bplot(longevity, treatment,int "CI", ylab "Male
longevity (days)", xlab "Treatment"))
```

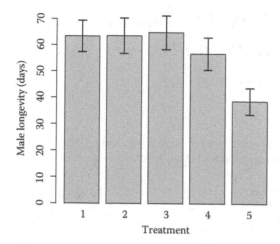

FIGURE 5.1

Barplot of male fly longevity with respect to mating treatments. Bar heights are arithmetic means. Error bars are 95% confidence intervals for the true mean (see also Example 10.20).

Sections 5.4 and 5.5 present techniques that will be of interest, and less frustrating, to more advanced practitioners. In particular, Section 5.4 discusses inferential methods that are useful when the sampling distribution of a statistic is unknown. These include bootstrap and jackknife methods that create distributions of statistics based on resampling from a dataset. Section 5.5 introduces Bayesian approaches for interval estimation. Specialized techniques are presented that allow the evaluation of the posterior distribution (e.g., identify mode(s) and obtain interval bounds for those points) when the posterior cannot be derived analytically. Of particular interest are Markov chain Monte Carlo (MCMC) procedures. Biological examples and **R**-code are provided to provide insight into some of these approaches.

All of the topics in this chapter are built on earlier concepts of probability (Chapter 2), pdfs (Chapter 3), and parameters (and their estimators) (Chapter 4).

5.2 Sampling Distributions

A *sampling distribution* is composed of an infinite number of independent statistical estimates (e.g., sample means) each calculated from random samples of the same sample size, from the same infinite *parent population* (sampled population).

Picture for a moment a stalwart biological researcher who randomly samples the weights of individual mountain goats from a large population 100 times with the same sized sample, for example, $n = 10$. The researcher then calculates the mean weight of each sample. While the parent population and the number of samples are both finite, the resulting collection of 100 means illustrates the underlying principles of a true sampling distribution.

For most parent populations the likelihood that any two statistical estimates based on random samples will be the same is minute. Note that in Figure 5.2, $\bar{x}_1$ is much smaller than $\bar{x}_{100}$ because lighter goats are randomly selected in the first sample. In most sampling distributions, certain estimates will be more likely than others. For example, in Figure 5.2, sample means are most likely to be between 87.5 and 95 kg.

The sampling distributions of many statistics will have specific forms (e.g., normal, t, and F) given a particular parent distribution. The sampling distributions of other statistics will obtain recognizable forms as sample sizes grow large, regardless of the parent distribution.

5.2.1 `samp.dist`

The function `samp.dist` from *asbio* is a useful tool for examining sampling distributions. It has two general purposes. First, it can create simultaneous "snapshots" of a sampling distribution at several sample sizes. For instance, the sampling distribution of the sample mean can be simultaneously viewed for $n = 3$, 10, and 50. Second, it can be used to "animate" the process of acquiring estimates for a particular estimator at a single sample size. That is, estimates are obtained for many random samples of size n, and are added one at a time to a density distribution over a specified number of iterations. Presets for both of these approaches have been programmed for a large number of statistics. These can be easily modified using three interactive GUIs. These are

- `samp.dist.snap.tck1`, which provides distributional snapshots for simple estimators requiring a single statistic and a single parent distribution
- `samp.dist.snap.tck2`, which provides snapshots for complex estimators requiring more than one statistic and/or multiple parental distributions
- `samp.dist.tck`, which provides sampling distribution animations

These three functions can be called from a single GUI, `samp.dist.method.tck`. For more information, see Examples 5.3 through 5.6 and/or type `?samp.dist`.

5.2.2 Sampling Distribution of $\bar{X}$

A sampling distribution of arithmetic means will have an expected value equal to the parent population mean, and a variance equal to the parent population variance divided by n. That is, let the parent population be a random variable, X, with mean, μ, and finite variance, σ^2. Then

$$E(\bar{X}) = \mu, \quad \text{and} \quad \text{Var}(\bar{X}) = \frac{\sigma^2}{n}. \tag{5.1}$$

$\text{Var}(\bar{X})$ is generally denoted $\sigma_{\bar{X}}^2$. From Equation 4.3, it immediately follows that

$$\text{Var}(\bar{X}) = \sigma_{\bar{X}}^2 = E(\bar{X}^2) - \mu^2. \tag{5.2}$$

The standard deviation of any sampling distribution is called the *standard error*. Thus, the standard error for a sampling distribution of arithmetic means is $\sigma/\sqrt{n}$.

```
samp.dist.mech(1)
samp.dist.mech(100)
```

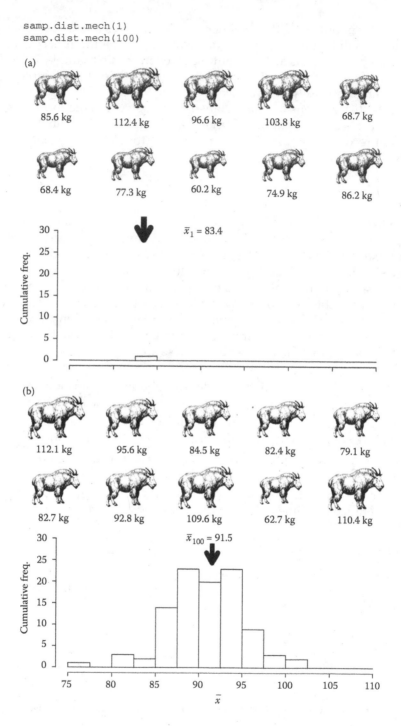

FIGURE 5.2

Illustration of the basic principles of a sampling distribution. Mountain goats are sampled 10 at a time and weighed, a mean weight is calculated from these measures, and this mean weight is added to collection of mean weights in the form of a continuously updated histogram. (a) The first mean, and a histogram with only this value. (b) The 100th mean, and a histogram containing all 100 means.

EXAMPLE 5.2

Assume that a parent distribution has a mean μ and a variance σ^2. The simple proofs below show that the mean of $\bar{X}$ equals μ, and that $\mathrm{Var}(\bar{X}) = \sigma^2/n$.

$$E(\bar{X}) = E\left(\frac{1}{n}\sum_{i=1}^{n} X_i\right) \quad \text{From Equation 4.8}$$

$$= E\left[\frac{1}{n}(X_1 + X_2 + \cdots + X_n)\right] \quad \text{Definition of } \Sigma$$

$$= \frac{1}{n}E(X_1 + X_2 + \cdots + X_n) \quad \text{Constants go outside expectations}$$

$$= \frac{1}{n}\left[E(X_1) + E(X_2) + \cdots + E(X_n)\right] \quad \text{Distribute expectations}$$

$$= \frac{n\mu}{n} = \mu.$$

Thus, μ defines $E(\bar{X})$. The proof also demonstrates that the arithmetic mean is unbiased for $E(X)$.

$$\mathrm{Var}(\bar{X}) = \mathrm{Var}\left(\frac{1}{n}\sum_{i=1}^{n} X_i\right) \quad \text{From Equation 4.8}$$

$$= \mathrm{Var}\left[\frac{1}{n}(X_1 + X_2 + \cdots + X_n)\right] \quad \text{Definition of } \Sigma$$

$$= \left(\frac{1}{n}\right)^2 \mathrm{Var}(X_1 + X_2 + \cdots + X_n) \quad \text{Constant} \times \text{variance; Equation 4.40}$$

$$= \left(\frac{1}{n}\right)^2 \left[\mathrm{Var}(X_1) + \mathrm{Var}(X_2) + \cdots + \mathrm{Var}(X_n)\right] \quad \text{Distribute expectations}$$

$$= \frac{n\sigma^2}{n^2}$$

$$= \frac{\sigma^2}{n}.$$

Thus, σ^2/n defines $\mathrm{Var}(\bar{X})$. Further, since the variance of the sampling distribution of $\bar{X}$ decreases with sample size, then $\bar{X}$ is Fisher consistent for $E(X)$. This result follows directly from the *weak law of large numbers* developed by Jakob Bernoulli (1654–1705). The law stipulates that because $\mathrm{Var}(\bar{X}) = \sigma^2/n$, and because σ^2 is fixed, $\lim_{n\to\infty} \sigma^2/n = 0$. That is, the sampling error of the statistic goes to zero if the sample includes the entire population. This is also in agreement with the definition of $E(X)$ given in Chapter 4. Specifically, the expected value of X is the arithmetic mean of the random variable given all possible outcomes.

5.2.2.1 Central Limit Theorem

Mathematical tenets called *limiting theorems* describe how a distribution converges to a particular form as sample size increases. Because they allow distributional approximations, limiting distributions are particularly important when the derivation of a pdf is *intractable* (unknown or unknowable). Limiting distributions have been described for the sampling distributions of many statistics, including the arithmetic mean, sample median, sample standard deviation, median absolute deviation, the estimator for the binomial parameter π, and many test statistics (Hall and Welsch 1985, Hall and Wang 2004).

The *central limit theorem* describes the limiting distribution of the sample mean. It can be summarized with two statements:

- First, if the parent population has the distribution $N(\mu, \sigma^2)$, then the sampling distribution of $\bar{X}$ is

$$\bar{X} \sim N(\mu, \sigma^2/n). \tag{5.3}$$

  From Equations 3.10 and 5.3, it follows that if the parent distribution is normal, then

$$Z = \frac{\sqrt{n}(\bar{X} - \mu)}{\sigma}. \tag{5.4}$$

- Second, regardless of the form of the parental distribution, $\bar{X}$ will converge to a normal distribution as n becomes large, with mean, μ, and variance, σ^2/n. This can be stated summarily as

$$\bar{X} \xrightarrow{d} N(\mu, \sigma^2/n). \tag{5.5}$$

  where the term $\xrightarrow{d}$ means "*converges in distribution.*"

The importance of the central limit theorem is clear. If the sample size is sufficiently large, then we can safely assume that the sampling distribution of $\bar{X}$ is approximately normal. This will hold for any parent population (discrete or continuous). Thus, while the parent distribution will almost never be known, the distribution of $\bar{X}$ will always be known given large sample sizes. It will be normal if X is normally distributed, or approximately normal, if X is not normal.

The sample size $n = 30$ is often used as a cutoff for a sufficiently large sample size to achieve an approximate normal distribution for $\bar{X}$. However, the appropriate sample size will depend on the parent distribution. The more normal the parent distribution (with respect to support and parameters, e.g., skew and kurtosis), the smaller the sample size required to obtain a normal sampling distribution for $\bar{X}$.

EXAMPLE 5.3

To demonstrate the principles of the central limit theorem, we will assume an exponential parental distribution with a rate of 1, that is, $X \sim EXP(1)$, and sample from it using six different sample sizes $n = \{1, 3, 7, 10, 20, 50\}$. Recall that the exponential distribution is strongly leptokurtic (peaked) and positively skewed (Chapters 3 and 4).

To be delivered to a sampling distribution GUI with these (modifiable) presets, I would type

```
samp.dist.method.tck()
```

and choose "mean," or simply type

```
samp.dist.snap.tck1("mean")
```

For each sample size, 10,000 samples (by default) will be drawn, and a mean for each sample will be calculated. Thus, each histogram in Figure 5.3 represents a distribution of 10,000 means.

A sample size of one is used to create the first plot in Figure 5.3. Because each mean is equivalent to an individual observation from X, the plot simply redisplays the parent distribution. As predicted by the central limit theorem, the sampling distribution of $\bar{X}$ becomes increasingly normal as sample size increases. In accordance with the weak law of large numbers, the tails of the distributions also contain less density (the sampling distributions are less variable) as n increases. In the context of a single sample, $\bar{X}$ will

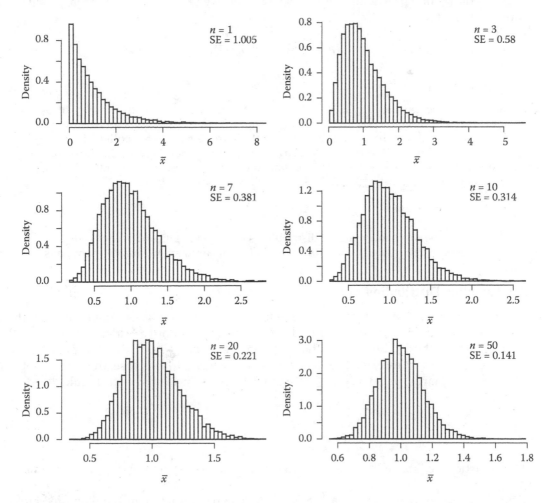

FIGURE 5.3
Empirical distributions of $\bar{X}$ for sample sizes of 1, 3, 7, 10, 20, and 50 from the parent distribution EXP(1). Each histogram represents 10,000 means.

decrease as n increases because extreme random observations will have a greater influence on $\bar{X}$ if the sample size, n, is small. As a result, $\bar{X}$ is more efficient if n is large.

It is important to remember that while the randomly generated histograms in Figure 5.3 illustrate the principles of central limit theorem, they cannot represent true sampling distributions. No simulation can perfectly characterize a continuous distribution with a nonzero variance. Thus, the sampling distribution is yet another frequentist idealization, albeit a very useful one.

5.2.3 Sampling Distribution of S^2

The sampling distribution of the sample variance is vital to a number of important procedures, including *variance components analysis* in conjunction with *random* and *mixed effect models* (Chapter 10).

Assuming a normal parent distribution, the sampling distribution of S^2 can be related to a χ^2 distribution. In particular, if $X \sim N(\mu, \sigma^2)$, and we sample infinitely from X with a sample size n, then

$$(n-1)S^2/\sigma^2 \sim \chi^2(n-1).\tag{5.6}$$

EXAMPLE 5.4

By default, the `samp.dist` GUI depicts the distribution of $(n-1)S^2/\sigma^2$, by first calculating sample variances for 10,000 random samples of sizes $n = 3, 7, 10$, and 20, from a standard normal parent population. These values are fed into an auxiliary function representing Equation 5.6 to obtain 10,000 $(n-1)S^2/\sigma^2$ outcomes (Figure 5.4). The correct presets for this demonstration can be obtained by typing `samp.dist.method.tck()`, and choosing the statistic "$(n-1)S^2/\sigma^2$," or by simply typing

```
samp.dist.snap.tck2("(n-1)S^2/sigma^2").
```

As part of the snapshot default, χ^2 distributions with $n-1$ degrees of freedom are overlaid. We see that the distribution of $(n-1)S^2/\sigma^2$ *does* appear to follow these distributions.

5.2.4 Sampling Distribution of t^*

Estimators known as *test statistics* are crucial to inferential procedures called null hypothesis tests. The test statistic, t^*, (pronounced t-star) is used to quantify evidence *against* a *null hypothesis* (a hypothesis of no effect) that the mean of a normally distributed population, with an unknown variance, is equal to a specified value, μ_0. In this case, the null hypothesis can be summarily stated as $\mu - \mu_0 = 0$ (see Chapter 6). The test statistic is calculated as

$$t^* = \frac{\sqrt{n}(\bar{X} - \mu_0)}{S}.\tag{5.7}$$

Test statistics are "starred" to distinguish them from their sampling distributions under null.

Oftentimes, several estimators (each specific to its own parameter) must be combined into a single estimator to obtain estimates for a parameter for which the individual statistics are not sufficient. This is true for Equation 5.7. To obtain t^* we require both $\bar{X}$ and S.

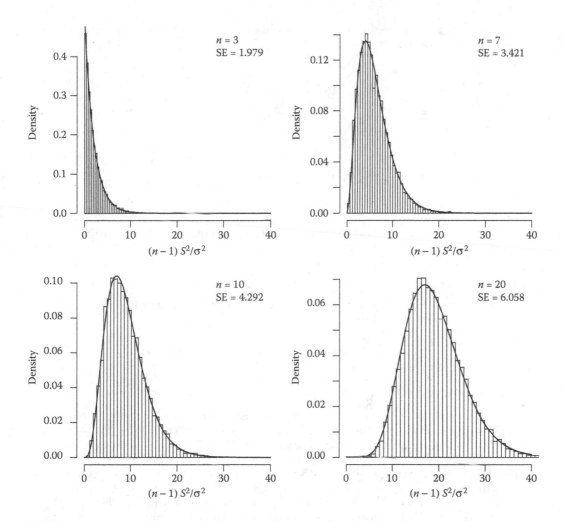

FIGURE 5.4
The conceptual sampling distribution of $(n-1)S^2/\sigma^2$, for sample sizes of 3, 7, 10, and 20, given a normal parent distribution with $\sigma^2 = 1$. Distributions of 10,000 values of $(n-1)S^2/\sigma^2$ are shown in each histogram. χ^2 distributions with $n-1$ degrees of freedom are overlaid.

If the null hypothesis is true, then the sampling distribution of t^* can be represented by a random variable T that will follow a t-distribution with $n-1$ degrees of freedom. In particular, if n samples are repeatedly taken from $N(\mu, \sigma^2)$, then it can be shown that

$$T = \frac{Z}{\sqrt{V/(n-1)}} \sim t(n-1), \tag{5.8}$$

where $Z \sim N(0, 1)$ and results from standardization of a normal distribution using Equation 3.10, and $V \sim \chi^2 (n-1)$. Thus, T is a ratio of two random variables.

The major motivation for the development of t^* was the need for a test statistic that made inference to population means, while not requiring knowledge of σ^2 (Student 1908). It is straightforward to show that t^*s are random outcomes from T under null, and thus satisfy this condition.

Under null, $\mu = \mu_0$, a normal parental distribution, and independent samples of size n, we have

$$T = \frac{\sqrt{n}(\bar{X} - \mu_0)}{\sigma} \bigg/ \sqrt{\frac{(n-1)S^2/\sigma^2}{n-1}},$$

$$= \frac{\sqrt{n}(\bar{X} - \mu_0)}{\sigma} \bigg/ S/\sigma,$$

$$= \frac{\sqrt{n}(\bar{X} - \mu_0)}{S} \sim t(n-1). \quad \text{See Equations 5.4 through 5.8}$$

Thus, the t-distribution allows inferences about the null mean without requiring knowledge of σ^2.

EXAMPLE 5.5

To be delivered to a GUI with the correct presets for the sampling distribution of t^* discussed above, we would type

```
samp.dist.method.tck(),
```

and choose: "t^* (1 sample)," or simply enter

```
samp.dist.snap.tck2("t* (1 sample)").
```

For the sample sizes, 3, 7, 10, and 20, 10,000 paired estimates of $\bar{X}$ and S are fed into an auxiliary function representing Equation 5.7 to calculate 10,000 t^* s (Figure 5.5).

As a default for this snapshot, the standard normal distribution (black dashed lines) and t-distributions with $n - 1$ degrees of freedom (gray solid lines) are overlaid in each plot. We see that the thicker tails of the t-distribution provide a better representation of the sampling distribution of t^* than the standard normal distribution, particularly for smaller sample sizes (Figure 5.5). For larger sample sizes, the sampling distribution of t^* and the standard normal distribution are essentially indistinguishable.

5.2.5 Sampling Distribution of F^*

All the statistics discussed so far in this text make inference to a single parent population. However, many statistics compare or summarize two or more populations. The test statistic F^* is often used to quantify evidence against the null hypothesis that two normal parent populations have equal variances (σ_1^2, σ_2^2). The test statistic has the form

$$F^* = \frac{S_1^2}{S_2^2}. \tag{5.9}$$

If the null hypothesis is true, then F^* will be a random outcome from an F-distribution with $n_1 - 1$ numerator degrees of freedom and $n_2 - 1$ denominator degrees of freedom.

A proof that a ratio of sample variances from normal distributions will follow $F(n_1 - 1, n_2 - 1)$ under null is shown below.

In Section 3.3.2.3, we learned that if $V_1 \sim \chi^2(v_1)$ and $V_2 \sim \chi^2(v_2)$, and $X = (V_1/v_1)/(V_2/v_2)$, then $X \sim F(v_1, v_2)$.

```
samp.dist.snap.tck2("t* (1 sample)")
```

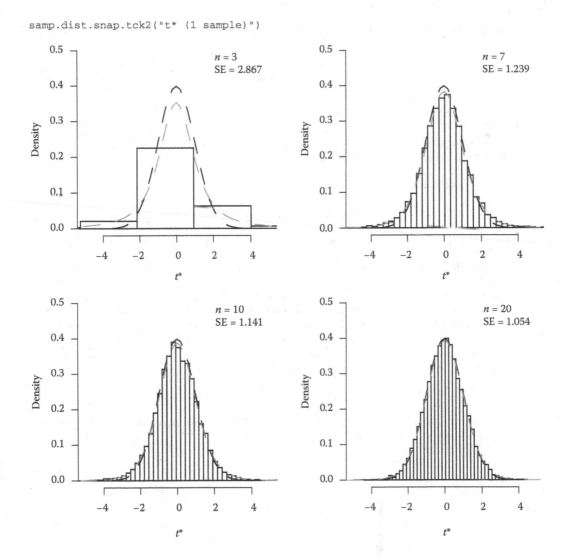

FIGURE 5.5
Sampling distribution of t^* under a true null hypothesis for sample sizes of 3, 7, 10, and 20. Each histogram is constructed from 10,000 t^*s. Standard normal and $t(n-1)$ distributions are overlaid with black and gray lines respectively.

Assuming that the parent distributions are normal, then from Equation 5.6 we have $V_1 = (n_1 - 1)S_1^2/\sigma_1^2 \sim \chi^2(n_1 - 1)$ and $V_2 = (n_2 - 1)S_2^2/\sigma_2^2 \sim \chi^2(n_2 - 1)$.

Thus, $S_1^2/\sigma_1^2 = V_1/\nu_1$ and $S_2^2/\sigma_2^2 = V_2/\nu_2$.

From division, we have

$$\frac{S_1^2 \sigma_2^2}{S_2^2 \sigma_1^2} = \frac{V_1/\nu_1}{V_2/\nu_2} \sim F(\nu_1, \nu_2). \tag{5.10}$$

Finally, assuming that the null hypothesis, $\sigma_1^2 = \sigma_2^2$, is true, we have

$$\frac{S_1^2}{S_2^2} \sim F(\nu_1, \nu_2). \tag{5.11}$$

EXAMPLE 5.6

The sampling distribution of F^* will be demonstrated using animation. To be delivered to a GUI with correct presets, I would type

```
samp.dist.method.tck(),
```

choose "animation," and the statistic "F^*," or simply type

```
samp.dist.tck("F*").
```

We will (arbitrarily) use sample sizes of 10 and 8 from standard normal parent distributions.

The animation in Figure 5.6 illustrates the process of accumulating statistical estimates. This is pedagogically useful because it demonstrates that sampling distributions assume an infinite number of statistical estimates. Clearly, a collection of 10 means will not resemble a normal distribution regardless of their sample size. Only when the number of estimates grow large do they collectively begin to approximate a "true" sampling distribution.

We overlay the correct distribution using

```
curve(df(x, 9, 7), from = 0, to = 20, add = TRUE, lwd = 2, col = gray(.4))
```

The test statistic F^* *does* appear to follow an $F(\nu_1, \nu_2)$ distribution assuming that the null hypothesis, $\sigma_1^2 = \sigma_2^2$, is true.

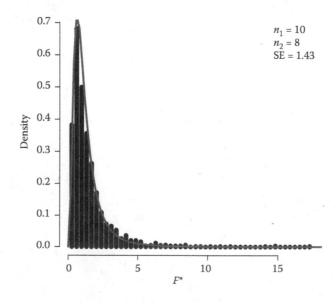

FIGURE 5.6

Sampling distribution of F^* under a true null hypothesis for sample sizes $n_1 = 10$, $n_2 = 8$. The histogram is constructed from 2000 F^*s. The distribution $F(9, 7)$ is overlaid.

5.3 Confidence Intervals

By itself, a point estimate provides no information concerning its effectiveness in estimating a parameter, θ. Of interest then is a measure to quantify the precision and accuracy of point estimates. One solution is to estimate a *random interval* (an interval with random endpoints) that contains θ at a particular probability, γ. The interval will be random because the estimator used to create the interval, $\hat{\theta}$, will be a random variable. At a fixed value of γ narrow intervals around the estimate indicate that the estimate is probably close to θ, while wide intervals indicate that the estimate is potentially far from θ. Conventionally, interval bounds will be quantiles (Section 3.1.2) from the sampling distribution of $\hat{\theta}$. The bounds containing $\gamma \times 100\%$ of this distribution will be C_L and C_U in

$$P[C_L \leq \theta \leq C_U] = \gamma. \tag{5.12}$$

After estimation, the random interval will be fixed. At this point, it would be incorrect to say that the interval contains θ with probability γ. This is because the estimated interval now either contains the parameter θ or it does not. On the other hand, because the random interval is based on probability γ, we might assert that we are $\gamma \times 100\%$ "confident" that the interval contains θ. Based on this rationale, an estimate of such a random interval is called a *confidence interval*, while the upper and lower limits of the interval are called *confidence bounds*.

Confidence intervals are strongly tied to significance testing in null hypothesis tests described in Chapter 6. Confidence, denoted above as γ, is equal to $1 - \alpha$, where α is the *significance level*: the probability of rejecting a null hypothesis when it is actually true (Neyman and Pearson 1928, 1933). Conventional values of α, are 0.05 and 0.01 (Chapter 6), corresponding to 95% and 99% confidence intervals, respectively. Larger values of γ result in wider confidence intervals. This is because, preceding estimation, they will have a higher probability of containing θ. Confidence intervals can be estimated for most conventional parameters (e.g., μ, σ^2, and σ) using the sampling distributions of their estimators. Of particular interest is the *confidence interval for* μ, which allows us to quantify how confident we are in a particular estimate of μ.

5.3.1 Confidence Interval for μ

A common way of thinking about confidence intervals is based on the frequentist view of probability (Chapter 2). Assume that we have taken an infinite number of samples, calculated the mean (and variance) of each sample, and found 95% confidence intervals for μ based on each of these summaries, then 95% of those confidence intervals would contain μ (Figure 5.7). That is, a confidence interval provides "...one interval generated by a procedure that will give the correct answer 95% of the time" (Antelman 1997).

On average, 95% of the confidence intervals generated by the algorithm in Figure 5.7 will contain the true mean. *Exactly* 95% of the intervals would contain μ if an infinite number of confidence intervals were calculated.

The method of calculation for the confidence interval of μ will vary depending on whether or not the variance of the parent population is known.

```
anm.ci.tck()
```

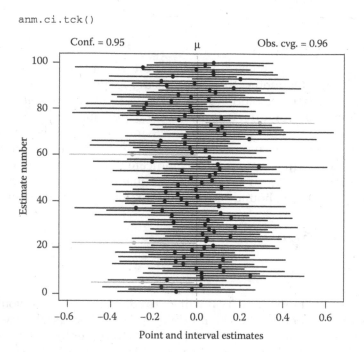

FIGURE 5.7
Animated depiction of confidence interval calculation.

5.3.2 Confidence Interval for μ,σ Known

We first consider a confidence interval for μ based on random sampling from a normal distribution whose variance, σ^2, is known. In this case, a $100(1 - \alpha)\%$ confidence interval can be found using

$$\bar{X} \pm z_{1-(\alpha/2)} \times \frac{\sigma}{\sqrt{n}}, \tag{5.13}$$

where $z_{1-(\alpha/2)}$ is the z-quantile function at probability $1 - (\alpha/2)$. We identify the z-quantile at probability $1 - (\alpha/2)$ to define the bounds of the central $1 - \alpha$ proportion of the standardized sampling distribution of $\bar{X}$. For example, let $\alpha = 0.05$, resulting in 95% confidence. We have $1 - (\alpha/2) = 1 - 0.025 = 0.975$. Thus, 2.5% of the Z-distribution will be above the $z_{1-(\alpha/2)}$ quantile, and 2.5% of the distribution will be below its additive inverse, $z_{\alpha/2}$. Between these values will lie, after the linear transformation using Equation 5.13, 95% of the estimated sampling distribution of $\bar{X}$.

The derivation of Equation 5.13 clarifies the purpose and mechanism of confidence intervals.

To calculate a confidence interval, we require bounds that contain the parameter with probability $1 - \alpha$. That is, we wish to estimate C_L and C_U in

$$P[C_L \leq \mu \leq C_U] = \gamma = 1 - \alpha.$$

We will do this by defining a pivotal quantity for μ. A *pivotal quantity* is a function that may include unknown parameters of interest (like μ), but does not *depend* on them for the expression of its distribution. Pivotal quantities allow the construction of *exact confidence intervals*, meaning that they have the exact stated (i.e., *nominal*) confidence, γ, for any sample size. This is in contrast to "asymptotic" or "large-sample" confidence intervals that at best approximate γ when n is large. The pivotal quantity for a confidence interval will include both the parameter of interest and its estimator.

Assume that the parent distribution is normal. This requires that the distribution of $\bar{X} - \mu$ is normal (Equation 4.47). Further, if σ is known, then $Z = (\bar{X} - \mu)/(\sigma/\sqrt{n})$. Given this conceptualization, C_L and C_U will be the outcomes from the standard normal quantile function at probabilities $\alpha/2$ and $1 - (\alpha/2)$. That is

$$P[z_{\alpha/2} < Z < z_{1-(\alpha/2)}] = 1 - \alpha.$$

Z is a pivotal quantity for μ. This is because Z includes μ, but does not *depend* on it, since μ is fixed. Expanding Z, we have

$$P\left[z_{\alpha/2} < \frac{\bar{X} - \mu}{\sigma/\sqrt{n}} < z_{1-(\alpha/2)} \right] = 1 - \alpha.$$

Rearrangement of the inequality gives

$$P\left[z_{\alpha/2}\sigma/\sqrt{n} < \bar{X} - \mu < z_{1-(\alpha/2)}\sigma/\sqrt{n} \right] = 1 - \alpha,$$

$$P\left[\bar{X} - z_{1-(\alpha/2)}\sigma/\sqrt{n} < \mu < \bar{X} - z_{\alpha/2}\sigma/\sqrt{n} \right] = 1 - \alpha.$$

Because of the symmetry of normal distributions, $-z_{\alpha/2} = z_{1-\alpha/2}$, and we have

$$P\left[\bar{X} - z_{1-(\alpha/2)}\sigma/\sqrt{n} < \mu < \bar{X} + z_{1-(\alpha/2)}\sigma/\sqrt{n} \right] = 1 - \alpha.$$

which gives the confidence interval in Equation 5.13:

$$\left(\bar{X} - z_{1-(\alpha/2)}\sigma/\sqrt{n}, \quad \bar{X} + z_{1-(\alpha/2)}\sigma/\sqrt{n} \right).$$

EXAMPLE 5.7

An alpine vegetation study using 16 samples at alpine late snowbank sites found that the mean cover of the grass *Agrostis variabilis* was 14.6%. Assume that we know $\sigma = 4$, and calculate a 95% confidence interval for μ.

Famously, the z-quantile for the probability of $1 - (\alpha/2) = 0.975$ equals approximately 1.96.

```
qnorm(0.975)
[1] 1.959964
```

Thus, we have

$$14.6 \pm 1.96\frac{4}{\sqrt{16}} = 14.6 - 1.96 = 12.64, \quad \text{and} \quad 14.6 + 1.96 = 16.56.$$

```
shade.norm(from = 12.64, to = 16.56, mu = 14.6, tail = "middle")
```

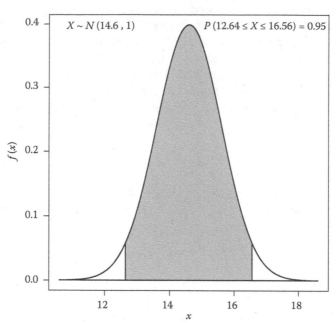

FIGURE 5.8
Sampling distribution of $\bar{X}$ from Example 5.7: $\bar{X} \sim N(14.6, 16/1)$.

The 95% confidence bounds are {12.64, 16.56}. These will contain the middle 95% of the sampling distribution of $\bar{X}$ with mean 14.6 and variance 1 (Figure 5.8).

5.3.2.1 Interpreting Confidence Intervals

There are correct and incorrect ways to interpret confidence intervals. The following are *correct* interpretations for the result in Example 5.7:

- By definition, we are 95% confident that the true mean cover (i.e., μ) of *A. variabilis* lies in the interval (12.64, 16.56).
- Assume that we sampled the *A. variabilis* parent population an infinite number of times, with a sample size 16, and calculated an infinite number of 95% confidence intervals from these samples. Then, 95% of those intervals will contain μ (see Figure 5.7).
- The confidence interval for μ comprises the central 95% of the estimated sampling distribution of its ML estimator, $\bar{X}$, for a sample size of 16 (see Figure 5.8).

The following are common *incorrect* interpretations:

- There is a 95% probability that the confidence interval contains μ. This interpretation is incorrect because, under the frequentist paradigm, μ is a constant. Therefore, once a confidence interval for μ has been calculated, it either contains μ or it does not; that is, $P = 1$ or $P = 0$.

- We are 95% confident that the sample mean cover is in the confidence interval. This is also incorrect. We are completely certain that the sample mean is in the center of the interval because we used it to obtain the confidence interval.

5.3.3 Confidence Interval for μ, σ Unknown

We now consider a confidence interval for μ based on random sampling from a normal distribution whose variance is unknown (the most likely scenario). If we do not know σ, then we must estimate $\sigma_{\bar{X}}$. A logical estimator is

$$\hat{\sigma}_{\bar{X}} = \sqrt{\frac{S^2}{n}} = \frac{S}{\sqrt{n}}. \tag{5.14}$$

The estimator $\hat{\sigma}_{\bar{X}}$ is the *sample standard error of the mean*, although it is often called the standard error of the mean (resulting in confusion of $\hat{\sigma}_{\bar{X}}$ and $\sigma_{\bar{X}}$). The estimator $\hat{\sigma}_{\bar{X}}^2$ is unbiased for $\sigma_{\bar{X}}^2$. However, because the sample standard deviation is biased low (Section 4.3.7), $\hat{\sigma}_{\bar{X}}$ will be biased low for $\sigma_{\bar{X}}$, although this effect will be negligible for sample sizes larger than 20.[*]

In Section 5.2.3, we learned that *t*-distribution does not require knowledge of σ. From Chapter 3 we also know that $t(\nu) \rightarrow N(0, 1)$ as $\nu \rightarrow \infty$. This reflects the fact that S is consistent for σ (Chapter 4). Assuming a normal (but unknown) parent distribution, a 100(1 − α)% confidence interval for μ is given by

$$\bar{X} \pm t_{(1-(\alpha/2), df=n-1)} \frac{S}{\sqrt{n}}, \tag{5.15}$$

where $t_{(1-(\alpha/2), df=n-1)}$ is the *t* quantile function with $n-1$ degrees of freedom for the probability 1 − (α/2), and S is the sample standard deviation. It is left as an exercise to find the pivotal quantity for μ when σ is unknown, and thus derive Equation 5.15.

Sokal and Rohlf (2012) describe other sample standard errors, including standard errors for the sample median, sample variance (whose square is the variance of the sample variance sampling distribution), the sample skewness, and sample kurtosis. However, while derivation of $\hat{\sigma}_{\bar{X}}$ only requires that σ^2 is finite, all other estimators require normal parent distributions.

5.3.3.1 Expressing Precision

The sample standard error of any estimator quantifies the precision of an estimate. Thus, standard errors serve a similar function to confidence intervals. Generally speaking, measures of precision and/or dispersion should always be reported with point estimates (although see below). This is commonly done by reporting the parameter estimate plus or minus the error estimate, for example, a standard error or confidence interval. For instance, if the sample mean is 15 and the standard error is 5, this can be written as 15 ± 5.

---

[*] See Sokal and Rohlf (2012) for standard error corrections for small sample sizes.

The method used to compute the margin of error *must also be reported* (e.g., confidence margin, standard error, standard deviation, IQR, etc.).

Sample standard errors of the mean and confidence intervals for μ are somewhat exchangeable as statements of precision. This is because $\bar{X} \pm 2\hat{\sigma}_{\bar{x}}$ will be approximately equal to a 95% confidence interval for μ for any n when σ is known (because $z_{1-(\sigma/2)} \approx 2$), and will be approximately equal to a 95% confidence interval for μ given large sample sizes when σ is unknown. In a situation where the sampling distribution of a statistic is asymmetric [e.g., the sampling distribution of $(n-1)S^2/\sigma^2$], standard errors tend to be less straightforward to interpret than confidence intervals.

In many cases, it will be more appropriate to report the sample standard deviation instead of the standard error or a confidence interval. For instance, if one wishes to describe the variability in the parent population (and not the precision of an estimate), then the point estimate should be reported with the sample standard deviation (e.g., $\bar{x} \pm s$) or a robust analog (Altman and Bland 2005).

Finally, it should be noted that biologists in many fields use interval summaries for observations that are not independent (Vaux 2012). This is not a valid practice (Hurlbert 1984, Chapter 7). If observations are not the result of random sampling, or another process that ensures independence (cf. Hurlbert 1984), then these summaries will have no inferential meaning. Similarly, if a sample comprises the entire population, a confidence interval will be meaningless because the standard error will be 0 after finite population correction (Chapter 7). This will make the confidence interval bounds equal the point estimate, resulting in a confidence interval that will *always* contain the true value.

EXAMPLE 5.8

Twenty-five cover measurements for the alpine grass *Agrostis variabilis* from alpine late snowmelt sites in Yellowstone National Park are included in the *asbio* dataframe agrostis. Using these data, we will calculate a 95% confidence interval for the true mean cover, μ.

```
data(agrostis)
```

From our sample, we find that $s = 12.25$ and $\bar{x} = 14.6$. The $t(24)$ quantile function for a probability of 0.975 equals 2.064.

```
qt(0.975, 24)
[1] 2.063899
```

We have

$$14.6 \pm 2.064 \frac{12.25}{\sqrt{25}} = 14.6 \pm 5.0568 = (9.543, \ 19.657).$$

We are 95% confident that the true mean cover of *A. variabilis* lies in the interval (9.543, 19.657).

The *asbio* functions ci.mu.z and ci.mu.t can be used to calculate confidence intervals for μ using the Z-distribution and *t*-distribution, respectively.[*] Here are the con-

---

[*] The functions ci.mu.z and ci.mu.t can also be used to calculate confidence intervals from summarized data. This requires that a user specify summarized = TRUE and provide the required summary statistics [i.e. $\bar{X}$, S^2 (for *t*-confidence), and n]. The functions can also account for finite population size corrections (see Chapter 7). This requires that fpc = TRUE, and for n and N to be specified.

fidence interval results using the `agrostis` dataset using the Z-distribution (with σ known).

```
ci.mu.z(agrostis, conf =.95,sigma = 4)
95% z Confidence interval for population mean
Estimate      2.5%     97.5%
   14.60     13.03     16.17
```

Here are our results using the *t*-distribution (σ unknown).

```
ci.mu.t(agrostis, conf =.95)
95% t Confidence interval for population mean
 Estimate      2.5%     97.5%
    14.60      9.54     19.66
```

The latter interval is much wider, and is therefore more conservative with respect to inferences concerning μ.

EXAMPLE 5.9 PIKAS AS ECOSYSTEM ENGINEERS

American pikas (*Ochotona princeps*) are small lagomorphs (an order that includes rabbits) that inhabit alpine talus areas in Western North America. Despite their inhospitable climate, pikas do not hibernate. Instead, to ensure sufficient winter food supplies, pikas harvest plants from surrounding meadows and use these to build food caches (haypiles). Aho (1998) hypothesized that pikas worked as ecosystem engineers (organisms that cause physical, chemical, or structural changes in the environment) by building relatively rich soils (via decomposing haypiles and fecal accumulations) in otherwise barren scree. Soils from 21 paired on-haypile and off-haypile sites were gathered from Rendezvous Peak in Grand Teton National Park in to determine if the habitats differed in total soil nitrogen.

```
data(pika)
```

Confidence intervals and standard errors are conventionally displayed with *error bars* overlaid on bar graphs. In this format, bar heights indicate a location measure (typically a mean), while error bars represent confidence bounds or the location measure ± a precision or dispersion estimate. The function `bplot` from *asbio* is helpful for creating such graphs (Figure 5.9).

The function `bplot` can also handle multiple categories that is, an x vector describing multiple treatments (see Examples 5.1 and 10.1 and Section 10.3.7).

Because zero is not contained in the confidence interval for the true difference of on-haypile minus off-haypile soils (this difference in soils is much greater than zero) there is strong evidence that on-haypile soils have higher %N than off-haypile soils.

5.3.3.2 One-Sided Confidence Intervals

In some situations, we may wish to quantify confidence in only the region above or below a mean estimate. For instance, a biologist working with an endangered species might be interested in saying "I am 95% confident that the true mean number of offspring this year is above (or below) a particular threshold." We calculate $(1 - \alpha)100\%$ lower and upper one-sided confidence bounds for μ using

$$C_L = \bar{X} - t_{(1-\alpha, df=n-1)} S/\sqrt{n}, \tag{5.16}$$

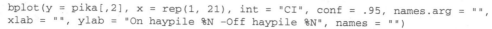

```
bplot(y = pika[,2], x = rep(1, 21), int = "CI", conf = .95, names.arg = "",
xlab = "", ylab = "On haypile %N -Off haypile %N", names = "")
```

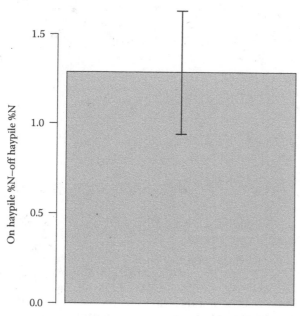

FIGURE 5.9

Mean difference of on and off pika haypile soil for %N. The error bar is a 95% confidence interval for the true mean difference.

$$C_U = \bar{X} + t_{(1-\alpha, df=n-1)} \, S/\sqrt{n}, \tag{5.17}$$

where $t_{(1-\alpha, df=n-1)}$ is the t quantile function with $n-1$ degrees of freedom, at probability $1-\alpha$.

EXAMPLE 5.10

Below we use the function `ci.mu.oneside` to calculate an upper one sided 95% confidence limit for μ using the *A. variabilis* data.

```
ci.mu.oneside(agrostis, tail = "upper")
95% One Sided Confidence Interval for Population Mean
Estimate 5% 95%
14.60    NA 18.79
```

We are 95% confident that the true mean is below 18.79. Note that the one-tailed 95% upper confidence bound is smaller (less conservative) than the conventional two tailed *t*-confidence interval of Example 5.8. Indeed, a one-tailed, 95% upper confidence limit will be identical to the upper confidence bound for a 90% confidence interval.

5.3.4 Confidence Interval for σ^2

In many cutting-edge statistical procedures, the estimation of confidence intervals for σ^2 is vital. From earlier work, we know that, given a normal parent distribution

$$(n-1)S^2/\sigma^2 \sim \chi^2 \, (n-1).$$

We will use this expression as a pivotal quantity for σ^2. We have

$$P[\chi^2_{\alpha/2} < (n-1)S^2/\sigma^2 < \chi^2_{1-(\alpha/2)}] = 1-\alpha,$$

$$P\left[\frac{(n-1)S^2}{\chi^2_{\alpha/2}} < \sigma^2 < \frac{(n-1)S^2}{\chi^2_{1-(\alpha/2)}}\right] = 1-\alpha,$$

where $\chi^2_{1-(\alpha/2)}$ is the χ^2 quantile function with $n-1$ degrees of freedom, for the probability $1-(\alpha/2)$. Thus, the $(1-\alpha)100\%$ confidence bounds for σ^2 are given by

$$C_L = (n-1)S^2/\chi^2_{1-(\alpha/2)}, \tag{5.18}$$

$$C_U - (n-1)S^2/\chi^2_{\alpha/2}. \tag{5.19}$$

Confidence bounds for σ are obtained by taking the square roots of the confidence bounds for σ^2.

EXAMPLE 5.11

Using the function ci.sigma on the *A. variablis* data, we have

```
ci.sigma(agrostis)
95% Confidence Interval for Population Variance
Estimate 2.5%   97.5%
150.17  91.56 290.62
```

Calculation of this interval by hand is left as an exercise.

We note that the confidence interval for σ^2 is highly asymmetric around S^2. This is because the χ^2 distribution is right-skewed. This asymmetry is clearly illustrated with the function anm.ci.tck by specifying confidence intervals for σ^2.

5.3.5 Confidence Interval for the Population Median

The sample mean is not resistant to outliers, and will have low efficiency compared to the median in contaminated distributions (Chapter 4). Of interest then is a confidence interval for the population median.

A large number of methods for calculating confidence intervals for the true median have been proposed (e.g., Bloch and Gastwirth 1968, McKean and Schrader 1984, Price and Bonett 2001). One simple method uses the binomial inverse cdf. With this approach the lower and upper $100(1-\alpha)\%$ confidence bounds for the median are given by

$$L = x_{\text{BIN}(\alpha/2,n,0.5)}, \tag{5.20}$$

$$U = x_{n-\text{BIN}(\alpha/2,n,0.5)+1}, \tag{5.21}$$

where L is the ordered observation number representing the lower confidence bound, and U is the ordered observation number representing the upper confidence bound. $\text{BIN}(\alpha/2,n,0.5)$ is the binomial quantile function for the probability $\alpha/2$ given a sample size, n, and probability of success, 0.5.[*]

---

[*] We note that if $\text{BIN}(\alpha/2,n,0.5) = 0$, then the confidence interval is not calculable.

EXAMPLE 5.12

In this example, we will calculate a 95% confidence interval for the true population median, $X_{0.5}$, from the *Agrostis variabilis* dataset. The sample median of these data is

```
median(agrostis)
[1] 10
```

The quantile associated with L is

```
qbinom(0.025, 25, 0.5)
[1] 8
```

Since $L = x_{BIN(\alpha/2,n,0.5)} = 8$, we use the 8th ordered observation to serve as the lower confidence bound. The 8th ordered observation is 5. This is the lower confidence bound. We find that U is the $25 - 8 + 1 = 18$th ordered observation, which is 20.

```
x[order(agrostis)]
 [1]  0  0  0  3  5  5  5  5  5  7 10 10 10 10 15 20 20 20 20 25 30 30 30 40 40
```

This method is guaranteed not to underestimate the coverage of the exact nominal level. However, the actual coverage of the interval will generally be well above the nominal level because the binomial distribution is discrete (see Agresti 2012). Several methods for addressing this problem have been proposed. These include interpolation procedures and studentizing the sample median (dividing by its standard error), and then considering this value in the context of the standard normal or *t*-distribution (see Sheather and McKean 1987 for more information).

We can compute the exact upper and lower level of *coverage* of the confidence interval using the binomial distribution cdf and inverse cdf, $F(x)$ and $F^{-1}(x)$ below:

$$\text{Coverage} = 1 - 2\{F[F^{-1}(\alpha/2 \,|\, n, 0.5) - 1 \,|\, n, 0.5]\} \tag{5.22}$$

```
1-2* (pbinom(qbinom(.025,25,.5)-1, 25, 0.5))
[1] 0.9567147
```

Since $0.957 > 0.95$ we are *at least* 95% confident that the true median cover of *Agrostis variabilis* is between 5% and 20%.

The function `ci.median` from *asbio* does the work for us.

```
ci.median(agrostis)
95% Confidence Interval for Population Median
Estimate 2.5% 97.5%
10        5    20
ci.median(agrostis)$coverage
[1] 0.9567147
```

Assuming roughly equal sample sizes, a 95% confidence interval for the difference of two population medians can be formulated (McGill et al. 1978, p. 16) as

$$1.57 \times \text{IQR}/\sqrt{n}, \tag{5.23}$$

where IQR is the interquartile range. The function `boxplot` in **R** depicts margins of error using this method. Nonoverlap of the confidence intervals provides evidence that population medians are not equal (see descriptions of nonparametric methods for comparing two populations in Chapter 6). Equation 5.23 is based on properties of the sampling distribution of the median. Specifically, if $X \sim N(\mu, \sigma^2)$ then the sample median follows $N(\mu, 1.57\sigma^2/n)$. This parameterization demonstrates that the median is asymptotically only 64% as efficient as the mean, given normality and large sample sizes. That is, a sampling distribution of means will have a variance that is 64% as large as the sampling distribution of medians because $1/1.57 = 0.64$. As an empirical demonstration, we can create sampling distributions for the mean and median using a normal parent distribution and compare the variance estimates.

```
samp.dist.snap(parent = expression(rnorm(s.size)), stat = mean,
s.size = c(10, 20, 40, 50), R = 10000)
dev.new(xpos = -750, ypos = 0)
samp.dist.snap(parent = expression(rnorm(s.size)), stat = median,
s.size = c(10, 20, 40, 50), xlab = "Median", R = 10000)
```

After squaring the standard errors to obtain the sampling distribution variances, we see that for the largest sample size, the depicted distribution of means is approximately 64% as variable as the distribution of medians.

It is also interesting to see what happens when normality goes away. The function `dirty.dist` creates contaminated sampling distributions. The default settings provide random outcomes for a standard normal parent distribution with 10% contamination from $N(10, 1)$.

```
samp.dist.snap(parent = expression(dirty.dist(s.size)), stat = mean,
s.size = c(10, 20, 40, 50))
dev.new(xpos = -750, ypos = 0)
samp.dist.snap(parent = expression(dirty.dist(s.size)), stat = median,
s.size = c(10, 20, 40, 50), xlab = "Median")
```

In this situation, the sample median is much more efficient than the sample mean.

5.3.6 Confidence Intervals and Sample Size

When considering a confidence interval for θ, the margin of error, m, represents how far the confidence bounds will be displaced above or below $\hat{\theta}$. Given this, we can compute the sample size required to create a confidence interval for μ for a particular margin of error. Assuming the underlying parent distribution is normal, and we know (or have a good estimate for) σ, the margin of error for μ is

$$m = z_{1-(\alpha/2)} \times \frac{\sigma}{\sqrt{n}}. \tag{5.24}$$

From this equation, we can derive a simple formula for calculating appropriate sample size.

$$\sqrt{n} = \frac{z_{1-(\alpha/2)}\sigma}{m}, \quad \text{thus}$$

$$n = \left(\frac{z_{1-(\alpha/2)}\sigma}{m} \right)^2. \tag{5.25}$$

Thus, if we find that $n = 10$ for $\alpha = 0.05$, then we are 95% confident that the margin of error will contain μ given a sample size of 10.

5.3.7 Assumptions and Requirements for Confidence Intervals

A confidence interval, like a point estimate, is an inferential statement concerning a population. Such statements will only be valid, however, if underlying assumptions are met.

1. All confidence interval methods discussed in Section 5.3 assume that data points are independent. For instance, they are the result of random sampling from the population(s) of interest.

2. The confidence intervals for μ and σ^2 assume that the underlying parent distributions are normal.

3. The z-confidence interval for μ assumes that σ^2 is known.

4. Outliers may dramatically affect interval estimates for μ and especially σ^2, resulting in invalid inferences.

From Section 5.2.3, we know that the sampling distribution $(n - 1)S^2/\sigma^2$ will only follow $\chi^2(n - 1)$ if the parent distribution is normal. Relatively minor deviations from normality may have dramatic effects on the validity of intervallic inferences for σ^2. As with confidence intervals for σ^2, confidence intervals for μ will be exact if underlying parent distributions are normal. However, confidence intervals for μ will be less affected by nonnormality, given larger sample sizes, because of the asymptotic normality of $\bar{X}$. Nonetheless, if sample sizes are small, and the parent population is nonnormal, or the data contain outliers, one should consider the use of a robust estimator of location (e.g., the sample median) along with confidence intervals based on that estimator.

★ ## 5.4 Resampling Distributions

There are many situations where the sampling distribution of a statistic will be unknown. This may occur because sample sizes prevent asymptotic convergence to a known distribution, or because the sampling distribution of the statistic is poorly understood or intractable. Biological statistics with unknown sampling distributions include estimators of growth rate and mark/recapture population sizes, and indices for diversity and resource selection. For these sorts of measures, *resampling* (i.e., randomly drawing from an existing sample) is required to characterize the sampling distribution. Resampling allows adjustments for potential bias in the statistic, estimation of the standard error of the statistic, and calculation of confidence intervals for the true value. Two general approaches for this are bootstrapping and jackknifing.

5.4.1 Bootstrapping

The name "bootstrap" describes the way this technique allows extrication from "the mud" (data analysis problems) using one's own "bootstraps" (data). With *bootstrapping*, we resample with replacement from a sample of independent observations of size n. The number of observations in a resample will be equal to n because the resampling distribution can then be used to describe the sampling distribution of a statistic with respect to the original sample size.

Two general approaches for bootstrapping can be used. With *nonparametric bootstrapping*, we make no assumptions regarding the underlying distribution of the data, and as a result all observations have an equal probability of being resampled. Conversely, with *parametric bootstrapping*, probabilistic weights from a pdf are used to ensure that certain observations are more likely to be resampled than others. Regardless of the approach, a statistic of interest is calculated for each bootstrap sample, and this procedure is repeated R times. The ith calculated bootstrap statistic, $i = \{1,2,\ldots,R\}$, is denoted as $\hat{\theta}_i^*$.

The capacity of bootstrapping for producing unique randomized datasets may seem limited; however, the probability of randomly recreating the original dataset when sampling once with replacement is $1/(n^n - n!)$. Note that $10^{10} - 10! = 9{,}996{,}371{,}200$. Thus, when n is greater than 10, it is reasonable that R should be $\gg 1000$.

An exact bootstrap distribution for a dataset can be created although $n^n - n!$ distinct data configurations will be required. Because of this issue, relatively simple algorithms are generally used for resampling with replacement, with the small potential for identical data configurations.

The value $\hat{\theta}_B$ is the sample mean of the R bootstrap statistics. That is

$$\hat{\theta}_B = \frac{1}{R} \sum_{i=1}^{R} \hat{\theta}_i^*. \tag{5.26}$$

The *bootstrap standard error* is simply the standard deviation of the bootstrap statistics (Efron 1982, Robertson 1991):

$$\hat{\sigma}_{\hat{\theta}_B} = \sqrt{\frac{1}{R-1} \sum_{i=1}^{R} \left(\hat{\theta}_i^* - \hat{\theta}_B \right)^2}. \tag{5.27}$$

The bootstrap method can be used to estimate bias for an estimator. This requires obtaining a conventional estimate, $\hat{\theta}$, based on the raw (nonbootstrapped) data. Bias is calculated by subtracting $\hat{\theta}$ from $\hat{\theta}_B$:

$$\widehat{\text{bias}}_{\hat{\theta}_B} = \hat{\theta}_B - \hat{\theta}. \tag{5.28}$$

We see that in the bootstrap worldview $\hat{\theta}$ is a parameter for the data population that is used to generate bootstrap estimates. Thus, if $\hat{\theta}_B$ is less than $\hat{\theta}$, this indicates negative bias; if $\hat{\theta}_B$ is greater than $\hat{\theta}$, this indicates a positive bias. An estimate of θ, corrected for bias, is

$$\hat{\theta}_{\text{corrected}} = \hat{\theta} - \widehat{\text{bias}}_{\hat{\theta}_B} = 2\hat{\theta} - \hat{\theta}_B. \tag{5.29}$$

5.4.1.1 `sample`

Because of its stripped down structure and computational power, **R** is well suited to resampling procedures. The workhorse for these approaches is the function `sample`. It has four arguments:

- The first argument, `x`, is generally a vector containing the data one wishes to resample.
- The second argument, `size`, specifies the sample size (the number of elements that are to be chosen from x). If it is not specified `size` is assumed to be of length n.

- The third argument, `replace`, tells the function whether or not to sample with replacement. By default `replace = FALSE`.
- If specified, the fourth argument, `prob`, will be a vector of probability weights for obtaining elements in x. By default `prob = NULL`.

EXAMPLE 5.13

Consider the dataset

```
x <- c(1, 2, 3, 5, 4, 6)
```

Here, we resample twice *without replacement* from x with a sample size of 3.

```
set.seed(1)
sample(x, 3); sample(x, 3)
[1] 2 6 3
[1] 6 2 5
```

Unless you used the same random seed, your answers are likely to be different than mine since there are $\binom{6}{3} = 20$ possible combinations.

Here are two random reorderings of x, a technique useful in permutation tests (Chapter 6):

```
sample(x); sample(x)
[1] 3 1 2 5 6 4
[1] 3 5 1 6 4 2
```

Here are two conventional nonparametric bootstrap samples:

```
sample(x, replace = TRUE); sample(x, replace = TRUE)
[1] 4 2 1 3 4 3
[1] 1 2 1 4 5 2
```

Note that because we are sampling with replacement, observations can reoccur in a bootstrap sample. For instance, in the first bootstrap sample, each of the outcomes 4 and 3 occur twice, while in the second sample, 1 and 2 occur twice.

Here are parametric bootstrap samples using density weights from a normal distribution with a mean and variance estimated from the data.

```
sample(x, replace = TRUE, prob = dnorm(x, mean(x), sd(x)))
[1] 3 4 5 3 6 4
```

In this case, we would expect most selections to be near the arithmetic mean of the data, 3.5.

5.4.1.2 *Bootstrap Confidence Intervals*

Confidence intervals for parameters can also be calculated using the distribution of bootstrap statistics. A confusing myriad has been proposed (Carpenter and Bithell 2000). Several of these are briefly described below.

5.4.1.2.1 *Normal Approximation*

The *normal approximation method* calculates a $(1 - \alpha)100\%$ confidence interval using

$$\hat{\theta} \pm z_{1-(\alpha/2)} \times \hat{\sigma}_{\hat{\theta}B}, \tag{5.30}$$

where $z_{1-(\alpha/2)}$ is the standard normal quantile function at probability $1 - (\alpha/2)$. The normal approximation approach assumes

1. A normal underlying sampling distribution for $\hat{\theta}$.

2. $\hat{\theta}$ is unbiased for θ.

3. $\hat{\sigma}_{\hat{\theta}_B}$ provides a good estimate for $\sigma_{\hat{\theta}}$.

Unfortunately, these assumptions may often be unrealistic or difficult to test.

5.4.1.2.2 Basic Bootstrap

The *basic bootstrap* method is based on a pivotal quantity (see Section 5.3.2).[*] We seek to estimate C_L and C_U in

$$P[C_L < \hat{\theta} - \theta < C_U] = 1 - \alpha,$$

which we can rearrange as

$$P[\hat{\theta} - C_U < \theta < \hat{\theta} - C_L] = 1 - \alpha.$$

Conventionally, we would use the sampling distribution of the pivot to estimate C_L and C_U; however, in bootstrap applications, this will be unknown. As a result, we let the collection of bootstrap statistical estimates stand in for the true sampling distribution. Specifically, in the notation below, we let $\hat{\theta}_j^*$ be the outcome from the bootstrap inverse ecdf corresponding to the jth probability, assuming R bootstrap iterations. Under the assumption that the sampling distribution and the bootstrap distribution are very similar, we now have

$$P[\hat{\theta}_{\alpha/2}^* - \hat{\theta} < \hat{\theta} - \theta < \hat{\theta}_{1-\alpha/2}^* - \hat{\theta}] \approx 1 - \alpha, \quad \text{and rearranging we have}$$

$$P[2\hat{\theta} - \hat{\theta}_{1-\alpha/2}^* < \theta < 2\hat{\theta} - \hat{\theta}_{\alpha/2}^*] \approx 1 - \alpha,$$

resulting in the confidence bounds $(2\hat{\theta} - \hat{\theta}_{1-(\alpha/2)}^*, 2\hat{\theta} - \hat{\theta}_{\alpha/2}^*)$.

The basic bootstrap method provides simple confidence intervals for the true median whose coverage is generally close to nominal; however, the method may result in substantial coverage error because the distributions of $\hat{\theta}_j^*$s will generally not effectively represent the true sampling distribution of $\hat{\theta}$ (Carpenter and Bithell 2000).

5.4.1.2.3 Studentized Bootstrap

The *studentized bootstrap* or *bootstrap-t* method (Efron 1981) reconciles the differences in the variance of the distribution of bootstrap statistics and the true sampling distribution variance by standardizing $\hat{\theta}_j^*$s using

$$\frac{\left(\hat{\theta}_j^* - \hat{\theta}\right)}{\hat{\sigma}_{Bj}^*}, \tag{5.31}$$

---

[*] Hall (1992) called this approach the "other" percentile method.

where for $\hat{\sigma}^*_{Bj}$ is the standard error for the jth bootstrap iteration. This, of course, requires the estimation of $\hat{\sigma}^*_{Bj}$ for each ordered value of $\hat{\theta}^*_j$. This is accomplished using a two-step nested process. The first step involves a conventional bootstrap of size n with R iterations; while in the second step, each bootstrap sample from the first step is itself bootstrapped using a sample size of n, but with a number of iterations much less than R. Lunneborg (2000) recommends 100 iterations for the second step. This process is repeated for each bootstrap sample. The sample standard deviation is obtained from the statistics calculated for each second-order (second step) bootstrap of each first order sample (first step), resulting in R values of $\hat{\sigma}^*_{Bj}$. The studentized bootstrap method performs well in many situations (Davison and Hinkley 1997); however, it is often a time-consuming and computationally intensive process.

5.4.1.2.4 *Percentile Method*

In the *percentile method* (Efron 1979), the $(1 - \alpha)100\%$ confidence bounds are simply the values delineating the middle $(1 - \alpha)100\%$ of the ordered R bootstrap statistics. The percentile confidence interval is simple and intuitive, and unlike the methods discussed so far, cannot include invalid values (i.e., values outside the support of $\hat{\theta}$). The method can, however, result in substantial coverage errors if the true sampling distribution of $\hat{\theta}$ is not nearly symmetric (Carpenter and Bithell 2000). Furthermore, the method implicitly assumes that a transformation exists that will convert the sampling distribution of $\hat{\theta}$ into a normal distribution (Manly 2007), and for many situations such a transformation will not exist (Carpenter and Bithell 2000).

5.4.1.2.5 *Bias Corrected and Accelerated*

The *bias corrected and accelerated bootstrap* (*BCa*) attempts to shift and scale percentile method intervals to compensate for bias. Details are not given here, but are provided in Dixon (1993), Manly (2007), Lunneborg (2000), and Carpenter and Bithell (2000). The advantages of the BCa method include the fact that the method can account for asymmetric distributions of $\hat{\theta}$, while often obtaining smaller intervals for θ than the percentile method. Disadvantages include the fact that the intervals can be computationally difficult to obtain, and coverage issues may arise as $\alpha \to 0$ (Carpenter and Bithell 2000).

> **EXAMPLE 5.14**
>
> To demonstrate the bootstrap process, we consider the heights of 20 salsify plants (*Tragapogon dubius*) from Example 4.15 contained in the dataframe `trag`. Assume that the statistic of interest is the trimmed mean with $\kappa = 0.2$ (Section 4.3.6.2). We calculate our initial estimate, $\hat{\theta}$, using the original data.
>
> ```
> data(trag)
> mean(trag, trim = 0.2)
> [1] 60.58333
> ```
>
> We use nonparametric resampling to bootstrap the *Tragapogon* data 1000 times.
>
> ```
> resamp <- matrix(nrow = 1000, ncol = 20, data = sample(trag, 1000,
> replace = TRUE), byrow = TRUE)
> ```
>
> To complete the bootstrap process, we calculate 1000 trimmed means from the resampled data. Because bootstrap iterations were placed in the rows of the object `resamp`, we accomplish this by typing
>
> ```
> boot.stats <- apply(resamp, 1, function(x) mean(x, trim = 0.2))
> ```

Point Estimates

The sample mean and standard deviation of the bootstrapped trimmed means are $\hat{\theta}_B$ and $\hat{\sigma}_{\hat{\theta}_B}$ respectively. The estimated bias of the statistic is $\hat{\theta}_B$ minus the original estimate.

```
theta.hat.B <- print(mean(boot.stats))
[1] 60.22139
bias <- print(theta.hat.B - 60.58333)
[1] -0.36194
sd(boot.stats)
[1] 3.465784 # standard error
```

The statistic appears to be negatively biased, although the bias is small in magnitude compared to the standard error. The bias-adjusted estimate is the original estimate minus the bias.

```
60.58333 - bias
[1] 60.94524
```

Confidence Intervals

- The normal approximation 95% confidence interval is

$$60.58 \pm 1.96 \times 3.47 = (53.79, 67.38).$$

Adjusting this interval for negative bias, we have

$$(53.79, 67.38) + 0.36 = (54.15, 67.74).$$

Given the validity of assumptions required for this approach, we are 95% confident that the true trimmed mean is in the interval (54.15, 67.74). This interpretation also holds for the results below.

- For calculating the basic bootstrap 95% confidence interval, we have

$$[1 - (\alpha/2)](R) = 0.025(1000) = 25 \quad \text{and} \quad (\alpha/2)(R) = 0.975(1000) = 975.$$

The 25th and 975th ordered bootstrap trimmed means are

```
quantile(boot.stats, 0.025); quantile(boot.stats, 0.975)
2.5%
53.66667
97.5%
66.83333
```

These are the values for $\hat{\theta}^*_{\alpha/2}$ and $\hat{\theta}^*_{1-(\alpha/2)}$ respectively. The 95% basic bootstrap confidence interval is

$$(2\hat{\theta} - \hat{\theta}^*_{0.975}, \ 2\hat{\theta} - \hat{\theta}^*_{0.025}) = (2(60.58) - 66.83, \ 2(60.58) - 53.67) = (54.33, \ 67.50).$$

- To calculate the percentile method 95% confidence interval bounds, we simply use the values $\hat{\theta}^*_{0.025}$ and $\hat{\theta}^*_{0.975}$ from the basic bootstrap approach. Thus, the percentile 95% confidence interval is (53.67, 66.83).

5.4.1.3 `bootstrap`

A number of **R** packages contain built-in functions for bootstrap analyses. These include the relatively simple function `bootstrap` from *asbio.** The function has five arguments:

- The first argument, `data`, specifies what sample data will be used for bootstrapping. It can be a vector or a multidimensional matrix or dataframe.

- The second argument, `statistic`, will be a statistical function with a single argument, a call to `data`.

- The third argument, `R`, specifies the number of bootstrap replicates. The default is 1000.

- An optional fourth argument, `prob`, can be used to specify a vector of probabilities for parametric bootstrapping.

- The fifth argument, `matrix`, tells the function whether the first argument, `data`, is multidimensional. If `matrix = TRUE`, then rows are treated as multivariate observations. The default is `matrix = FALSE`.

EXAMPLE 5.15

Of great interest to community ecologists are measures of *alpha diversity* that quantify the richness (number of species) and usually the evenness (degree of monodominance) within individual plots in a dataset. High evenness and high richness result in high α-diversity.

A famous measure of alpha diversity, the Shannon Weiner index, is calculated as

$$H' = \sum_i p_i \ln p_i, \tag{5.32}$$

where p_i is the proportional abundance of the ith species.

Consider a single site (site 18) in which the percent ground cover of 44 Scandinavian bryophyte, vascular plant, and lichen species were measured (Väre et al. 1995).

```
data(vs)
site18 <- t(vs[1,])
```

The function `SW.index` from *asbio* can be used to calculate Shannon Weiner diversity. We will call it using the `statistic` argument in `bootstrap`.

```
bs.SW <- print(bootstrap(site18, SW.index, R = 10000))
Bootstrap summary :

original  theta.hat.B      bias         SE
2.017763     2.040785  0.0230212  0.2389707
```

The `print` function for `bootstrap` gives the original estimate, $\hat{\theta}$, the mean Shannon Weiner diversity from 10,000 bootstrapped diversity estimates, $\hat{\theta}_B$, the bias, and the standard error.

The *asbio* function `ci.boot` obtains confidence intervals for output from `bootstrap`. Up to five different interval estimation methods can be called simultaneously:

---

* Much more flexible (but much more difficult to use) is the function `boot` from library *boot* (Canty and Ripley 2012).

the normal approximation, the basic bootstrap, the percentile method, the BCa method, and the studentized bootstrap method.

```
ci.boot(bs.SW)
95% Bootstrap confidence interval(s)

                2.5%     97.5%
Normal       1.545060 2.490466
Basic        1.557244 2.503286
Percentile   1.532241 2.478283
BCa          1.630324 2.566165
Studentized        NA       NA
Bootstrap SEs req'd for studentized intervals
```

Note that studentized confidence intervals are not calculated by the function as they require user specification of bootstrap variances (see Section 5.4.1.2.3).

5.4.2 Jackknifing

Jackknife resampling predates bootstrapping, and is computationally less intensive. The analogy here is of a pocket-knife with a large number of useful built-in tools. Like bootstrapping, jackknifing can be used to estimate the sampling distribution of a statistic based on resampling. Jackknifing is well suited to many applications in biology, including the derivation of species accumulation curves. It is also preferred over bootstrapping for estimating statistical bias (Efron and Stein 1981). However, unlike bootstrapping, jackknifing will provide consistent estimates of standard errors only when an estimator's mathematical function is smooth and differentiable (Krewski and Rao 1981). Differentiable statistics include totals, means, proportions, ratios, odd ratios, and regression coefficients, but not medians or other quantiles (see Exercise 18 at the end of the chapter).

With the *first-order jackknife*, a statistic $\hat{\theta}$ is calculated using all n independent observations. It is then calculated with the first observation removed ($\hat{\theta}^*_{-1}$), with *only* the second observation removed ($\hat{\theta}^*_{-2}$), and so on. Following these calculations, *pseudovalues* corresponding to observations in the original data are calculated from the partial estimates as

$$\tilde{\theta}_i = n\hat{\theta} - (n-1)\hat{\theta}^*_{-i}, \tag{5.33}$$

where $\hat{\theta}^*_{-i}$ is the statistic calculated with the ith observation omitted. We can see that the ith pseudovalue, $\tilde{\theta}_i$, is a weighted version of the original statistic. The *jackknife estimate*, $\hat{\theta}_J$, is simply the mean of the pseudovalues. That is

$$\hat{\theta}_J = \frac{1}{n}\sum_{i=1}^{n}\tilde{\theta}_i. \tag{5.34}$$

Unlike bootrapping, there is only one possible jackknife estimate for θ, given a particular order of jackknifing. The common method for estimating the standard error of $\hat{\theta}_J$ is (Miller 1974):

$$\hat{\sigma}_{\hat{\theta}_J} = \sqrt{\frac{n-1}{n}\sum_{i=1}^{n}\left(\tilde{\theta}_i - \hat{\theta}_J\right)^2}. \tag{5.35}$$

The inflation of the squared deviations ($(n - 1)/n$ compared $1/(R - 1)$ for the bootstrap) is necessary because of the inherent similarity of the jackknife sample to the original data, which will cause $\sigma_{\hat{\theta}_J}$ to be underestimated.

The jackknife estimator of bias is

$$\widehat{\text{bias}}_{\hat{\theta}_J} = (n - 1)(\hat{\theta}_J - \hat{\theta}).$$ (5.36)

Again, this measure is inflated by $n - 1$ (compared to the bootstrap estimator of bias). As with bootstrapping, a corrected estimate for θ is made by subtracting the bias from the original estimate.

It is important to note that while pseudovalues are assumed to be independent in the formulae described above, in reality, they will be correlated. Because of this, the variance in the jackknife estimator will tend to be biased upward or downward (Manly 2007). It follows that the jackknife procedure performs poorly in estimating confidence intervals (Lohr 1999, Manly 2007). As a result, no methods for jackknife confidence intervals are given here.

Second-order and higher jackknife resampling procedures are also possible. In general, whenever n/p is an integer, the *pth order jackknife* is performed by calculating a statistic using all n samples, with the first p observations removed, with observations $p + 1$ to $2p$ removed, and so on. Pseuodovalues are then calculated using Equation 5.33.

EXAMPLE 5.16

We will demonstrate first-order jackknifing with a subset of five observations from the *Tragapogon* dataset used in Example 5.14 . Once again we will synthesize the sampling distribution of a trimmed mean with $\kappa = 0.2$.

```
set.seed(1)
sub <- sample(trag, 5)
sub
[1] 68 78 31 73 43
```

We first calculate the trimmed mean using all the data. This is $\hat{\theta}$.

```
mean(sub, trim=0.2)
[1] 61.33333
```

We then calculate partial trimmed mean estimates, $\hat{\theta}_i^* s$, by leaving out single observations:

```
theta.hat.i <-c(
mean(sub[2:5], 0.2),
mean(sub[c(1, 3:5)], 0.2),
mean(sub[c(1:2, 4:5)], 0.2),
mean(sub[c(1:3, 5)], 0.2),
mean(sub[1:5], 0.2))
```

Next, we calculate the pseudovalues by weighting $\hat{\theta}_i^* s$ and $\hat{\theta}$, and finding their difference.

```
pseudo <-(61.33333 * 5)-(theta.hat.i * 4)
```

Here are the resulting jackknife estimate, bias, and bias-adjusted estimate.

```
mean(pseudo)
[1] 73.19998
bias <- print(4 * (mean(pseudo)- mean(sub, 0.2)))
[1] 47.4666
mean(sub, 0.2)- bias
[1] 13.86673 # bias adjusted estimate
```

The function se.jack from *asbio* can be used to calculate the jackknife standard error from a set of pseudovalues.

```
se.jack(pseudo)
[1] 35.20253
```

5.4.2.1 pseudo.v

The function pseudo.v in *asbio* calculates jackknife pseudovalues of any order. It requires four arguments:

- The first argument, data, is a call to the data to be used for jackknifing.
- The second argument, statistic, requires a function with one argument: a call to data.
- The third argument, order, is the order of jackknifing to be used.
- The fourth argument, matrix, specifies whether object data is multidimensional. If matrix = TRUE (FALSE is the default), then resampling from within the data matrix is conducted using rows as multivariate observations.

EXAMPLE 5.17

Critical to many biological investigations is the effective estimation of the true number of species in an area. One approach is to perform a first- or second-order jackknife of a site by species presence/absence matrix, and take the total species counts for the collections of sites (often called *gamma diversity*) at each step. The jackknife estimate from the resulting pseudovalues can be used as an estimate of total richness. Many experts consider this approach to be superior to all other methods for estimating richness, including Preston's lognormal distribution demonstrated in Example 3.16 (Palmer 1990, 1991, Rosenberg et al. 1995, Melo and Froehlich 2001).

We will use first-order jackknifing to estimate the richness of the tropical tree survey from Barro Colorado Island in Panama discussed in Example 3.16.

Here is an algorithm for finding the observed number of species for a set of sites:

```
#- - - - - - - - - - - - - - - - - - - - - - - - - - - - - - - - - - - #
# data = A site by species matrix with cells containing species abundance
# (e.g., counts), or presence absence (1, 0) information.
#
gamma.div <- function(data){
s <- apply(t(data) >0, 1, sum) # Gives a matrix of presences/absences
length(s[s > 0])                # Number of species present
}
#- - - - - - - - - - - - - - - - - - - - - - - - - - - - - - - - - - - #
```

We have

```
data(BCI.count)
pseudo <- pseudo.v(BCI.count, matrix = TRUE, statistic = gamma.div)
```

Note that we specify in the function that BCI.count is a (site by species) matrix in the argument matrix = TRUE. Output from pseudo.v is a two-column matrix with partial estimates for richness, $\hat{\theta}$*s in the first column and richness pseudovalues in the second column. The jackknife estimate of richness is

```
mean(pseudo[,2])
[1] 245.58
```

This number is quite a bit lower than the richness estimate of 254 species given in Example 3.16.

A unique standard error estimator is commonly used in combination with first-order jackknife estimates of species richness (Heltshe and Forrester 1983, Smith and van Belle 1984). This is

$$\hat{\sigma}_{\hat{\theta}/1} = \sqrt{\frac{1}{n(n-1)} \sum_{i=1}^{n} (\tilde{\theta}_i - \hat{\theta}_J)^2}. \tag{5.37}$$

The function se.jack1 performs the calculation

```
se.jack1(pseudo[,2])
[1] 4.867895
```

★ 5.5 Bayesian Applications: Simulation Distributions

Crucial to Bayesian statistical methods is the ability to adequately describe the variability in the posterior distribution. To accomplish this, central intervals of the posterior probability distribution are generally derived. These are called *credible intervals* (Gelman et al. 2003). While a credible interval serves a similar function to a confidence interval, it is important to realize that they are not the same thing. Confidence intervals quantify how close we would expect a point estimate of a parameter to be to the true value, θ. However, Bayesians assume that θ does not have a fixed true value. They hold that θ is a random variable with its own density distribution. Thus, a 95% credible interval generally contains the central 95% of the posterior distribution of $\theta | data$.

Credible intervals will be equivalent to confidence intervals for location parameters given a uniform prior, and will be equivalent to confidence intervals for scale parameters given Jefferys' prior (Section 4.6, Jaynes 1976, p. 175). In all other cases, however, these intervals will be nonequivalent.

Before reading this section, it will be helpful to revisit Bayesian approaches described in Section 2.6 and Section 4.6, and to become familiar with the procedures for linear algebra described in the Appendix.

EXAMPLE 5.18

If the posterior distribution is recognizable because its priors and likelihood function are conjugate (Chapter 4), then credible intervals will be easy to derive. For instance, in Example 4.19, we modeled the probability of male births in the United States using a Bayesian approach with a binomial likelihood function, and uniform priors bounded at [0, 1]. The resulting unnormalized posterior was beta distributed, $\theta | data \sim \text{BETA}(x + 1, n - x + 1)$, where x is the number of male births and n is the sample size. Four different sample sizes (5, 50, 100, and 1000) were used and 3, 26, 52, and 512 male births were observed, respectively (Table 4.3).

Below are the central 95% credible intervals for θ (the probability for a male birth) calculated directly from the posterior beta distribution (Table 3.5).

```
x <- c(3, 26, 52, 512)
n <- c(5, 50, 100, 1000)
lower <- qbeta(0.025, x+1, n-x+1)
E.theta <-(x+1)/((x+1)+(n-x+1))
upper <- qbeta(0.975, x+1, n-x+1)
cred.int <- t(cbind(lower, E.theta, upper))

colnames(cred.int)<- c("x=3, n=5", "x=26, n=50", "x=52, n=100",
"x=512, n=1000")

cred.int
          x = 3, n = 5 x = 26, n = 50 x = 52, n = 100 x = 512, n = 1000
lower    0.2227781      0.3845873       0.4229612        0.4810254
E.theta  0.5714286      0.5192308       0.5196078        0.5119760
upper    0.8818828      0.6524720       0.6155258        0.5428814
```

We note that the intervals become narrower as sample sizes increase. The interpretation for the largest sample size ($n = 1000$) is that 95% of possible values for the probability of male birth, given prior and current data, lie between 0.481 and 0.543.

5.5.1 Direct Simulation of the Posterior

When the priors and likelihood are nonconjugate, then there may be no analytical solution for the posterior distribution. In these cases, computer simulations will be necessary for posterior summaries. There are two general approaches: direct simulation and indirect simulation (Gelman and Rubin 1996). With *direct simulation*, random draws are obtained from the posterior distribution directly in one step using independent and identically distributed samples. With *indirect simulation*, the posterior is described through an iterative random walk process.

EXAMPLE 5.19

To demonstrate direct simulation, we will rework Exercise 4.19 using a piecewise triangular prior density distribution centered at 0.5, and equaling zero for values of θ below 0.4 or above 0.6 (Figure 5.10).

$$f(\theta) = \begin{cases} 0 & \text{if } \theta < 0.4 \\ 5\theta - 2 & \text{if } 0.4 \leq \theta < 0.5 \\ -5\theta + 3 & \text{if } 0.5 < \theta \leq 0.6 \\ 0 & \text{if } \theta > 0.6 \end{cases}.$$

This distribution is not conjugate to the binomial likelihood function. Nonetheless, as before, we multiply the likelihood function by the prior densities to get the unnormalized posterior distribution.

```
theta <- seq(0.4, 0.6, by =.001)# support
prior <- tri.pr(theta)# see Ch. 5 code for function tri.pr
lk <- theta^512 * (1-theta)^(1000-512)
post <- prior * lk
```

```
Fig.5.10() ## see Ch. 5 code
```

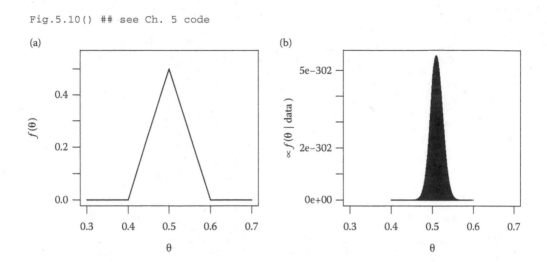

FIGURE 5.10
Reanalysis of the probability of male birth example in Example 4.19. (a) Piecewise triangular prior distribution. (b) The resulting nonnormalized posterior density distribution.

The posterior distribution cannot be summarized analytically. As a result, we describe it using strictly computational procedures. For instance, the maximum density (mode) of the posterior occurs at

```
theta[post == max(post)]
[1] 0.509
```

Credible intervals can be derived by randomly sampling from the vector theta, with replacement, using probability weights from the posterior distribution. Here, we take 1000 random draws.

```
post.samp <- sample(theta, 1000, replace = TRUE, prob = post)
lower <- quantile(post.samp, 0.025)
lower
    2.5%
0.480975
upper <- quantile(post.samp, 0.975)
upper
    97.5%
0.539025
```

The 95% credible interval for the posterior is (0.481, 0.539). Despite having different priors than Example 5.18, the posterior distribution here is highly similar for the largest sample size, $n = 1000$. This is because the large sample size causes the likelihood function to overwhelm the prior distribution.

5.5.2 Indirect Simulation: Markov Chain Monte Carlo Approaches

Describing posterior distributions using direct simulation is possible for simple density surfaces like the previous example. Often, however, the response surface will be topologically complex (e.g., multimodal and asymmetric), multidimensional (i.e., multiple variables are modeled simultaneously), and/or hierarchical (i.e., the distributions of parameters are conditional on other parameters, each with their own distributions). In these situations, direct simulation may be impossible, and/or may poorly represent the posterior distribution.

Markov chain Monte Carlo (MCMC) algorithms model probability distributions through the use of a random walk process called a *Markov chain.*[*] A Markov chain consists of *n steps*, indexing a sequence of random variables, $\{X_0 \ldots ,X_n\}$. The range of possible values for the variables is called the *state space*, and is made up of distinct outcomes called *states*. A random variable will move between states based on *transition probabilities*, p_{ij}, that define the probability that a process at state s_i will move to state s_j in a single step. In a Markov chain, the transition probabilities for the next variable in the sequence, X_t+1, will be defined solely by the outcome from the current random variable, X_t. That is

$$P(X_{t+1} = s_j \mid X_0 = s_k, \ldots, X_t = s_i) = P(X_{t+1} = s_j \mid X_t = s_i). \tag{5.38}$$

Thus, to predict X_{t+1}, one only needs to consider the outcome from X_t. Although a Markov process is nondeterministic, it will tend to wander into higher density states, allowing the posterior distribution to be described.

The starting state(s) in a Markov chain are defined by the user. In theory, these may consist of any outcome in state space. In practice, however, the specification of unusual/outlying states may slow or even prevent convergence to a target *stationary distribution* in which the estimated probability/densities for states are independent of the starting conditions. The number of steps required for a Markov chain to achieve stationarity is called the *burn-in period*. The burn-in period is generally not considered in summaries of the posterior.

A Markov chain is said to be *irreducible* if all states can be reached from any other state (although it may take more than one step). *Accessible* states can be reached in a single step because they have nonzero transitional probabilities states in the current step. In contrast, *nonaccessible* states have transitional probabilities of zero in the current step.

Markov transition probabilities can be summarized with a *transition matrix*:

$$\mathbf{T} = \begin{bmatrix} p_{11} & p_{12} & \cdots & p_{1n} \\ p_{21} & p_{22} & \cdots & p_{2n} \\ \vdots & \vdots & \ddots & \vdots \\ p_{41} & p_{42} & \cdots & p_{nn} \end{bmatrix}, \tag{5.39}$$

where the first subscript element indicate the starting point for a step, and the second subscript represents the end point. For instance, p_{11} is at the intersection of row one and column one, and represents the probability of staying at state 1. The probability p_{12}, is at the intersection of row one and column two and represents the probability of moving from state 1 to state 2, and so on. We see that a single Markov step must take place within a single row of $\mathbf{T}$, and as a result the transitional probabilities for each row must sum to one.

5.5.2.1 Simple Applications

A *countably finite* set has a finite number of elements. For example, {1, 2, 3, 4.5, 7} is a countably finite set. A *countably infinite* set will have the same number of elements as some subset

---

[*] Named after the Russian mathematician Andrey Markov.

of natural numbers. Examples include integers and natural numbers. Examples of *uncountably infinite* sets include real and irrational numbers.

The expected probability for a countably finite Markov state space, for a given number of steps, t, and a particular set of starting probabilities, $\pi(0)$, can be obtained from

$$P(X_t \mid X_0) = \pi(0)\mathbf{T}^t, \tag{5.40}$$

where $\pi(0)$ is a row vector containing probabilities for the state space at step 0, and $\mathbf{T}$ is the transition matrix. All elements of $\pi(0)$ will generally be zeroes except for a single element of one, corresponding to the starting state of the chain. That is, at X_0, the probability the chain will be at the starting state will be 1, and the probability it will be at any other state will be zero. Matrix algebra, required for computation of Equation 5.40, is described in the Appendix.

EXAMPLE 5.20

Consider a Markov random process with four possible states: 1, 2, 3, and 4, and assume that only numerically adjacent states are accessible to each other. For instance, because state 1 is only adjacent to state 2 on the number line, then state 2 is the only accessible state from state 1, and conversely state 1 is the only accessible from state 2. We will use the transition matrix below.

$$\mathbf{T} = \begin{vmatrix} 0.5 & 0.5 & 0 & 0 \\ 0.25 & 0.5 & 0.25 & 0 \\ 0 & 0.25 & 0.5 & 0.25 \\ 0 & 0 & 0.5 & 0.5 \end{vmatrix}.$$

In the matrix $\mathbf{T}$ probabilities of zero indicate nonaccessible states. We note that this Markov chain is irreducible because all states can communicate with each other.

A starting outcome of one would result in $\pi(0) = (1, 0, 0, 0)$. Thus, the expected probabilities for the state space after 10 steps, given a starting value of one, would be

$$P(X_{10} \mid X_0 = 1) = (1, 0, 0, 0)\mathbf{T}^{10}$$

```
T <- matrix(nrow=4, ncol=4, c(0.5, 0.5, 0, 0, 0.25, 0.5, 0.25, 0, 0,
0.25, 0.5, 0.25, 0, 0, 0.5, 0.5), byrow=T)
pi.0 <- c(1, 0, 0, 0)
```

To find the solution to $\mathbf{T}^{10}$, we can use the function `mat.pow` from *asbio*, which calculates the cumulative product of a matrix with itself. The function requires a symmetric matrix, and the power it is to be raised to. Thus, we have

```
Tp10 <- mat.pow(T, 10)
```

In **R**, we use the notation `%*%` to specify the multiplication of two conformable matrices, the multiplication of a matrix with q columns to a column vector with p rows, or, as in this current case, the multiplication a row vector with q columns to a matrix with p rows (see Appendix).

```
pi.0 %*% Tp10
          [,1]      [,2]      [,3]      [,4]
[1,] 0.1854382 0.3521042 0.3145618 0.1478958
```

Thus, the expected probabilities for $P(X_{10} \mid X_0 = 1) = \{0.185, 0.352, 0.315, 0.148\}$.

The Markov process reaches a stationary distribution within 100 steps.

```
c(1, 0, 0, 0) %*% mat.pow(T, 100)# starting state = 1
          [,1]      [,2]      [,3]      [,4]
[1,] 0.1666667 0.3333333 0.3333333 0.1666667

c(0, 1, 0, 0) %*% mat.pow(T, 100)# starting state = 2
          [,1]      [,2]      [,3]      [,4]
[1,] 0.1666667 0.3333333 0.3333333 0.1666667
```

From Equation 4.1, our expected value (state) is

```
c(1, 0, 0, 0) %*% mat.pow(T, 100) %*% c(1, 2, 3, 4)
      [,1]
[1,]   2.5
```

5.5.2.1.1 A Markov Chain Monte Carlo Algorithm

To build a single *random* Markov chain, we require an MCMC algorithm. In the current context of a known transition matrix with countably finite states, such an approach would be unnecessary. This is because the expectations and other parameters of the transition matrix can be obtained analytically, and thus do not need to be estimated through simulation. Nonetheless, for pedagogic purposes, we will create an MCMC algorithm for just such scenarios.

```
#- - - - - - - - - - - - - - - - - - - - - - - - - - - - - - - - - #
# T = transition matrix.
# start = starting value.
# length = length of chain.
#
MC <- function(T, start, length){
states <- ncol(T)
m <- seq(1 : length) # The vector m; will hold the MCMC results
m[1] <- start # The 1st element in m; value specified in "start".
for(i in 2 : length){
m[i] <- sample(1 : states, size = 1, prob = T[m[i-1],])
# randomly acquire a new number, 1 through 4, based on
# the probabilities in the current row of T.
}
m                       # MCMC results
}
#- - - - - - - - - - - - - - - - - - - - - - - - - - - - - - - - - #
```

Here is a random Markov chain with 10 steps. I define a starting value of 1.

```
res <- MC(T, 1, 10)
res
[1] 1 1 2 1 1 2 3 4 3 4
```

The function Rf from *asbio* gives proportions of outcomes. We have the following proportions for states 1 through 4:

```
Rf(res)
[1] 0.4 0.2 0.2 0.2
```

Of course, we get a different answer by rerunning the algorithm.

```
Rf(MC(T, 1, 10))
[1] 0.1 0.2 0.6 0.1
```

The averages of an infinite number of random MCMC walks of length 10 using the same starting state, 1, would give the expected probabilities we calculated earlier, that is $P(X_{10}|X_0 = 1) = \{0.185, 0.352, 0.315, 0.148\}$.

5.5.2.1.1.1 MCMC Convergence   The degree of convergence of a random Markov chain can be estimated using a statistic, $\hat{R}$, which is based on the stability of outcomes between and within m chains of the same length, n (see Gelman and Rubin 1992 and details in R.hat for algorithm description). $\hat{R}$ values close to one indicate convergence to the underlying probabilistic framework, while outcomes greater than 1.1 indicate nonconvergence (Gelman et al. 2003).

The function R.hat in *asbio* provides calculation of $\hat{R}$. To allow burn-in time the function only uses the second half of chains to calculate the statistic.

```
a   <- matrix(nrow=200, ncol = 10)
for(i in 1:10) a[,i] <- MC(T, 1, 200)
R.hat(a)
[1] 1.015689
```

Convergence of random chains to the true probability structure is reached after approximately 200 steps. We combine all mn (10×200) observed states from the second half of each chain to get an estimate of the true probability distribution.

```
b <- a[100 : 200,]
Rf(stack(data.frame(b))[,1])
[1] 0.1841584 0.3277228 0.3198020 0.1683168
```

The MCMC estimate of the expected value (state) is

```
mean(b)
[1] 2.472277
```

5.5.2.2 Advanced Applications

For complex multivariate processes, infinite sets, and/or hierarchical structures, more sophisticated MCMC approaches will be necessary. A number of techniques have been devised for examining Bayesian posterior distributions resulting from such scenarios. These include Gibbs sampling, and the Metropolis and Metropolis–Hastings algorithms.

A *Gibbs sampler** draws random observations from conditional univariate distributions derived from a joint multivariate posterior distribution. New outcomes are randomly sampled, conditional on the outcomes from the other variables at the previous Markov step.

---

* Named after Josiah Willard Gibbs, an American physicist, chemist, and mathematician. The algorithm was first described by the brothers Stuart and Donald Geman in the context of image restoration (Geman and Geman 1984).

Gibbs sampling has three steps:

1. Choose a starting point or points.
2. Derive conditional distributions from the proposal distribution. A multivariate proposal distribution $f_1(\theta_1, \theta_2)$ can be based on crude estimates (summary statistics), or more sophisticated approximation methods (see Gelman et al. 2003, Chapter 12).

   For instance, given the bivariate normal proposal distribution

$$\theta_1, \theta_2 \sim MVN\left(\begin{bmatrix} \mu_1 \\ \mu_2 \end{bmatrix}, \begin{bmatrix} \sigma_1^2 & \tau \\ \tau & \sigma_2^2 \end{bmatrix}\right),$$

   and the starting points $\theta_1 = 3$, $\theta_2 = 4$, we have the following univariate conditional distributions for the next Markov step:

$$\theta_1 \mid \theta_2 \sim N(\mu_1 + \tau \times 4, \sigma^2 - \tau^2),$$

$$\theta_2 \mid \theta_1 \sim N(\mu_2 + \tau \times 3, \sigma^2 - \tau^2).$$

   See Gelman et al. (2003) for a description of the conditional distributions resulting from a multivariate normal distribution. See Chapter 8 for more information on the bivariate normal distribution.

3. Randomly draw from the conditional distributions, and use these points to repeat step 2.

The *Metropolis* and *Metropolis–Hastings* algorithms[†] each use an acceptance/rejection region, based on the ratios of densities, to converge to the target posterior distribution. For both algorithms, Markov chains are created by generating *proposal values* (θ^*s) from a *jumping distribution* based on the previous state in the chain. At time t, the proposal value is either accepted, allowing θ^* to be used as θ^t, or rejected, in which case θ^t will be given the previous value, θ^{t-1}.

The Metropolis algorithm has four steps:

1. Choose a starting point or points and a univariate or multivariate starting distribution, $f_1(\theta)$, as described for the Gibbs sampler above.
2. Create a proposal value θ^* from a jumping distribution, $j_t(\theta)$. The jumping distribution will often be centered at the previous parameter value, θ^{t-1}. For the Metropolis algorithm, the jumping distribution is required to be symmetric. Normal and uniform distributions satisfy this condition, and are frequently used as jumping distributions. For instance, a common jumping distribution is $N(\theta^{t-1}, \sigma_1^2)$, where σ_1^2 is the variance from the starting distribution.

[†] Named after Nicholas Constantine Metropolis, a Greek-American physicist, and a member of the original staff of 50 scientists at the Los Alamos atomic bomb project, and W. K. Hastings, a Canadian mathematician and computer scientist.

3. Calculate the ratio of the densities for the previous and proposed values θ^{t-1}, and θ^*, respectively, with respect to the starting distribution:

$$r = \frac{f_1(\theta^*)}{f_1(\theta^{t-1})}. \tag{5.41}$$

4. Set

$$\theta^t = \begin{cases} \theta^* & \text{with probability } \min(r, 1) \\ \theta^{t-1} & \text{otherwise} \end{cases}. \tag{5.42}$$

The Metropolis–Hastings algorithm is identical to the Metropolis algorithm except that the ratio in Equation 5.41 is replaced by

$$r = \frac{f_1(\theta^*)/j_t(\theta^*)}{f_1(\theta^{t-1})/j_t(\theta^{t-1})}. \tag{5.43}$$

This change allows the use of nonsymmetric jumping distributions.

The Gibbs, Metropolis, and Metropolis–Hastings algorithms can be interactively demonstrated with the function `anm.mc.bvn`. The function examines MCMC walks in a bivariate normal posterior space. For this distribution, density is conical across the intersections of variable values. Bivariate normal jumping kernels are also used for the Metropolis and Metropolis–Hastings algorithms (Figure 5.11).

EXAMPLE 5.21 A NORMAL HIERARCHICAL MODEL FOR CUCKOO EGGS

To demonstrate a real-world MCMC application, we use a relatively simple Bayesian *hierarchical model* (one in which the hyperparameters have modeled distributions). In particular, we will use a *normal hierarchical model* (Gelman et al. 2003) to compare the distributions of the true means from a series of groups. This model assumes

1. The groups have underlying normal distributions with the same variance.
2. The group means are normally distributed with an unknown mean and variance.

As with other Bayesian analyses in this book, this example is intended to illustrate underlying mathematical and computational approaches. It does not include model refinement and model checking, vital to effective and responsible use of these methods. For guidance in these matters, see Gelman et al. (2003).

Many species of cuckoos (family Cuculidae) are obligate brood parasites. That is, they only reproduce by laying their eggs in the nests of other birds. Cuckoo eggs hatch earlier than the host's, cuckoo hatchlings are fed by the hosts, and in most cases cuckoo chicks evict the host eggs or young from the nest. Tippett (1952) listed the lengths of cuckoo eggs found in the nests of other birds. A subset of this data, describing cuckoo eggs in the nests of three species, tree pipits (*Anthus trivialis*), hedge sparrows (*Prunella modularis*), and robins (*Erithacus rubecula*), are in the dataset `cuckoo` in `asbio`.

For the purpose of normal hierarchical modeling, we will assume that the data from each host are independent and from normal populations with means θ_j, $j = \{1, \ldots, J\}$, and a common variance, σ^2. The total sample size combining all groups is denoted by n. Thus, $n = \sum_{j=1}^{J} n_j$. The j host means, θ_js, are normally distributed (prompting visions of

```
anm.mc.bvn.tck()
```

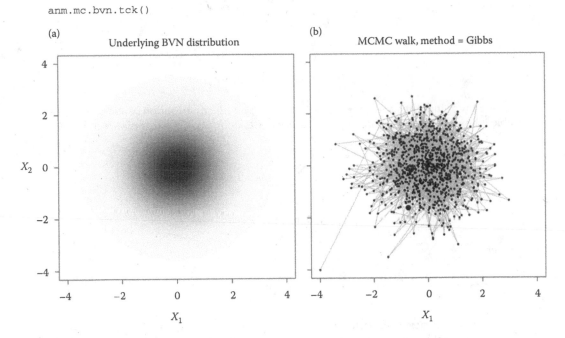

FIGURE 5.11

Random walk through a bivariate standard normal distribution. (a) A defined underlying probability space $X_1 X_2 \sim \mathrm{MVN}\left(\begin{bmatrix} 0 \\ 0 \end{bmatrix}, \begin{bmatrix} 1 & 0 \\ 0 & 1 \end{bmatrix}\right)$. (b) MCMC walk through (a) using Gibbs sampling for a Markov chain of length 1000.

an infinite number of hosts) with an unknown mean, μ, and an unknown variance, τ^2. Uniform prior distributions will be used for μ, log σ, and τ.

The posterior distribution will describe the joint distribution of θ_1, θ_2, θ_3, μ, σ, and τ. This will be difficult to obtain because the joint posterior distribution is built upon a system of complex conditional relationships. The only reasonable approach is MCMC modeling.

We could draw directly from the complex joint posterior likelihood using the Metropolis or Metropolis–Hastings algorithms. It turns out, however, that a series of conjugate conditional distributions can be derived from the posterior, allowing straightforward Gibbs sampling. Specifically

$$\theta_j | \mu, \sigma, \tau, \text{data} \sim N(\hat{\theta}_j, V_{\theta_j}),$$

where

$$\hat{\theta}_j = \frac{1/\tau^2\, \mu + n_j/\sigma^2\, \bar{x}_j}{1/\tau^2 + n_j/\sigma^2} \quad \text{and} \quad V_{\theta_j} = \frac{1}{1/\tau^2 + n_j/\sigma^2}$$

$$\mu | \theta, \sigma, \tau, \text{data} \sim N(\hat{\mu}, \tau^2/J),$$

where

$$\hat{\mu} = \frac{1}{J}\sum_{j=1}^{J}\theta_j$$

$$\sigma^2 | \theta, \mu, \tau, \text{data} \sim \text{Inv } \chi^2(n, \hat{\sigma}^2),$$

where Inv χ^2 indicates a scaled inverse χ^2 distribution (the distribution of a random variable whose reciprocal divided by its degrees of freedom, is a χ^2 distribution, see Section 9.23), and

$$\hat{\sigma}^2 = \frac{1}{n}\sum_{j=1}^{J}\sum_{i=1}^{n_j}(x_{ij} - \hat{\theta}_j)^2.$$

$$\tau^2 | \theta, \mu, \sigma, \text{data} \sim \text{Inv } \chi^2(J-1, \hat{\tau}^2),$$

where

$$\hat{\tau}^2 = \frac{1}{J-1}\sum_{j=1}^{J}(\theta_j - \mu)^2.$$

The function `mcmc.norm.hier` provides a Gibbs sampling comparison of groups given a normal hierarchical model. For each chain, starting points for θ_js are randomly drawn from empirical observations within groups. The mean of these values is then used as the starting point for μ. User-defined starting points for σ^2 and τ^2 are not necessary because their conditional proposal distributions are based on the starting points for θ_j and μ.

Below I specify 10 chains, each 1000 steps long.

```
data(cuckoo)
nhm.results <- mcmc.norm.hier(cuckoo, length = 1000, n.chains = 10)
```

The function `mcmc.summary` provides credible intervals, and calculations of $\hat{R}$ for each parameter for an output array object from `mcmc.norm.hier`.

```
mcmc.summary(nhm.results)
                 2.5%       Median       97.5%     R.hat
theta1 2.262828e+01  23.0322816   23.486492  1.002781
theta2 2.264146e+01  23.0479135   23.521209  1.002834
theta3 2.222260e+01  22.6787944   23.110390  1.002370
mu     2.026260e+01  22.9247506   26.171767  1.071839
s.sq   5.426111e-01   0.8047594    1.286276  1.000449
tau.sq 7.447745e-04   0.2535635  137.307170  1.007803
```

The 95% credible intervals show little evidence that cuckoo egg length is different in the nests of tree pipits, hedge sparrows, and robins.

5.6 Summary

Chapter 5 is concerned with the process of interval estimation for parameters, including resampling and simulation methods.

- Conventional interval estimation for a parameter requires knowledge of the sampling distribution of the estimator for the parameter. A sampling distribution represents a collection of statistics from a conceptual infinite number of samples using the same sample size. Estimates of the sampling distribution variance can be used to construct confidence intervals. Confidence intervals quantify the reliability of a parameter estimate.

- Resampling can be used to synthesize sampling distributions when the underlying distribution of a statistic is unknown. Two methods for resampling were introduced. Bootstrapping consists of sampling with replacement using a sample of size n. A statistic of interest is calculated from the bootstrap data, and this is repeated a large number of times. With jackknifing, a statistic is calculated using all n samples. It is then calculated with the first observation removed, with only the second observation removed, and so on. A jackknife estimate is based on the weighted average of these measures.

- Credible intervals describe central regions of posterior probability distribution, and are the Bayesian equivalent of confidence intervals. Credible intervals are easily computed for well-understood posterior distributions. However, posterior distributions that cannot be derived analytically require simulation procedures for description. Two methods were introduced: direct simulation and indirect simulation. With direct simulation, random draws are obtained from the posterior distribution directly in one step using independent and identically distributed samples. Indirect simulation includes iterative processes that change over time like MCMC sampling. Examples and descriptions at the end of the chapter are given concerning three important MCMC approaches: Gibbs sampling, and the Metropolis and Metropolis–Hastings algorithms.

EXERCISES

1. Bob and Jimmy Joe are middle-distance runners. The 1600 m (metric mile) times for Bob can be represented by a random variable, B, which is normally distributed with a mean, μ_B, of 260 s and a variance, σ_B^2 of 20 s^2; that is, $B \sim N(260, 20)$. The 1600 m times for Jimmy Joe can be represented by a random variable, J, which is normally distributed with a mean, μ_J, of 265 s and a variance, σ_J^2, of 17 s^2; that is, $J \sim N(265, 17)$.

   a. If we record the 1600 m times from 35 of Bob's races, what is the distribution of $\bar{X}_B$?

   b. If we record the 1600 m times from 42 of Jimmy Joe's races, what is the distribution of $\bar{X}_J$?

   c. Given the results from (a) and (b) what is the distribution of $\bar{X}_J - \bar{X}_B$?

   d. What is the probability that Jimmy Joe beats Bob, that is, what is $P(\bar{X}_J - \bar{X}_B) < 0$?

2. Define, in your own words, the central limit theorem and the weak law of large numbers.

3. Play around with the central limit theorem formula in **R** using snapshots from the function `samp.dist.method.tck()`. First run using the default exponential parent distribution. Now rerun using WEI(3, 3) and UNIF(0,1) by specifying: `expression(rweibull(s.size, 3, 3))` and `expression(runif(s.size, 0, 1))` in the *Parent* widget. In which distribution (Weibull, exponential, or uniform)

does the distribution of means appear to be approaching normality more quickly? Does 30 appear to be a good cutoff for establishing a normal distribution for $\bar{X}$? Attach all figures.

4. Compare the sampling distributions of $\bar{X}$ and the sample median using snapshots from `samp.dist.method.tck()`. For both statistics specify a standard normal parental distribution by typing `expression(rnorm(s.size))` in the *Parent* widget. Attach the resulting figures and compare the standard errors of the sampling distributions.

5. Compare the sampling distributions of $\bar{X}$, the sample median, and Huber's estimator of location using `samp.dist.method.tck()` using snapshots. For the parent distribution, specify a standard normal parent distribution with 10% contamination from a second distribution $N(10, 1)$. Do this by typing `expression(dirty.dist(s.size))` in the *Parent* widget. Use sample sizes of 3, 7, 10, 20, and 50 for all three statistics. The Huber estimator will require significantly more time to run since it requires an iterative computational process.

6. This exercise examines the sampling distribution of the F^* test statistic.

   a. Demonstrate the sampling distribution of the F^* statistic using `samp.dist.method.tck()` with the snapshot defaults. Note that the default setting is the standard normal parent distribution. Thus, normality is satisfied and the distributions have equal variances, so H_0 is true. The overlaid lines are F-distributions with the "correct" $n_1 - 1$, and $n_2 - 1$ degrees of freedom. Attach figure to homework.

   b. Reopen the two sample F^* test statistic GUI. Replace the first parent distribution (*Parent1*) with `expression(rexp(s.size, 0.5))`. This results in an exponential distribution with a variance of 4. Replace the second parental distribution (*Parent2*) with `expression(runif(s.size, 0, 6))`. This results in an uniform distribution with a variance of 4. Thus, the null hypothesis of the F-test is true, but the normal parent distribution assumption is violated. Attach the resulting figure, and compare to the earlier results.

7. Given that $X \sim N(\mu, \sigma^2)$ and that $Z \sim N(0, 1)$, define the resulting distributions:

   a. $X - X$

   b. $X + 4X$

   c. Z^2

   d. $\dfrac{\sqrt{n}(\bar{X} - \mu_0)}{\sigma}$

   e. $\dfrac{\sqrt{n}(\bar{X} - \mu_0)}{S}$

   f. $Z/\sqrt{Z^2}$

8. Given that $V_1 \sim \chi^2 (2)$, $V_2 \sim \chi^2 (4)$, and $Z \sim N(0, 1)$, use **R** to find

   a. $P(V_1 V_2 > 4)$

   b. $P(Z/\sqrt{V_1/2} > 2)$

   c. $P\left(\dfrac{V_1/v_1}{V_2/v_2} < 3\right)$

9. A researcher knows that the variance σ^2 in height for a large population of 20-year-old farmed pine trees is 3.3 m². She finds that the sample mean of 20-year-old pine tree heights from a sample of 15 is 7.6 m.

   a. Calculate the 95% confidence interval for μ by hand (without using `ci.mu.z`).

   b. Verify your results using `ci.mu.z`.

   c. Correctly interpret the meaning of the confidence interval.

10. Assume that the researcher in Question 9 does not know σ^2 but finds that that the sample variance, $s^2 = 3.3$.

   a. Calculate the 95% confidence interval for μ by hand (without using `ci.mu.t`).

   b. Verify your results using `ci.mu.t`.

   c. Correctly interpret the meaning of the confidence interval.

   d. Do your results differ from those in Question 9? Why?

   e. Calculate a one-sided upper 95% confidence limit for μ. Does it differ from the upper confidence bound in (a)? Why?

   f. Calculate the 99% confidence interval for σ^2 by hand (without using `ci.sigma`). Interpret the result correctly.

11. For the salsify dataset (dataframe `trag`), calculate 95% confidence intervals for μ and the true median by hand. Verify your results using `ci.mu.t` and `ci.median`. Discuss your results given that sample mean and the sample median are both unbiased for μ.

12. Find the pivotal quantity for μ when σ is unknown, and use this to derive Equation 5.15.

13. Use the function `amn.ci.tck()` to demonstrate the mechanism of confidence interval calculation.

   a. Describe what is happening.

   b. In general, does the observed coverage approach the theoretical coverage as iterations increase? Why?

14. Redo Example 5.11 by hand (without using `ci.sigma`).

15. A biologist wishes to estimate the effect of an antibiotic on the growth of a particular bacterium by examining the mean amount of bacteria present per plate of culture when a fixed amount of an antibiotic is applied. Previous experimentation with the antibiotic on this type of bacteria indicates that the population standard deviation, σ, is 11/cm². Use this information to determine the number of cultures that need to be developed to estimate the mean number of bacteria, with 99% confidence, and a margin of error of ±3/cm².

★ 16. Create bootstrap and jackknife analyses for the Simpson diversity index. The function below will be useful:

```
simpson.p <- function(x){
p.i <- x/sum(x)
1-sum(p.i^2)
}
```

where x is a vector of species abundances. Use the abundances `x <- c(20, 4, 3, 3, 2, 1, 2, 1, 1, 1, 1, 2)` for the following questions:

a. Find the original bootstrap observation, bias, and standard error using `bootstrap`.

b. Calculate a 95% normal, basic, and percentile bootstrap confidence interval for the true population diversity, θ, by hand (without using `ci.boot`).

c. Check your results using `ci.boot`.

d. Calculate the jackknife estimate, bias, and conventional standard error by hand (without `pseudo.v` or `se.jack`).

e. Verify your results using the function `pseudo.v`.

★   17. Use `pseudo.v` to get a first-order jackknife estimate of species richness for the area sampled in the classic dataset dune from package *vegan*.

★   18. Using `pseudo.v`, perform a jackknife analysis for the true median using a random sample of 100 from a standard normal distribution. Discuss potential problems with the pseudovalue output.

★   19. Describe, in your own words, a Markov chain process.

★   20. The transition probabilities below represent a state space with three outcomes {1, 2, 3}.

$$T = \begin{vmatrix} 0.75 & 0.25 & 0 \\ 0.1 & 0.8 & 0.1 \\ 0 & 0.5 & 0.5 \end{vmatrix}$$

a. Calculate the expected probabilities given a chain with 10 steps and a starting value of 3.

b. Repeat (a) using a starting value of 2. Describe differences in results.

c. What chain length is required to achieve stationarity?

d. Randomly create a Markov chain of length 10 using the function MC in the chapter. Are the proportions of states equivalent to the expected probabilities? Why or why not?

★   21. How do starting values and jump distance effect convergence of chains in `anm.mc.bvn`?

★   22. Describe a series of coin flips as a Markov chain process and create a transition matrix for the chain.

★   23. For an MCMC process, assume a starting distribution $N(0, 1)$, and a starting value of 3.

a. Create a proposal value, using a normal jumping distribution, centered at 3, with a variance of 2. This will require use of the function `rnorm`.

b. Using the Metropolis and Metropolis–Hastings methods, determine if this proposal point will be utilized next in the chain.

6

Hypothesis Testing

...it must be possible for an empirical scientific system to be refuted by experience.

K. Popper (1959)

6.1 Introduction

Hypothesis testing is essential to the scientific method (Chapter 1). In this process, an investigator quantifies the strength of a hypothesis using evidence obtained from data.

Statistical hypothesis testing procedures can be placed into one of two general groups: those that use a null hypothesis (formally introduced in Section 6.2) and those that do not. Null hypothesis procedures can, in turn, be subdivided into three types: (1) parametric, (2) permutational, and (3) rank-based permutational. *Parametric* methods assume particular distributional characteristics for the population(s) underlying the data. Methods 2 and 3 are often referred to as *nonparametric* because they have no *a priori* distributional assumptions, or as robust because they are generally resistant to outliers. Topics associated with these approaches are considered in Sections 6.2 through 6.6. Statistical hypothesis tests/comparisons that do not require a null hypothesis generally involve Bayesian and/or likelihood-based methods. I discuss these approaches in Section 6.7.

6.2 Parametric Frequentist Null Hypothesis Testing

As an introduction to this section, two statements should be made: (1) the most common use of statistics by biologists is for hypothesis testing (cf. Quinn and Keough 2002), and (2) biologists generally use null hypothesis tests based on parametric assumptions and a frequentist interpretation of probability. An adherence to this framework can be traced to three causes: (1) there is a strong need for hypothesis testing in the biological sciences, (2) these methods generally produce plausible results, and (3) this approach is strongly emphasized in biometric and introductory statistics textbooks. However, while frequentist null hypothesis tests have become deeply enmeshed in biological research, justifications for their use are seldom considered by biologists (Gill 1999). This chapter addresses this discrepancy by describing null hypothesis testing (including a large number of simple tests) and then considering both criticisms and alternatives.

6.2.1 Null Hypothesis and Its Motivation

In Chapter 1, we learned that we can never prove a deductive argument is "true" using empirical evidence (Hempel 1966). This is because the proof requires the impossibility that all possible observations concerning the hypothesis are made (cf. Hume 1740). Instead, we can only reject (*modus tollens*) or fail to reject a hypothesis. This condition has given rise to the useful idiom: "nothing in science is proven until it is disproven."

By rejecting a hypothesis, we have taken a decisive action: we have eliminated a particular line of reasoning. However, this does not clarify how one would *support* a hypothesis. One way around this difficulty is the *null hypothesis*, denoted H_0. The null hypothesis is often a statement of no effect or no difference, and is generally constructed to encompass all possible outcomes except an expected effect. As a result, the rejection of H_0 conceptually provides support for a hypothesis expressing the expected effect. The logical form of *reductio ad absurdum* arguments is clearly followed in this dichotomy (Chapter 1).

We concern ourselves with H_0 and not with a research hypothesis directly for two reasons. First, as noted above, we cannot prove that a hypothesis is true; however, it may be possible to prove it is false, and H_0 is often a hypothesis we do not mind rejecting. Second, it is simply easier to consider statistical evidence from the perspective of H_0. This is because the research hypothesis will generally be no more than an inexact supposition that there is "some effect" (Quinn and Keough 2002). The null hypothesis, on the other hand, can often be expressed in exact mathematical terms. For instance, a common null hypothesis is that the effect of a treatment is exactly zero. This characteristic allows an investigator to more clearly distinguish between experimental outcomes leading to falsification or nonfalsification for H_0.

The notion of the null hypothesis can be historically linked to R.A. Fisher and Karl Popper who independently proposed that severe testing and falsification should be the goal of hypothesis testing (Howson 2001). To quote Popper (1959, p. xix): "the point is that, whenever we propose a solution to a problem, we ought to try as hard as we can to overthrow our solution, rather than defend it." The statistical null hypothesis (invented by Fisher) can be regarded as a mathematical response to the philosophical need for falsification in hypothesis testing.[*]

A large number of null hypothesis testing procedures have been developed to address particular applications and assumptions of empirical science. Despite this variety, all such methods take the same approach. They address the question: *how probable are the data if the null hypothesis is true?*

EXAMPLE 6.1

Hansen et al. (2011) vaccinated 24 rhesus monkeys against a powerful form of simian immunodeficiency virus (SIV) (a simian cousin of HIV). The vaccine was unique in that it used a long-lived delivery vehicle, a herpesvirus called cytomegalovirus (CMV), to carry AIDS proteins that conferred immune responses. The researchers believed that the vaccine would provide additional protection from SIV infection. Thus, their null hypothesis was that the vaccine would provide no additional protection. The vaccine was found to protect half of the tested monkeys. This result would be highly unlikely if H_0 were true. Thus, the investigators rejected the null hypothesis of no effect, which provided implicit support for the efficacy of the new vaccine.

---

[*] A justification for null hypothesis testing can be made by linking Popperian severe falsification to the use of probability (e.g., Hillborn and Mangel 1997, pp. 15–16). It should be noted, however, that Popper advised caution in the use of probability since explicit falsification may be "delayed" as sufficient trials are acquired to adequately discern the frequentist framework of a phenomenon (Popper 1959, p. 185; Mayo 1996).

6.2.2 Significance Testing

In *significance testing*, we are limited to one of two possible decisions:

1. Reject H_0 or
2. Fail to reject H_0.

These choices follow directly from the philosophical coupling of deduction and empirical data discussed at length in Chapter 1 .

In a null hypothesis we might predict that a parameter describing the difference between two populations is zero, or that a parameter equals a particular number (often zero). Thus, when H_0 is true, we would expect an estimate of the parameter to take a value near the value of the parameter specified in H_0, making this difference close to zero. An estimator called a *test statistic* is used to quantify the difference between the parameter value specified in H_0 and the estimate of the parameter based on data. Generally speaking, the sampling distribution of the test statistic under H_0 will be known (Chapter 5). As a result, an investigator can calculate probabilities based on test statistic outcomes assuming that H_0 is true. These are called *probability values* or *P-values*. In particular, a *P*-value is the probability of seeing a test statistic as or more extreme than the test statistic observed if H_0 is true. This somewhat complex definition can be summarized succinctly as: $P(\text{data}|H_0)$ if we remember that this probability will comprise the tail(s) of the null distribution defined in the hypothesis statements. Smaller *P*-values designate stronger evidence against H_0. The choice between rejecting and failing to reject H_0 is based on a *decision rule*. The decisive criterion in the decision rule is the *P*-value.

EXAMPLE 6.2

To understand the concept of a *P*-value, consider a researcher who is interested in distinguishing the effect of two soil nutrient treatments, X and Y, on crop yield. The researcher stipulates the following null hypothesis:

H_0: The true mean difference of X and Y is zero.

Data are gathered concerning the groups and a test statistic is calculated estimating the standardized (adjusted for variability) mean difference between X and Y.

To test the validity of H_0, we define a random variable called the *null distribution*. In the null distribution, the true difference of X and Y is assumed to be zero (H_0 is assumed to be true in all null hypothesis tests). Nevertheless, its probabilistic form acknowledges that nonzero outcomes may be observed due to variability in the sampled populations of X and Y. This variability is estimated from the samples and incorporated into the null distribution. The calculated test statistic is assumed to be a random outcome from this distribution.

On one hand, suppose that evidence indicated that the difference in the crop yield of X and Y was small. In this case, the test statistic will have a value similar to the highest likelihood value (zero) of the null distribution. Thus, test statistic outcomes close to zero will result in *P*-values that are large. Recall that the *P*-value is the probability of seeing an outcome in the null distribution equal to or more extreme than the test statistic. The values −0.5 or 0.5 are not unusual if the null distribution is standard normal; large portions of the distribution are more extreme than these values (Figure 6.1). As a result, a *P*-value for these test statistics will (1) be a proportion comprising much of the null distribution, (2) provide essentially no evidence against H_0, and (3) prompt *failure to reject* H_0.

On the other hand, assume that the evidence showed that X was very different from Y. In this case, the test statistic will be an outcome far from the expected value (zero) of the

```
Fig.6.1() ## see Ch. 6 code
```

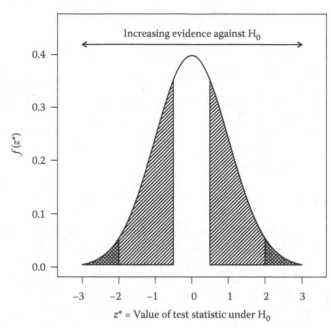

FIGURE 6.1

Illustration of a test statistic whose sampling distribution is standard normal under H_0. More extreme values (those farther from the mean of the sampling distribution) will result in smaller P-values and stronger evidence against the null hypothesis of no difference. Hatched areas under the curve indicate the magnitude of P-values for test statistics of -0.5 or 0.5 and -2 or 2.

null distribution and evidence supporting H_0 will come only from the extremes of the null distribution. For instance, the values -2 or 2 are fairly unusual if the null distribution is standard normal (Figure 6.1). The P-value, calculated by finding the area under pdf equal to or more extreme than -2 or 2 will (1) be small, (2) provide resultant strong evidence against H_0, (3) and prompt *rejection of* H_0.

6.2.3 Models for Null Hypothesis Testing

A number of different methods have been proposed for null hypothesis testing. Fisher (1954, 1956) suggested a five-step process:

1. State H_0.

2. Conduct an investigation that produces data concerning H_0.

3. Choose an appropriate test and test statistic, and calculate the test statistic.

4. Determine the P-value.

5. Reject H_0 if the P-value is small; otherwise, retain H_0.

Fisher (1925) formally introduced the idea of using P-values as the criterion for rejecting H_0. In particular, he proposed that a threshold be established describing outcomes that would be extremely improbable if H_0 were true, allowing rejection of H_0. An outcome that meets this criterion is said to be *statistically significant* (Figure 6.2). The decisive P-value at the cutoff between what is and what is not significant is known as the *significance level* and is denoted by the symbol α (Figure 6.2). Fisher initially proposed that a *P*-value of 0.05 be used for α, although he later recommended that fixed significance levels would be too restrictive and that these probabilities should depend on the characteristics of the scientific investigation. The *critical value* is the value that a test statistic must exceed in order to reject H_0 at a particular significance level (Figure 6.2).

If we let $\alpha = 0.05$ (still the common default, *à la* Fisher 1925), then the test statistic will exceed the critical value no more than 5% of the time if H_0 is true. If we choose $a = 0.01$, we are insisting on even stronger evidence against H_0. In this case the test statistic will exceed the critical value no more than 1% of the time if H_0 is true. Thus, we can also define a *P*-value as the smallest possible significance level at which H_0 can be rejected.

Neyman and Pearson (1928, 1933) proposed a model that differed from Fisher's model in three important respects (cf. Quinn and Keough 2002):

1. A significance level should be established before beginning data collection. This step is particularly important because in the Neyman–Pearson system, a *P*-value designates only whether one should reject H_0 or fail to reject H_0. Specifically, we reject H_0 if the *P*-value is less than or equal to α, and we fail to reject H_0 if the *P*-value is greater than α. Fisher (in his later publications) disagreed with this conceptualization by arguing that *P*-values should be viewed as a continuous measure of the strength of evidence against H_0 (Oakes 1986, Huberty 1993).

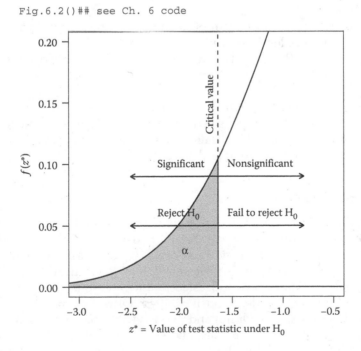

```
Fig.6.2()## see Ch. 6 code
```

FIGURE 6.2
The left tail of a conventional null distribution, $N(0, 1)$. The potential test statistic scores comprise the X-axis. A conventional significance level, $\alpha = 0.05$, is equal to the proportion of area under the curve indicated by the shaded area. The lower-tailed critical value is –1.64. Thus, lower-tailed test statistic values that are more "extreme" (more negative) than –1.64 would result in the rejection of H_0.

2. Neyman–Pearson explicitly incorporated an *alternative hypothesis* (specified as H_A) into their method. The alternative hypothesis generally makes a statement an investigator suspects is true, and ideally encompasses all outcomes except H_0. Neyman–Pearson argued that the rejection of H_0 requires acceptance of H_A. Fisher strongly opposed this idea on the grounds that evidence against H_0 does not necessarily imply evidence for H_A (Quinn and Keough 2002).

3. Neyman–Pearson proposed that there were two types of long-run error associated with significance testing. These are the possibility of rejecting H_0 when H_0 is true and the possibility of failing to reject H_0 when H_0 is false. Neyman–Pearson referred to these as type I and type II errors, respectively, and denoted their probabilities as α and β. The significance level is equivalent to the probability of type I error, α. It should be noted that Fisher also opposed the concept of type I and II errors on the grounds that H_0 is never explicitly true or false (Oakes 1986).

Both the Fisher and the Neyman–Pearson approaches have desirable attributes; today, most biologists use a hybrid schema for null hypothesis testing (Gill 1999). This system can be summarized in five steps (cf. Gill 1999, Quinn and Keough 2002):

1. Specify H_0, H_A, and the significance level, α, to be used.
2. Conduct an investigation that will produce data (evidence) concerning H_0.
3. Choose an appropriate test statistic with a sampling distribution that measures deviations from H_0 and calculate the test statistic.
4. Calculate the *P*-value.
5. The decision rule consists of a pair of statements. If the *P*-value is greater than the specified significance level, conclude that there is insufficient evidence to reject H_0 and retain H_0. If the *P*-value is less than or equal to the specified significance level, reject H_0.

Below, I discuss the interpretation for null hypothesis tests given nonsignificant and significant results. It should be emphasized that both significant and nonsignificant results offer important information if a study is designed and executed properly (Chapter 7). Both provide insight into the phenomenon under study.

6.2.3.1 Nonsignificant Results

There are several important things to remember when interpreting a nonsignificant result. First, failure to reject H_0 does not mean that H_0 is true. Recall that accepting H_0 as true results in the logical fallacy *affirming the consequent* (Chapter 1). On the other hand, if we have insufficient evidence to reject H_0, then we will have very little evidence to support H_A. Because of this, some authors have suggested that it is reasonable to conclude that H_A is false given nonsignificance (Underwood 1990, 1999). Others prefer a less severe approach with respect to H_A. For instance, Quinn and Keough (2002) recommend withholding judgment on H_A, given nonsignificant results, and conclude only a failure to reject H_0. They note that an exception would be in the case where we fail to reject H_0 and the probability of type II error is low (i.e., the probability of detecting an effect specified in H_A is high; see Section 6.3).

Given nonsignificance, we could reformulate or revise our hypotheses, re-evaluate our underlying model(s) and assumptions, regather data, and thus restart the steps of the scientific method.

6.2.3.2 Significant Results

If we have sufficient evidence to reject H_0, this does not mean that H_0 is untrue, but it does mean that the outcome from our experiment would be unlikely if H_0 were true, permitting *probabilistic* falsification of H_0. Divergent interpretations have been ascribed to H_A given a significant result. A strict Fisherian view is that H_A merely provides a context by which evidence against H_0 is judged and that rejection of H_0 does not imply anything about H_A (von Storch and Zwiers 2001, p. 99). Most texts, however, take a less severe (Neyman–Pearson) view of H_A. They hold that if H_0 represents all other outcomes except H_A, then the rejection of H_0 corroborates H_A. From this perspective, statements following a significant result have the form: "we reject H_0 and conclude in favor of H_A." It should be emphasized, however, that just as one cannot identify H_0 as true given an insignificant result, one assuredly cannot define H_A as true (despite this claim in some introductory texts) given significance.

After a significant test, an investigator may wish to test the research hypothesis in the context of other scenarios or datasets to confirm his or her results, or to identify its conceptual limitations.

6.2.4 Upper-, Lower-, and Two-Tailed Tests

It may be possible to anticipate directionality in the measured effect of the phenomenon under study. For instance, we might expect that plants given a nutrient supplement would be larger than control plants given no supplements. Such scenarios can be addressed with upper- and lower-tailed tests.

Upper- and lower-tailed tests are specified in the alternative hypothesis. An upper-tailed test will contain a "greater than" sign in H_A and a lower-tailed test will contain a "less than" sign in H_A. Two-tailed tests are concerned with any effect or difference, and will contain a "not equal" sign in the alternative. For instance, if we anticipate that parameter X will be greater than Y, we denote H_A: $X > Y$; if we believe that X will be less than Y, we would designate H_A: $X < Y$; and if we suspect that X and Y are different, but are not sure how they are different (either $X > Y$ or $X < Y$), then we would specify H_A: $X \neq Y$.

Two schools of thought exist for the specification of H_0 in one-tailed tests. Many experts recommend that H_0 should have the same *exact* form, regardless of the alternative (Fisher 1949, Moore and McCabe 2004, Ott and Longnecker 2004). For example, in an upper-tailed test, in which we expect that parameter X is greater than parameter Y, we would have

$$H_0: X = Y$$

versus

$$H_A: X > Y.$$

The statement: H_0: $X = Y$, would also be used in lower- and two-tailed tests. This approach *defines* the distribution under H_0 and emphasizes that the null distribution does not change with the alternative (Fisher 1949).

Other experts, particularly biometricians, specify the null as the *directional opposite* of the alternative (Quinn and Keough 2002, Zar 1999). Thus, for the example above, we would have

$$H_0: X \leq Y$$

versus

$$H_A: X > Y.$$

Only the upper tail from the null distribution is considered in an upper-tailed test (see below). Thus, the possibility that X is less than Y is ignored. The latter formulation of H_0 acknowledges this condition. Stating the null as the mathematical opposite of the alternative also provides a clear conceptual connection to *reductio ad absurdum*. In this text, when required to state H_0 in a one-tailed test, I will use the latter method.

Evidence in a one-tailed test comes only from the tail of the null distribution specified in the alternative. Specifically, a P-value from an upper-tailed test is the proportion of the null distribution *equal to or above* a test statistic outcome, whereas a lower-tailed P-value is the proportion of the null distribution *equal to or below* the outcome. In the simplest statistical tests (those using a symmetric null distribution), a two-tailed test quantifies the proportion of the null distribution both greater than or equal to the absolute value of the test statistic, for example, $|z^*|$, *and* less than or equal to its additive inverse, $-|z^*|$. Recall that an asterisk is assigned to a test statistic so that it can be distinguished from its null distribution (Chapter 5). With one-tailed tests it is extremely important to specify the treatments in the same order that they were given in the null and alternative hypothesis. Misspecification will result in the calculated P-value being 1 minus the correct P-value.

EXAMPLE 6.3

Let the distribution under H_0 be $N(0, 1)$. Now, suppose that we have specified a lower-tailed alternative hypothesis and calculated a test statistic of -1.2. Evidence concerning H_0 comes from the tail of the distribution given in H_A. Thus, the P-value is simply $P(Z \leq -1.2) = 0.11507$ (Figure 6.3a). If we had specified an upper-tailed alternative, we would use the upper tail of the null distribution to calculate the P-value. In this case, the P-value is $P(Z \geq -1.2) = 0.88493$ (Figure 6.3b). Finally, if we had required a two-tailed alternative, we would quantify evidence concerning the null distribution using both tails. In this case, since the null distribution is symmetric, the P-value is $2[P(Z \leq |-1.2|)] = 0.23104$ (Figure 6.3c). Note that, in contrast to the approach used here, only one type of test, lower-, upper-, or two-tailed, should be defined for an analysis, and this specification should be made *a priori* (prior to the start of the experiment).

One-tailed (lower- or upper-tailed) tests are useful since they result in more evidence against H_0 if the direction of the effect is correctly stipulated in H_A. For instance, compare the areas under the curve for Figure 6.3a and c. The probability $P(Z \leq -1.2)$ is half as large as $2[P(Z \leq |-1.2|)]$ and therefore provides twice as much evidence against H_0. Nonetheless, investigators should use such tests with caution. This is because after the alternative hypothesis is defined, we are obliged to ignore effects in the other direction. For example, we would expect soil nitrogen to increase plant biomass, and because of this expectation, we might define an upper-tailed alternative hypothesis. Unfortunately, if soil N turns out to decrease

```
par(mfrow = c(2, 3), mar = c(0, 0, 0, 0))
shade.norm(x = -1.2, show.dist = F, show.p = F)
shade.norm(x = -1.2, tail = "upper", show.dist = F, show.p = F)
shade.norm(x = -1.2, tail = "two", show.dist = F, show.p = F)
```

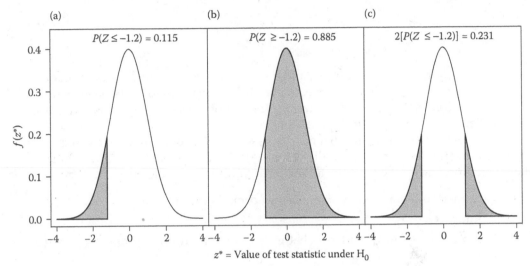

FIGURE 6.3
Illustration of *P*-values from lower-, upper-, and two-tailed tests for test statistic with a standard normal distribution under H_0. (a) A lower-tailed test with an observed test statistic of -1.6. (b) An upper-tailed test with the test statistic observed in (a). (c) A two-tailed test with the same test statistic observed in (a) and (b).

biomass, we cannot use this potentially valuable information to make any formal conclusions. We can only report that the upper-tailed test was insignificant.

6.2.5 Inferences for a Single Population Mean

In null hypothesis tests for a single population mean, we assume that the mean for the population of interest, μ, will equal a null value, μ_0. Thus, a two-tailed null hypothesis would be specified as

$$H_0: \mu = \mu_0,$$

and the two-tailed alternative hypothesis would be

$$H_A: \mu \neq \mu_0.$$

The lower- and upper-tailed alternatives would be specified as

$$H_A: \mu < \mu_0,$$

$$H_A: \mu > \mu_0.$$

Generally, we let $\mu_0 = 0$.

6.2.5.1 One-Sample z-Test

We will begin our demonstrations using a *one-sample* z-test. The test statistic is calculated as

$$z* = \frac{\bar{X} - \mu_0}{\sigma/\sqrt{n}}. \tag{6.1}$$

If test assumptions hold (see below), and if H_0 is true, then $z*$ will be a random outcome from a standard normal distribution (Chapter 3). The test statistic, when calculated, will indicate the number of standard deviations that an observed mean, $\bar{x}$, is away from its hypothesized expectation, μ_0 (Chapters 3 and 4).

All parametric tests (including all those discussed in this section) have particular assumptions. Meeting these assumptions will determine if the test results are trustworthy and will allow valid inferences. We have the following assumptions for the one-sample z-test:

1. The underlying population of interest is normally distributed.
2. We know σ.
3. Observations are independent. That is, they are based on random samples from the underlying population.

Normality can be visually assessed using a number of different methods, including histograms (Chapter 3, ecdfs (Section 3.4.1)) and normal quantile plots (see Examples 6.6 and 6.17). A number of diagnostic hypothesis testing procedures are also possible including the Shapiro–Wilk normality test (Royston 1982; Example 6.6). The assumption of normality will be unimportant for large sampling sizes (n greater than around 30) because $\bar{X}$ (the variable whose outcomes are considered in the test statistic) will generally be approximately normal. This property is true for all tests in Sections 6.2.5 through 6.2.7. The assumption of independence allows straightforward calculation of probability (Chapter 2) and inferential statements with respect to the underlying population (see Chapter 7).

EXAMPLE 6.4

Port et al. (2000) found that for males 45–54 years of age, systolic* blood pressure was normally distributed with $\mu = 131$ mm Hg (millimeters of mercury) and $\sigma = 12$ mm Hg. A medical administrator at a state university examines the records of 85 male faculty members in this age group and finds that $\bar{x} = 128$ mm Hg. The administrator is concerned that the blood pressure of faculty members at his university may differ from that of the overall population of males 45–54 years of age.

Step 1: State H_0, H_A, and α

$$H_0: \mu = 131 \text{ mm Hg,}$$

$$H_A: \mu \neq 131 \text{ mm Hg,}$$

$$\alpha = 0.05.$$

Note that the hypotheses are expressed in terms of the parameter μ. This is because *we are interested in making inferences concerning the true difference* from

---

* The blood pressure when the heart is contracted. The units "millimeters of mercury" means the pressure exerted at the base of a column of mercury fluid, exactly 1 mm high, at 0°C.

a systolic blood pressure of 131. The "not equal" ($\neq$) sign in the alternative hypothesis indicates a two-sided test (i.e., we are interested in any difference from μ_0).

Steps 2 and 3: Decide on an appropriate test statistic and calculate the test statistic.

$$z* = \frac{128 - 131}{12/\sqrt{85}} = -2.305.$$

Step 4: Calculate the *P*-value. For a two-tailed test this requires that we find the proportion of the Z-distribution greater than or equal to 2.305, and add it to the proportion of the Z-distribution that is less than or equal to −2.305. That is, we find $P(-2.305 \geq Z \geq 2.305)$. This probability will represent the proportion of the population of middle-aged males more than 2.305 standard deviations above or below a systolic blood pressure of 131 mm Hg. Since normal distributions are symmetric, we can find the probability associated with either tail and simply multiply it by 2 to get the two-tailed probability.

$$2[P(Z \geq |z*|)] = 2[P(Z \geq 2.305)] = 2[P(Z \leq -2.305)] = 0.02117.$$

Step 5: State a conclusion based on the decision rule. Because $0.02117 \leq 0.05$ (the specified significance level), we reject H_0 and conclude that the blood pressure of middle-aged faculty at the university is different from that of the overall population.

A plot of the example is shown in Figure 6.4.

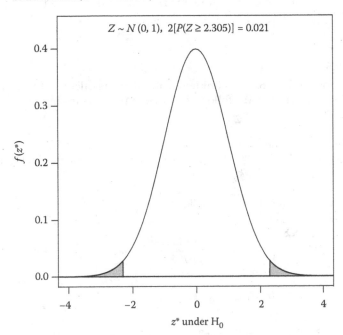

FIGURE 6.4
Standard normal distribution, with the probability $2[P(Z \leq -2.305)]$ overlaid.

We can run the entire test using the function `one.sample.z` from *asbio*.

```
one.sample.z(null.mu = 131, xbar = 128, sigma = 12, n = 85,
alternative = "two.sided")
$test.statistic
[1] -2.304886
$p.val
[1] 0.02117295
```

6.2.5.2 One-Sample t-Test

If σ is unknown (and this will be the general case), then we must estimate σ with the sample standard deviation (Chapter 4). Because of this step, the test statistic will no longer follow a Z-distribution under H_0. Instead, if the null hypothesis is true, and if assumptions for the test are valid (see below), then the test statistic

$$t* = \frac{\bar{X} - \mu_0}{S/\sqrt{n}}, \tag{6.2}$$

will be a random outcome from a t-distribution with $n - 1$ degrees of freedom (see Section 5.2.4).

The null and alternative hypothesis statements for a one- sample t-test will have the same form as for the one-sample z-test described above.

We have the following assumptions:

1. The underlying population of interest is normally distributed.
2. Observations are independent.

EXAMPLE 6.5

Assume that for Example 6.4, σ is unknown and that s is calculated from sample data and found to be 12 mm Hg. We follow the same hypothesis testing process as we did for Example 6.4.

Step 1: State H_0, H_A, and α.
We will use the two-tailed hypotheses H_0: $\mu = 131$ mm Hg versus H_A: $\mu \neq 131$ mm Hg, and the significance level $\alpha = 0.05$.

Steps 2 and 3: Calculate the test statistic

$$t* = \frac{128 - 131}{12/\sqrt{85}} = -2.305.$$

Step 4: Let $T \sim t(n-1)$. The resulting P-value is

$$2[P(T \geq |t*|)] = 2[P(T \geq 2.305)] = 2[P(T \leq -2.305)] = 0.0236.$$

We note that the t-test gives a more conservative (larger) P-value than the z-test.

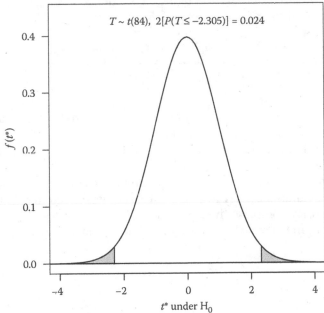

FIGURE 6.5

A $t(n - 1 = 84)$ distribution, with the probability $2[P(T \le -2.305)]$ overlaid.

> *Step 5*: Because $0.0236 \le 0.05$, we reject H_0 and again conclude that the blood pressure of middle-aged faculty at the university is different from that of the overall population.

A plot of the example is shown in Figure 6.5.

We can run the test using the function one.sample.t from *asbio*.

```
one.sample.t(xbar = 128, null.mu = 131, s = 12, n = 85, alternative = "two.
sided")
$test.statistic
[1] -2.304886
$p.val
[1] 0.02363993
$Df
[1] 84
```

In null hypothesis tests, one should report the distribution under H_0, the test statistic, the P-value, and some measure of *effect size*: the magnitude of the phenomenon under study in the context of H_0. A logical estimator of the true effect size in one-sample tests is $\bar{X} - \mu_0$. In Example 6.5, we would denote the null distribution and test statistic simultaneously as $t_{84} = -2.304$, and the estimated effect size as -3. Methods for summarizing effect size in t-tests are discussed further in Section 6.4.1.

6.2.6 Confidence Intervals and Hypothesis Testing

There is a fundamental connection between two-tailed confidence intervals using $\gamma = 1 - \alpha$, and a two-sided significance test using a significance level of α. Specifically, if we can reject H_0 for a two-sided hypothesis test using a significance level of α, then μ_0 will not be contained in a confidence interval for μ using a confidence level of $1 - \alpha$. For instance, the 95% confidence interval for μ from Example 6.5 is

```
ci.mu.t(n=85, xbar=128, s=12, summarized=TRUE)
95% t Confidence interval for population mean
Estimate       2.5%      97.5%
128.0000 125.4117 130.5883
```

If we reject H_0 at $\alpha = 0.05$, then μ_0 will not be contained in the 95% confidence interval for μ. Note that in the interval above, $\mu_0 = 131$ is *not* contained in the 95% confidence interval for μ. Conversely, if the *P*-value was >0.05, then 131 (the value of μ_0) *would* be in the 95% confidence interval.

It is also clear that the t quantiles used to calculate the confidence interval, that is, $\pm t_{(1-(\alpha/2), df=n-1)}$ are critical values (i.e., they are the values t^* must exceed to reject H_0 at probability α). This is because they comprise the 2.5th and 97.5th percentiles of the null distribution. That is, to reject null in Exercise 6.5, we require that t^* must be greater than

```
qt(0.975, 84)
[1] 1.98861
```

or less than

```
qt(0.025, 84)
[1] -1.98861
```

Because t^* was more extreme than the lower critical value, the *P*-value is less than the significance level, prompting the rejection of H_0.

One-tailed confidence intervals are analogous to one-tailed null hypothesis tests. That is, if μ_0 is not contained in a 95% lower-tailed confidence interval for μ, we would also reject H_0: $\mu \geq \mu_0$ at $\alpha = 0.05$.

6.2.7 Inferences for Two Population Means

It is also possible to inferentially compare the means of two populations using the family of *t*-hypothesis tests. Tests involving even more than two population means are possible using a procedure called ANOVA (discussed in Chapter 10).

In null hypothesis tests comparing two population means, we assume that the differences in populations will follow a distribution with mean, D_0. Thus, a two-tailed null hypothesis comparing populations X and Y would be specified as

$$H_0: \mu_X - \mu_Y = D_0,$$

and the two-tailed alternative hypothesis would be

$$H_A: \mu_X - \mu_Y \neq D_0.$$

We specify a lower-tailed test that assumes that the mean of X is smaller than the mean of Y with the alternative

$$H_A: \mu_X - \mu_Y < D_0.$$

Finally, we specify an upper-tailed alternative that assumes that the mean of X is larger than the mean of Y as

$$H_A: \mu_X - \mu_Y > D_0.$$

Generally, we let $D_0 = 0$.

We will examine three scenarios for comparing two population means where the family of t-tests may be applied. These are when (1) samples are paired, (2) population variances are equal, and (3) population variances are unequal. Before demonstrating these tests individually, however, we examine some underlying theory.

Let $X \sim N(\mu_X, \sigma_X^2)$ and $Y \sim N(\mu_Y, \sigma_Y^2)$, then from Chapter 4

$$X - Y \sim N(\mu_X - \mu_Y, \sigma_X^2 + \sigma_Y^2). \tag{4.47}$$

If $\bar{X}$ estimates μ_X and $\bar{Y}$ estimates μ_Y, then from Chapter 5, we have

$$\bar{X} \sim N(\mu_X, \sigma_X^2/n_X) \quad \text{and} \quad \bar{Y} \sim N(\mu_Y, \sigma_Y^2/n_Y). \tag{5.3}$$

We can combine Equations 4.47 and 5.3 to derive the sampling distribution of $\bar{X} - \bar{Y}$:

$$\bar{X} - \bar{Y} \sim N\left[\mu_X - \mu_Y, (\sigma_X^2/n_X + \sigma_Y^2/n_Y)\right]. \tag{6.3}$$

These principles form the theoretical basis for t-tests comparing two populations because they describe the sampling distribution of the mean difference of two normal random variables.

6.2.7.1 Paired t-Test

Paired t-tests are appropriate for analyzing experimental designs in which pairs of identical or highly similar experimental units (units to which treatments are applied, Chapter 7) are assigned to two distinct treatments. Matched pairing helps to account for confounding variables because other variables are held constant, while levels in the variable of real interest are varied (see Chapter 7). Treatment responses within an experimental unit are blocked and are not necessarily independent. As a result, it is incorrect to analyze paired designs with a t-test that assumes independence, for example, a pooled variance t-test or the Welch approximate t-test (see Sections 6.2.7.2 and 6.2.7.3).

Examples of paired designs include

- Testing for the effect of contamination on eggshell density at the *same nests* before and after contamination by dichlorodiphenyltrichloroethane (DDT)
- Comparing the effect of drug and placebo on the *same patients*
- Comparing biomass of vegetation with and without fertilizer for pairs of sites that are spatially adjacent

For a paired *t*-test, we use the test statistic

$$t* = \frac{\bar{X}_D - D_0}{S_D / \sqrt{n}},$$

(6.4)

where $\bar{X}_D$ is the mean of the paired differences, S_D is the standard deviation of the paired differences, and n is the number of paired differences.

Our assumptions are

1. The distribution underlying the differences in paired treatments is normally distributed.

2. Observations within treatments are independent, resulting in independent differences.

If these assumptions hold, and if H_0 is true, then $t*$ will be a random outcome from a *t*-distribution with $n - 1$ degrees of freedom.

EXAMPLE 6.6

Scientists have long been concerned with identifying physiological characteristics that result in a disposition for schizophrenia. Early studies suggested that the volume of particular brain regions of schizophrenic patients may differ from nonafflicted individuals. However, these studies did not address confounding variables (e.g., socioeconomic status and genetics) that might obfuscate brain volume/schizophrenia relationships (Ramsey and Schafer 1997; see Chapter 7). To control for confounding variables, Suddath et al. (1990) examined 15 pairs of monozygotic twins in which one twin was schizophrenic and the other was not. The twins were located from an intensive search throughout the United States and Canada. The authors used magnetic resonance imaging to measure the brain volume of the left hippocampus region in the twin's brains. The researchers were interested in any difference in hippocampus volume. We will use $\alpha = 0.05$.

The data are located in *asbio* under the name `sc.twin`.

```
data(sc.twin)
D <- with(sc.twin, unaffected - affected)
```

While it will not affect the results of the two-tailed test (such as the one we have here), the ordering of variables in calculation of the difference is vital for one-tailed paired *t*-tests since misspecification will reverse the correct sign of the test statistic.

Before running the hypothesis test, we will perform some diagnostics. A *normal quantile plot* compares the observed quantiles to theoretical normal quantiles[*]. If the data are normally distributed, there should be a strong linear relationship between these variables (all points should be near a linear fit line). This allows inference to the normality of the parent population.

We can create a graph that includes a normal quantile plot and a histogram with the following commands:

```
dev.new(height = 3.5)
par(mfrow = c(1,2), mar = c(4.4, 4, 0.5, 0.5))
```

---

[*] To create a normal quantile plot by hand: (1) Arrange observations from smallest to largest. (2) Find what percentile each ranked observation occupies. For instance, the smallest ranked observation in a dataset of 20 occupies the 5% point, 10% point, and so on. (3) Find the *z*-scores (theoretical quantiles) at these percentiles. (4) Plot the *z*-scores versus the observed quantiles.

```
qqnorm(D, main = "")
qqline(D)
hist(D, main = "",xlab = "Difference in brain volume")
```

We note that points in the normal quantile plots do not fall on the fit line and that the sample distribution appears right skewed in the histogram (Figure 6.6).

We can also perform hypothesis tests to check for normality. The *Shapiro–Wilk test* is the most frequently used null hypothesis test for this purpose. It has the following hypotheses:

$$H_0: \text{The underlying population is normally distributed,}$$

versus

$$H_A: \text{The underlying population is not normally distributed.}$$

We run the Shapiro–Wilk test using

```
shapiro.test(D)
        Shapiro-Wilk normality test
data: D
W = 0.9041, p-value = 0.1099
```

The diagnostic plots and hypothesis test indicate that there are issues with nonnormality (i.e., scatter of points in the quantile plot is nonlinear and the Shapiro–Wilk *P*-value is small). We will run an analogous hypothesis test on these data using the Wilcoxon sign rank test (a method for paired samples that does not assume normality) in Exercise 6.16. Rerunning this test after log transformation of the data is left as an exercise at the end of this chapter.

We proceed with the hypothesis test.

Step 1: We have the following hypotheses and significance level:

$$H_0: \mu_A - \mu_U = 0,$$

versus

$$H_A: \mu_A - \mu_U \neq 0; \quad \alpha = 0.05.$$

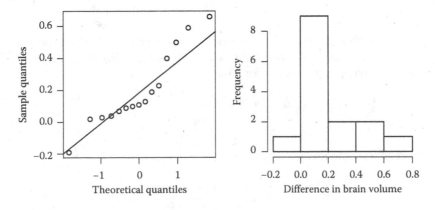

FIGURE 6.6
Normal quantile plot and histogram for the data in Example 6.6.

Steps 2 and 3: Next, we calculate the test statistic. To do this, we first calculate the mean and standard deviation of the pairwise brain volume differences for the affected and unaffected twins.

```
c(mean(D), sd(D))
[1] 0.1980000 0.2368906
```

$$t* = \frac{0.198 - 0}{0.2369/\sqrt{15}} = 3.237151.$$

Step 4: We use the test statistic to compute the *P*-value.

```
2*pt(abs(3.237151), df=14, lower.tail=F)
[1] 0.005963199
```

Step 5: Because $0.006 \leq 0.05$, we reject the null hypothesis and conclude that the hippocampus volumes are different for schizophrenic and nonschizophrenic twins.

In two sample tests, a common estimator of effect size is $(\bar{X} - \bar{Y}) - D_0$ For a paired *t*-test, the analogous measure is $\bar{X}_D - D_0$ (see Section 6.4.1 for other approaches). Thus, we can numerically summarize the results for this example as $t_{14} = 3.24$, estimated effect size $= \bar{x}_D = 0.198$, and *P*-value $= 0.006$.

We can easily run the entire test in **R**.

```
t.test(unaffected, affected, alternative="two.sided", mu=0,
paired=TRUE)
Paired t-test
data: unaffected and affected
t=3.2372, df=14, p-value=0.005963
alternative hypothesis: true difference in means is not equal to 0
```

6.2.7.2 Pooled Variance t-Test

The pooled variance *t*-test is the simplest statistical method for comparing the means of two populations. If the random variables X and Y are independent and the variances of X and Y are equal, then the standard error of the sampling distribution of $\bar{X} - \bar{Y}$ is

$$\sigma_{\bar{X}-\bar{Y}} = \sqrt{\frac{\sigma^2}{n_X} + \frac{\sigma^2}{n_Y}} = \sqrt{\sigma^2\left(\frac{1}{n_X} + \frac{1}{n_Y}\right)} = \sigma\sqrt{\frac{1}{n_X} + \frac{1}{n_Y}}. \tag{6.5}$$

We estimate σ^2 with a *pooled variance estimator*. Another name for the pooled variance estimator is the *mean squared error* or *MSE* (see Chapters 9 and 10). MSE is a weighted average of the sample variances corrected for bias:

$$MSE = \frac{(n_X - 1)S_X^2 + (n_Y - 1)S_Y^2}{n_X + n_Y - 2}. \tag{6.6}$$

The test statistic has the form

$$t* = \frac{(\bar{X} - \bar{Y}) - D_0}{\sqrt{MSE}\sqrt{\frac{1}{n_X} + \frac{1}{n_Y}}}. \tag{6.7}$$

If assumptions for the test hold (see below) and if H_0 is true, then t^* will be a random outcome from a t-distribution with $n_X + n_Y - 2$ degrees of freedom.

For a pooled variance t-test, we assume

1. The parent distributions underlying the treatments are normally distributed.
2. Observations are independent.
3. The parent distributions underlying the treatments have equal variances.

Heteroscedasticity (unequal population variances) may result in misleading pooled variance t-test results, especially if sample sizes are unequal (Glass et al. 1972). In particular, if the smaller sample has the larger population variance, then type I error will be higher than the stipulated significance level (see Section 6.3). Conversely, if the larger sample has the larger population variance, then the rate of type II error will be higher than expected (see Section 6.3; Coombs et al. 1996). Heteroscedasticity assumptions can be checked with *side-by-side box plots*, summary statistics, and diagnostic tests such as the Fligner–Killeen test. The *Fligner–Killeen test* ranks the absolute values of the difference of data from corresponding sample medians and uses this information to create a test statistic that will be asymptotically χ_1^2 distributed if the null hypothesis of equal population variances is true (see Conover et al. 1981). The test is not sensitive to violations of normality, allowing normality and homoscedasticity to be verified independently.

Like nonnormality, heteroscedasticity can often be corrected with data transformations (e.g., square root and log transformations) (Chapter 7). If heteroscedasticity problems persist after transformations and the underlying populations are normal, one can still perform a two-sample t-test using Welch's procedure (see below).

EXAMPLE 6.7

Monteith et al. (2009) examined the maternal life-history characteristics of white-tailed deer (*Odocoileus virginianus*) originating from two regions: the Black Hills in southwestern South Dakota, and eastern South Dakota. Because litter size and dam size affect offspring mass, the investigators used proportional birth mass (total litter mass/dam mass) as a measure of reproductive investment by deer. We wish to test whether the proportional birth mass differs at the two sites. We will use $\alpha = 0.05$.

The data are located in *asbio* in the dataframe deer.

```
data(deer)
```

Before beginning hypothesis testing, we will perform a diagnostic test for homoscedasticity. The Fligner–Killeen test will have the following hypotheses:

$$H_0: \sigma_{BH}^2 = \sigma_{ER}^2,$$

versus

$$H_A: \sigma_{BH}^2 \neq \sigma_{ER}^2,$$

where BH indicates Black Hills and ER indicates Eastern Region. We run the Fligner–Killeen test using

```
with(deer, fligner.test(Prop.mass ~ Region))

      Fligner-Killeen test of homogeneity of variances
data: Prop.mass by Region
Fligner-Killeen:med chi-squared = 0.97, df = 1, p-value = 0.3247
```

In `fligner.test`, the proportional offspring mass of deer is defined to be a function of region by using the argument `Prop.mass ~ Region`. This allows the response variable (`Prop.mass` in this case) to be a function of an arbitrary number of treatments in a categorical variable (in this case `Region`). We fail to reject the null hypothesis of equal variances at $\alpha = 0.05$ and proceed with the pooled variance testing procedure.

Step 1: We have the following hypotheses and significance level:

$$H_0: \mu_{BH} - \mu_{ER} = 0$$

versus

$$H_A: \mu_{BH} - \mu_{ER} \neq 0; \, \alpha = 0.05$$

Steps 2 and 3: Using the sample means and variances, we calculate the MSE and the test statistic.
First, here are some summary stats:

```
with(deer, tapply(Prop.mass, Region, mean))#treatment means
        BH           ER
0.11418734 0.09934436
with(deer, tapply(Prop.mass, Region, var))#treatment variances
         BH           ER
0.0003434640 0.0002660172
```

Thus, we have

$$\sqrt{MSE} = \sqrt{\frac{11(0.00034) + 23(0.00027)}{34}} = 0.01706.$$

$$t* = \frac{(0.114 - 0.099) - 0}{0.0171\sqrt{2/24}} = 2.4607.$$

Step 4: Next, we calculate our *P*-value.

```
2* pt(2.4607, df = 34, lower.tail = F)
[1] 0.01910169
```

Step 5: Because $0.019 \leq 0.05$, we reject H_0 and conclude that proportional birth weights are not equal at the two sites.

The result of the analysis can be summarized as $t_{34} = 2.46$, estimated effect size $= \bar{x}_{BH} - \bar{x}_{ER} = 0.015$, and *P*-value $= 0.019$.
Again, we can run the entire test in **R**.

```
with(deer, t.test(Prop.mass ~ Region, var.equal = TRUE))

Two Sample t-test

data: Prop.mass by Region
t = 2.4607, df = 34, p-value = 0.0191
alternative hypothesis: true difference in means is not equal to 0
95 percent confidence interval:
0.00258462 0.02710134
```

The 95% confidence interval in the output above is for the parameter $\mu_{BH} - \mu_{ER}$.

I stipulate the assumption of equal variances with var.equal = TRUE. As in the function fligner.test, I specify that proportional birth mass to be a function of region (BH or ER) using the tilde character (~). The function t.test can also be run by inputting vectors of observations from the hypothesized populations separately as the first two arguments. Regardless of the approach, care should be taken to place populations into t.test in the same order that they are given in the hypotheses. While these steps would not affect the results in the current example (since a two-tailed test is being used), it is good practice to specify BH as the first argument in t.test because it is the first population given in the hypotheses. If the observations are specified as a function of character vector denoting the treatments, then the first alphanumeric population will always be placed first in computations by t.test. That is, BH will be treatment X, while ER will be Y in Equation 6.7.

6.2.7.3 Welch's Approximate t-Test

If observations from experimental units for treatments X and Y are independent (e.g., not paired) and we cannot assume that their underlying variances are equal, a number of options are still available for comparing population means. The most commonly used is an approximation of the pooled variance t-test called the *Welch test* or the *Satterthwaite adjusted t-test* (Welch 1947).

Test statistics calculated using the pooled variance procedure will be random outcomes from a t-distribution with $n_X + n_Y - 2$ degrees of freedom if H_0 is true. Welch test statistics, however, will not exactly follow t-distributions under H_0. Instead, an approximate null t-distribution is identified by using the *Satterthwaite procedure* to compute the degrees of freedom. This will be v in the equation below.

$$v = \frac{\left((S_x^2/n_X)+(S_Y^2/n_Y)\right)^2}{\left((S_X^2/n_X)^2/(n_X-1)\right)+\left((S_Y^2/n_Y)^2/(n_Y-1)\right)}. \tag{6.8}$$

Equation 6.8 will generally produce a noninteger solution for v.
The Welch test statistic has the form

$$t* = \frac{(\bar{X}-\bar{Y})-D_0}{\sqrt{\dfrac{S_X^2}{n_X}+\dfrac{S_Y^2}{n_Y}}} \tag{6.9}$$

For a Welch approximative test, we assume

1. The parent distributions underlying the treatments are normally distributed.
2. Observations are independent.

If these assumptions are valid and if H_0 is true, then $t*$ will be a random outcome from a t-distribution with approximately v degrees of freedom.

EXAMPLE 6.8

Magnets have long been used as an alternative medicine for speeding the recovery of broken bones and for pain relief. Vallbona et al. (1997) tested whether chronic pain experienced by postpolio patients could be treated with magnetic fields applied directly to pain trigger points. The investigators identified 50 subjects who not only had postpolio syndrome but also experienced muscular or arthritic pain. Magnets were applied to pain trigger points in 29 randomly selected subjects, while the other 21 patients were treated with a magnet-shaped placebo. The experiment was *double blind*, that is, neither the patient nor the investigator knew what treatment the patient was receiving after the experiment. This step decreased bias by preventing investigators from "leading" patients to report their pain reactions in a particular way (see Chapter 7). The patients were asked to subjectively rate their pain on a scale from 1 to 10 before and after application of the magnet or placebo. The authors were interested in whether the pain reduction in the magnet group was greater than for the placebo group. Hence, we will specify an upper-tailed hypothesis test.

Step 1: We have the following hypotheses and significance level:

$$H_0: \mu_M - \mu_P \leq 0,$$

versus

$$H_A: \mu_M - \mu_P > 0; \quad \alpha = 0.05.$$

The data are located in *asbio* in a dataset called `magnets`.

`data(magnets)`

We are interested in the difference in pain ratings before and after application of the magnets. Here are the treatment means for these differences:

```
mean.diffs <- with(magnets, tapply(Score_1 - Score_2, Active, mean)
mean.diffs
  Magnet  Placebo
5.241379 1.095238
```

The variable `Active` distinguishes whether the device applied to the subject was active (a magnet) or inactive (a placebo). The treatment variances are

```
var.diffs <- with(magnets, tapply(Score_1 - Score_2, Active, var)
var.diffs
   Magnet   Placebo
10.475369 2.490476
```

We note that the sample variances have more than a threefold difference, prompting the use of Welch's procedure. The Fligner–Kileen test also supports the use of Welch's test.

```
with(magnets, fligner.test(Score_1 - Score_2, Active))

        Fligner-Killeen test of homogeneity of variances

data: Score_1 - Score_2 and Active
Fligner-Killeen:med chi-squared = 6.5756, df = 1, p-value = 0.01034
```

Steps 2 and 3: Using summary output, we can calculate the Satterthwaite degrees of freedom and the test statistic.

$$v = \frac{\left(\dfrac{10.475}{29} + \dfrac{2.490}{21}\right)^2}{\dfrac{(10.475/29)^2}{29-1} + \dfrac{(2.490/21)^2}{21-1}} = 42.92598.$$

$$t* = \frac{(5.24 - 1.10) - 0}{\sqrt{(10.475/29) + (2.490/21)}} = 5.986.$$

We are careful to put the magnet treatment before the placebo treatment in the numerator of the test statistic because this was the order of treatments given in the hypotheses.

Step 4: Here is the *P*-value

```
pt(t.star, df = 42.92598, lower.tail = FALSE)
[1] 1.930065e-07
```

Step 5: Because 0.0000002 is less than 0.05, we reject H_0 and conclude that patients treated with magnets experience decreased pain compared to patients treated with a placebo.

The result of the analysis can be numerically summarized as $t_{42.93} = 5.99$, estimated effect size $= \bar{x}_M - \bar{x}_P = 4.15$, and *P*-value < 0.001.

Again, we can run the entire test in **R**.

```
t.test(Score_1 - Score_2 ~ Active, alternative = "greater", mu = 0,
paired = FALSE

Welch Two Sample t-test

data: Score_1 - Score_2 by Active
t = 5.9856, df = 42.926, p-value = 1.93e-07
alternative hypothesis: true difference in means is greater than 0
95 percent confidence interval:
2.981643 Inf
```

In the line `Score_1 - Score_2 ~ Active`, I specify that I want the difference of before and after reactions to be considered as a function of the treatment (magnet or placebo). Magnet is the first alphanumeric treatment. This means that `t.test` will consider the magnet group to be the first population in its computations. This fits with ordering of treatments in the null and alternative hypotheses.

6.3 Type I and Type II Errors

The significance level, like other frequentist methods, describes how our analytic approach would perform with repeated use. By definition, if we use $\alpha = 0.05$ repeatedly when H_0 is in fact true, we will be wrong (the test will reject a true H_0) 5% of the time.

Biologists in certain fields should be particularly aware of testing error. For instance, in a genetic microarray experiment, tens of thousands of significance tests may be run. The concerns associated with testing error should be obvious. Given 10,000 tests and a significance level of 0.05, we would expect to reject null hypotheses 500 times when H_0 was actually true.[*] Much research in genomics has been devoted to dealing with this problem (e.g., Craig et al. 2003).

If we reject H_0 when in fact H_0 is true, this is a *type I error*. It will occur with a probability equal to the significance level of the test, α. Conversely, if we fail to reject H_0 when in fact H_0 is false, then this is a *type II error*, which occurs with probability β. Recall that type I and

---

[*] If the results for tests are not independent and/or they are conducted for the same experiential units, it is generally recommended that *P*-values be adjusted so that the probability of type I error actually agrees with the significance level being used. This topic is readdressed with greater emphasis in Chapter 10.

TABLE 6.1

Summary of Type I and Type II Error

	H_0 True	H_0 False
Reject H_0	α	$1 - \beta$
	Type I error	Correct decision, power
Retain H_0	$1 - \alpha$	β
	Correct decision	Type II error

II errors are implicit to the Neyman–Pearson hypothesis testing procedure (Section 6.2). The probability that we will reject a false H_0 at a fixed alpha for a particular effect size is called *power*. It occurs with a probability of $1 - \beta$. Summaries of type I error, type II error, and power can be obtained interactively using the function see.typeI_II.

The columns of Table 6.1 comprise the states of nature for a null hypothesis (it is either true or false) and will sum to 1; that is, $\alpha + (1 - \alpha) = 1$ and $\beta + (1 - \beta) = 1$. However, the rows of Table 6.1 *do not* have to sum to 1. For instance, investigators conventionally use $\alpha = 0.05$ and $1 - \beta = 0.8$.

EXAMPLE 6.9

Many nutrients are essential to organisms, but are fatal at higher dosages. Assume that extensive testing indicates that the correct recommended daily allowance (RDA) of selenium (an essential nutrient but one that is fatal at high doses) for an endangered animal is 87.5 mcg. A researcher wishes to test if the RDA of selenium is greater than 87.5 mcg (1000 mcg = 1 mg). He quantifies the effect of selenium by measuring the fitness (number of offspring of the endangered animal) at different dosages. The null and alternative hypotheses are

$$H_0: \mu \leq 87.5 \text{ mcg,}$$

$$H_A: \mu > 87.5 \text{ mcg.}$$

What if we reject null, but we should not have? (i.e., the RDA *should be* ≤87.5 mcg, but our test indicates that the RDA is >87.5 mcg). In this case, a type I error occurs. Because the dosage is too high we could poison the animals.

What if we fail to reject null, but we should have rejected it? (i.e., the RDA *should be* >87.5 but our test indicates that the RDA is ≤87.5). In this case, a type II error occurs. The animals may become ill from selenium deprivation.

As we will see in the next section, one way to decrease type I and II errors is to increase sample size. Power analyses are often used to define appropriate sample size with respect to both types of error.

6.4 Power

In a *power analysis*, desired experimental characteristics (effect size and sample size) are considered in the context of both α and β. Power analyses are generally conducted *a priori* (before starting an experiment) to aid in defining the proper replication for an experiment. *Post hoc* power analysis are generally considered inappropriate (Hoenig and Heisey 2001), but are occasionally used to convince a journal editor that nonsignificant results are worth publishing. Such actions are a response to *publication bias* by scientific journals that are more apt to publish only significant results (Dickersin 1990). Power has the basic form

$$\text{Power} = 1 - \beta \propto \frac{E\alpha\sqrt{n}}{\sigma}, \tag{6.10}$$

where E indicates the true effect size. The proportion sign in Equation 6.10, $\propto$, is indicative that power will depend in part on the measure of effect size being used, the type of statistical analysis one is conducting, and the particular testing procedure used in the analysis. Note that Equation 6.10 requires that we know σ or have a good estimate for it preceding analysis.

EXAMPLE 6.10 POWER ANALYSIS, SMOKING, AND ALZHEIMER'S

Surprisingly, the incidence of Alzheimer's disease has been negatively associated with moderate amounts of smoking (Van Duijn and Hoffman 1991, Graves et al. 1991, Brenner et al. 1993, Salib and Hillier 1997). Smoking may offer some protection because nicotine may reduce apoptosis (programmed cell death) of neurons (Larrick 1993).

Let us assume that previous to the start of the experiment concerning the effect of smoking on Alzheimer's, researchers were interested in being able to detect a 7% decrease in Alzheimer's incidence for subjects smoking 10–20 cigarettes a day, given that $\sigma = 45\%$. The investigators wanted to know if a sample size of 200 was sufficient to detect this effect (produce a significant result) given $\alpha = 0.05$.

Because we know σ, we will run the power analysis assuming that a one-sample z-test will be used for hypothesis testing. We are interested in the lower-tailed alternative hypothesis that smoking decreases the occurrence of Alzheimer's. Our hypotheses are $H_0: \mu \geq 0$ and $H_A: \mu < 0$. We note that we have a sample size large enough to assume a normal sampling distribution for $\bar{X}$, regardless of its underlying parent distribution.

A one-sample lower-tailed z-test would reject null at $\alpha = 0.05$ whenever z^* is less the lower-tailed critical value -1.645.

```
qnorm(.05)
[1] -1.644854
```

Solving for $\bar{x}$ in Equation 6.1, we have

$$\frac{\bar{x} - 0}{45/\sqrt{200}} < -1.644854 \quad \bar{x} < (45/\sqrt{200})\,1.644854 \quad \bar{x} < 5.234684.$$

The observed decrease in Alzheimer's as a result of smoking (effect size) has to be greater than 5.234684% for us to reject H_0 at $\alpha = 0.05$.

Given a significance level of 0.05, effect size of -7%, variance of 45^2, and a sample size of 200, power is obtained by finding

$$P(\bar{X} \leq -5.234684), \quad \text{where } \bar{X} \sim N(-7, 2025/200).$$

In **R**, we have

```
pnorm(-5.234684, -7, 45/sqrt(200))
[1] 0.7104792
```

Thus, $1 - \beta = 0.7104792$. In other words, there is a probability of 0.71 of rejecting the null hypothesis $\mu \geq 0$ if the true effect of smoking is a 7% decrease in Alzheimer's. This is the power of the experiment.

The relationship of type I error and power can be visualized in a number of ways. In one approach, a plot containing the null distribution is placed atop a plot containing a distribution that assumes H_A is true. The two distributions are then connected with a dashed vertical line representing the boundary for α and β. This is demonstrated for the Alzheimer's analysis in Figure 6.7.

```
Fig.6.7()##See Ch. 6 code
```

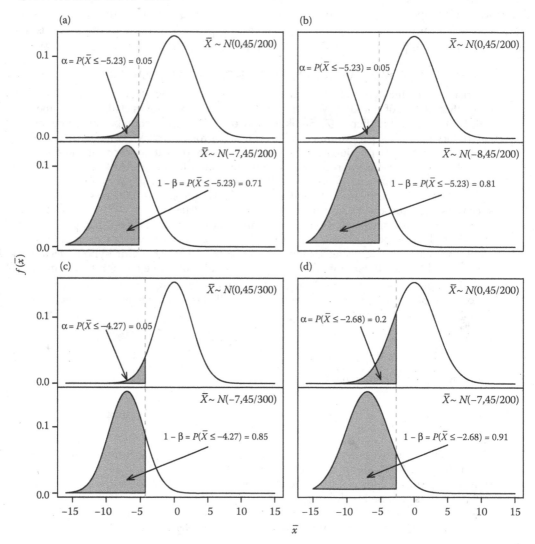

FIGURE 6.7

Illustration of type I error and power for Example 6.10. For (a) through (d), the top figure illustrates the sampling distribution for $\bar{X}$ assumed by the null hypothesis and shows (shades) α, while the bottom figure plots the distribution for a particular value of the alternative and shows power, $1 - \beta$. (a) provides an initial example (effect size $= -7$, $\alpha = 0.05$, $\sigma = 45$, $n = 200$, and power $= 0.71$). Figures (b) through (d) demonstrate ways to increase power: (b) increase effect size (effect size $= -8$, $\alpha = 0.05$, $\sigma = 45$, $n = 200$, and power $= 0.81$), (c) increase n (effect size $= -7$, $\alpha = 0.05$, $\sigma = 45$, $n = 300$, and power $= 0.85$), and (d) increase α (effect size $= -7$, $\alpha = 0.2$, $\sigma = 45$, $n = 200$, and power $= 0.91$).

There are three ways to increase power. First, power will always increase with increasing effect size. For instance, we increase power from 0.71 to 0.81 by increasing effect size from –7% to –8% (Figure 6.7b). Second, one can increase sample size since this will drive down the standard error for the mean. For the Alzheimer's example, power increases from 0.71 to 0.85 by increasing sample size from 200 to 300 (Figure 6.7c). Finally, we can increase the level of significance. While α is not the inverse of β, increasing α will decrease our probability for a type II error and increase power (Figure 6.7d).

The *asbio* library contains an interactive GUI useful for demonstrating the effect of changing parameter values on power. Simply type

```
see.power.tck()
```

6.4.1 Sample Adequacy

Power analyses can be used to find adequate sample sizes given a particular effect size along with prescribed levels of α and $1 - \beta$.

In Chapter 5, we derived a simple formula to find the sample size n necessary for a particular significance level and margin of error for estimating μ

$$n = \left(\frac{z_{1-(\alpha/2)}\sigma}{m} \right)^2, \tag{6.11}$$

where m is the margin of error that is desired or anticipated.

It is possible to consider the probability of both power $(1 - \beta)$ and type I error (α) when estimating the sample sizes necessary to detect particular effects. Below are equations for calculating sample sizes for one-sample z-tests. Slightly different formulae are required for one-tailed (Equation 6.12) and two-tailed tests (Equation 6.13). Note, however, that except for replacement of m with $\mu - \mu_0$ and the additional consideration of power, these equations are identical to Equation 6.11.

$$n = \frac{(z_{1-\alpha} + z_{1-\beta})^2 \sigma^2}{(\mu - \mu_0)^2}, \tag{6.12}$$

$$n \approx \frac{(z_{1-(\alpha/2)} + z_{1-\beta})^2 \sigma^2}{(\mu - \mu_0)^2}. \tag{6.13}$$

EXAMPLE 6.11

In quantifying the effects of logging on insects, an entomologist would be particularly concerned if insect species richness dropped by 5% or more. She wants to use $\alpha = 0.05$ and $\beta = 0.8$ (both conventional values) and thinks that $\sigma = 10\%$. We have

$$n = \frac{(1.645 - 0.842)^2 100}{(-5 - 0)^2} = 24.7.$$

We round up to 25 and conclude that a sample size of 25 is necessary to detect 5% drop in richness given $\alpha = 0.05$, $\beta = 0.8$, and $\sigma = 10\%$. This answer can be obtained with **R** using the function `power.z.test` from *asbio*. To get the result above, I would type

```
power.z.test(sigma = 10, power = 0.8, alpha = 0.05, effect = 5, test = "one.
tail")
```

6.4.1.1 Power in t-Tests

Methods for performing power analyses for *t*-test are conceptually analogous to those described above for *z*-tests. In this case, however, the form of the null distribution will be dependent upon sample size (and vice versa) via the degrees of freedom. As a result, iterative procedures are required for computation using the *noncentral t-distribution* (a *t*-distribution whose expected value can be nonzero). The function `power.t.test` runs power analyses using this approach.

> **EXAMPLE 6.12**
>
> For Example 6.11, assume that σ is unknown, but that previous work indicates that its value may be close to 10. In **R**, we have
>
> ```
> power.t.test(sd = 10, power = 0.8, sig.level = 0.05, delta = 5, type = "one.
> sample", alternative = "one.sided")
>         One-sample t test power calculation
>                 n = 26.13751
>             delta = 5
>                sd = 10
>         sig.level = 0.05
>             power = 0.8
>       alternative = one.sided
> ```
>
> A slightly larger sample size is required to detect the effect, compared to Example 6.11, due to the more conservative nature of the *t*-test.

6.4.1.2 Effect Size in t-Tests

Other than the differences $\mu - \mu_0$ and $(\mu_X - \mu_Y) - D_0$, other measures of effect size commonly used with *t*-tests include variants on *Cohen's d* (Cohen 1988)

$$d = \frac{|\mu_1 - \mu_2|}{\sigma},$$

(6.14)

where σ^2 is the common group error variance, that is, $E(\text{MSE})$. Cohen (1988) recommended examining power from the perspective of small, medium, and large effect sizes if the true effect size was unknown. On the basis of his work in the behavioral sciences, he suggested that $d \leq 0.2$ denoted small effect sizes, that medium effect sizes were between 0.2 and 0.8, and that large effects were ≥ 0.8. This approach is conceptually useful; however, it has been criticized on the grounds that cutoffs for the magnitude of effect sizes should be considered within the context of the study system (Quinn and Keough 2002). The function `pwr.t.test` in library *pwr* uses the definition of effect size given in Equation 6.14. *Post hoc* estimators analogous to Cohens *d* include a measure recommended by Hedges (1981):

$$\hat{d} = \frac{|\bar{X}_1 - \bar{X}_2|}{\sqrt{\text{MSE}}}.$$

(6.15)

6.5 Criticisms of Frequentist Null Hypothesis Testing

The frequentist null hypothesis testing paradigm (Section 6.2) has been criticized by many individuals including philosophers of science (Edwards 1972), statisticians (Salsburg 1985), and biologists (Johnson 1999). I have subset these criticisms into six classes:

1. *Sample space of the test statistic exceeds what could be sampled.* The sample space of the test statistic distribution is generally larger than what could possibly be observed from sampling. For instance, the support for a z-test statistic is defined on the interval $(-\infty, +\infty)$. Because this range represents outcomes beyond what could possibly be observed (e.g., female heights approaching $-\infty$ cm), tests assuming these distributions for H_0 may overstate evidence against H_0 (Royall 1997). A Bayesian approach (Chapter 2) addresses this problem to some degree. This is because the Bayesian posterior density distribution is the result of conditioning using only sample data.

2. $P(\text{data}|H_0)$ *not* $P(H_0|\text{data})$. Frequentist null hypothesis tests find the probability of the data if H_0 is true. However, it has been argued (Cohen 1994) that the result of greatest interest to scientists is the probability of H_0 being true, given the data. These two conditional probabilities are not interchangeable (Chapter 2).

   Associated with this criticism is the *likelihood principle* (Fisher 1956), which states that (1) all evidence for an unknown parameter θ is contained in the likelihood function, which, if the data are fixed while θ is varied, can be specified as $\mathcal{L}(\theta|\text{data})$ (see Section 4.4.2) and (2) two likelihood functions are equivalent with respect to inferential evidence concerning θ if they are proportional to each other (Edwards et al. 1963). The footnote below[*] contains a famous illustration

---

[*] Lindley and Phillips (1976) provided the following demonstration of the likelihood principle. Consider a test for a fair coin, that is, H_0: $\pi = 0.5$. In the experiment, we record 12 coin flips and observe nine heads and three tails. We use two probability density functions (binomial and negative binomial) to quantify evidence concerning H_0.

   Approach 1 (binomial). The number successes (heads) will follow a binomial distribution, $X \sim \text{BIN}(n, \pi)$. The upper tailed P-value is the probability of seeing nine or more heads if H_0 is true. Thus, we have

$$P(X \geq 9 \mid H_0) = \sum_{x=9}^{12} \binom{12}{x}(0.5)^x(0.5)^{12-x} = 0.073$$

```
pbinom(8, 12, 0.5, lower.tail = FALSE)
[1] 0.07299805
#Note that I stipulate P(X > 8) to get P(X ≥ 9)
```

   Approach 2 (negative binomial). The number of failures (heads) required to achieve three successes (tails) will follow the negative binomial distribution, $X \sim \text{NB}(3, 1 - \pi)$. Now, the upper tailed P-value is the probability of seeing nine or more successes after three failures if H_0 is true. Thus, we have

$$P(X \geq 9 \mid H_0) = \sum_{x=9}^{\infty} \binom{3+x-1}{3-1}(0.5)^x(0.5)^3 = 0.0327$$

```
pnbinom(8, 3, 0.5, lower.tail = FALSE)
[1] 0.03271484
```

This is unexpected. The P-values are different although the data and hypotheses are the same. This offends our frequentist sensibilities but the likelihood principle is satisfied. The likelihood function for both frameworks is $c(\pi)^9(1 - \pi)^3$. The two functions will differ only with respect to multiplication by a constant, c, and are therefore proportional. Because of this, the likelihood principle presumes that both approaches contain identical inferential information about π.

that frequentist inference violates the likelihood principle. Bayes rule incorporates likelihood functions, and as a result, the likelihood principle has been used as a justification for Bayesian over frequentist inference.

3. *Evidence against H_0 does not imply evidence for H_A.* An assumption of the Pearson–Neyman approach is that hypotheses can be designed so that evidence against H_0 provides resultant evidence *for* H_A. However, statistical tests are designed assuming H_0 is true (not that H_A is false). Thus, at best, evidence supporting H_A is obtained indirectly. As a result, one should be careful in overstating the relevance of H_A given a significant result. We note that Bayesian and likelihood-based approaches allow one to explicitly quantify the weight of evidence for and against particular hypotheses. For instance, evidence *in favor of* H_0 can be quantified (see Section 6.7).

4. *Significance levels are arbitrary.* The significance level $\alpha = 0.05$ has been recommended for statistical hypothesis testing (Fisher 1936), but this value should not be considered particularly magical. Because of this arbitrary cutoff, many of the criticisms of frequentist hypothesis testing have been leveled at significance testing. Of course, scientists should not feel compelled to make any formal statements about the significance of the test. We can simply present P-values as evidence against H_0 and let those who study our results make their own decisions with respect to significance (cf. Oakes 1986).

5. *Dependence on sample size.* Biologists may be particularly prone to finding significant effects as a result of their large datasets. For instance, in some time series analyses (e.g., radio collar transmissions) and spatial analyses (e.g., per pixel data in a digital map), thousands or even millions of observations might be acquired for a single test. Holding all other factors constant, increasing the sample size will always decrease a frequentist P-value (see Section 7.4). Thus, while a hypothesis of no effect can often be rejected in a large sample test, this effect may be very small. This prompts a distinction between statistical significance and *biological significance*. In particular, "a scientific fact should be regarded as experimentally established only if a properly designed experiment rarely fails to give this level of significance" (Fisher 1926, p. 504). This problem can be addressed by using care with sampling and experimental designs (Chapter 7). For instance, by basing sample sizes on power analyses.

6. *Asymmetry in error rate criteria.* Hypothesis decisions are based on type I error rates (type II error rates are rarely even mentioned in scientific papers). Because of this, demanding restrictions are usually placed on type I error (e.g., $\alpha = 0.05$). A number of authors have argued that type I error constitutes a more serious scientific mistake than type II error since it emphasizes the importance of falsification and stresses H_0 over H_A (Shrader-Ferchette and McCoy 1992).[*] However, this convention makes it difficult to maintain similarly restrictive levels for β unless sample sizes or effect sizes are exceptionally large. The importance of type II error for biological studies has been acknowledged by a number of authors, particularly those involved with environmental monitoring or human health, leading to proposals of scalable decision criteria that incorporate both types of error (Mapstone 1995).

---

[*] By applying type I and II error to "real-world" scenarios, one could argue that attention to type I error (over type II error) has been favored by *natural selection*. Assume you are a small mammal and that you hear a noise. You immediately consider two hypotheses: (1) it is a predator and (2) it is not a predator. The first hypothesis has been selected to be the default position (it is the null hypothesis). Here, a type I error (rejecting the fact that a predator is present) may have calamitous consequences, whereas a type II error (reacting to an imagined predator) merely results in unnecessary stimulation of your flight or fight response.

6.5.1 A Final Word

My purpose in listing criticisms of frequentist null hypothesis tests is not to make biologists terrified of using them, but to make clear the limitations and issues involved. A large number of authors have argued that while these tests can be misused and are often misunderstood, this does not change the fact that they can also provide an objective and consistent method of evaluating research hypotheses (Dennis 1996, Quinn and Keough 2002).

6.6 Alternatives to Parametric Null Hypothesis Testing

In this section, I introduce tests that rely on a null hypothesis paradigm and a frequentist interpretation of probability, but utilize nonparametric approaches to quantify evidence against H_0. Because of their lack of distributional assumptions and because they are generally resistant to outliers, these methods are often called *nonparametric* or *robust*.

6.6.1 Permutation Tests

Permutation tests do not rely on assumed *a priori* distributions for random variables. Instead, empirical distributions are created from randomization of observed data. Consider a null hypothesis that two drug treatments are equal in efficacy. If H_0 is true, then random assignment of treatments to observations should produce the similar differences among treatments. Randomization tests differ from bootstrapping (Chapter 5) in that sampling (or subsampling) with replacement does not occur (cf. Ter Braak 1992). A single observation from the response vector will always occur exactly once, although it may, in a particular permutation, be assigned to a different treatment than the one to which it was originally assigned.

Permutation tests are robust to violations of normality. However, like rank-based permutation procedures (discussed next), tests based purely on randomization are sensitive to differences in treatment variances (Boik 1987). As a result, these procedures should not be looked upon as a cure-all for heteroscedasticity (Quinn and Keough 2002).

Permutation tests have the advantage of being valid when sampling is nonrandom, since inference to a parent population may be dubious anyway (Manly 2007). Examples include situations when samples are unavoidably opportunistic (e.g., museum specimens), or situations where samples are difficult or dangerous to gather (Potvin and Roff 1993). Generally speaking, inference to populations only makes sense if clearly defined populations are being randomly sampled (see Chapter 7). If this is not true, then permutation tests may be more appropriate than conventional parametric tests (Quinn and Keough 2002).

A major problem with permutation tests is that their scope of inference is limited to the sample (Manly 1997). The importance of this difficulty, however, has been questioned by a number of authors, since permutation tests and parametric tests have similar power when conventional parametric assumptions are valid (Crowley 1992, Manly 1997). The function MC.test in *asbio* provides permutation tests for inferentially comparing the means of two populations.

EXAMPLE 6.13

To investigate pollination patterns among bumblebee queens and honeybee workers, Harder and Thompson (1989) recorded the proportion of pollen removed from glacier lilies (*Erythronium grandiflorum*) by 35 bumblebee queens (*Bombus ternarius* and *Bombus terrricoa*) and 12 honeybee (*Apis melifera*) workers near Kebler Pass, Colorado. The data are available under the name bombus in *asbio*. We will test the two-tailed hypotheses:

$$H_0: Q = W$$

versus

$$H_A: Q \neq W.$$

Here is the absolute value of the difference of the means:

```
data(bombus)
abs(diff(tapply(bombus[,1], bombus[,2], mean)))
0.2395476
```

Here are 10,000 permutated absolute differences:

```
perm<-rep(1,10000)
for(i in 1:10000){
perm[i]<- abs(diff(tapply(bombus[,1], sample(bombus[,2], replace = F),
mean)))}
```

How many times is a random difference equal to or larger than the observed difference? This number plus one (the original observed difference is included in the count) divided by the total number of permutations, is equivalent to the *P*-value.

```
(length(perm[perm>=0.2395476])+1)/10001
[1] 3e-04
```

This proportion is equivalent to the probability of seeing an outcome as or more extreme than the outcome observed by chance if the null hypothesis of no effect is true.

6.6.2 Rank-Based Permutation Tests

The attraction of permutation-testing procedures predated software (such as **R**) that made them possible (cf. Fisher 1925) . A precursor to permutation tests are *rank-based permutation tests*. In this approach, data are rank transformed and all possible permutations of the ranks are derived to create an empirical distribution.

Rank permutation distributions have the appeal of only requiring a single null distribution (representing all possible permutations) given a particular sample size. Because the frequencies of all possible outcomes under H_0 can be determined, distributional parameters can also be directly computed. As a result, tables of rank-based permutation distributions for small sample sizes are conventionally included in most introductory statistics texts. In addition, test statistics for these procedures are often asymptotically normal. Thus, for large sample sizes, the standard normal distribution has traditionally been used for inferences. Note, however, that computational procedures to derive all rank permutations distributions are now available in **R** even for large datasets (Hothorn et al. 2006).

Rank-based permutation procedures have three advantages over parametric and non-ranked procedures. First, they are much less sensitive to outliers compared to parametric methods (and permutation tests based on mean differences). Second, because their rank-based empirical distributions include all possible outcomes, the scope of inference is generally considered to be less of an issue than for purely permutational procedures. Finally, rank-based permutation procedures are only slightly less powerful than parametric methods if their parametric assumptions hold and may be more powerful than parametric methods if parametric assumptions do not hold (Pitman 1948).

There are four disadvantages to rank-based tests. First, the presence of ties adds an additional level of complexity to analyses. Because a permutation distribution that includes ties is much more complex than a permutation distribution excluding ties, most software packages use normal approximations for the distribution under H_0 given ties, resulting in approximate P-values. Second, a rank transformation amounts to throwing out a large amount of data as we convert a continuous variable into a discrete ordinal variable. Third, with only a few exceptions (see Wilcox 2005), rank-based permutation tests, such as parametric tests, are sensitive to heteroscedasticity. That is, similar distributional shapes are assumed for the populations being compared, although they need not be normal (Boik 1987). Finally, a traditional problem with both rank and nonrank permutation methods is an inability to quantify interactions in multifactor studies (Chapter 10). This issue, however, has been addressed to a large degree by recent advances in robust estimation and hypothesis testing (see Chapter 10, Wilcox 2005).

EXAMPLE 6.14

Consider a comparison of two hypothesized populations, X_1 and X_2. Assume that we have two observations for X_1, three observations for X_2, and are interested in the following lower-tailed test: H_0: $X_1 \geq X_2$ versus H_A: $X_1 < X_2$. In the absence of ties, there will be $(n_1 + n_2)!/(n_1!n_2!) = 10$ possible ranks of X_1 or X_2. The possible ranks for X_1 are shown in Table 6.2. The ranks for X_2 are those *not given* for X_1.

Since we know everything about the distribution, we can easily determine probabilities. For instance, we can ask: "What is the smallest possible P-value?" That is, what is the probability of seeing ranks equal to or smaller than 1 and 2 for X_1? The answer is 0.1. One tenth of the outcomes in Table 6.2 meet this criterion.[*]

A large number of rank-based permutation methods have been designed to replace particular parametric tests when parametric assumptions are not met. Be warned, however, that (1) these tests are not free from assumptions, and (2) hypotheses statements will generally differ from parametric tests. Nonparametric analogs for t-tests include the *Kolmogorov–Smirnov* test, which can be used to replace a one-sample z or t-test, the *Wilcoxon sign rank*

TABLE 6.2

Table of Possible Ranks for X_1 in Example 6.14

X_1-Ranks	Probability	X_1-Ranks	Probability
1,2	1/10	2,4	1/10
1,3	1/10	2,5	1/10
1,4	1/10	3,4	1/10
2,3	1/10	3,5	1/10
1,5	1/10	4,5	1/10

---

[*] This results in a Wilcoxon rank-sum test statistic of 0. The probability that $W \leq 0$, if H_0 is true, is 0.1.

```
pwilcox(0, n = 2, m = 3, lower.tail = TRUE)
[1] 0.1
```

test, which is often used as a nonparametric substitute for one-sample *t*-test or a paired *t*-test, and the *Wilcoxon rank sum test*, which can be used in place of a two-sample *t*-test.

6.6.2.1 Kolmogorov–Smirnov Test

The Kolmogorov–Smirnov procedure performs a goodness-of-fit test using a rank-based permutation distribution. The method is often used to address one of two questions: (1) Do two distributions differ in some unspecified way (e.g., with respect to location, scale, skewness, or kurtosis) and (2) does a sample come from a particular hypothesized distribution?

For the latter question, we compare an empirical cdf (ecdf) derived from the sample to a hypothesized cdf. Our hypotheses are

$$H_0: \text{ecdf} = F_0(X)$$

versus

$$H_A: \text{ecdf} \neq F_0(X),$$

where ecdf denotes the true empirical cumulative distribution and $F_0(X)$ is the cdf under H_0.

We test the hypothesis using the Kolmogrov–Smirnov test statistic:

$$D* = \max_{i=1}^{n} |F_0(X_i) - \text{ecdf}_i|. \tag{6.16}$$

Our only assumption is that observations are independent.

The test statistic $D*$ is the maximum difference between the empirical cumulative probability for a particular quantile of ranked observed data (ecdf) and the cumulative probability for the same quantile for a distribution specified by H_0 (see Example 3.18 for additional information). If H_0 is true and the assumptions for the test are valid, then $D*$ will be a random outcome from the Kolmogrov distribution. Ties in data will result in conservative *P*-values. Hollander and Wolfe (1999, p. 183) discuss methods for computation of exact Kolomogrov–Smirnov *P*-values in the presence of ties.

EXAMPLE 6.15

In this example, we test if counts for a plant species follow a uniform distribution. The maximum difference in cumulative probabilities between the ecdf and UNIF(0, 8) is shown in Figure 6.8. This difference is the test statistic.

```
plant.count <- c(1, 3, 4, 1, 1, 2, 1, 6, 6, 6, 1, 4, 7, 3, 1, 3, 2, 3,
0, 4, 1, 3, 5, 1, 1, 4, 1, 3, 2, 4)

ks.test(plant.count,"punif", min=0, max=8)
One-sample Kolmogorov-Smirnov test
data: plant.count
D=0.3333, p-value=0.002545
alternative hypothesis: two-sided

Fig.6.8() ##See Ch. 6 code
```

At $\alpha = 0.05$, we reject H_0 and conclude that the plant counts are not uniformly distributed.

Fig.6.8 ## see Ch. 6 code

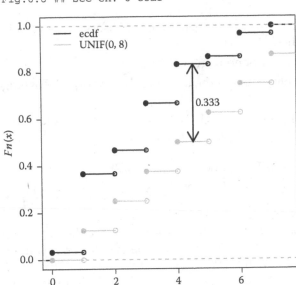

FIGURE 6.8
Empirical cdf for the plant count data compared to a hypothetical uniform cdf, *UNIF*(0, 8). The maximum difference in cumulative probabilities is 0.333. Thus, the statistic D^* occurs at a plant count of 4.

6.6.2.2 *Wilcoxon Sign Rank Test*

The Wilcoxon sign rank test is a nonparametric method for one sample or two paired samples. Thus, it is often used as a substitute for the paired *t*-test. Like the paired *t*-test, the Wilcoxon sign rank test utilizes the differences between paired observations. In the sign rank test, however, the absolute values of these differences are ranked and then the original signs (positive or negative) of the differences are reassigned. In a one-sample setting, ranks of nonzero differences from the value hypothesized in the null are used in the analysis.

Let τ be the true shift in location (Chapter 4) of one population with respect to a second population or hypothesized value.

A two-tailed null hypothesis and alternative hypotheses would be specified as[*]

$$H_0: \tau = 0,$$

$$H_A: \tau \neq 0.$$

The lower and upper-tailed alternatives would be

$$H_A: \tau < 0,$$

$$H_A: \tau > 0.$$

---

[*] Note that while Wilcoxon tests have been described as a method for comparing marginal medians (cf. Minitab), it is inadequate for this purpose (Hettmansperger 1984) and hypotheses should not be stated in terms of population medians. Bauer (1972) describes point and confidence interval procedures for *pseudomedians* that are analogous to results from one- and two-tailed Wilcoxon tests.

The calculation of the Wilcoxon sign rank test statistic, V^*, requires four steps:

1. Calculate the differences between r pairs of observations, or between observations and the hypothesized value.

2. Let n be the number of nonzero differences.

3. Rank the absolute values of the n differences. Assign the average rank in the case of ties.

4. Reassign the signs from step 1. The sum of positive ranks is denoted as T_+, while the sum of negative ranks is denoted as T_-.

Note that $T_- = (n(n + 1)/2) - T_+$ and $T_+ = (n(n + 1)/2) - T_-$.

For a two-tailed test, the Wilcoxon sign rank test statistic, V^*, will be the smallest of the two values T_+ or T_-. In an upper-tailed test, the test statistic V^* is T_-. In a lower-tailed test, the test statistic V^* is T_+.

If H_0 is true and the assumptions for the test are valid, then V^* will be a random outcome from the Wilcoxon sign rank distribution. The two-tailed P-value will be $2 \times P(V \le V^*)$, the lower-tailed P-value will be $P(V \le V^*)$, and the upper-tailed P-value will be $P(V \ge V^*)$, where $V \sim WSignRank(n)$. Ties in data will prevent the calculation of exact P-values under this approach. Hollander and Wolfe (1999, p. 46) discuss methods for computation of exact P-values in the presence of ties.[*]

The sampling distribution of the test statistic V^* will be approximately normal, $N(\mu_V, \sigma_V^2)$, given large sample sizes (>25 pairs of observations, Siegel 1956). Because the permutation distribution will be known (Example 6.13), the parameters for the normal distribution can be directly computed. This is a dramatic departure from conventional frequentist methods, which assume fixed and unknown values for parameters. The population mean of V is

$$\mu_V = \frac{n(n + 1)}{4} \tag{6.17}$$

and the population variance is

$$\sigma_V^2 = \frac{n(n + 1)(2n + 1) - 1/2 \sum_{i=1}^{k} t_i(t_i - 1)(t_i + 1)}{24}. \tag{6.18}[\dagger]$$

where k is the number of tied groups and t_i denotes the number of tied observations in the ith group of ties.

The Wilcoxon sign rank test has the following assumptions:

1. The differences between pairs have a parental distribution that is symmetric (e.g., normal, uniform, and t).

2. Observations within treatments are independent, leading to independent differences among pairs of observations.

---

[*] The package *exactRankTest* provides exact Wilcoxon tests in the presence of ties.
[†] If no ties are present, σ_V^2 reduces to $\sigma_V^2 = (n(n + 1)(2n + 1))/24$.

EXAMPLE 6.16

Diagnostic plots and tests generated from the schizophrenia data in Example 6.6 suggested that the underlying population of paired differences is nonnormal (Figure 6.6). As a result, we will reanalyze these data using the Wilcoxon sign rank test.

We first find the paired differences:

```
data(sc.twin)
D <- with(sc.twin, unaffected - affected)
```

We then rank these differences:

```
rD <- rank(abs(D))
```

And reassign the original signs:

```
srD <- rD*sign(D)
```

The sum of positive ranks is T_+, while the sum of negative ranks is T_-.

```
T.plus <- sum(rD[srD > 0])
T.minus <- sum(rD[srD < 0])
```

For a two-tailed test (the kind we are interested in), the test statistic, V^*, will be whichever is smaller, T_+ or T_-.

```
V.star <- min(T.minus, T.plus)
V.star
[1] 9
```

The two-tailed P-value is $2[P(V < 9)]$, where $V \sim WSignRank(15)$.

```
2 * psignrank(9, n = 15)
[1] 0.00201416
```

We reject H_0 and conclude that brain volumes for schizophrenic and nonschizophrenic twins differ.

We can run this entire test in **R** using the function `wilcox.test`.

```
with(sc.twin, wilcox.test(unaffected, affected, alternative = "two.
sided", correct = FALSE, paired = TRUE))
Wilcoxon signed rank test

data: affected and unaffected
V = 9, p-value = 0.002014
alternative hypothesis: true location shift is not equal to 0
```

The command `correct = FALSE` concerns the Yates correction for discontinuity discussed in Chapter 11.

6.6.2.3 *Wilcoxon Rank Sum Test*

The Wilcoxon rank sum test (Wilcoxon 1945) is often used as the nonparametric equivalent of a two-sample t-procedure for independent samples. The test was introduced at about the same time as another rank-based procedure, the *Mann–Whitney U*-test (Mann and Whitney 1947). Although they derive their test statistics in different ways, the tests are inferentially equivalent. That is, they provide the same P-values given the same hypotheses. Hypothesis statements have the same form as the Wilcoxon sign rank test in the previous example.

The calculation of the Wilcoxon rank sum test statistic, W^*, requires three steps:

1. Rank the data values for *both* samples from the smallest to the largest. If there are ties (duplicated values), the ranks in the data are taken to be the average of the ranks for those observations.
2. Let T_1 denote the sum of ranks from the sample representing population 1 and let T_2 denote the sum of ranks from population 2.
3. Calculate W_1 and W_2:

$$W_1 = T_1 - \frac{n_1(n_1+1)}{2} \tag{6.19}$$

$$W_2 = T_2 - \frac{n_2(n_2+1)}{2}. \tag{6.20}^*$$

Note that $W_2 = n_1 n_2$ and $W_1 = n_1 n_2 - W_2$.

For a two-tailed test, the Wilcoxon rank sum test statistic W^* will be the smallest of the two values W_1 or W_2. In an upper-tailed test, the test statistic W^* will be W_1. In a lower-tailed test, the test statistic W^* will be W_2.

If H_0 is true and assumptions for the test are valid, then W^* will be a random outcome from the Wilcoxon rank sum distribution. The two-tailed P-value will be $2 \times P(W \le W^*)$, the lower-tailed P-value will be $P(W \le W^*)$, and the upper-tailed P-value will be $P(W \ge W^*)$, where $W \sim WRankSum(n_1, n_2)$. P-values will be inexact in the presence of ties under this approach. Hollander and Wolfe (1999, p. 115) discuss methods for the computation of exact P-values in the presence of ties.

The sampling distribution of W^* will be approximately normal $N(\mu_W, \sigma_W^2)$ under null given moderately large sample sizes (>20 for each group, Siegel 1956). The population mean of W will be

$$\mu_W = \frac{n_1 n_2}{2} \tag{6.21}$$

and the population variance of W (adjusted for ties) will be

$$\sigma_W^2 = \frac{n_1 n_2}{12} \left((n_1 + n_2 + 1) - \frac{\displaystyle\sum_{i=1}^{k} t_i(t_i^2 - 1)}{(n_1 + n_2)(n_1 + n_2 - 1)} \right), \tag{6.22}^\dagger$$

where k is the number of tied groups and t_i denotes the number of tied observations in the ith group of ties. For the Wilcoxon rank sum test, we have the following assumptions:

---

[*] Note that in some statistical texts, the unadjusted sum of the ranks is used for W^*; that is T_1 and T_2 are used as the test statistic (S-Plus and Minitab use this method). **R** uses the adjusted ranks as shown in Equations 6.19 and 6.20. This agrees with the most commonly used definition of the test statistic: the number of observations in one sample that are exceeded by each datum in the other sample. The test statistic associated with the first population, W_1 in the hypothesis is always shown as the test statistic, W, in the **R** functions `wilcox.test`, `wilcox _ test` (from *coin*), and `wilcox.exact` (from *exactRankTest*).

[†] If no ties are present, σ_W^2 reduces to $\sigma_W^2 = (n_1 n_2/12)(n_1 + n_2 + 1)$.

1. The parent distributions underlying the samples have similarly shaped distributions (e.g., both right skewed or left skewed).
2. Observations are independent.

EXAMPLE 6.17

Serum β-2 microglobulin is produced in the body as a result of myelomas, and can therefore be used as an indicator of the severity of disease. Murakami et al. (1997) randomly assigned 20 multiple myeloma patients identified through the School of Health Sciences at Gunma University, Japan, to treatment and control groups. The treatment patients received two types of drugs: malphalan and sumerifon, while the control group received only sumerifon. The data are in the dataframe `myeloma`.

Normal probability plots and histograms indicate nonnormal parent populations, particularly for the control subjects (Figure 6.9). This characteristic, in combination with small sample sizes, prevents trustworthy analyses with *t*-tests.

The researchers were interested in the possibility that the control group would have elevated levels of β-2 microglobulin compared to the treatment group. Thus, letting the control treatment represent hypothesized population 1, we have

$$H_0: \tau \leq 0$$

versus

$$H_A: \tau > 0.$$

We first rank the control and treatment serum levels:

```
data(myeloma)
r <- with(myeloma, rank(mglobulin))
```

The sums of the ranks within drug treatments are

```
T.1 <- print(sum(r[myeloma$drug =="Control"]))
[1] 121
T.2 <- print(sum(r[myeloma$drug =="Trt"]))
[1] 89
```

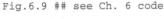

Fig.6.9 ## see Ch. 6 code

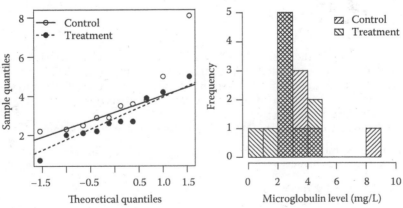

FIGURE 6.9
Normal diagnostics for the β-2 microglobulin data in Example 6.17.

For the Wilcoxon test statistic, we have

$$W_1 = 121 - \frac{10(10+1)}{2} = 66,$$

$$W_2 = 89 - \frac{10(10+1)}{2} = 34.$$

Because this is an upper-tailed test, we use W_1 as the test statistic.

We can get the lower tailed P-value directly by finding $P(W \geq 66)$, where $W \sim WRankSum(10, 10)$. The function pwilcox provides the Wilcoxon rank sum distribution cdf.

```
pwilcox(65, 10, 10, lower.tail = FALSE) #Subtract 1 from W* in upper-
  tailed test since H0 dist. is discrete
[1] 0.1237253
```

The presence of ties makes the P-value inexact.
Using the normal approximation, we have

$$\mu_W = \frac{10 \times 10}{2} = 50$$

and accounting for four ties, each among two observations, we have

$$\sigma_W^2 = \frac{10 \times 10}{12}\left((10+10+1) - \frac{2(2^2-1)+2(2^2-1)+2(2^2-1)+2(2^2-1)}{(10+10)(10+10-1)} \right) = 174.4737.$$

Recall from Chapter 3 that we can convert any quantile in any normal distribution into a standard normal quantile using a standard normal transformation. Our resulting z-score is

$$z* = \frac{W* - \mu_W}{\sigma_W} = \frac{66-50}{13.20885} = 1.2113.$$

The upper-tailed P-value associated with this test statistic is

```
pnorm(1.2113, lower.tail = FALSE)
[1] 0.1128902
```

Because $0.113 > 0.05$, then using the standard significance level and a strictly Neyman–Pearson perspective, we fail to reject H_0 and conclude that serum levels are the same or lower for the control. From a Fisherian perspective, however, the relatively small P-value suggests a potentially meaningful difference between the groups that may be apparent with larger sample sizes.

To obtain an approximate asymptotic P-value, we can use the function wilcox.test.

```
wilcox.test(mglobulin ~ drug, alternative = "greater", correct = FALSE)
Wilcoxon rank sum test

data: mglobulin by drug
W = 66, p-value = 0.1129
alternative hypothesis: true location shift is greater than 0

Warning message: cannot compute exact p-value with ties
```

Again, the command `correct = FALSE` concerns the Yates correction for discontinuity (see Chapter 11).

★ ### 6.6.3 Robust Estimator Tests

As a conclusion to this section, I present a robust method for comparing two groups that does not use rank transformation. This procedure is based on two topics introduced in earlier chapters: M-estimators of location (Chapter 4) and bootstrapping (Chapter 5). Recall that M-estimators of location and scale are robust to outlier contamination. Coombs et al. (1996) found that Welch's approximate t-test was inadequate for comparing the means of heteroscedastic populations if the underlying populations were not normal, or contained outliers, and recommended a test utilizing M-estimators for skewed heteroscedastic distributions.

A bootstrap M-estimator test for comparing two groups is provided by the function `boot.ci.M` in *asbio*. In this procedure, a bootstrap distribution of differences between one-step Huber location estimates is created (Chapter 4). A $(1 - \alpha)100\%$ confidence interval of the true difference in location values is then calculated using the percentile method (Chapter 5) and permutation of P-values for a two-tailed alternative hypothesis are obtained.

EXAMPLE 6.18

Reanalysis of the microglobulin data in Example 6.17 with bootstrapped M-estimators, using a two-tailed alternative hypothesis, gives the result:

```
ci.boot.M(mglobulin[drug=="Control"],mglobulin[drug=="Trt"])

95% Confidence interval for true difference of location measures:
     2.5%        97.5%
-0.5574997   2.3218891

Test results:
        SE     P.value
0.6265163  0.2866000
```

As might be expected, the two-tailed P-value is slightly more than double the Wilcoxon one-tailed P-value.

6.7 Alternatives to Null Hypothesis Testing

This section addresses methods that do not explicitly require a null hypothesis. Two conceptual frameworks are described: Bayesian and likelihood-based.

6.7.1 Bayesian Approaches

Bayesian hypothesis tests represent a fundamental departure from classical frequentist methods since parameters are viewed as nonfixed entities and hypotheses are evaluated with the degrees-of-belief conception of probability (Chapter 2).

Consider a research problem with c-competing hypothesis $\{H_1, H_2, \ldots, H_c\}$. To evaluate H_1 using a Bayesian approach, we compute the posterior probability $P(H_1 \mid \text{data})$ using

$$P(H_1 \mid \text{data}) = \frac{P(\text{data} \mid H_1)P(H_1)}{P(\text{data})}. \tag{6.23}$$

We can compare competing hypotheses by creating a ratio of their posterior probabilities. For instance, to quantify the degrees of belief for H_2 compared to H_1, we could obtain the *posterior odds ratio* (see Sections 2.4 through 2.6)

$$\frac{P(H_1 \mid \text{data})}{P(H_2 \mid \text{data})}. \tag{6.24}$$

We can derive posterior odds by multiplying the prior odds by an entity called the *Bayes factor* (Jeffreys 1935, Kass and Raftery 1995)

$$\frac{P(H_1 \mid \text{data})}{P(H_2 \mid \text{data})} = B_{1,2} \times \frac{P(H_i)}{P(H_2)}, \tag{6.25}$$

where

$$B_{1,2} \text{ is the Bayes factor for hypotheses 1 and } 2 = \frac{P(\text{data} \mid H_1)}{P(\text{data} \mid H_2)}. \tag{6.26}$$

When hypotheses are simple (i.e., $P(H_1)$ and $P(H_2)$ correspond to single values), $B_{1,2}$ is a *likelihood ratio* (see Example 6.20). Likelihood ratios are frequently used to inferentially compare candidate models or to compare a candidate model to a null model. When hypotheses are complex (i.e., $P(H_1)$ and $P(H_2)$ correspond to probability distributions with unknown parameters), then probabilities for the numerator and denominator in Equation 6.26 are obtained by integrating over the range of values specified in the hypotheses using

$$P(\text{data} \mid H_i) = \int f(x_1, \ldots, x_n \mid \theta_i, H_i) P(\theta_i \mid H_i) d\theta_i, \tag{6.27}$$

where θ_i is the parameter under consideration by H_i, $f(x_1, \ldots, x_2 \mid \theta_i, H_i)$ is the probability density function for the data evaluated across the range of data values: $x_1, \ldots, x_n$ (i.e., the range of the definite integral), and $P(\theta_i \mid H_i)$ is the prior probability density function for θ_i assumed by H_i.

EXAMPLE 6.19

Hemophilia may dangerously decrease the capacity of an affected individual to clot blood if blood vessels are broken. The disease is a sex-linked recessive disorder that is carried on the X chromosome. Thus, the male offspring of a female carrier will have a 50% chance of inheriting the disease, and a male offspring of a female who is affected will always express the disease (the male Y-chromosome will not contain an allele to allow heterozygosity). Consider a situation where a woman (let us call her Mary) has a mother and father who do not express hemophilia, but has a brother who does express hemophilia. Because of this, we know that the mother was a carrier and that Mary has a 50% chance of being a carrier. Consider further that Mary has three sons, none of whom express hemophilia. We have two competing hypotheses: H_1: Mary is a carrier, and H_2: Mary is not a carrier. We have

$$P(\text{data} \mid H_1) = (0.5)(0.5)(0.5) = 0.125,$$

$$P(\text{data} \mid H_2) = (1)(1)(1) = 1.$$

The first probability is computed assuming that Mary is a carrier. If this assumption is true, then the probability that a son *would not* inherit the allele is 0.5 and the probability that no sons inherit the allele (which is what was actually observed) is 0.5^3. The second probability is computed assuming that Mary is not a carrier. In this case, the probability that a son would not inherit the allele is 1 and the probability that no sons inherit the allele (which is what was observed) is 1. The Bayes factor is

$$B_{1,2} = \frac{P(\text{data} \mid H_1)}{P(\text{data} \mid H_2)} = \frac{0.125}{1} = 0.125.$$

We get the posterior odds by multiplying the Bayes factor by the prior odds [the odds of Mary being a carrier (or not) before considering her offspring]. Because the priors are equal (both = 0.5), the Bayes factors will equal the posterior odds.

$$\frac{P(H_1 \mid \text{data})}{P(H_2 \mid \text{data})} = B_{1,2} \times \frac{P(H_1)}{P(H_2)} = 0.125 \frac{0.5}{0.5} = 0.125.$$

The posterior odds in favor of H_1 is 0.125.[*] Performing the appropriate calculations, we find that the posterior odds in favor of H_2 is 8. Therefore, we choose H_2 over H_1.

Bayesian alternative approaches have been proposed for a large number of frequentist analyses (Box and Tiao 1973, Berry and Stangl 1996, Gelman et al. 2003, see Chapters 9 and 10). In this capacity, P-values from conventional frequentist test are often smaller than the posterior probability of comparable Bayesian models, even when H_0 and H_A are given equal (noninformative) priors (Kass and Raferty 1995). One explanation is that the probability space of the frequentist test statistic representing H_0 can often greatly exceed what can possibly be sampled, exaggerating the evidence against H_0 (see criticism 1 in Section 6.5; Casella and Berger 1987). In addition, Bayes factors are consistent (Section 4.3) for identifying the true model or hypothesis in a number of different settings (Dass and Lee 2004), and can identify parsimonious models (Chapter 9) without explicit inclusion of penalty terms (Jefferys 1961).

While a Bayesian approach addresses several criticisms of frequentist hypothesis testing, it has its own drawbacks. First, priors (required for any Bayesian analysis) may be subjective and/or difficult to acquire (Edwards 1996). Second, a frequentist approach with fixed parameters often makes more sense. For instance, the variable height measured within a biological population for a particular point in time will have a single (if unknown) parameter value. Third, as evident in Markov chain models from Chapter 5, Bayesian analyses can be complex. Given the frequent misapplication of classical statistical tests by biologists

---

[*] We can also directly derive the posterior odds ratio (and in this case Bayes factors) using Bayes theorem:

$$P(H_1 \mid \text{data}) = \frac{P(\text{data} \mid H_1)P(H_1)}{P(\text{data} \mid H_1)P(H_1) + P(\text{data} \mid H_1')P(H_1')} = \frac{0.125 \times 0.5}{(0.125 \times 0.5) + (1 \times 0.5)} = 0.1111111$$

$$P(H_2 \mid \text{data}) = \frac{P(\text{data} \mid H_2)P(H_2)}{P(\text{data} \mid H_2)P(H_2) + P(\text{data} \mid H_2')P(H_2')} = \frac{1 \times 0.5}{(1 \times 0.5) + (0.125 \times 0.5)} = 0.8888889$$

$$\frac{P(H_1 \mid \text{data})}{P(H_2 \mid \text{data})} = \frac{0.1111111}{0.8888889} = 0.125$$

$$\frac{P(H_2 \mid \text{data})}{P(H_1 \mid \text{data})} = \frac{0.8888889}{0.1111111} = 8.000001$$

(Heffner et al. 1996), the misuse of Bayesian approaches seems even more likely (e.g., Van Dongen 2006).[*] Fourth, Bayesians are not freed from parametric assumptions. The proper use of distributions for priors and likelihood functions requires these to be legitimate reflections of reality. Checks of these assumptions may involve diagnostic procedures, similar to those used for conventional frequentist approaches (Petit 1986). Because of these problems Bayesian methods are generally used for defining the probability space of parameters, and not for hypothesis testing (Kass and Raftery 1993, p. 43).

6.7.2 Likelihood-Based Approaches

Leading scientific and statistical figures have long argued for the inferential use of likelihood over frequentist *P*-values (Barnard 1951, Hacking 1965, Edwards 1972, Royall 1997, Lindsay 1997, Bain and Engelhart 1992).[†] Many of the desirable characteristics of a Bayesian approach, for example, nonviolation of the likelihood principle, insensitivity to sample size (at least for likelihood ratios), and comparisons of nonnested models (Chapter 9), are also implicit in likelihood analyses. Likelihood ratios can be created using proportional model likelihoods. The likelihood ratio of the ith model or hypothesis compared to the jth model or hypothesis is simply

$$\mathcal{L}(H_i | \text{data}) / \mathcal{L}(H_j | \text{data}) \qquad (6.28)$$

As with Bayesian methods, likelihood-based methods require an understanding of the working of probability density functions and do not universally preclude the usefulness of frequentist null hypothesis tests (Burnham and Anderson 2002, p. 83).

EXAMPLE 6.20

Bayes factors will be equivalent to likelihood ratios in the case that hypotheses are simple and processes are discrete. Further, if the priors for the hypothesis are equal, then the Bayes factors will also equal the posterior odds. Thus, for the hemophilia example, the likelihood ratios equal the Bayes factors and the Bayes factors equal the posterior odds. The likelihood ratio of H_1 relative to H_2 is 0.125 and the likelihood ratio of H_2 relative to H_1 equals 8. Thus, again, we choose H_2 over H_1.

6.8 Summary

This chapter introduces topics associated with hypothesis testing.

- The null hypothesis was developed because a hypothesis cannot be deductively proven with data, although it can be falsified. Through falsification of H_0, the research hypothesis can be supported.

---

[*] As Burnham and Anderson (2002, p. 104) noted: "... investigators should understand the methods leading to the results of their work."

[†] John Nelder (1999) stated: "at least once a year I hear someone at a meeting say that there are two modes of inference: frequentist and Bayesian. That this sort of nonsense should be so regularly propagated shows how much we have to do. To begin with there is a flourishing school of likelihood inference, to which I belong."

- Two progenitors of modern null hypothesis testing frameworks are the Fisher and Neyman–Pearson approaches. With the former, a P-value (the probability of seeing an outcome as or more extreme than the one observed if H_0 is true) is regarded as a continuous measure of the strength of evidence against H_0. The latter considers type I and II errors, and power (i.e., the capacity to reject a false null hypothesis). Most modern scientists use a hybrid of their approaches.

- A large number of frequentist tests for making inferential statements about one or two populations are described in this chapter, including the family of t-tests and nonparametric analogs.

- The frequentist approach to null hypothesis testing has been criticized by a large number of individuals, giving rise to alternative procedures. Alternatives to frequentist null hypothesis testing include permutation tests using ranked-transformed or nonranked data. Alternatives to null hypothesis testing in general include Bayesian and likelihood-based approaches.

EXERCISES

1. Define the following terms:
   a. Null hypothesis
   b. Alternative hypothesis
   c. Test statistic
   d. P-value
   e. Significance level
   f. Critical value
   g. Decision rule
   h. Type I error
   i. Type II error
   j. Power

2. It is likely that you will use null hypothesis testing constantly in your career as a biologist. To increase your familiarity with null hypothesis testing, describe (in your own words) major criticisms to this approach.

3. Contrast what the word falsification meant to philosophers of science Popper, Kuhn, and Lakatos (see Chapter 1).

4. The height (in inches) of female freshmen at Poky High has the distribution $N(64.5, 16)$. Of interest is whether the height of honors students is different from the overall population. With this in mind, I take a random sample of 20 female freshmen and find that the mean height is 61 inches.
   a. State the null and alternative hypotheses.
   b. Conduct a null hypothesis test following the steps described for the hybrid method and verify your result using the function `one.sample.z`. State your conclusions correctly.

5. Ott and Longnecker (2004) describe an investigation in which a pollution control inspector suspects that a riverside community is releasing semitreated sewage into a river, causing the river to become eutrophic. He records dissolved O_2 readings in parts per million for 15 random locations above and below the riverside community.

Test if the O_2 below the town is lower (more eutrophic) than the O_2 above the town. Assume that the variances are equal. Use $\alpha = 0.05$. The data are in the dataframe dO2.

a. Correctly state the hypotheses in terms of population parameters.
b. Use **R** to calculate sample means and variances.
c. Calculate the pooled variance.
d. Calculate the test statistic.
e. Calculate the P-value.
f. Verify your test result by running the whole test in **R**. Attach this output to homework.
g. State your conclusions.

6. Polychlorinated biphenyls (PCBs) are a group of synthetic oil-like chemicals whose toxicity was first recognized in the 1970s. Until then, they were widely used as insulation in electrical equipment, particularly transformers. PCB concentrations in bird eggs helps researchers quantify bioaccumulation of PCBs in ecosystems. Thirteen sites in the Great Lakes were selected for a study to quantify PCB concentrations in 1982 and 1996 (Hughes et al. 1998). At each site, 9–13 American herring gull (*Larus smithsonianus*) eggs were randomly collected and tested for PCB content. Test to see if 1996 levels were lower than 1982 levels using a two-sample t-test that assumes unequal variances. Use $\alpha = 0.01$. The data are in the dataframe PCB.

a. Correctly state the hypotheses in terms of population parameters.
b. Use **R** to calculate sample means and sample variances.
c. Calculate the degrees of freedom, ν, by hand.
d. Calculate the test statistic.
e. Calculate the P-value.
f. Verify your answer by running the appropriate test in **R**. Attach this output to homework.
g. State your conclusions.
h. Does the independence assumption for this test appear to be violated? Why? What can you do about it?

7. To compare two different weight loss programs (X and Y) and to control for the confounding potential of genetics, 12 pairs of identical twins of similar weight were studied. Each pair of twins was randomly assigned to program X or Y and the amount of lost weight in pounds was recorded. Test if program X is superior to program Y.

X	12.42	9.31	6.83	11.51	10.42	8.87	6.73	9.53	8.8	8.01	7.01	9.69
Y	13.8	10.0	8.51	11.95	10.66	8.76	7.93	11.81	11.62	9.76	9.20	11.20
$X - Y$	−1.38	−0.69	−1.68	−0.44	−0.24	0.11	−1.2	−2.28	−2.82	−1.75	−2.19	−1.51

a. State your hypotheses.
b. Calculate the appropriate test statistic by hand.
c. Calculate the P-value.
d. State your conclusions.
e. Verify your answer by running the appropriate test in **R**. Attach this output to homework.

8. Repeat Example 6.6 after log transforming the data.

9. Type `see.typeI_II()` or open the book menu [by typing `book.mneu()`] and go to *Ch6 > Type I and II error.* Click the *More info* widget to learn more about type I and II errors and power (you will need to click at the edge of the widgets).

   a. What is considered a more serious type of error, type I or II? Use information from this chapter to explain why this is true.

   b. What are the conventional (most frequently used) levels for α, β, and $1 - \beta$?

10. Type `see.power()` or go to *Ch6 > Power.* The top graph shows the null distribution, which assumes that H_0 is true. This is the one we use to compute our *P*-values. The lower graph shows power, that is, the probability of rejecting H_0 if H_0 is false.

    a. Does power $= 1 - \alpha$?

    b. Manipulate the slider widgets. What are the four ways to increase power?

11. Type `see.ttest.tck()` or click on *t-test mechanics* button from Chapter 6 pull-down menu in the `book menu()`.

    a. What do the normal distributions represent? What do the points under the curves represent? Why are the degrees of freedom for the *t*-distribution nonintegers when you turn off the *Variance equal* widget?

    b. Set the populations to have equal population means. In this case, the null hypothesis is true. Resample from the populations repeatedly (>30 times) by clicking on the *Refresh* button. Is it still possible to reject H_0 at $\alpha = 0.05$? Why? What is this called?

12. Describe how hypothesis testing involves both *reductio ad absurdum* and *modus tollens* (Chapter 1) arguments.

13. Test to see if the following data are normally distributed using the Kolmogrov–Smirnov test (note that this is a poorer test for assessing normality than the Shapiro–Wilk test):

    ```
    Q13 <- c(6.466, 3.254, 2.869, 3.781, 10.861, 2.084, 2.431, 4.841,
    0.269, 5.024, 5.500, -1.300, 5.674, 6.791, 8.867)
    ```

14. Lloyd et al. (1969) studied the transferal of tritiated water (water containing a radioactive isotope of hydrogen with detrimental health effects) across the human chloroamnion (placental membrane). Permeability to tritium was measured for two groups of membranes, A and B. The data are

    ```
    A <- c(0.80, 0.83, 1.89, 1.04, 1.45, 1.38, 1.91, 1.64, 0.73, 1.46)
    B <- c(1.15, 0.88, 0.90, 0.74, 1.21)
    ```

    a. Check for normality using normal probability plots and histograms.

    b. Conduct a Wilcoxon rank sum test by hand (without using `wilcox.test`) for the alternative hypothesis that A is shifted above B. Confirm your results using `wilcox.test`.

    c. Calculate a *z*-score from your Wilcoxon test statistic and calculate the *P*-value. How does this compare to the Wilcoxon *P*-value calculated above?

    d. Repeat the test of the Mann–Whitney hypotheses using the randomization test `MC.test`.

     i. What are your results? Do they agree or disagree with the Mann–Whitney results? Why or why not? Attach the results.

     ii. Run the randomization test again (and again, *ad infinitum* if you wish). Are your results the same? Why or why not? If they *are* different, how different *are* they? Attach the new results.

15. Hollander and Wolfe (1999) presented Hamilton depression scale-factor measurements for nine patients with mixed anxiety and depression. The measurements were taken at two times; a time preceding administration of the tranquilizer and a time after tranquilizer administration. Conduct the appropriate rank-based permutation analysis by hand (using **R** to help) and confirm your result in **R**. Use a two-tailed alternative with a significance level of 0.05. The data are in the dataframe `depression`.

16. Give scenarios when we would want to minimize type I and type II error. Come up with other examples than those used in Example 6.9. Give a scenario for each type of error.

17. You are a reviewer on a study plan detailing sample sizes. You find that the power for the experiment to be conducted = 0.2. What are your comments?

18. You are writing a large grant proposal to the National Science Foundation (NSF) to find if high amounts of cheat grass (*Bromus tectorum*) decrease sage grouse (*Centrocercus urophasianus*) percent survivorship. On the basis of previous work, you conclude a reasonable value for σ is 5%. Decreases in sage grouse of greater than 3% would be particularly relevant. What sample size should you use given $\alpha = 0.05$ and $1 - \beta = 0.8$? Perform calculations by hand and confirm your results using `power.z.test`.

19. Address the following questions:

    a. What are the advantages and disadvantages of frequentist parametric tests?

    b. What are the advantages and disadvantages of permutation tests?

    c. What are the advantages and disadvantages of rank-based permutation tests?

20. What are some alternatives to null hypothesis testing in general? What are the factors supporting and disparaging these approaches?

21. Define the likelihood principle. Describe how it can be used to support or disparage a Bayesian approach to statistical inference.

22. Describe what is happening at each line of code in the `for` loop in Example 6.13.

23. Consider an experiment with three samples from X_1 and three samples from X_2. Excluding ties, what are all the possible ranks for X_1 and X_2?

24. Three recent surveys of Lake Victoria have failed to turn up a rare species of cichlid that persisted in the lake 5 years ago. The electrofishing technique used to survey the lake has a probability of 0.78 for detecting the cichlid species when it is really there; that is, $P(Det^+|C^+) = 0.78$, $P(Det^-|C^+) = 0.22$.

    Compare the hypothesis that the cichlid is extinct versus the hypothesis that it is not extinct using a Bayesian approach. Use uninformative uniform priors, that is, $P(C^+) = 0.5$ and $P(C^-) = 0.5$. Compare the hypotheses using posterior odds and Bayes factors.

25. Consider the binomial occurrence of male and female births where a male is a success. We observed 11 males and nine females and considered two hypotheses H_1: $\pi = 0.5$ and H_2: $\pi = 0.25$.

    a. Calculate $P(\text{data}|H_1)$ and $P(\text{data}|H_2)$.

    b. Calculate $\mathcal{L}(H_1|\text{data})$ and $\mathcal{L}(H_2|\text{data})$.

    c. Calculate $\ln\left[\dfrac{\mathcal{L}(H_1|\text{data})}{\mathcal{L}(H_2|\text{data})}\right]$. This is the log-likelihood ratio.

    d. Is our approach here Bayesian? Why or why not?

7

Sampling Design and Experimental Design

Theories are nets we cast to catch "the world."

K. Popper (1959)

7.1 Introduction

In the preceding chapters, we have been introduced to foundational concepts in biostatistics, including historical and philosophical views of science (Chapter 1), probability (Chapter 2), probability density functions (Chapter 3), parameters and point estimation (Chapter 4), sampling distributions and interval estimation (Chapter 5), and hypothesis testing (Chapter 6). In this chapter, we conclude the consideration of foundational topics by presenting information to allow effective planning for research and analysis. Of particular importance, sampling and experimental design are formally introduced. We will see that careful consideration of these processes is required to meet two important goals of inferential statistics: (1) extending inference to the population(s) of interest and (2) providing evidence of cause and effect.

7.2 Some Terminology

Before addressing the core topics of this chapter, it will be helpful to define some terms.

7.2.1 Variables

In Chapter 2, we learned that a variable is a measurable phenomenon whose outcomes vary from measurement to measurement, while a *random* variable will have outcomes that cannot be predicted with certainty in advance. Scientific *data* consist of empirical observations (e.g., measurements) from variables. Data are observed and recorded for entities called *experimental units* or *sampling units*. For instance, to acquire data concerning plant height (a variable), we measure the heights of individual plants (experimental units). Experimental units are generally called *subjects* if measurements taken are on humans. An experimental unit can be measured with respect to one or many variables, and can be measured at one or at many points of time.

The methods and tools to collect data are strongly associated with particular fields of biology. A stream ecologist may use a Surber sampler[*] to disturb random areas of stream bottoms and identify the diversity of benthic insects. A fisheries biologist may use a rotary screw trap,[†] along with mark–recapture methods to monitor fish populations in a river. A molecular biologist may use polymerase chain reactions[‡] to obtain many copies of a gene and determine its allelic variability. A plant ecologist may use special sampling frames to estimate plant species abundance. A plant physiologist might use an infrared gas analyzer[§] to measure atmospheric carbon flux at the leaf surface, and so on.

Despite this diverse array of tools and techniques for observing and measuring biological phenomena, the goal of data analysis is always the same: *make valid inferences about the system(s) that produced the data.*

7.2.1.1 Explanatory and Response Variables

Scientists are often concerned with observing a variable and then trying to explain it. *Explanation*, in turn, is often accomplished by identifying a cause for the variable. As a result, scientists are often interested in analyses that make logical connections between cause and effect, that is, *causality*.

If two variables are *associated*, then some measures of one variable tend to occur more often with certain values of the second variable than with other values of the second variable. If these variables are quantitative, then their association requires that they are *correlated*. That is, one variable will increase or decrease as the other variable increases or decreases (Chapter 8). An association may be causal, as defined above, or it may not. For instance, the distance of the earth to the moon may be correlated with the number of traffic accidents in downtown Los Angeles, but this does not require that this distance is somehow causing the car accidents to happen. Such circumstances explain the useful dictum: "correlation (mere association) does not imply causation." More specifically, causation requires correlation, but correlation does not require causation. The distinction of causal and noncausal associations is often difficult, but extremely important for valid inference.

In the context of causality, we can distinguish two types of variables. The first is called the *explanatory variable* or *predictor variable* and is hypothesized to exert influence on (i.e., determine) a second type of variable, the *response variable*. The explanatory and response variables are often called *X and Y variables*, respectively.

Examples include

- Parental genotype (X) → Offspring genotype (Y)
- Temperature (X) → Enzymatic activity (Y)
- Mutagenic agent (X) → Number of mutations (Y)

---

[*] A Surber sampler is a small cylinder with a sock-like net extension permeable to water. A benthic area is disturbed and objects liberated from the benthos (including aquatic organisms) are filtered into the net.

[†] A rotary screw trap is a bladed cone that funnels fish into a collection area where they can be counted and released without harm.

[‡] An assay in which a portion of DNA representing a genetic region of interest is replicated many times by adding a primer specific to the region, and cycling temperatures rapidly in the presence of heat-resistant DNA polymerase.

[§] Because it is a greenhouse gas, CO_2 is opaque to infrared radiation. As a result, CO_2 fluxes due to photosynthesis can be measured using the absorption of infrared radiation in a leaf chamber.

In some cases, the assignment of X and Y variables is unclear, and could be reversed given a different scientific question. For instance, consider annual precipitation and net primary productivity (NPP) at some site. At first glance, it is most reasonable to see precipitation as the X variable and NPP as the Y variable, because plants require water for CO_2 fixation, which is used to create biomass. However, given that a terrestrial ecosystem and a lake cover the same planar area, the amount of water released by terrestrial plants to the atmosphere via transpiration may far exceed the amount of water evaporated from the lake's surface. Indeed, tropical rain forests largely recycle their precipitation back to the atmosphere, and effectively determine their own climate. From this large spatial perspective, the assignment of X and Y variables could easily be reversed.

In other cases, an investigator may be unwilling to ascribe explanatory and response labels, or such labels may be unimportant. For instance, the variables' height and weight are generally correlated in organisms, but their causal relationship (if any) may be difficult to deduce. In these situations, the relationship of variables can be studied with methods that test for independence, or that quantify correlation without considering causation (see Chapters 8 and 11).

7.2.1.2 Categorical, Ordinal, and Quantitative Variables

Variables can also be classified as categorical, ordinal, or quantitative.

With *nonquantitative* or *categorical variables*, outcomes are either nonnumeric or are numeric indices with no quantitative meaning. For instance, the sex of a deer could be recorded as "M" or "F," as "XY" or "XX," or as 1 and 0. In all cases, variable outcomes will have no quantitative meaning. That is, a deer of class 1 is not necessarily larger (in any sense) than a deer from class 0. Categorical variables were tacitly used with hypothesis testing examples in Chapter 6. For instance, in Example 6.7, we examined a quantitative variable: proportional deer birth weight (dam mass/total litter mass) with respect to a categorical variable: geographical region. The first was a response variable (Y), while the second was an explanatory variable (X) (see Table 7.1).

Outcomes from *ordinal variables* have some quantitative meaning, although it is imprecise. Suppose a scientist records a qualitative soil water index (from 1 to 10 indicating dry to wet) at sites she is studying. While this system affords a relative measure of water in the soil (10 is wetter than 1), it does not provide an exact measure of outcomes with respect to each other. An index score of 10 probably does not indicate that the soil is exactly 10 times wetter than a soil with a score of 1. Ordinal variables are often referred to as *categorical variables with ordered classes*. This definition intimates that rank-transformed variables (Chapter 6) are ordinal.

In contrast to the preceding types of variables, outcomes from a *quantitative variable* will have a precise numerical meaning. For instance, if we record temperatures of 10°C and

TABLE 7.1

Three Observations from a Dataset Describing White-Tailed Deer Reproduction

Y Proportional Birth Mass	X Region
0.81995	BH
0.11546	ER
0.13356	BH

Source: Adapted from Monteith et al. 2009. *Journal of Mammalogy* 90(3): 651–660.

20°C, we know that the second measurement is exactly twice as warm as the first, with respect to the baseline 0°C.

7.2.1.3 Discrete and Continuous Variables

Quantitative variables will be either discrete or continuous (see Chapter 3). *Discrete variables* have outcomes that are discontinuous. An example would be counts of mountain goats at a particular location and time (Example 3.1). Here, outcomes will always be *natural integers*, that is, $X = \{0, 1, 2, \ldots\}$. *Continuous variables* have no conceptual breaks. As a result, the capacity to distinguish outcomes depends solely on the resolution of the measuring device used to gather the data. An example would be crop yield, measured in kilograms, from an agricultural experiment. Within the range of its support, this variable will not have any conceptual breaks. That is, any interval bounded by two distinct yields could contain an infinite number of other distinct yield outcomes.

7.2.1.4 Univariate and Multivariate Analysis

Another important distinction can be made between univariate and multivariate analyses. These are distinguishable by the number of response variables. In a *univariate analysis*, there will be only one response variable, although it may be modeled to be the simultaneous function of several explanatory variables. In a *multivariate analysis*, several response variables will be examined simultaneously with respect to one or more explanatory variables.[*]

EXAMPLE 7.1

In Example 6.7, we used two variables from a dataset that described life history characteristics of white-tailed deer (*Odocoileus virginianus*) in South Dakota (Monteith et al. 2009). Table 7.1 shows the first three observations from this dataset. There is only one *Y* variable; therefore, a univariate analysis is required.

Müller and Hothorn (2004) studied the habitat of tree pipits (*Anthus trivialis*), a small passerine bird common to Europe and temperate western and central Asia. The authors established eighty-six 1-hectare grids in oak forests of the Franconian region of northern Bavaria, and measured bird densities and environmental variables at each grid. Table 7.2 shows data from three grids. We note that there are four explanatory variables, but only one *Y* variable. Thus, an analysis of these data would again be univariate.

In 1982, Dutch researchers examined the plant community structure and abiotic characteristics of dune meadows on Terschelling, a small island in the North Sea (Batternik and Wijffels 1983). The authors were interested in the way the entire plant communities responded to land management practices and natural gradients. Some of these data are shown in Table 7.3. There are three response variables: the cover of the vascular plant species *Achillea millefolium*, *Agrostis stolonifera*, and *Elymus repens*. There are also two predictors: ordinal variables that describe soil moisture and manure application. Because there are multiple *Y* variables, an analysis will require a multivariate approach.

While this book does not emphasize multivariate analyses, several pertinent applications are discussed in Chapters 5, 10, and 11. Multivariate approaches were used with complex Bayesian applications in Section 5.5. Multivariate approaches for hypothesis

---

[*] This definition is in disagreement with some authors who define any analysis involving more than two variables as multivariate (e.g., Everitt 2005). It *is* in agreement, however, with most modern biometric texts (cf. Gotelli and Ellison 2004, Quinn and Keough 2002, etc.).

TABLE 7.2

Subset of Data Describing the Habitat of European Tree Pipits

Grid	Y Bird Count	X_1 Stand Age (years)	X_2 Oak Overstory Cover (%)	X_3 Number of Dead Trees	X_4 Distance to Forest Edge (m)
6	2	200	80	1	30
7	0	200	70	0	40
8	0	200	75	0	60

Source: Adapted from Müller J., and Hothorn, T. 2004. *European Journal of Forest Research* 123: 219–238.

Note: The complete dataset is contained in the *coin* package under the name `treepipit`.

testing will also be considered in the context of repeated measures designs (Section 10.12) and tabular designs (Chapter 11).

It is important to note that while the explanatory and response labels for the variables in the deer, tree pipit, and dune plant community investigations discussed in Example 7.1 are appropriate, *none of these studies will generate causal evidence with respect to X on Y*. This is because the experimental designs are nonrandomized (see Sections 7.6 and 7.6.2). Thus, inferences depend not only on explanatory/response variable designations but also on the underlying characteristics of the sampling and experimental design. These topics are the focus of this chapter.

7.2.1.5 *Lurking Variables and Confounding Variables*

It is often easier to decide on which response variable(s) to measure than to identify a set of important predictors. This is because a particular response variable will generally be central to a research hypothesis, while a large number of predictors may influence the response. As a consequence, important predictors may also go unmeasured (they may remain *lurking*). Furthermore, if predictors are correlated (X_1 increases or decreases as X_2 increases or decreases), then their effects on the response may be difficult to distinguish (they may be *confounded*).

Lurking variables are unmeasured variables that may be important in explaining the response. A lurking variable will often result in false claims about causality of X on Y. This is because, although X may appear to determine Y, the true causal variable remains unmeasured. Consider a case where a response variable Y and an explanatory variable X

TABLE 7.3

Dune Meadow Biotic and Abiotic Data

Site	Y_1 ACMI (% Cover)	Y_2 AGST (% Cover)	Y_3 ELRE (% Cover)	X_1 Moisture (1–5)	X_2 Manure (1–4)
2	3	0	4	1	2
13	0	5	0	5	3
4	0	8	4	2	4

Source: Adapted from Batternik, M., and Wijffles, G. 1983. Een verglijkend vegetatiekundig onderzoek naar de typologies en invloeden van het beheer van 1973 tot 1982 in de Duinweilanden op Terschelling. Report of the Agricultural University, Department of Vegetation Science, Plant Ecology and Weed Science, Wageningen.

Note: The complete dataset is available from the package *vegan* under the names `dune` and `dune.env`.

```
Fig.7.1()  ##See Ch. 7 code
```

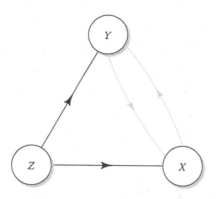

FIGURE 7.1
A situation in which both the response variable, Y, and the explanatory variable, X, are causally associated with a lurking variable, Z. The association between X and Y is merely correlative.

may each be associated with a third lurking variable, Z. Assume further that Z has a causal association with Y, while X does not. Although X does not "cause" Y, the association may appear causal because of the association of Y with Z (Figure 7.1).

EXAMPLE 7.2

For cities in the United States, a researcher notices that there is a strong association between the number of taverns and the number of churches. The investigator concludes that the number of churches is somehow determining the number of taverns. However, she has not considered that the number of churches and the number of taverns are both likely to be causally associated with the unmeasured (lurking) variable population size.

Confounded variables are predictors that are indistinguishable with respect to their effect on the response. As a result, we may be unsure about their individual causal effects. Confounding may result from measured predictors being associated (Figure 7.2), poor

```
Fig.7.2() ##See Ch. 7 code
```

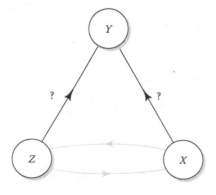

FIGURE 7.2
A situation in which two predictors X and Z are confounded. As a result, we are unsure about whether the relationship of these variables to Y is causal.

control of unmeasured (lurking) predictors, or from inadequate consideration of explanatory variable interactions (Sections 7.6 and 10.8).

EXAMPLE 7.3

The optimal environments of plants vary with respect to a suite of soil variables, including rockiness, pH, and water availability. These variables tend be highly correlated. This is because as soil rockiness increases, the water-holding capacity of the soil decreases. This, in turn, causes soil pH to increase because leachable bases are not lost in saturated soils. Because these variables are confounded, the causal effect (if any) of an individual soil variable on a plant response will often be difficult to discern.

7.3 The Question Is: What Is the Question?

Before gathering data, a researcher should be able to define a series of working research hypotheses concerning his or her study system. A *research hypothesis* is a general, usually nonquantitative statement describing a scientifically testable process. In contrast, a *statistical hypothesis* (Chapter 6) is a mathematical abstraction that (hopefully) represents key aspects of the research hypothesis. Statistical hypotheses are quantitative, and (in a parametric context) make inferential statements concerning parameters. A number of logical problems arise when research hypotheses for an investigation are identical to statistical hypotheses (James and McCulloch 1985, Quinn and Keough 2002).

EXAMPLE 7.4

In 2008, autism spectrum disorders (ASDs) were estimated to affect 1 in 88 children in the United States, representing a 2100% increase in the last 25 years (McDonald and Paul 2010). While a single causal agent for ASDs does not exist, it is likely that this escalation is due to genetically susceptible individuals exposed to novel environmental triggers (Baio 2012).

Thomas and Klaper (2012) hypothesized that the presence of unmetabolized psychoactive pharmaceuticals (UPPs), including prescription antidepressants in streams, may cause changes in neurological processes linked to autism. This statement constitutes a general *research hypothesis*. The investigators randomly assigned fathead minnows (*Pimephales promelas*) to UPP tanks or control tanks (free of pharmaceuticals). After 18 weeks, the fish were euthanized and the expression profiles of human ASD-associated genes (and 11 other gene sets) were examined. The investigators tested the *statistical null hypothesis* that the expression of genes was identically enriched (up- or downregulated) in the UPP and control fish. They rejected this null hypothesis, supporting the premise that UPPs cause changes in the products of genes associated with ASDs.

7.3.1 Four Types of Questions

The formulation of compelling research hypotheses and questions is among the most challenging and important tasks undertaken by a scientific investigator. There are at least four general types of questions that may arise in biological studies.

1. *Does X cause Y to happen or does Y vary with X?* Much of science involves observing a phenomenon and then trying to explain it, and explanation is often accomplished by identifying the cause of the phenomenon. Consider a reasonable hypothesis

that current evapotranspiration patterns on earth are a function of solar incident radiation (short-wave solar radiation hitting the earth's surface). If evidence can be acquired supporting this hypothesis, then progress has been made in explaining the phenomenon of interest, evapotranspiration. Often, however, causal variables may be unidentifiable, or the experimental design will not allow causal inference. In this case, useful questions concerning the noncausal association of variables can commonly be addressed.

2. *Is phenomenon under study consistent with an existing model or hypothesis?* Not exclusive of other questions is the possibility that an observed process may be consistent (or inconsistent) with a proposed or hypothetical model. Of course, such considerations require that an investigator is familiar with previous work in his field including theoretical and mathematical models. The support, refinement, and refutation of such frameworks are fundamental to the progress of science (Chapter 1).

3. *Given data from Y, what is the best estimate of the parameter* θ*?* Parameter estimation (Chapters 4 and 5) is often the chief concern of biological investigations. For instance, a biologist may be interested in estimating the true growth rate for a population of mountain goats (Example 4.6), the true variability in soil elemental characteristics (Example 4.11), or the Bayesian parameter space for the probability of a male birth in the United States (Example 4.19).

4. *What configuration of X is the best for predicting Y?* This is the central question for the biological modeler. Often, a researcher will measure a relatively large number of explanatory variables, but be unaware of which configuration of variables provides the best predictive model for Y. By comparing the efficacy of a set of candidate models, inferences about an optimal model may be derived. A number of issues are associated with this topic, including the central question: "What constitutes a good model?" While some scientists insist that more variables are always better (Gelman 2009), others prefer models that balance uncertainty, caused by too many parameters, and bias, caused by too few parameters (Burnham and Anderson 2002).

7.4 Two Important Tenets: Randomization and Replication

The primary goal of this chapter is to introduce techniques for sampling and experimental design. As a final preparation for this discussion, we consider two fundamental concepts: randomization and replication.

7.4.1 Randomization

Randomization is a selection or allocation process that produces outcomes that cannot be known in advance. Random outcomes can be obtained by an investigator using random number generators (**R** contains many of these), or some other nondeterministic process, for example, die throws and coin tosses.

With *random sampling*, we randomly select experimental units from a population of interest.

Random sampling is used to ensure independence among experimental units, which allows *inference to the sampled population(s)* (see Section 7.5). All inferential procedures discussed in Chapters 5 and 6 assume that data were obtained from random samples. This assumption will continue to hold for essentially all of the analytical techniques discussed in Chapters 8 through 11. Nonrandom sampling will often lead to invalid population inferences. For instance, with *samples of convenience*, we select experimental units that are easy to collect, or that are particularly obvious, leading to *anecdotal* evidence (i.e., data based on biased or nonrigorous observation) that is poorly representative.

In a *randomized experiment*, we randomly assign treatments to experimental units.

Randomized experiments allow for the control of confounding or lurking variables. As a result, these experiments allow *causal inference* with respect to the effect of X on Y (see Section 7.6). Conversely, the nonrandom assignment of treatments may introduce investigator bias, and generally does not allow for the control of confounding and/or lurking variables.

7.4.2 Replication

Replication is the process of obtaining sufficient observations to precisely and accurately describe the characteristics of a population, or the effects of an experimental treatment. Each repetition (e.g., application of a treatment) is called a *replicate*. Thus, an experimental unit assigned to a treatment is a replicate of that treatment.

Larger sample sizes (increased replication) will always increase the precision and accuracy of frequentist estimators. This is because increasing n will decrease sampling distribution variances. Conversely, inadequate replication will often result in invalid inference with estimates poorly representing true values. While not often considered, replication is also important for Bayesian applications since it will decrease the influence of potentially misleading prior distributions.

Holding other factors constant, replication will always increase the power of null hypothesis tests. Too much replication, however, may be inferentially dangerous. P-values can become meaningless in the presence of excessive replication (Chapter 6). Indeed, any effect will be statistically significant if sample sizes are large enough. This prompts a potential distinction between *biological significance* (which indicates the phenomenon under investigation may be important to the function of organisms/populations/communities/ecosystems) and statistical significance (which may be strongly influenced by replication, while being biologically unimportant).[*]

Power analyses can help researchers decide on appropriate levels of treatment replication. These procedures, however, require specification of anticipated effect size and system variability (Section 6.4.1). In the absence of such information, some biologists recommend general guidelines like the *Rule of 10* (Gotelli and Ellison 2004). Under this approach, at least 10 replicates are obtained for each treatment or unique combination of treatments. Ten replicates, however, may be far too many when effect sizes are large and treatment variances are low, may be unachievable in situations with many treatment combinations, or at large spatial scales, and may be wholly inadequate when effect sizes are small and treatment variances are high. Thus, it is much better to establish sample sizes based on known or plausible characteristics of the study system, and to employ, whenever possible, objective tools such as power.

---

[*] For a graphical illustration of this in the context of correlation (Chapter 8) and linear regression (Chapter 9), type see.adddel(). Note that statistically significant relationships can be obtained for very small effect sizes (small regression slopes) if sample sizes are large enough.

7.5 Sampling Design

Sampling is necessary in empirical science because of the practical impossibility of recording all possible outcomes from a variable of interest. For instance, a soil scientist cannot possibly measure soil characteristics for the entire surface of a mountain range. Instead, he or she randomly selects experimental units to represent a larger clearly defined population that he or she wishes to make inference to. *Sampling design* refers to the way these experimental units are selected from a population.

7.5.1 Randomized Designs

There are three basic types of randomized sampling designs (Figure 7.3).

7.5.1.1 Simple Random Sampling

A simple random sample is the simplest application of a *probability sample*; one in which every experimental unit has a nonzero probability of being selected (Lohr 1999). In this process, the investigator mixes up (randomizes) the population before selecting an experimental unit (Figure 7.3b).

7.5.1.2 Stratified Random Sampling

In a *stratified random sample*, a population is divided into logical subunits called strata. A simple random sample is then taken from each stratum (Figure 7.3c). In a well-designed stratified random sample, the elements in strata tend to be more similar to each other than to randomly selected units from the entire population. Stratification will allow the representation of rare groups that may be missed by simple random sampling, and as a result will often increase both the precision and accuracy of parameter estimates (Lohr 1999). The number of samples taken within strata can be based on a number of criteria, including *proportional allocation*. Using this approach, the number of sampled units in each stratum will be proportional to the size of the stratum.

7.5.1.3 Cluster Sampling

In a *cluster sample*, experimental units are aggregated into larger sampling units called clusters. A sample of clusters will be randomly selected from a population of clusters. A *census* (a sample of every possible experiment unit) is then taken from each selected cluster (Figure 7.3d). This process allows efficient acquisition of data, often in the context of human demographics.

7.5.2 Other Designs

An alternative to randomized sampling is *systematic sampling* (e.g., Hurlbert 1984). Under this approach, a population is partitioned at particular intervals to maximize independence. For instance, observations can be systematically separated by a certain time interval to ensure that they are not temporally autocorrelated, or placed at a sufficient distance to ensure that they are not spatially autocorrelated (see Section 7.5.5). In general, a systematic sample gives results similar to a simple random sample, and methods assuming random sampling are often used in the analysis of these data (Lohr 1999).

```
anm.samp.design.tck()
```

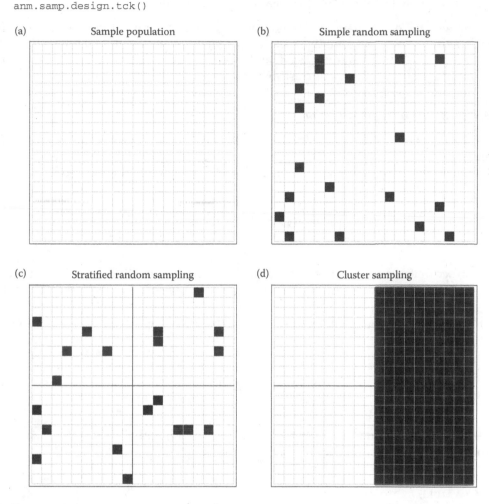

FIGURE 7.3
Demonstration of sampling designs. (a) An entire population of 400 experimental units (i.e., elements in a 20×20 grid). (b) A simple random sample of 20 from the population. (c) A stratified random sample; four strata are established and five samples are randomly taken from within each strata. (d) Cluster sampling; four clusters are established, two of these are randomly selected and a census is taken from each selected cluster.

7.5.3 Comparison of Designs

All three designs in Section 7.5.1 involve random selection of experimental units. In a simple random sample, units are randomly selected from the population. With stratified random sampling, experimental units within each stratum are randomly selected. With cluster sampling, the clusters are randomly selected from a population of clusters.

One concern with systematic sampling is that if a gradient coincides with the systematic pattern of sampling, then the sampling method will strongly confound the characteristics of the phenomenon under investigation. Random sampling allows inferences to underlying population(s) because it helps insure to independence among experimental units, and allows for the control of confounding variables and investigator bias. This may not be assured with systematic sampling.

7.5.4 Adjustments to Estimators to Accounting for Sampling

The estimator for $\sigma_{\bar{X}}$ (Equation 5.14) assumes that (1) the sampled parent population is essentially infinite in size, and (2) a sample was taken using a simple random sample. If simple random sampling was not used, and/or the parent population size is small, then Equation 5.14 should not be used to estimate $\sigma_{\bar{X}}$.

7.5.4.1 Finite Population Correction

When the sample size, n, approaches the population size, N, then the standard error will require a *finite population correction*. This adjustment will be particularly relevant to wildlife biologists who often work with populations that are small and restricted.

$$\hat{\sigma}_{\bar{X}} = \sqrt{\left(1 - \frac{n}{N}\right)\frac{S^2}{n}}. \tag{7.1}$$

We can see that as n approaches N, $\hat{\sigma}_{\bar{X}}$ goes to 0, and that as the difference between n and N increases, the corrected standard error approaches the conventional sample standard error, $S/\sqrt{n}$.[†]

A $100(1 - \alpha)\%$ confidence interval for μ using the finite population corrected standard error is

$$\bar{X} \pm t_{(1-\alpha/2, n-1)}\sqrt{1 - \frac{n}{N}} \cdot \frac{S}{\sqrt{n}} \tag{7.2}$$

where $t_{(1-\alpha/2, n-1)}$ is the quantile function for a t-distribution with $n - 1$ degrees of freedom, at probability $1 - (\alpha/2)$.

EXAMPLE 7.5

A study of the demography of Yellowstone National Park grizzly bears (*Ursus arctos*) was conducted using 20 years of data. One of the goals of the study was to obtain an estimate of the *rate of population growth*, λ. The researchers obtained the estimate $\hat{\lambda} = 1.01$ with a standard error of 0.04. The resulting 95% confidence interval was given as (0.93, 1.09). The authors concluded that because a Gaussian distribution with a mean of 1.01

---

[*] *Question:* What is the standard deviation of $\bar{X}$ (i.e. the standard error of the mean) if $n = N$? Is this reasonable? *Answer:* $\hat{\sigma}_{\bar{X}} = 0$. In this situation, the sample means will all be equal, and there will be no variation among sample means (because the "sample" comprises the entire population).

[†] For large populations, the size of the sample, not the proportion of the population sampled, determines the precision of the sample mean. For instance, a sample size of 100 from a population with 100,000 units has almost the same precision as a sample size of 100 from a population of 10 million.

$$\hat{\sigma}_{\bar{X}} = \sqrt{\left(1 - \frac{100}{100,000}\right)\frac{S^2}{100}} = \frac{S}{10}\sqrt{\frac{99,900}{100,000}} = \frac{S}{10}\sqrt{0.999} = S(0.09995)$$

$$\hat{\sigma}_{\bar{X}} = \sqrt{\left(1 - \frac{100}{10,000,000}\right)\frac{S^2}{100}} = \frac{S}{10}\sqrt{\frac{9,999,900}{10,000,000}} = \frac{S}{10}\sqrt{0.99999} = S(0.0999995)$$

```
Fig.7.4()## see Ch.7 code
```

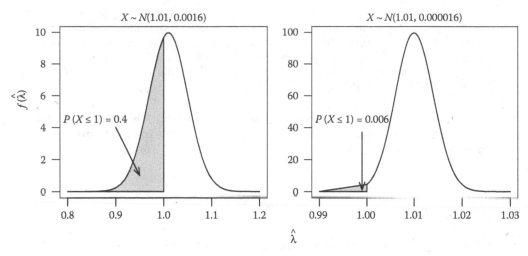

FIGURE 7.4

Two distributions for growth rate: $\hat{\lambda} \sim N(1.01, 0.0016)$, and $\hat{\lambda} \sim N(1.01, 0.000016)$. While observations of $\hat{\lambda} \leq 1$ are fairly likely in the left-hand distribution, that is, $P(\hat{\lambda} \leq 1) = 0.40129$ (viz., Pease and Mattson 1999), they are extremely unlikely in the right-hand distribution, which has a smaller standard error: $P(\hat{\lambda} \leq 1) = 0.00620$. In the right-hand diagram, the probability that the population is actually growing is $P(\hat{\lambda} > 1) = 1 - 0.0062 = 0.9937$. This is the opposite of what the authors are proposing.

and a standard deviation of 0.04 has about 40% of its probability mass below 1.00, that there was a probability of approximately 0.4 that the number of animals had declined over the 20 years. It was apparent that no finite population size corrections were made by the authors. Are their conclusions valid?

To answer this question, we note that the Yellowstone grizzly population consisted of less than 500 individuals at the time of the study. As a result, the influence of the finite population correction could be very large, and the standard error was likely overestimated. Thus, the difference between 1 and 1.01 could easily be significant; that is, the population was probably growing, not shrinking* (Figure 7.4).

7.5.4.2 Adjustments for Sampling Design

The conventional formulae for sample standard errors assume simple random sampling. If different sampling methods are used, then a different estimation approach will be required. Only corrections for stratified random sampling are described here. For analogous adjustments for cluster sampling, see Lohr (1999).

For a stratified random sampling design, let N be the known total number of units in the defined population of interest, and assume that the population can be logically divided into k strata; $N = N_1 + N_2 + N_3 + \cdots + N_k$. That is, we assume that we know both the total population size, and the population size in each stratum. Strata are sampled with $n = n_1 + n_2 + n_3 + \cdots + n_k$ experimental units.

---

* The Greater Yellowstone Ecosystem (GYE) grizzly bear provides a success story for the endangered species act. During the period 1975–2005, the size of the GYE grizzly population rose from 136 individuals to around 500 individuals in 2005. This trend allowed the U.S. Fish and Wildlife Service, which listed the GYE grizzly population as threatened in 1993, to delist the population in 2007 (USFWS-DOI 2007).

EXAMPLE 7.6

Consider an experiment to measure the population size of dall sheep (*Ovis dalli*) on a mountain in Alaska. Assume that the population is divided into subsets of three geographic strata (representing catchments). Each subset is again divided into a grid of 60 hexagons. Thirty of these hexes are randomly chosen from each stratum, and counts of sheep are then made in each hex (experimental unit). In this case, $N_1 = N_2 = N_3 = 60$, $N = 180$, $n_1 = n_2 = n_3 = 30$, and $n = 90$.

Reflecting conventional usage of stratified random sampling with population surveys, the formulae given below assume that the responses are counts of organisms.

We estimate the true population total (total number of organisms) with

$$\hat{T} = \sum_{h=1}^{k} N_h \bar{X}_h. \tag{7.3}$$

We estimate the true strata mean (organisms per strata) with

$$\bar{X}_{str} = \frac{\hat{T}}{N}. \tag{7.4}$$

We estimate the variance in the hth stratum as

$$S_h^2 = \frac{1}{n_h - 1} \sum_{i=1}^{n_h} \left(X_{(h)i} - \bar{X}_h \right)^2. \tag{7.5}$$

where $X_{(h)i}$ is the ith observation nested in the hth strata and $\bar{X}_h$ is the hth stratum sample mean.

An unbiased estimator for the standard error of $\hat{T}$ is

$$\hat{\sigma}_{\hat{T}} = \sqrt{\sum_{h=1}^{k} \left(1 - \frac{n_h}{N_h} \right) N_h^2 \frac{S_h^2}{n_h}}. \tag{7.6}$$

While an unbiased estimator for the standard error of $\bar{X}_{str}$ is

$$\hat{\sigma}_{\bar{X}_{str}} = \sqrt{\sum_{h=1}^{k} \left(1 - \frac{n_h}{N_h} \right) \left(\frac{N_h}{N} \right)^2 \left(\frac{S_h^2}{n_h} \right)}. \tag{7.7}$$

Note that Equations 7.6 and 7.7 have finite population corrections built into them.

Assuming that sample sizes within each stratum are large, or that the sampling design has a large number of strata, a $100(1 - \alpha)\%$ confidence interval for true strata mean, μ, and the true total, T can be constructed using

$$\hat{T} \pm z_{1-(\alpha/2)} \hat{\sigma}_{\hat{T}} \tag{7.8}$$

and

$$\bar{X}_{str} \pm z_{1-(\alpha/2)}\hat{\sigma}_{\bar{X}_{str}}, \tag{7.9}$$

where $z_{1-(\alpha/2)}$ is the z-quantile function at probability $1 - (\alpha/2)$. In situations where sample sizes or the number of strata are small, a $t(n - k)$ distribution should be used for the calculation of confidence intervals, where $n = n_1 + n_2 + \cdots + n_k$.

EXAMPLE 7.7

Stratified random sampling was used to estimate the size of the Nelchina herd of Alaskan caribou (*Rangifer tarandus*) in February 1962 (Siniff and Skoog 1964). The total population of sample units (for which responses would be counts of caribou) consisted of 699 areas of 4 mi² each. This population was divided into six strata, and each of these was randomly sampled. Results are shown in Table 7.4. The first row of the table should be interpreted as follows: 98 4 mi² sampling units were randomly chosen from a population of 400 in strata A. An average of 24.1 caribou per 4 mi² was found at those sampling units.

We let `ci.strat` from *asbio* do the work for us. Performing these calculations by hand is left as an exercise at the end of the chapter.

```
data(caribou)
with(caribou, ci.strat(strat = stratum, N.h = N.h, conf = 0.95,
summarized = TRUE, use.t = FALSE, n.h = n.h, x.bar.h = x.bar.h,
var.h = var.h))
$strat.summary
  N.h    n.h   x.bar.h    var.h
A 400     98     24.1      5575
B  30     10     25.6      4064
C  61     37    267.6    347556
D  18      6    179.0     22798
E  70     39    293.7    123578
F 120     21     33.2      9795
$CI
          estimate        2.5%          97.5%
CI.mu     77.96366     61.58854       94.33879
CI.T   54497.60000  43050.38811    65942.81189
```

The resulting 95% confidence interval for the true number of caribou per 4 mi² area is {61.59, 94.34}. The 95% confidence interval for the true total number of caribou in the $699 \times 4 = 2796$ mi² area of interest is (43050, 65943).

TABLE 7.4

Data Describing Caribou Counts

Stratum	N_h	n_h	$\bar{x}_h$	s_h^2
A	400	98	24.1	5575
B	30	10	25.6	4064
C	61	37	267.6	347,556
D	18	6	179.0	22,798
E	70	39	293.7	123,578
F	120	21	33.2	9795

7.5.5 Lack of Independence in Samples

Inferential methods for parameters in statistical models become more complicated when sampled outcomes are not independent. That is, when one outcome affects the probability that another outcome will occur (Chapter 2). For instance, when two random variables X_1 and X_2 are dependent, then specification for the variance of their joint probability distribution requires an additional covariance term (Chapters 4 and 8).

A lack of independence among samples may arise from nonrandom sampling, hierarchical experimental designs (see Section 7.6.5.3) or when data are acquired over time, or from spatially adjacent or identical areas. As Tobler (1970) noted: "...everything is related to everything else but near things are more related than distant things." Lack of independence due to spatial proximity is called *spatial autocorrelation*, while a similar lack of independence with time series data is called *temporal autocorrelation*. Autocorrelation can often be prevented by using existing knowledge of a study system to define observations that are well separated in time or space, and then using this criterion to establish a sampling interval. Temporally or spatially autocorrelated data require specialized analytical approaches, including time series modeling (see the following text), repeated measures analyses (Chapter 10), and spatially explicit models (not discussed here).

7.5.5.1 Time Series Models

Time series models acknowledge and model a lack of independence in observations due to temporal autocorrelation. One approach is to make observations at time t a function of observations at a particular time in the past called a *lag*. For instance, observations at time t are often modeled as a linear function of $t - 1$. That is, the response will be a linear function of itself at time $t - 1$,

$$Y_t = \beta_0 + \beta_1 Y_{t-1} + \varepsilon_t, \tag{7.10}$$

where β_0 is a constant, β_1 is a linear coefficient associated with lag $t - 1$, and ε_t is random error at time t. The form of Equation 7.10 is in marked contrast to conventional statistical models that assume that Y is a linear function of distinct explanatory variables (see Section 7.6.3, Chapters 9 and 10).

A time series analysis attempts to estimate three model components: trend, seasonality, and error. *Trend* represents systematic (often linear) nonrepeating changes over time. For instance, an overall increase in biological population numbers over time, or a plateau in population growth followed by a linear increase. *Seasonality* describes repeating patterns at particular intervals over time. For example, high population numbers may occur each spring, and low numbers may occur at other seasons. *Error* represents random variability in the response that is not due to either trend or seasonality. This is the term ε_t in Equation 7.10.

A representation of the response will often be a combination of trend and seasonality. For example, the population size of an organism may fluctuate seasonally over the course of a year, while showing a trend of increase over several years. Trend and seasonality are also a matter of temporal perspective. That is, what is a mere trend over a single year (high temperature in the summer, low temperature in the winter) will be a seasonal effect over several years.

A frequently used procedure for diagnosing temporal autocorrelation is the *Durbin–Watson* test. Its null hypothesis is that the true autocorrelation among time intervals is zero, and that the error in a linear model of Y as a function of time consists merely of random *white noise*. For details, see Shumway and Stoffer (2000). Other methods for assaying and addressing nonindependence are discussed in Chapter 9.

EXAMPLE 7.8

PM 2.5 pollutants (those less than 2.5 microns in diameter) can be produced in conflagrations like forest fires, or can form when gases discharged from power plants, industries, and automobiles react in the air. Once inhaled, these particles can affect the heart and lungs and cause serious health problems. The Idaho Department of Environmental Quality (DEQ) began monitoring PM 2.5 pollutants in Pocatello, Idaho in November 1998. A resulting time series dataset, consisting of 65 monthly measurements, can be found in *asbio* under the name PM2.5.

We use the Durbin–Watson test to diagnose autocorrelation.

```
data(PM2.5)
install.packages("lmtest"); library(lmtest)
dwtest(PM2.5 ~ seq(1:65), data = PM2.5)

        Durbin-Watson test
data: PM2.5 ~ seq(1:65)
DW = 1.278, p-value = 0.0007898
```

We reject the null hypothesis of temporal independence and conclude that particulate concentrations are the result of an autocorrelative process.

An exploratory time series analysis is provided by the function stl.

```
pmts <- ts(PM2.5[,2], start = c(1998, 11),frequency = 12)
plot(stl(pmts, s.window = "periodic"))
```

With respect to seasonality, there is a large spike in 2.5 PM pollutant that occurs annually during winter months (November–January) (Figure 7.5). This is almost certainly due to increased burning of fossil fuels for heating coupled with inversion effects. A second, smaller peak occurs during the fall and may be due to particulate production from local potato harvesting. With respect to trend, there is a slight improvement (decline) in PM 2.5 pollutants from particularly high levels in the late early 2000s (Figure 7.5).

7.5.5.2 *Psuedoreplication*

Pseudoreplication indicates a lack of independence among sample entities assumed to be replicates. It can result in a situation where treatments are not replicated (although samples may be) and "testing for treatment effects with an error term inappropriate to the hypothesis being considered" (Hurlbert 1984). Pseudoreplication can be directly or indirectly tied to spatial or temporal sampling, or to hieratical experimental designs (see Section 7.6.5.3).

EXAMPLE 7.9

We are interested if oak (*Quercus*) leaves decompose more quickly than maple (*Acer*) leaves at a 1-m isobath in a lake (an isobath is a line representing a consistent depth in a body of water (Figure 7.6)). We place eight bags of oak leaves randomly within a 0.5 m² plot on the isobath (A), and eight bags of maple leaves randomly within an "identical" plot (B). We remove the bags after several weeks and determine biomass loss via decomposition by drying and weighing each leaf bag, and comparing this to the weights preceding immersion. What is wrong with this sampling design?

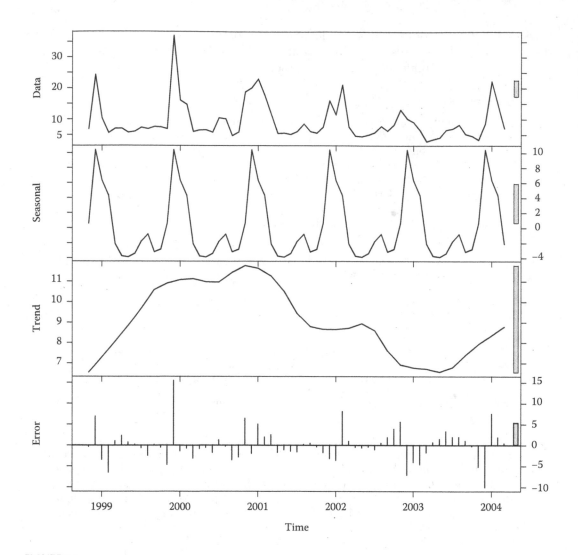

FIGURE 7.5
A time series analysis of PM 2.5 particulate levels in Pocatello, Idaho. Estimation of seasonal and trend components are found here using a combination of lowess smoothing and moving averages (Chapter 9). Heights of gray bars at the right show a common scale measure for the four figures.

Because treatments are segregated into two locations, and are not interspersed randomly (or even systematically), there is no replication of the oak and maple leaf "treatments." The bags within A and B are not replicates; they are pseudoreplicates. As a result, we can only evaluate whether oak leaves at site A decompose more quickly than maple leaves at site B.

In itself, pseudoreplication is not a bad thing. Pseudoreplicates can be extremely helpful in accurately describing a true replicate. For instance, in the aforementioned example, we are likely to have good descriptions of what is happening to oak leaves at site A, and maple leaves at site B. We cannot, however, make inference to isobath/species associations because there is no replication. One should never treat pseudoreplicates as independent samples.

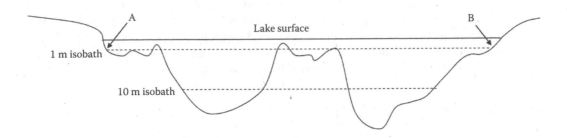

FIGURE 7.6
Cross section of a lake showing the 1 m and 10 m isobaths. (Adapted from Hurlbert, S.H. 1984. *Ecological Monographs* 54(2): 187–211.)

One way to account for pseudoreplication is to use the average of the pseudoreplicates as the value of the replicate. This approach is often used with autocorrelated data, and in nested vegetation and zoological surveys (Mueller-Dombois and Ellenberg 1974, Gotelli and Ellison 2004). A disadvantage of this method is that the information about subsample variability will be discarded. Another alternative is a model that explicitly estimates both the treatment effects associated with independent replicates and the variance components of nested subsamples (see discussion of nested designs in Section 7.6.5.3).

7.5.6 Other General Sampling Concerns

There are several other general concerns with sampling and management of resultant data that researchers should be aware of. These include the detection and handling of outliers, errors in the measurements of experimental units, sampling bias, missing data, and data transformation.

7.5.6.1 Outliers

Outliers are outcomes in data that fall well outside the overall observed patterns. Such outcomes will generally decrease power and increase type II error in conventional analyses.

Common methods for identifying outliers include flagging observations greater than 1.5(*IQR*) from the median, or identifying those that satisfy (Wilcox 2005):

$$\frac{|X_i - M|}{MAD/0.6745} > 2.24, \qquad (7.11)$$

where M is the sample median, and MAD is the median absolute deviation (Chapter 4). The impact of outliers in conventional analyses using sample variances will be amplified because variability will be measured using squared deviations from the sample mean. Thus, the relationship between outlier size and outlier effect is exponential, not linear.

Some researchers delete outliers as a matter of course. However, this is poor scientific practice. In fact, deleting data just because it is messy, or does not show what you want, can be considered scientific fraud. Outliers should only be discarded if it can be verified that they were recorded or measured in error. For instance, the datum is the result of a false entry in a field notebook, an incorrect keystroke on a computer, or a transient malfunction in measurement apparatus. Default deletion of outliers may be inferentially dangerous. For instance, we noted in Chapter 4 that the detection of the Antarctic ozone hole was delayed for several years because a sensor ignored low O_3 levels as outliers (Kandel 1991). Alternatives exist for most conventional statistical analyses that remain robust to the effect of outliers without requiring their deletion (Chapters 6, 8 through 10).

7.5.6.2 Measurement Error and Precision

Sampling error occurs as a result of the fact that a variable will vary naturally, and complete sampling of the population is generally not possible (Chapter 4). It should not be confused with *measurement error*, that is, variability in observations because of imprecision or inaccuracy in measurement. In practice, however, these errors may be very difficult to distinguish. If errors in measurement are random, then they will tend to cancel each other out although they may cause the sampling error to appear larger as they are silently incorporated into model error estimates (Kutner et al. 2005). If measurement errors are correlated or systematic, this may lead to measurement bias (see below) and prevent valid inferences.

7.5.6.3 Bias

Bias describes a predisposition for an experiment, or measurement, or researcher, to arrive at a particular result. Because it will result in systematically incorrect measures, bias will almost always impede valid inference. *Selection bias* (also called *sampling bias or ascertainment bias*) occurs when certain individuals or groups are more likely to be chosen as experimental units than others and will generally be prevented by random sampling. Unexpectedly, unbalanced designs (unequal replications among treatments) may also result in selection bias because the characteristics of one treatment will be estimated with greater precision than another. *Measurement bias* describes when a variable is systematically measured incorrectly. Such an outcome may occur when subjective data measures are used (e.g., ocular estimates of plant cover), or with defective or noncalibrated data recorders.

7.5.6.4 Missing Data and Nonresponse Bias

Difficulty in getting to the field, sensor failure, subjects dropping out of cohort studies (with repeated measures on the same subjects), and many other factors may result in missing data. This will generally result in an unbalanced design, which may result in bias and complicate analyses. For this reason, experimental designs should be balanced whenever possible. Particularly damaging are cases in which missing observations result in no replicates for a combination of treatments. The experimental is now said to have *missing cells.* Such data are particularly difficult to analyze (see Chapter 10).

   The mechanisms underlying missing data will determine if any action needs to be taken by the investigator (Little and Rubin 1987). If data are *missing completely at random (MCAR)*, then the missing data are not associated with the response variable, Y, or the structure of the experimental design or sample selection process, X. Thus, the missing data will be a random subsample of the population, and can be ignored without concerns about selection bias. If a technician drops and breaks a test tube representing a sample of white blood cell counts, there is no reason to suspect that the white blood cell count caused the tube to be dropped. Thus, this sample can be considered MCAR. If data are *missing at random given covariates*, then the nonresponse depends on X, and not Y. In these cases, because Y is a function of X, then a missing response can be *imputed* (assigned a value) using nonmissing data (see Lohr 1999). Finally, if the probability of nonresponse depends on the value of the response variable, then it is a *nonignorable nonresponse*. For instance, response measurements in a particular range of a standard curve (see Figure 9.1b) may be especially difficult to obtain, but will prevent derivation of valid predictive models if missing. In this case, missing values will undoubtedly require rerunning at least part of the experiment.

7.5.6.5 Transforming Data

By *transforming* a variable, we change its units of measurement to a form that is more useful or interpretable. Such an action is justifiable because variable units are conceptually arbitrary. That is, it really should not matter if plant height is measured in millimeters, kilometers, or for that matter the inverse sine of (mm).* Transformations can often dramatically improve the appropriateness of the data for inferential procedures including confidence interval estimation (Chapter 5), and hypothesis testing (Chapters 6, 8 through 11).

For instance, in a linear regression model (Chapter 9), we assume that error terms are independent, with a constant variance, and normally distributed, and that the relationship between X and Y is linear. After collecting and analyzing the data, we may find that these assumptions are being violated, and that transformation will be required if the model is to be used.

In Chapter 4, we discussed methods of linear transformation and their effects on parameters and statistics. A linear transformation has the form

$$Y' = a + bY,$$

where Y is a random variable, Y' is the transformed variable, and a and b are constants. A linear transformation to variables X and/or Y will not change their linear association/correlation (Chapter 8). For instance, consider the correlation of outcomes from a sample x, and $3x + 5$.

```
x <- c(1, 2, 3, 4, 5)
cor(x, 3 * x + 5) #correlation of x and a linear transformation of x
[1] 1
```

An important type of linear transformation is *standardization*. This technique will result in a new set of observations with a sample mean of zero and unit sample variance.†

$$Y'_i = \frac{Y_i - \bar{Y}}{S} \tag{7.12}$$

---

* A few additional words are warranted. First, while it is true that units of measurement are generally arbitrary, it also true that the original units made particular sense to the researcher. As a result, an intuitive understanding of the analysis may be lost as a result of a data transformation. It is simply easier to interpret a variable measured in millimeters than a variable measured in the inverse square root of millimeters. As a consequence model results (e.g., point and interval estimates) should always be back transformed and presented in the original units. Second, while our linear model assumptions now apply to the transformed model, a researcher should not forget that, in the pretransformation units, the data will still have the same "problems" they had in the first place (e.g., nonlinearity and heteroscedasticity).

† Centering a variable with respect to its mean (i.e., $Y'_i = Y_i - Y$) is often called *translation* (Legendre and Legendre 1998). This technique is often used to reduce multicollinearity in polynomial regression (Section 9.14), or to increase the interpretability of results from ANCOVA (Section 10.13). *Relativizing* variable outcomes with respect to a maximum response, that is, $Y'_i = Y_i/\max(Y)$, or measure of location, that is, $Y'_i = Y_i/Y$, also allows unitless comparability of variables and can be used to address bias in sampling (see McCune and Grace 2002).

Standardization facilitates comparison of particular outcomes among variables since each variable will have the same mean and variance, and will be expressed as a unitless quantity (the number of standard deviations from the mean). As we know, standardization involving observations and parameters of a normal random variable results in a z-score (Chapter 3).

While linear transformations are often useful (e.g., for standardizing or relativizing variables or for changing units of measurement), they will be inadequate for resolving problems with heteroscedasticity, nonlinearity, and nonnormality. To address these concerns, nonlinear transformations will be necessary.

Nonlinear transformation describes alterations of a variable that cannot be expressed as a linear function of individual components. Examples include transformations in which variables are used as exponents or are multiplied or divided by each other. By definition, nonlinear transformation will change (increase or decrease) measures of linear association. Several frequently used nonlinear transformation methods and their applications are summarized in Table 7.5.

TABLE 7.5

Nonlinear Transformations Commonly Used in Statistics

Name	Example	Application and Comments
Log	$Y' = \log(Y)$	Used for heteroscedastic data when the treatment standard deviations are proportional to the treatment means. May also correct nonnormality in positively skewed data.
Square root	$Y' = \sqrt{Y}$, or $Y' = \sqrt{Y} + \sqrt{Y+1}$	Used for heteroscedastic data when the treatment variances are proportional to the treatment means. May also correct nonnormality in positively skewed data. May be particularly helpful with count data. A more rigorous analytical approach to response data which are counts involves Poisson GLMs (Chapter 9).
Inverse	$Y' = \dfrac{1}{Y}$	Used for heteroscedastic data when the treatment variances are proportional to the treatment means squared. May also correct nonnormality in positively skewed data.
Square	$Y' = Y^2$	May correct nonnormality in negatively skewed data.
Antilog	$Y' = e^Y$	May correct nonnormality in negatively skewed data.
Arcsine	$Y' = 2 \times \arcsin(Y)$	Frequently used when data are proportions. Proportional data are frequently nonnormal, and treatment variances will be unstable if treatment proportions are different. The transformation can be used to address both issues. Percentages need to be transformed to proportions before usage. A more rigorous analytical approach to proportional response data involves logistic GLMs (Chapter 9).
Rank	$Y' = \text{Rank}(Y)$	Useful for handling outliers. Will transform a monotonic association into a linear association (Chapter 8).

If a transformation is used in an analysis, then one should report the results, for example, parameter estimates, predicted values, and confidence intervals, in back-transformed units. For instance, assume that natural log-transformation was used on plant height measures (taken in centimeters) to meet inferential assumptions of normality, and to allow confidence interval calculation for μ. Suppose we then calculate a 95% confidence interval with the confidence bounds (2, 3). The units for this confidence interval are now in logged centimeters. To convert back to original units we have $(e^2, e^3) = (7.4 \text{ cm}, 20.1 \text{ cm})$. Note that while the original confidence bounds are symmetric around the mean of the logged data, $2.5 \log_e(\text{cm})$, the back-transformed interval is not symmetric around the back-transformed mean, 12.2 cm. Transformation in the context of general linear models is discussed in Chapter 9.

7.5.6.6 Altering Datasets

After examining a dataset, it may be apparent that corrective measures are needed. For instance, it may be necessary to delete an erroneous measurement, interpolation may be required to replace a missing observation, or a variable might require transformation to meet the assumptions of an analytical model. In such cases, care should be taken to ensure that the original data are not lost, and changes to the raw data should be carefully documented so that (if necessary) the raw data can be recreated from the altered data.

7.6 Experimental Design

Experimental design is defined by four factors:

1. Whether or not a predictor variable is consciously altered or defined by the investigator.
2. Whether or not experimental units were randomly assigned to treatments.
3. The forms of response and explanatory variables (categorical, quantitative, or ordinal).
4. The arrangement of categories in categorical predictors (e.g., blocked and nested).

The characteristics will determine both the general analytical approach, and the types of inferences that can be drawn from the investigation.

7.6.1 General Approaches

7.6.1.1 Manipulative Experiments versus Observational Studies

In a *manipulative experiment*, an investigator consciously alters the levels of a predictor variable and then measures change in a response variable as a result of these alterations. The resulting Y outcomes are then analyzed, often to test a null hypothesis of no predictor effect. Conversely, in an *observational study* or *mensurative study*, we describe the natural variation associated with the studied phenomena. The system is not manipulated; as a result "treatments" or explanatory gradients of interest exist in place naturally.

7.6.1.2 *Randomized versus Nonrandomized Experiments*

Experimental designs can also be distinguished by whether randomization was used to assign levels of the predictor(s) to the response. In observational studies (and some manipulative experiments), experimental units are not randomly assigned to treatments, while in a *randomized experiment* they are. Randomized experiments (which are by definition manipulative) provide causal evidence of X on Y. This is because, through the process of randomization, the effect of confounding and lurking variables is "averaged out" (see below). As a result, we can generally assume that the effect of X on Y is the only phenomenon being examined. Conversely, nonrandomized manipulative and observational studies generally do not control for lurking or confounding variables, preventing causal inferences. Therefore, while randomized experiments and observational studies superficially generate the same sorts of data, and may be analyzed using the same sorts of statistical analyses, the interpretation of results from these designs will differ dramatically.

> **EXAMPLE 7.10**
>
> An important advocate of randomized experiments, R. A. Fisher (see Section 1.9) worked as a statistician at the Rothamsted Agricultural Experimental Station near London around 1920. Fisher noticed that plants he was studying would respond differently to treatments (e.g., fertilizer and water) depending on where they were placed in experimental plots, and that this would often obscure the effect of the treatment. He noted that this variability could be due to a number of factors. For example, in certain cases, an unknown fertility gradient could be present causing plants in some parts of the plot to grow well or poorly regardless of the treatment. In other cases, certain plants may have had genetic characteristics that caused them to respond strongly (or weakly) to the treatment. Fisher reasoned that if experimental units were randomly assigned to treatments, the effect of such confounding variables would be "averaged out" allowing one to quantify the true overall effect of the treatment. For instance, in testing the effectiveness of fertilizer, some plants in the fertile parts of a plot would randomly receive fertilizer, while others would not. Conversely, some plants in infertile parts of the field would receive fertilizer, while others would not. Application of fertilizer might cause dramatic increases in productivity in infertile parts of the field, and a less pronounced increase in fertile parts of the field (where plants were already doing fairly well). In either case, the confounding effect of field fertility would be canceled by random application of treatments across the gradient.

Drawing causal inference from an observational study or nonrandomized experiment is only possible if one can somehow account for potential researcher bias and confounding variables (Gelman et al. 2003, pp. 226–227). Some control over these factors can be obtained through matching, and blocking of experimental units in specialized designs (discussed in Section 7.6.5.3). In fact, these procedures may increase precision and power in both randomized experiments and observational studies. In addition, confounding variables, even if they cannot be controlled, can often be accounted for in analyses by including them as covariates in models (see below). With many observational studies, however, these steps are not taken, are not possible, or still may not provide adequate controls.

There exist two caveats to the general premise that causal evidence cannot be generated by observational studies. First, multiple corroborating observational studies may provide support for causal statements, or at least partially address the concern that lurking variables may be influencing the response. An important example involves smoking and lung cancer.

It would be unethical to randomly assign smoker and nonsmoker groups over long periods of time if we suspected that smoking was truly dangerous. As a result, evidence of the risk of lung concern due to smoking has been observational.[*] Nonetheless, the agreement of many observational studies, conducted in many different settings, has prompted most medical experts to view smoking as a causal factor for cancer (e.g., Cornfield et al. 1959).

Second, some statisticians have argued that inferential statements can be based on subject matter expertise alone (Ramsey and Schafer 1997). For instance, to derive his "big bang theory," Edwin Hubble quantified the association between distance (as measured by the apparent luminosity of cepheid variable stars) and the velocity (as measured by the red shift) of 24 nearby galaxies. These galaxies did not constitute random samples, and the data were observational. Nonetheless, Hubble inferred his results to the universe and inferred causality based on his expertise, and the assumption that there is nothing particularly special about our location in the universe.

Observational studies may provide useful information, and in many cases may be more appropriate than randomized experiments. For example, because they do not involve artificial manipulations, observational studies may be more representative of the complexities of some biological systems, and may be more ethical to conduct, particularly in ecology (Quinn and Keough 2002). Many ecological field studies are observational because of the difficulty of randomly assigning treatments to experimental units. For instance, transporting populations of invasive mountain goats to randomly selected mountains to determine the causal effect of goats on alpine vegetation is probably unrealistic, and moreover, unethical.

7.6.1.3 Controls

A *control* ideally represents the combined effect of all unmeasured factors minus the treatment. As a result, a control can be used to account for lurking variables. Consider a researcher who wishes to quantify the effect of nutrient additions on plant growth. This person might randomly assign 20 plants to a supplemental soil nutrient treatment, while 20 other plants would receive no additional nutrients, but would otherwise be handled in exactly the same way. In this case, the plants not receiving fertilizer could be considered a *control group*.[†]

Controls are generally used to help quantify the true effect of a treatment after accounting for background conditions. Lack of controls will often lead to false inferences concerning the true effect size, and incorrect causal inferences.

---

[*] R. A. Fisher, a lifelong smoker, argued that unmeasured genetic factors that predispose one to smoke may also be responsible for higher levels of lung cancer in smokers. Thus, people predisposed to smoke would contract lung cancer at higher rates whether they smoked or not (Fisher 1958). Cornfield et al. (1959), however, argued that if an agent S is not causal, but because of correlation with a causal agent X, shows apparent risk, r, then $(X|S)/(X|S')$ must be greater than r, where $S' =$ "not S." That is, if cigarette smokers have nine times the risk of smokers of developing lung cancer, and this is not the result of smoking, but because smokers are genetically predisposed to have cancer because they produce protein X, then the amount of X must be at least nine times higher in smokers compared to nonsmokers.

[*] This is an example of a *negative control*, that is, a control when one expects "no response" from the control group. Conversely, a *positive control* would be a treatment with a "real response" that is known. Both negative and positive controls are frequently used in spectrophotometric enzyme assays. In this case, the negative control will be a cuvette containing all parts of the reagent except the enzyme. The positive control will be a cuvette containing the reagent and enough enzyme to elicit a spectrophotometric response. If there is no difference between the positive and negative controls, then the reagent is probably faulty.

EXAMPLE 7.11

In many medical experiments, patients will respond positively to any treatment, even a *placebo* (i.e., a dummy treatment). This may be due to psychosomatic effects associated with patients' trust of doctors, or the fact that many illnesses improve without treatment.

"Gastric freezing" was proposed in the 1960s as means for controlling pain in individuals suffering from ulcers in the upper intestine (Wangensteen et al. 1962). In the treatment, a balloon is attached to a nasogastric tube and pushed down the pharynx into the stomach. Refrigerated liquids are then pumped through the balloon in the belief that cooling the stomach would decrease acid production, and relieve ulcer pain. Artz et al. (1977) reported that gastric freezing appeared to reduce pain in patients with peptic ulcers. The study, however, did not account for the placebo effect by assigning a control treatment to some of the patients. To address this concern, a later study randomly divided a group of ulcer patients into two groups. The first group received the gastric freezing treatment, as described above, while a second group received a placebo treatment (a balloon was inserted, and water at body temperature was pumped through). The researchers found that 34% of the patients in the treatment group improved, but that 38% of the patients in the control group improved (Miao 1977). The experiment demonstrated that gastric freezing worked no better than a control, and prompted its discontinuation.

7.6.1.4 Measuring Appropriate Covariates

In a properly designed experiment, an investigator would be careful to account for all confounding variables so that the only differences between the treatment groups are levels in the explanatory variable of interest. For instance, to study the effect of soil nutrients on plants, the confounding variables (light, temperature, water, etc.) would be kept identical among treatments.

Covariates are explanatory variables that may affect the response variable, are not variables of central interest, and are generally not manipulated by the researcher.

In the event that confounding variables cannot be controlled by the experimenter (e.g., most observational studies), they should be included in statistical models as covariates, since such a framework allows for the predictor(s) of interest to be considered while *holding the covariates constant* (Section 9.5).

7.6.1.5 Prospective and Retrospective Studies

Several experimental design terms are generally associated with epidemiological and pharmaceutical investigations. These include the expressions prospective study and retrospective study.

In a *prospective study*, an investigator identifies two groups, one that has been exposed to a disease risk factor, and one that has not, and then follows the frequency of disease development in the groups over time. Thus, in a *prospective randomized experiment*, an investigator randomly designates treatment and control groups, and exposes the treatment group to the risk factor, and then measures disease outcomes in the groups over time.

In a *retrospective study*, an investigator looks into the past and contrasts groups or treatments. There are two types of retrospective studies. In a *cross-sectional study*, an investigator identifies several populations of interest, designates whether they came from risk, or nonrisk groups, and then finds the prevalence of disease in the groups. In a *case–control study*, an investigator selects specific individuals that already have the disease of interest. The investigator then identifies a second group strongly resembling the disease group, but without the disease, and attempts to identify risk factors. The latter framework allows a stronger case for causality because confounding factors are being addressed.

7.6.2 Summary: Inference in the Context of Both Experimental and Sampling Design

We can formatively summarize the materials presented so far on sampling and experimental design by considering the two types of inferential statements most frequently made in statistical analysis: (1) inference to the population(s) of interest, and (2) causal inference.

In Section 7.5, it was explained that in general, inference to a population can only be made if the experimental units representing that population were collected using random sampling.

In Section 7.6.1, it was explained that causal inference can be generated from randomized experiments, but (in general) not from observational studies. This information is synthesized in Table 7.6.

We note that in the "best case" inferential scenario, we have random sampling and a randomized experiment, allowing causal inference to the population, while in the "worst case" scenario, we have nonrandom sampling in an observational study, leading to non-causal inference to the sample.

7.6.3 Classification of Experimental Designs

We can distinguish experimental designs by classifying their variables as being explanatory or response, and then reclassifying the variables as either categorical or quantitative (cf. Gotelli and Ellison 2004, Whitlock and Schluter 2009). In Table 7.7, three general experimental designs result from the intersection of these categories. These are (1) regression designs (including specialized approaches when Y is categorical), (2) ANOVA designs, and (3) tabular designs. The approaches are briefly described in this section and are readdressed in greater detail in Chapters 9 through 11, respectively.

The designation of response and explanatory variables implies that all of the designs in Table 7.7 can be used to test causal relationships. It should be reemphasized here, however, that the evidence of causality will not be a function of the type of analysis used, but the way that confounding variables, lurking variables, and bias are controlled in the experimental design. That is, a significant P-value from any analysis in Table 7.7 only provides evidence of causality if the experimental design was handled appropriately with this goal in mind.

While it provides a useful simplification, a large number of designs do not fit cleanly into any of the categories in Table 7.7. First, *ANCOVA* requires at least two explanatory variables, one of which is quantitative, and the other categorical. Second, specification

TABLE 7.6

Statistical Inference Permitted by Different Sampling Procedures and Experimental Designs

	Randomized Experiment	Observational Study
Random sample	Causal inference to population	Inference to population
Nonrandom sample	Causal inference to sample	Inference to sample

Note: There are two types of randomization important to inference: (1) the process of selecting experimental units, and (2) the process of assigning treatments to experimental units (cf. Ramsey and Schafer 1997).

TABLE 7.7

Four General Classes of Experimental Designs as a Result of Variable Typology

	Response Variable	
Explanatory Variable	**Quantitative**	**Categorical**
Quantitative	Regression General linear model for continuous Y, GLM approaches for bounded or discrete Y	GLM regression approaches Includes models for dichotomous, or polytomous Y
Categorical	ANOVA t-tests and analogs (2 categories), ANOVA and analogs (>2 categories), GLM approaches for bounded or discrete Y	Tabular designs Includes GLM approaches, particularly loglinear models

Note: Regression analyses result from quantitative or categorical response and quantitative explanatory variables. ANOVA-type designs result when the response variable is continuous and the explanatory variables is (are) categorical. Tabular analyses are required when both explanatory and response variables are categorical.

of response or explanatory variables is pointless in correlation and often unnecessary in tabular association analyses (Chapters 8 and 11). Third, ordinal variables (including variables resulting from rank-transformation) are not addressed. Fourth, multivariate methods (which require multiple simultaneous Y-variables) are ignored.

7.6.3.1 General Linear Models

Many of the analysis approaches in Table 7.7, including simple linear regression, multiple regression, ANOVA, and ANCOVA, can be classified as *general linear models*. Such models assume that the response variable is a function of a linear transformation of the explanatory variable(s) plus an error term. All general linear models have the form

$$Y = \beta X + \varepsilon. \tag{7.13}$$

The terms βX represent the "fit" part of the model, while error (nonlinear noise) in Y is represented by ε. The terms in Equation 7.13 are capitalized and bolded because they will be described by vectors, or even matrices. General linear models all assume that components of ε are independent and identically normally distributed (see Chapters 9 and 10).

Recall that probabilistic (statistical) models contain both deterministic and stochastic elements (Chapter 2). This is clearly illustrated in the fit and error components of general linear models.

Permutational and rank-based permutational alternatives are possible for most of the general linear model analyses listed in Table 7.7. These approaches will not have all of the assumptions of general linear models. They will, however, have their own drawbacks (Sections 6.5 and 6.6). Bayesian and likelihood (information-theoretic) applications for general linear models are also possible and are discussed in Chapters 9 and 10.

7.6.3.2 Generalized Linear Models

GLMs (Nelder and Wedderburn 1972) allow nonnormal errors, and have a linear functional form as the result of the application of link functions (Chapters 9 through 11, Appendix). Because GLMs include general linear models, they provide an even more unifying framework for analysis. Indeed, GLM approaches can be used in all four classes of designs from Table 7.7.

7.6.4 Regression Designs

When the explanatory variable is quantitative, the result will be a regression design (Table 7.7, Chapter 9).

7.6.4.1 Simple Linear Regression

With *simple linear regression*, we will have a single response variable modeled as the linear function of a single explanatory variable. An example was shown in Figure 2.1. Here, spider web strand length was considered as a function of temperature. A negative association was demonstrated because of the elasticity of spider web polymers.

The chief advantage of simple linear regression over other regression approaches is its simplicity. A simple linear regression provides a straightforward representation of the linear relationship of the response as a function of the predictor (a scatterplot with a regression line superimposed), and allows the interpretation of parameter estimates without the additional consideration of holding other variables in the model constant. For instance, if tree height (in meters) was regressed as a function percent N in soils, then a slope coefficient of 2.1 would simply mean that we would expect a tree to grow, on average, 2.1 m for each 1% increase of N in soils.

There are three main experimental design concerns associated with simple linear regression (and other regression designs). First, one should be sure that the range of values recorded for the explanatory variable captures the full range of interest of the response variable. Consider Figure 7.7. Clearly, the strong overall association between Y and X would not be detected if we only used the central predictor values (and corresponding Y values) in a regression analysis.

Second, one should be sure that the explanatory values are approximately uniform within the sampled range. This is because outlying X observations may be unduly influential in determining the slope of the regression line, resulting in invalid inference. For a visual demonstration, type see.move(). By moving points in the scatterplot, it is apparent that the outlying outcomes on the X axis are the most influential with respect to the regression function (they result in more dramatic swings in the regression line).

Third, it should be emphasized that in conventional regression analyses, levels of X are assumed to be fixed, and measured without error. However, this assumption is likely to be invalid in many biological investigations, hampering both straightforward estimation of regression parameters and causal inferences. In fact, this scenario will be analogous to the nonrandom assignment of treatments given categorical explanatory variables.

7.6.4.2 Multiple Regression

With *multiple regression*, we model a single Y variable as a linear function of more than one quantitative explanatory variable. Multiple regression quantifies the effect of a predictor in the context of other predictors (when holding other predictors constant). In this way, the inclusion of covariates in a multiple regression model helps an investigator to account for confounding variables in observational studies.

In addition to the concerns associated with all regression models (see above), a particular problem with multiple regression is multicollinearity, which occurs when predictor

```
X - seq(1,20); Y - c(seq(1,7), rep(7,6), seq(8,14)) rnorm(20, sd 0.5)
plot(X, Y)
```

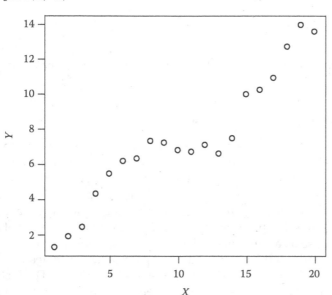

FIGURE 7.7
An illustration of the importance of considering the range of X values to measure in regression analysis. The relationship between Y and X flattens in the middle of the range of X values. To detect the overall true linear relationship of Y as a function of X, one would have to sample across the entire range of X.

variables are strongly correlated to each other. Given multicollinearity, the true effect of the explanatory variables on the response will be obscured (signs for regression coefficients may even be reversed) resulting in invalid inference.

7.6.5 ANOVA Designs

A common experimental design involves a continuous response variable and categorical explanatory variable(s) (Table 7.7). An omnibus approach called ANOVA that compares within-treatment to between-treatment variability has been developed for this scenario (Chapter 10). In Chapter 6, we learned how to make inferences to one and two populations using *t*-tests and their nonparametric analogs. ANOVAs can be used to extend inference to more than two populations. Furthermore, this framework can be used to examine the effect of more than one explanatory variable on the response, and to examine interactions among explanatory variables.

7.6.5.1 ANOVA Terminology

ANOVA designs have a specialized terminology, which is briefly described here. Categorical explanatory variables are called *factors*. It follows that a factor must have at least two categories (e.g., male or female). These are called *factor levels*. ANOVA designs can have a single factor, or they may have more than one factor. This distinguishes *single factor designs* (those with one explanatory variable) and *multifactor designs* (those with two or more explanatory variables).

For a single-factor ANOVA, *treatments* will (generally) be related categorical variations on a single factor. For instance, consider a factor called "Nitrogen" that has three factor levels, each representing a different amount of N applied to the soil ("High N," "Medium N," "Low N"). An investigator would test to see if the nutrient treatments had a different effect on a response variable (e.g., plant biomass). Another example might be a factor called "River," with factor levels distinguishing different catchments, for example, "Snake," "Columbia," and "Mississippi." Defining factors in this way (with logically related factor levels) facilitates rational comparisons in multifactor analyses.

In a multifactor design, the treatments will be combinations of factor levels from the factors. For instance, in a study with both the nitrogen application and river factors, we would have nine possible factor-level combinations. One would be the combination "Snake River" and "Low N."

7.6.5.2 Fixed and Random Effects

The way factor levels are chosen, and their underlying meaning to an investigation will affect the computation and interpretation of ANOVA results. These considerations require the differentiation of two types of factor levels: fixed and random.

Fixed factor levels have two characteristics: (1) they are not an obvious subset of a population of factor levels, and (2) the factor levels are informative, and are chosen by the investigator specifically because they have a unique and important meaning. Examples of fixed factor levels include

- "Male" and "female"
- "Predator" and "prey"
- "Drug A," "drug B," and "drug C"
- "Control" and "treatment"
- "Before" and "after"
- "Tree," "grass," "shrub," and so on
- Other important variants on a predictor, for example, "high," "medium," and "low"

Random factor levels have two characteristics: (1) the selected factor levels often can be considered a subset from a population of levels, and (2) the factor levels are not informative. Because of these characteristics, we are unlikely to care about particular differences among factor levels. Random factor levels include

- Time intervals in a time sequence
- A collection of seed types (or some other levels) chosen randomly from a population of seed types (or some other population)
- Nested groups in an observational study
- Subjects on whom repeated measurements are made

If all factors contain fixed factor levels, then one would use a conventional *fixed effect* model for hypothesis testing procedures. Conversely, if all factors are random, then one should use a *random effect model*. A design with both random and fixed factors will require a *mixed effect model*. Fixed, random, and mixed effect models are also called model I, model II, and model III ANOVAs, respectively.

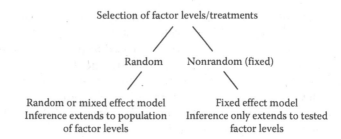

FIGURE 7.8
Inferential distinctions for random and fixed effect models.

Two inferential characteristics distinguish random and mixed effect models from fixed effect models. First, their null hypothesis tests concern different parameters. Fixed effect models make inferences concerning true factor-level means, while random effect models make inferences concerning the variance of a population of factor-level means. Second, random selection of factor levels (random effects) will allow inferences concerning the entire population of factor levels, while fixed factors, which are not concerned with such a population, will not. These concepts are summarized in Figure 7.8.

7.6.5.3 A Compendium of ANOVA Designs

There are a large number of experimental designs associated with ANOVAs. Six that are commonly used by biologists are shown in Figure 7.9. Of chief concern to introductory students is the first subsection describing one-way ANOVA. The five remaining (more sophisticated) designs will be of greater interest to more advanced students and users of statistics.

The animation in Figure 7.9 demonstrates that each design is ideally implemented as a randomized experiment. That is, randomization in assigning experimental units to treatments is possible and preferable for all of the frameworks. Observational studies may also incorporate these designs. In this case, however, the treatments as well as the blocks, nesting, matching, and so on will be the result of natural features in the studied system. Causal inference will, of course, be affected by this distinction (Table 7.6).

7.6.5.3.1 One-Way ANOVA

The simplest type of ANOVA design will have a single factor, and is called a *one-way ANOVA*. It follows that ANOVA designs with two factors result in *two-way ANOVAs*, and so on. In a *completely randomized design (CRD)*, experimental units are randomly assigned to factor levels without constraints like blocking (Figure 7.9a). This approach is commonly used in both one-way ANOVAs and more complex formats like factorial and hierarchical designs. In the simplest type of one-way design, the explanatory variable has only two factor levels (see Figure 7.9a). This format is best analyzed using a *t*-test or a parallel method that allows one-tailed tests, since these are not possible with ANOVA.[*]

---

[*] A one-way ANOVA with two factor levels will provide the same *P*-value as a two-tailed pooled variance *t*-test.

```
anm.ExpDesign(method c("CRD","factorial2by2","nested","RCBD","split",
"pairs"))
```

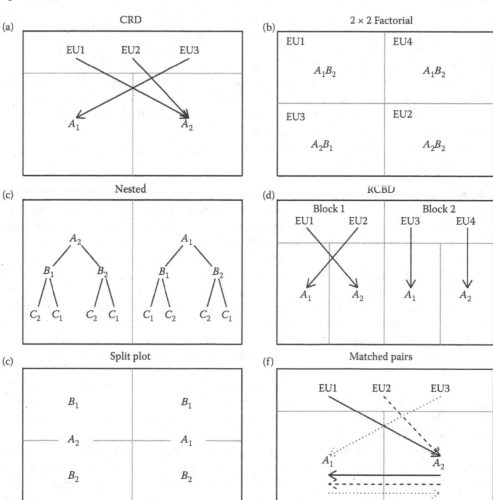

FIGURE 7.9

Demonstration of six types of ANOVA experimental designs. The letters *A*, *B*, and *C* represent factors, while subscripts to letters represent factor levels. EU = experimental unit. (a) A completely randomized design (CRD), in which EUs are randomly assigned to ≥2 factor levels without blocking or nesting. (b) A 2 × 2 factorial design. Here, every level of *A* occurs in every level of factor *B*, and vice versa. (c) A nested design. Here, particular levels of *B* occur only within a single level of *A*. The highest level of sampling represents the level of an independent experimental unit. (d) A randomized complete block design (RCBD). Here, treatments are randomly assigned to EUs within blocks. (e) A split-plot design. Here, one factor level from *A* is assigned to each whole plot. Whole plots are split creating split plots to which levels in *B* are randomly assigned. (f) A type of repeated measures design, matched pairs. Here, EUs are randomly assigned to either A_1 or A_2, and then are assigned to the treatment they did not initially receive.

There are two advantages to a one-way ANOVA. First, the design allows straightforward interpretation of factor-level effects. Second, computations will not be complicated by a lack of balance. This will not be true for unbalanced two-way or higher layouts.

A major disadvantage of one-way ANOVAs is that they may constitute an oversimplification of a system containing lurking variables, important interactions, and/or substantial background variability. The last circumstance may occur when treatments from a single factor are applied in a heterogeneous environment, resulting in background "noise" neutralizing the response "signal" from the treatments. Other ANOVA designs, including randomized block designs, can be used to address this issue.

A second disadvantage is that it may not be possible to rationally accommodate all factor levels of interest into a single factor. For instance, at an experimental site near Pocatello, Idaho, sagebrush steppe communities are often examined with respect to a single factor containing three levels: nitrogen addition, shrub removal, and a control. This system would probably be better analyzed using a two-factor approach (with N and shrub removal as separate factors, and the control as a factor level within each factor). This would allow the consideration of the confounding effects of shrub removal and N, and allow quantification of the interaction of the factors.

★ *7.6.5.3.2 Factorial Design*

Many ANOVA experiments will have more than one factor, and distinct treatments can be derived by combining factor levels from the multiple factors. This is called a *factorial design*. In a fully crossed factorial design with two factors, A and B, every level in factor A is contained in every level of factor B, and vice versa. As a result, if there are a factor levels in A and b factor levels in B, then there will be $a \times b$ distinct factor-level combinations (Figure 7.9b).

If a factorial design is not fully crossed, it will have confounded factors. For instance, a researcher recently measured bird diversity on two of the Aleutian Islands in Alaska, before and after removal of nonindigenous foxes. She measured diversity on both islands in year one (with foxes) and year two (after all foxes had been removed). However, because the experiment was not fully crossed, the effect of year and the effect of foxes were confounded. Thus, it was impossible to say whether changes in bird diversity were due to foxes, or year to year variation, or both.

A major advantage of factorial designs is that they allow quantification of both *main effects* (e.g., the effect of the factor A, holding factor B constant) and *interaction effects* between factors. Mathematically speaking, an interaction represents the portion of the effect of factors on the response that is nonadditive. That is, an interaction effect results in a response that cannot be obtained by simply adding the main effects (Chapter 10). The analysis of two factors using two separate one-way designs will not allow consideration of the interaction of factors.

A second advantage is that two-way designs (including factorial designs) may represent a more economical way of quantifying the effect of two explanatory variables on the response. That is, it may simply be more cost effective to run one experiment with two factors than to conduct two separate experiments, each with one factor.

A disadvantage of factorial designs is that it may be difficult to obtain adequate replication for a large number of factor-level combinations.[*] As a result of this (and the fact that it is extremely difficult to interpret the interactions of several factors), factorial designs with more than three factors are extremely rare.

---

[*] In a fully crossed factorial design, every experimental unit receives a factor level from each factor. In addition, there will (ideally) be an equal number of experimental units assigned to each factor-level combination. For instance, in Figure 7.9, there is one replicate for each of the four factor-level combinations. This is in contrast to a *fractional factorial design* in which not every experimental unit receives a factor level from each factor. Fractional factorial designs are often used in situations with large numbers of factors where adequate fully crossed replication is difficult or impossible to achieve.

7.6.5.3.3 *Randomized Block Design*

In a *randomized block design*, a researcher randomly assigns experimental units to treatments separately within units called blocks. If all treatments are assigned exactly once within each block, this is known as a *randomized complete block design (RCBD)* (Figure 7.9c).[*]

Blocks can be established randomly or systematically, but in either case, they should be placed to ensure that there is more variation between blocks than within blocks. Thus, blocks should be apportioned across a range of environmental conditions, and not merely at one point on a biological gradient (Figure 7.10). Blocks should be small enough to ensure that they are relatively homogenous. However, they should be large enough to ensure that the treatments within blocks are well separated in space (or time). The independence of experimental units should not be compromised by blocking.

The main advantage of a randomized block design is that it can be used to account for heterogeneity in a study system. As a result, a frequently used idiom is "block what you can; randomize what you cannot." A RCBD will be more efficient than a CRD if there is more heterogeneity among blocks than within blocks. That is, given appropriate application to a heterogeneous system, a randomized block design will require fewer replicates to have the same power as a CRD.

There are three disadvantages to using randomized block designs. First, there is a statistical cost to blocking. Thus, an RBD may be less efficient than a CRD if sample sizes are

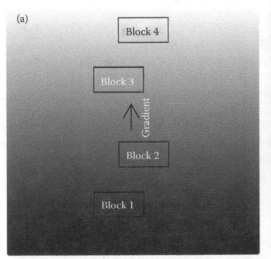 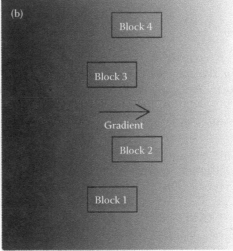

FIGURE 7.10
Correct and incorrect experimental blocking. (a) A correctly implemented randomized block design. The blocks are placed across the gradient. (b) An incorrectly implemented randomized block design. In this case, the blocks are placed perpendicular to the gradient. As a result, the difference between blocks will be small and variability within blocks will be high. This will result in weak blocking effects, and mischaracterization of main effects.

---

[*] Two designs closely related to RCBDs are randomized incomplete block designs and Latin squares designs. In a *randomized incomplete block design* or *RIBD*, treatments are assigned randomly but not all treatments are assigned within a block. Balanced RIBDs require careful planning. For a demonstration type: anm. ExpDesign("RIBD"). In a *Latin squares design*, experimental units are arranged into row and column blocks. A randomization strategy is used in which treatments are assigned no more than once in each row and column. The Latin square can be even more efficient than an RBD, provided that the blocking effects are sizable. To demonstrate a 3×3 Latin squares design, type anm.ExpDesign("latin").

small and block effects are weak. However, this does not mean that one should analyze a randomized block design as a CRD if block effects are nonsignificant. If an experiment was designed with blocks in place, then it should still be analyzed as a randomized block design (Kutner et al. 2005). Second, missing data pose a significant problem. This is because a missing observation results in an empty cell (no observation for a particular factor-level combination). Third, and most importantly, in a randomized block design we assume that the combined effect of the treatments and blocks is purely additive. This condition, which requires that the response is equal to the sum of the true main effects, is called *additivity*. Additivity entails that the rank order of responses to the treatments does not vary with block (Gotelli and Ellison 2004). Consider an experiment with three treatments: *a*, *b*, and *c*. Given additivity, blocks may differ with respect to the magnitude of the response across all treatments (the response to *a*, *b*, and *c* may be consistently high or low), but the ranking of *a*, *b*, and *c* should not vary from block to block. Nonadditivity will generally cause a randomized block design to incorrectly characterize main effects.

Because of these problems, a number of authors (e.g., Underwood 1997) have recommended that randomized block designs should not be used without within-block replication of treatments. An RBD framework allowing replication is called a *generalized randomized block design* (Wilk 1955), and results in a nested two-way ANOVA layout (see below). Such a design allows the consideration of main and additive effects, and helps alleviate problems with missing data. However, the additional replicates required may be too costly or logistically impossible to obtain for some investigations.

★ *7.6.5.3.4 Nested Design*

In a *nested design*, factor levels from one factor will be contained entirely within one factor level from another factor. Consider a design with two factors *A* and *B*. When every level of factor *A* appears with every level of factor *B*, and vice versa, then we have a fully crossed factorial design (see Section 7.6.5.3.2). Conversely, when levels of *B* only occur within a single level of *A*, then *B* is nested in *A* (Figure 7.9d).

Nesting can occur in three ways: (1) in a designed experiment with nested fixed factors, (2) in a designed experiment with nested random factors, or (3) as the result of a hierarchical sampling. The last two types of nesting, both of which will contain random effects, are much more prevalent in biological studies (e.g., Andrew and Underwood 1993, Letourneau and Dyer 1998).

EXAMPLE 7.12

Consider an experiment to compare the grain yield (kg/ha) in four counties in Idaho (factor *A*). Yield is measured within four randomly chosen fields at four randomly chosen farms in each county. The fields (call these factor levels in factor *C*) are nested within farms (call these factor levels in factor *B*) which are nested within *A*. Note that while there are four factor levels of *B* within each level of *A*, that farm B_1 in county A_1 is not the same as farm B_1 in county A_2. As a result, although we call the factor levels of *B* within each level of *A*, B_1, and B_2, this is only valid if we specify that *B* is nested within *A*. The fields within a farm within a county are not independent. In fact, they are more likely to be similar because they are much closer to each other spatially. In addition, the levels within county (the particular farms), or levels within farm (the particular fields) were chosen randomly, and have little meaning. Thus, B and C are random factors containing random effects.

There are three advantages to using a nested design. First, if nesting is the result of subsampling, then the nested subsamples (pseudoreplicates) will increase precision in estimating characteristics for the true replicates. Second, nested designs allow testing of hypotheses at both the treatment and subsample level. In particular, we can (1) use a one-way layout to test for treatment effects using the averages of the subsamples, and (2) use a one-way layout to test for differences among nested levels using values of the subsamples *within* the true replicates. Third, we can use subsampling to quantify variation in a biological system at different hierarchies.

The chief disadvantage of nested designs is that they are often analyzed incorrectly. This is because researchers often treat pseudoreplicate nested subsamples as independent samples, artificially inflating sample size, and increasing the probability of type I error (cf. Hurlbert 1984). This problem can be addressed by collapsing pseudoreplicates, using their averages, and analyzing treatment effects using a one-way layout. This will create independent replicates, however, it will prevent hierarchical analysis of a study system, using a nested ANOVA, as discussed above.

★ ### 7.6.5.3.5 Split-Plot Design

A *split-plot design* contains two nested levels of randomization. At the highest level are larger *whole plots* that are randomly assigned factor levels from one factor. At a lower (nested) level, whole plots are split to form *split plots* that are randomly assigned to factor levels from a second factor (Figure 7.9e). Blocks are frequently used to comprise a complete set of whole-plot treatments. Split-plot designs are often created inadvertently by researchers (Littell et al. 2006). As a result, one should examine factorial experiments carefully for split-plot features.

Split-plot designs have confusing similarities to both nested and randomized block designs. We can distinguish a split-plot design from a nested design because split plots contain informative (fixed) factor levels that mean the exact same thing in other whole plots. We can distinguish split-plot designs from randomized block designs because, unlike block levels, each whole-plot level will represent an informative and replicated treatment.

> **EXAMPLE 7.13**
>
> Consider an experiment to study the effect of two irrigation methods (factor *A*) and two fertilizers (factor *B*) using four fields as experimental units. Using a completely randomized design, we randomly assign one of the four treatments (A_1B_1, A_1B_2, A_2B_1, A_2B_2), to each field. However, because this is a one-way layout with four treatments and just four experimental units, there is insufficient replication for calculating the ANOVA test statistic (Chapter 10). One solution is to subdivide the fields into smaller experimental units. Allowing for this, assume that we can apply the fertilizer treatments to the smaller units, while logistical constraints require that irrigation treatments be applied to the entire field. As a result we designate irrigation as the whole-plot factor and fertilizer as the split-plot factor. Test statistics are now calculable allowing hypothesis tests concerning both main effects and interactions.

There are two advantages to split-plot designs. First, split plots can be used to effectively examine the effect of two factors when those factors have a hierarchical experimental or natural structure. Hypothesis tests can be run to separately examine the main effects (the whole- and split-plot factors) and to examine their interaction. Second, by establishing whole plots along a gradient, split-plot designs, like randomized block designs, can be used to account for "noise" in heterogeneous systems.

There are two disadvantages to split-plot designs. First, they are frequently misanalyzed as simple two-way ANOVAs. This will result in incorrect variances, and will increase the probability of a type I error. Second, in a conventional split-plot framework hypothesis tests are not possible for the interaction of blocks (containing whole-plot factor replicates) and the whole-plot factor. Thus, additivity assumptions are required for these factors, analogous to the RCBD assumption of no interaction between blocks and treatments.

★ *7.6.5.3.6 Repeated Measures Design*

With all previously described ANOVA designs, experimental units are measured at just one point in time. Conversely, in a *repeated measures design*, experimental units are measured more than once. Repeated measures designs can take on a variety of forms, and can be analyzed in a number of ways. One way to classify these approaches is to distinguish designs that use matched pairs from those based on longitudinal measures.

7.6.5.3.6.1 Matched Pairs   In a *matched pairs* design, treatments are compared by using the same (or highly similar) experimental units (Figure 7.9f). For simplest applications, it is generally assumed that outcomes within an experimental unit are independent; that is, there is no "carryover" effect from the previous treatment. In addition, we assume that the effects of experimental unit and treatment are purely additive. Under these assumptions, a conventional matched pairs design is a randomized block design with experimental units serving as blocks. Note that we have already encountered a very simple form of this approach for two factor levels, the paired *t*-test in Chapter 6.

7.6.5.3.6.2 Longitudinal Design   In another approach, an individual experimental unit receives only one treatment, and the response of the experimental unit is measured repeatedly over time. The analyses resulting from these sorts of studies are often called *longitudinal* (West et al. 2008).

The format of longitudinal data can be seen as a split-plot design. Specifically, the *between-subjects factor* describes the treatment associated with a particular experimental unit. These serve as the whole-plot treatment. The *within-subjects factor* is made up of time intervals at which the experimental unit was observed. These serve as split plots. A split-plot design in this context has been called a *split plot in time* (Milliken and Johnson 2009). A split plot in time assumes that the lack of independence among time frames is unchanging for all time differences.

7.6.5.3.6.3 Sphericity   Traditional repeated measures analyses assume that the variance of the difference of any two time sequences within an experimental unit measured repeatedly in time will be the same. This assumption is called *sphericity*, and violation of the assumption is called *asphericity*. The mechanics of sphericity estimation are shown in Table 7.8. The left hand of the table shows forced expiratory air volumes for three asthma patients

TABLE 7.8

Forced Expiratory Air Volumes for Three Asthma Patients at Three Time Intervals

Patient	T_1	T_2	T_3	$T_1 - T_2$	$T_1 - T_3$	$T_2 - T_3$
201	2.46	2.68	2.76	−0.22	−0.3	−0.08
202	3.5	3.95	3.65	−0.45	−0.15	0.3
203	1.96	2.28	2.34	−0.32	−0.38	−0.06
			$S^2 =$	0.0133	0.0136	0.0457

Source: Adapted from Littell, R. C., Stroup, W. W., and Fruend, R. J. 2002. *SAS for Linear Models.* Wiley, New York.

at three time intervals (see Exercise 10.17). The right side of the table shows differences in the time sequence, the variance of the differences is shown under these columns. While the variances of the $T_1 - T_2$ and $T_1 - T_3$ differences are similar, the variance for $T_2 - T_3$ is much larger, indicating a violation of sphericity. Sphericity is less restrictive than a condition called *compound symmetry* in which the variances within time segments are constant, and the covariances among time segments are also equal. While compound symmetry requires sphericity, the converse is not true.

Randomized block designs with random blocks (including those with repeated measures) assume sphericity, while a split plot in time design assumes that the dependence among observations is unchanging over time. If treatments are randomly assigned and are sufficiently separated in time, then this assumption will generally be valid (Quinn and Keough 2002). However, time segments are unlikely to be randomly assigned to experimental units, and the variance of treatment differences close together in time will generally be smaller than for those further apart in time. Indeed, the serial "memory" of a time sequence is the basis for the time series analysis described in Section 7.5.5.

Repeated measures designs have three advantages. First, in a matched pairs design, each experimental unit serves as its own control. This effectively accounts for confounding caused by lurking variability among experimental units. For instance, in an experiment to ascertain the effect of a drug on human subjects, variability in age, weight, and life history would all be controlled for by blocking by experimental unit. Second, longitudinal designs account for the fact that it is often more cost effective to repeatedly measure the same experimental units than to establish new units for each time by treatment combination. Third, repeated measures designs often allow quantification of time by treatment interactions. As with other multifactor designs, failure to measure these interactions may result in the misinterpretation of main effects. For instance, in a recent study, Gause (2012) examined the photosynthetic recovery of a moss species from two different levels of desiccation (high and low). The "high" treatment samples had higher photosynthetic rates than "low" treatment samples at the start of the experiment, but had lower photosynthetic rates after several hours. Thus, there was a significant interaction between time and treatment. Lack of knowledge of this interaction would have resulted in the false conclusion that high desiccation mosses recover more quickly than low desiccation mosses at each time interval.

The main disadvantage with repeated measures designs is the constraint of sphericity. Failure to address asphericity will inflate type I error. There are four solutions:

1. One can simply analyze and make inference to time frames separately (see Snedecor and Cochran 1989). Note that while this will prevent temporal autocorrelation, it will prevent quantification of time effects and time by treatment interactions.

2. In designing an experiment, one can use existing knowledge of the study system to determine when observations will be sufficiently separated in time so as to be independent. Note that this may not always be possible, particularly when time is a fixed factor.

3. One can use an analysis method that adjusts for or models asphericity (e.g., analysis of contrasts or a mixed model, see Chapter 10).

4. One can collapse repeated longitudinal measures for experimental units into averages, or use the slope of lines describing the response of subjects over time as a response variable (Gotelli and Ellison 2004). This approach will create independent observations, and allow a simplified one-way layout for analysis, but like

separate analyses of time segments it precludes description of time effects and the interaction of time and treatments.

7.6.6 Tabular Designs

In many cases, (1) the explanatory and response variables will both be categorical (ordered or not), or (2) all measured variables will be categorical, but will defy distinct X and Y classification. In both scenarios, the result is a *tabular design* (Table 7.7, Chapter 11). Examples include

- An experiment to determine whether or not AIDS has progressed, $Y = \{Yes, No\}$, for patients given either the drug AZT or a placebo, $X = \{A, P\}$ (Cooper et al. 1993).
- A study in which the presence/absence of a plant community, $Y = \{P, A\}$ is modeled as a function of particular topographic types $X = \{toe slope, ridgetop, drainage bottom\}$ (Aho et al. 2011).
- An experiment examining eastern chipmunk (*Tamias striatus*) trilling vocalization $Y = \{Yes, No\}$ as a function of distance released from home burrow $X = \{near far\}$ (Burke da Silva et al. 2002, McDonald 2009).

A *contingency table* (a tabular cross-classification of categorical variables) is often used to summarize data from tabular designs. The cells in the table contain counts for categorical intersections, and row and column sums (i.e., *marginal totals*). For instance, Table 7.9 shows a cross-classification of nucleotide sites on the alcohol dehydrogenase gene for three species of fruit flies (*Drosophila* sp.), with respect to synonymy and fixity. Note that in this case, the categorical variables (fixity and synonymy) cannot be clearly distinguished as X and Y.

The chief advantage of tabular designs is that they can accommodate and logically summarize ordered or unordered categorical X and Y data. There are two disadvantages. First, cell characteristics in tables will dictate not only the results of inferential tests, but the type of tabular tests that are possible. For instance, because they rely on asymptotic null distributions, tabular *chi-squared* and likelihood ratio tests (Chapter 11) should have no more than 20% of the cross-classified categories with expected frequencies less than five (Cochran 1954). A second disadvantage is that a reversal of trends

TABLE 7.9

Number of Replacement and Synonymous Alcohol Dehydrogenase Nucleotide Substitutions for Fixed Differences between Three Species of *Drosophila*, and Polymorphisms within a Single Species of *Drosophila*

	Fixed	Polymorphic	Total
Replacement	7	2	9
Synonymous	17	42	59
Total	24	44	68

Source: Adapted from McDonald, J. H., and Kreitman, M. 1991. *Nature* 351: 652–654.

Note: With respect to synonymy the "replacement" category indicates variation that resulted in a different amino acid sequence among species, while the "synonymous" category indicates that the amino acid sequences were the same. The "fixed" and "polymorphic" categories designate whether or not a particular nucleotide base was fixed across all individuals within one of the three species.

often occurs as a result of lurking variables and/or aggregating groups. This outcome, called *Simpson's paradox* (after Simpson 1951), may arise with both categorical and quantitative response variables, but is particularly problematic with causal hypotheses in tabular data (Pearl 2000).

7.6.6.1 Three Tabular Designs

Sokal and Rohlf (2012) distinguish three experimental designs for tabular analysis. They can be distinguished based on whether or not the investigator fixes marginal totals.

- In a model I format, marginal totals are not fixed. This is the most commonly used tabular design in the biological sciences (indeed all real-world examples given in this section follow this format). It will occur when experimental units are classified as they are acquired. Given random sampling, the count contained in each cell in a model I design will be a Poisson random variable.

- In a model II design, the marginal totals for one of the dimensions are fixed. Consider an experiment to test weed survivorship (Y) given four different herbicides (X). In model II designs, the marginal totals for the response variable are usually fixed, so we would use a fixed number of weed individuals, and assign one-fourth of these to each herbicide treatment to eliminate bias due to imbalance.

- In a model III design, the marginal total for all dimensions are fixed. While conceptually useful, this framework is seldom used by biologists (Quinn and Keough 2002). Given random sampling cell counts in a model III design can be considered hypergeometric random variables.

7.7 Summary

- This chapter introduces important terminology associated with statistical analysis, and the concepts associated with sampling design and experimental design. Variants in sampling and experimental designs exist to facilitate the attainment of inferential statistical goals. These include (1) extending inference to the population(s) of interest and (2) providing evidence for cause and effect relationships. Many issues affecting these goals were discussed, including lurking and confounding variables, nonindependence of samples, measurement error, bias, and controls. In general, valid inferences are greatly facilitated through thoughtful consideration and knowledge of the study system, by using adequate replication, and by using randomization in (1) the selection of experimental units, and (2) assignment of treatments to experimental units.

- Sampling design describes the methods used to select experimental units from a population one wishes to make inference to. A number of issues associated with sampling design were discussed, including adjustments to standard errors for both finite population size and sample selection methods.

- Experimental design describes how experimental units are assigned to treatments. Three general types of experimental designs can be distinguished: regression, ANOVA, and tabular. Regression and ANOVA are general linear models. That is, they define the response as a linear function of explanatory variables and assume identical normal errors for observations. Regression designs have continuous explanatory variables, and include approaches like simple linear regression and multiple regression. ANOVA designs have continuous response variables and categorical explanatory variables, and include completely randomized designs, randomized block designs, nested designs, split-plot designs, and repeated measures designs. Tabular designs have both categorical response and explanatory variables, and are used to test hypotheses of independence or causality among table dimensions.

EXERCISES

1. Differentiate between the terms in (a) through (f), and give an example for each.
   a. *Statistical* and *biological populations*
   b. *Research* and *statistical hypothesis*
   c. *Response* and *predictor* variables
   d. Types of data: *nonquantitative, quantitative,* and *ordinal*
   e. Types of quantitative data: *continuous* and *discrete*
   f. Types of analyses: *univariate* and *multivariate*

2. Provide definitions for the following terms:
   a. Experimental unit
   b. Variable
   c. Explanatory variable
   d. Response variable
   e. Lurking variable
   f. Confounding variable

3. For the hypothetical soils dataset in the given table, identify the experimental units, and decide whether data are quantitative, nonquantitative, or ordinal. If data are quantitative, decide whether they are discrete or continuous.

Site	Soil %Nitrogen	Soil % Carbon	Species Richness	Texture	Soil Water Class (0 = Dry, 10 = Wet)
1	20.1	40.3	26	Silty	7
2	17.3	30.0	22	Silty	7
3	17.2	27.1	21	Silty	6
4	13.9	24.2	20	Clayey	7
5	12.4	20.4	21	Clayey	5
6	15.5	26.6	20	Clayey	5
7	17.1	30.5	25	Sandy	3
8	14.0	24.1	20	Clayey	5
9	15.3	23.0	13	Sandy	1
10	10.1	15.1	10	Sandy	2

4. Decide whether the scenarios described below depict univariate or multivariate datasets/analyses by determining how many response and explanatory variables are present.

a. A plant ecologist is interested in how plant photosynthesis is affected by levels of CO_2. She measures plant photosynthesis (as O_2 produced) in chambers with 15 different levels of CO_2.

b. A wildlife biologist is interested in how the size and degree of isolation of patches with suitable small mammal habitat are associated with amount of distance traveled per day by rodents. She releases 100 marked voles (*Microtus* sp.) into each of 20 patches, which vary with respect to size and isolation. She recaptures the voles using live traps and determines how far they have traveled before releasing them again.

c. A community ecologist wonders how a community of 15 native plant species is affected by elk grazing. She randomly places 10 elk exclosure plots and 10 control plots onto a landscape that contains all 15 species and measures the ground cover of the 15 plant species at all 20 plots.

5. Decide whether the following examples describe simple random sampling, stratified random sampling, or cluster sampling.

a. An alpine plant ecologist designates three topographic areas (ridges, late snowmelt sites, and south faces). He randomly samples soils within these topographic areas to compare five different mountains.

b. A scientist randomly selects 50 people who suffer from chronic migraines from a large pool. She randomly assigns a new medicine to 25 of them and gives the rest a placebo.

c. A scientist establishes 40 four block areas within a city for demographic sampling. He randomly selects 10 of the blocks and takes a census inventory for each of these blocks.

6. Identify whether the experimental designs described here are completely randomized, randomized block, or matched pairs designs.

a. An agricultural scientist is interested in the yield of four species of corn. He divides a field up into four equally sized quadrangles and randomly assigns each species of corn to a quadrangle. He repeats this process in five fields.

b. Another agricultural scientist randomly assigns nitrogen supplements to four of eight fields. The other four fields are used as controls. She measures yield per acre for each of the fields.

c. A medical researcher is comparing the effects of drugs A and B on patients with sleep disorders. He randomly selects 100 patients from a large pool. He randomly assigns 50 of the patients to drug A and the rest to drug B for a 3-week period. During this time, the amount of REM sleep is measured for each patient. After an additional 2 weeks, the patients who were initially given drug A are now given B, and the patients initially given drug B are given drug A. Again the drugs are administered for 3 weeks and REM sleep is monitored.

8. Define the word *inference*. How do we establish inference to a population using a sample?

9. Define the word *causal*. Why is it difficult to establish causality? What general experimental method is used by scientists to generate causal evidence?

10. The following are terms important to establishing the scope of inference of a study. Contrast them.

a. *Experimental study* versus *observational study*

b. *Random* versus *nonrandom sampling* of experimental unit populations

c. *Random* versus *fixed effects*

11. Are the following examples observational studies or experiments?

a. A software company wants to compare the effectiveness of computer animation for teaching cell biology versus a textbook presentation of the same information. The company tests the biological knowledge of a group of 500 randomly selected first year college students, and then randomly assigns the students to one of two groups. One group uses the computer animation software while the other learns using a conventional textbook. The company retests all the students using a continuously scaled index and compares the increase in knowledge of cell biology in the two groups.

b. In an 1898 lecture at Woods Hole, Massachusetts, Herman Bumpus, a professor of zoology, presented measurements on house sparrows (*Passer domesticus*) brought to the anatomical laboratory at Brown University after a severe winter storm. Some of the birds survived the storm while others had died. Bumpus measured physical characteristics (e.g., humerus length) of the survivors and the mortalities and drew comparisons between survivorship and physical characteristics.

c. A researcher is interested in how the weight of gray wolves (*Canis lupus*) changes with latitude. She determines the average adult weight from ten randomly selected packs of wolves situated from northern Alaska to Southern Canada. She compared the pack weight to the average pack latitude.

12. Aerial surveys for mountain goats in Yellowstone National Park were constructed by dividing a 500 km² area into 4 strata (*A, B, C,* and *D*). A sample represents a 5 km². Thus, the total population of potential samples, $N = 100$. Furthermore, we know that $N_A = 20$, $N_B = 60$, $N_C = 10$, and $N_D = 10$. Individual entries in the following table describe goat counts for 5 km² samples.

Strata	$x_{(h)i}$	Strata	$x_{(h)i}$	Strata	$x_{(h)i}$	Strata	$x_{(h)i}$
A	2	B	4	C	7	D	1
A	2	B	5	C	7	D	2
A	1	B	6	C	6	D	0
A	0	B	4	C	3	D	2
A	3	B	3	C	5		
A	0	B	7				
A	1						

a. Calculate n_hs, $\bar{X}_h$, and S_h^2

b. Calculate $\bar{X}_{str}$

c. Calculate $\hat{T}$

d. Calculate $S_{\bar{Y}_{str}}$

e. Calculate $S_{\hat{T}}$

f. Calculate 95% stratified confidence intervals for μ and T

g. Check your results using `ci.strat` (also use `ci.strat` for (h) and (i))

h. Treat the data as if they were not stratified (but use a finite population correction) and calculate a 95% confidence interval for μ

i. Treat the data as if they were not stratified and do not use a finite population correction to calculate a 95% confidence interval for μ

j. Comment on the different confidence interval widths in (f), (h), and (i)

15. Explain how the following may affect inferences:
    a. Sampling bias
    b. Measurement bias
    c. Missing values
    d. Outliers
    e. Absence of controls
    f. Unmeasured covariates

16. A biological education researcher wants to compare the effectiveness of two teaching approaches using test scores. There are two classroom sections per semester. The researcher randomly assigns the teaching methods (one to each section), and repeats this for six semesters.
    a. Are the students in the classrooms pseudoreplicates? Why or why not?
    b. What is the experimental design? Why?

17. A well-intentioned plant ecologist wants to determine the effect of snow depth on plant biomass. He builds two snow fences, one 6-feet tall, and one 4-feet tall. On the lee side of each fence, he randomly establishes 20 plots to measure biomass. How many independent samples are there? Why?

18. Define the terms *additivity* and *sphericity*.

19. For Examples 6.6 through 6.8, 6.13, and 6.17, what is the scope of inference? In particular, consider the following:
    a. What is the population to which inference can be made?
    b. Will causal evidence be generated with respect to X on Y?
    Note that the answers, in all cases, may not be black/white.

20. With respect to Question 6
    a. What are the explanatory variables in examples (a), (b), and (c)?
    b. What are the response variables?
    c. As far as you can tell, what is the scope of inference and causality for examples (a), (b), and (c)?

21. Give biological examples, along with advantages and disadvantages, for each of the following regression designs/approaches:
    a. Simple linear regression
    b. Multiple regression

22. Give biological examples, and advantages and disadvantages, for each of the following ANOVA experimental designs/approaches:
    a. Completely randomized design
    b. Factorial design
    c. Randomized block design

    d. Nested design

    e. Split plot

    f. Matched pairs design

23. Find examples in the biological literature of completely randomized, factorial, randomized block, nested, split plot, and repeated measures designs. Provide citations and briefly summarize the experimental methods.

24. What characteristics allow us to distinguish nested and split-plot designs?

25. Give biological examples of model I and II tabular designs.

Section II

Applications

8

Correlation

The invalid assumption that correlation implies cause is probably among the two or three most serious and common errors of human reasoning.

Stephen Jay Gould (1981)

8.1 Introduction

It is commonplace for biologists to record more than one quantitative variable at an experimental unit. For instance, soil pH and percent nitrogen are often measured at each site where soils are sampled. We may be interested in how, or if, these variables are associated. For example, do pH and percent N increase together, does pH increase as percent N decreases, or does pH remain unchanged as percent N changes?

If two variables are _correlated_, then one variable will tend to increase or decrease as the other variable increases or decreases. Note, however, that this neither requires nor implies that these variables are causally associated (Chapter 7). _Correlation analysis_ does not test causal hypotheses, or distinguish X and Y variables. Instead it is concerned with independence/dependence of variables, and the direction and degree to which two variables change together.

EXAMPLE 8.1

Extremely important 21st-century correlations include those that can be demonstrated for atmospheric carbon components and time. Figure 8.1 shows a positive correlation between year and CO_2 concentration (left axis), and a negative correlation between year and the concentration of ^{14}C (right axis). CO_2 is an important greenhouse gas, whose concentration has increased by more than 40% since the onset of the industrial revolution in the late 18th century (Mann and Kump 2009). This relationship is shown for the years 1992–2008 in Figure 8.1 (white points). ^{14}C is an unstable carbon isotope with a half-life of 5730 years. Over time, ^{14}C will degrade into ^{14}N. Thus, fossil fuels that are often millions of years old have very little ^{14}C. Large anthropogenic contributions to atmospheric carbon can be linked to a linear decline in the atmospheric component of ^{14}C beginning around 1850 (Bacastow and Keeling 1974). This relationship for the years 1992–2008 is shown by the black points in Figure 8.1.

```
Fig.8.1()# See Ch. 8 code
```

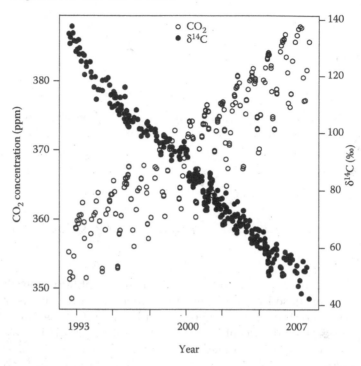

FIGURE 8.1
Daily atmospheric fluctuations for both CO_2 and $\delta^{14}C$ at La Jolla California (data from Graven et al. 2012. *Journal of Geophysical Research* 117: D02302: 1–14). The large (within year) variability in CO_2 is due to seasonal effects (CO_2 net assimilation by plants decreases during hemispheric winter months). $\delta^{14}C$ measures were derived from flask air samples and are calculated by comparing observed $^{14}C/^{13}C$ ratios to a standard. The recent decline in atmospheric $\delta^{14}C$ was first documented by Hans Suess and Harold Urey (1956) and is called the *Suess effect*. For both datasets, $n = 280$.

8.2 Pearson's Correlation

Recall from Chapter 4 that the *population covariance* describes how the *marginal variables* from a *bivariate distribution* covary. The parameter is defined as

$$\text{Cov}(Y_1, Y_2) = E(Y_1 Y_2) - E(Y_1)E(Y_2). \tag{8.1}$$

That is, the population covariance is the expectation of the product of Y_1 and Y_2 minus the product of their expectations. Note that because $E(Y_1 Y_2) = E(Y_2 Y_1)$, then $\text{Cov}(Y_1, Y_2) = \text{Cov}(Y_2, Y_1)$. Unlike variances, which must be ≥ 0, a covariance can be either positive or negative. If $\text{Cov}(Y_1, Y_2)$ is negative, then Y_1 tends to increase when Y_2 decreases (and vice versa). Conversely, if $\text{Cov}(Y_1, Y_2)$ is positive, then Y_1 and Y_2 tend to increase (and decrease) together.

The *population correlation* is simply the population covariance divided by the product of the standard deviations of Y_1 and Y_2.

$$\rho_{Y_1,Y_2} = \frac{\text{Cov}(Y_1, Y_2)}{SD(Y_1)SD(Y_2)} \tag{8.2}$$

As a result of this standardization, ρ will always be a number between -1 and 1. Values of ρ near 0 indicate a lack of linear association between Y_1 and Y_2, while values near -1 and 1 indicate strong negative and positive linear associations, respectively.

We note that

- ρ_{Y_1,Y_2} will have the same sign as $\text{Cov}(Y_1,Y_2)$.
- ρ_{Y_1,Y_2} is only defined if both population variances are finite and nonzero.
- Because covariances are symmetric, ρ_{Y_1,Y_2} must equal ρ_{Y_2,Y_1}.
- Because of the range of ρ, the absolute value of $\text{Cov}(Y_2,Y_1)$, must be less than or equal to $SD(Y_1)SD(Y_2)$.

Although the population correlation and the population covariance both quantify the degree to which two random variables change linearly together, the correlation coefficient is generally easier to interpret. This is because (1) the correlation clearly represents the strength of linear association with respect to the bounds $[-1, 1]$ and the center of its support, zero, and (2) linear transformation of Y_1 and/or Y_2 will generally have no effect on its value. The only exception will be when a multiplier changes the sign of only one of the variables, causing the sign of ρ to be reversed. Conversely, the covariance will have no strict upper and lower limits, and may change dramatically with linear transformation (for instance, changing the unit of measurement from grams to kilograms in one or both variables).

8.2.1 Association and Independence

It is important to be able to distinguish whether variables are independent or dependent. Recall from Chapter 2 that if two events A and B are independent, then

$$P(A \cap B) = P(A)P(B). \tag{2.11}$$

In other words, because A and B are independent, the product of their marginal probabilities is equivalent to the probability of their intersection.

Following these principles, if two random variables, Y_1 and Y_2, are independent, then

$$f(y_1,y_2) = f(y_1)f(y_2). \tag{8.3}$$

That is, the product of the marginal densities of Y_1 and Y_2 will be equivalent to their joint (bivariate) density.

If Y_1 and Y_2 are independent, then $\text{Cov}(Y_1,Y_2) = 0$. Furthermore, because $\text{Cov}(Y_1,Y_2) \neq 0$ will result in $\rho \neq 0$, then dependent (nonindependent) variables will have nonzero correlations. Note, however, that the converse is not true. That is, even if $\text{Cov}(Y_1,Y_2) = 0$ and $\rho = 0$, this does not necessarily mean that Y_1 and Y_2 are independent. Because the covariance and correlation can only quantify linear association, nonlinear forms of dependence will remain undetected. This is a rather difficult concept to grasp, and a mathematical/graphical example is provided below.

EXAMPLE 8.2

Consider a pair of discrete random variables Y_1 and Y_2 with the joint pdf $f(y_1, y_2) = 0.25$ for the joint outcomes $(y_1, y_2) = (0, 1), (1, 0), (0, -1), (-1, 0)$, and $f(y_1, y_2) = 0$ for all other outcomes. Let the marginal pdf of Y_1 considered alone be $f(-1) = 0.25, f(1) = 0.25, f(0) = 0.5$, and $f(y_1) = 0$ for all other outcomes, and let the marginal pdf of Y_2 be identical to the marginal pdf for Y_1 (Figure 8.2). We see that $E(Y_1) = E(Y_2) = 0$. This is because, for both marginal distributions, the expected value is $-1(0.25) + 0(0.5) + 1(0.25) = 0$. Furthermore, the product of joint quantiles will be 0 whenever $f(y_1, y_2) > 0$; thus, $E(Y_1Y_2) = 0$. It follows immediately that the covariance and correlation of Y_1 and $Y_2 = 0$, because $E(Y_1Y_2) - E(Y_1)E(Y_2) = 0$.

Nonetheless, Y_1 and Y_2 are not independent. This is because $f(y_1)f(y_2) \neq f(y_1, y_2)$. For instance, $f(0)f(0) \neq f(0, 0)$ because $(0.25)(0.25) \neq 0$. Thus, two variables can be dependent with a covariance and correlation equal to zero.

```
Fig.8.2() # See Ch. 8 code
```

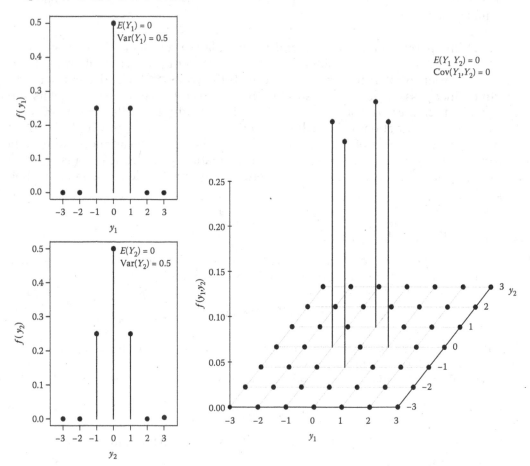

FIGURE 8.2
Example of discrete marginal and joint bivariate distributions from Example 8.2. Y_1 and Y_2 are not independent despite the fact that their covariance and correlation are zero. This is because the products of the marginal densities are not equal to their respective joint densities. Note that there are three nonzero marginal densities for each pdf, leading to nine nonzero bivariate combination densities under independence (the center of the bivariate grid above). However, there are only four nonzero bivariate densities.

8.2.2 Bivariate Normal Distribution

The only exception to the generality "$\rho = 0$ does not necessitate independence," is the bivariate normal distribution. In this case, if $\text{Cov}(Y_1, Y_2) = 0$ (which requires that $\rho = 0$), then Y_1 and Y_2 will be independent. This is because, if it exists, dependence in a bivariate normal distribution will be entirely linear.

From the Bayesian examples in Chapter 5, recall that a bivariate normal distribution is an extension of the normal distribution to two variables considered simultaneously. If Y_1 and Y_2 are bivariate normal, then their joint density can be expressed as

$$f(y_1, y_2) = \frac{1}{\sqrt{4\pi^2 |\sigma^2|}} \exp\left[-\frac{1}{2}(\mathbf{y} - \mu)' \sigma^2 (\mathbf{y} - \mu) \right] \tag{8.4}$$

where $\mathbf{y}$ is a two-element vector specifying quantiles for Y_1 and Y_2, μ is a two-element vector defining the population means for Y_1 and Y_2, σ^2 is a 2×2 *variance covariance matrix* containing the marginal population variances on the diagonal, and the covariance in the off-diagonal elements, and $|\sigma^2|$ is the determinant of σ^2 (see Appendix), that is

$$\mathbf{y} = \begin{bmatrix} y_1 \\ y_2 \end{bmatrix}, \quad \mu = \begin{bmatrix} \mu_1 \\ \mu_2 \end{bmatrix}, \quad \sigma^2 = \begin{bmatrix} \sigma_1^2 & \sigma_{1,2} \\ \sigma_{1,2} & \sigma_2^2 \end{bmatrix}.$$

Because covariances are symmetric (i.e., $\sigma_{1,2} = \sigma_{2,1}$), the off-diagonal elements in σ^2 will be equal.

A bivariate normal distribution is shown in Figure 8.3.

Note that the distribution is bell shaped in three dimensions instead of two. As a result, a planar slice of the joint distribution perpendicular to the Y_1 axis, or perpendicular to the Y_2 axis, will be a normal distribution. Figure 8.3 shows the *bivariate standard normal distribution*. It is defined by three characteristics: (1) the *centroid* (joint mean) is {0, 0}, (2) the covariance is zero, that is, $\sigma_{1,2} = \sigma_{2,1} = 0$, and (3) $\sigma_1^2 = \sigma_2^2 = 1$. In this setting, Y_1 and Y_2 are independent. As the covariance of the joint distribution increases compared to the marginal standard deviations, the distributional "mounds" become narrower and leptokurtic. In the case that $\rho = 1$ or -1, the entire bivariate population will be supported on a single line, allowing a planar projection of the joint pdf. For instance

```
bvn.plot(cv =.999, vr = c(1, 1))
```

EXAMPLE 8.3

For the bivariate normal distribution shown in Figure 8.3, we will calculate the joint density (the height of the bivariate surface) for $f(1.96, 2.3)$, relying entirely on **R**.

```
install.packages("mvtnorm")
library(mvtnorm)
sigma <- matrix(2, 2, data = c(1, 0, 0, 1))
mu <- c(0, 0)
dmvnorm(c(1.96, 2.3), mu, sigma) #f(1.96, 2.3)
[1] 0.001655459
```

```
bvn.plot(mu = c(0, 0), vr = c(1, 1), cv = 0, res = 0.1)
```

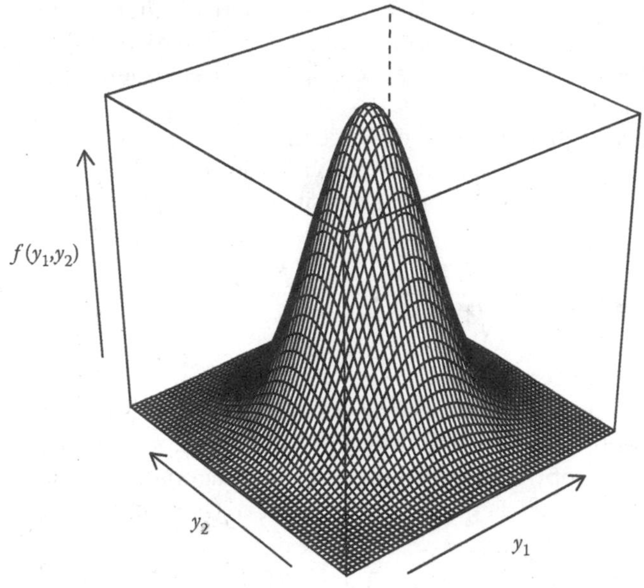

$f(y_1, y_2)$

y_2

y_1

FIGURE 8.3
A bivariate standard normal pdf, $Y_1, Y_2 \sim MVN(\mathbf{\mu}, \mathbf{\sigma}^2)$, $\mathbf{\mu} = \begin{bmatrix} 0 \\ 0 \end{bmatrix}$, $\mathbf{\sigma}^2 = \begin{bmatrix} 1 & 0 \\ 0 & 1 \end{bmatrix}$. Y_1 and Y_2 are independent because they are bivariate normal with a correlation of 0.

8.2.3 Estimation of ρ

The population covariance is estimated with the sample covariance.

$$S_{1,2} = \frac{\sum_{i=1}^{n} (Y_{i1} - \bar{Y}_1)(Y_{i2} - \bar{Y}_2)}{n-1}, \tag{8.5}$$

where n is the number of experimental units in which both variable 1 and variable 2 are measured.

Pearson's product moment correlation coefficient, r, named after Karl Pearson (Section 1.9), estimates the linear association between two random variables

$$r = \frac{S_{1,2}}{S_1 S_2}. \tag{8.6}$$

Assuming bivariate normality for the joint pdf, $S_{1,2}$ and r are the ML estimators for $Cov(Y_1, Y_2)$ and ρ, respectively. However, because the covariance and Pearson's correlation coefficient are defined by moments, they have no distributional requirements for point estimation.

Like the parameter it estimates, ρ, and like all other correlation estimators, r is unitless, and will lie in the range $[-1,1]$. Also, like ρ, a positive value of r indicates that Y_1 tends to increase as Y_2 increases, a negative value means Y_1 tends to decrease as Y_2 increases, and a value near 0 indicates a lack of linear association.

The statistic r will be affected by a number of factors (Figure 8.4).

- Maximum correlation will occur when all bivariate sample points lie along a straight line (Figure 8.4a). In this situation, the absolute value of covariance will equal the product of the marginal standard deviations, resulting in $r = 1$ or $r = -1$. The exception will be when the slope of the line either equals zero, or is infinite. In this case, r will be undefined because the conditional variance of one of the variables will equal 0.

- As bivariate scatter increases, resulting in decreased linearity, then r will also decrease (Figure 8.4b).

- Because it is based on sample variances and covariances, which are not resistant, r will be affected by outliers more than any other correlation estimator (Abdullah 1990). Even for datasets with an otherwise perfectly linear association, a single outlier will strongly influence estimates (Figure 8.4c).

- Like ρ, r does not detect nonlinear association. Because of this, dependence may exist even when $r = 0$. For instance, in Figure 8.4d, we see that Y_2 is entirely determined by Y_1. In fact, $Y_2 = Y_1^2 - Y_1$. However, because the dependence is nonlinear, the correlation is zero.

Bivariate data may have the same Pearson's correlation, while having dramatically different forms of association. The four scatterplots in Figure 8.5 represent *Anscombe's quartet* (Anscombe 1973, Chatterjee and Firat 2007). The four datasets all have the same correlation. Only in one, however, does r accurately represent the association between Y_1 and Y_2 (Figure 8.5a). In Figure 8.5b (as in Figure 8.4d), r provides a linear correlation for a

```
Fig.8.4() ## see Ch. 8 code
```

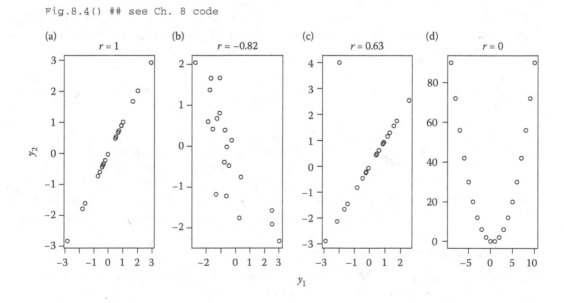

FIGURE 8.4

Examples of linear association as measured by Pearson's correlation coefficient: (a) a perfect linear association, (b) a negative linear association with random variability, (c) a perfect linear association with a single outlier, perhaps due to a data entry or measurement error, and (d) a concave parabolic association, for example, spring to fall soil water availability.

```
example(anscombe)
```

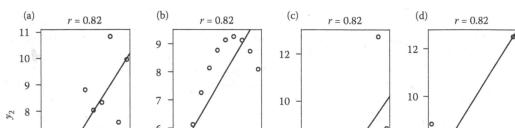

FIGURE 8.5

A demonstration of Anscombe's quartet. These data not only have the same Pearson's correlations, but also have the same conditional $(Y_1 | Y_2)$ sample means = 7.5, conditional sample variances = 14.97, and are described with the same regression line, $Y = 3 + 0.5X$ (Chapter 9).

nonlinear association. In Figure 8.5c (as in Figure 8.4b), an outlier strongly affects the measure of linear association. In Figure 8.5d, an outlier produces a high correlation coefficient for an association that should be undefined. The take-home message here is that summary statistics are often inadequate for describing an association (or any process), and should be accompanied with graphical examinations of the data (cf. Fisher and Switzer 2001).

The estimator for the standard deviation of the sampling distribution of r is

$$\hat{\sigma}_r = \sqrt{\frac{1 - r^2}{n - 2}}, \tag{8.7}$$

where r^2 is simply the correlation coefficient squared. Like all standard errors, $\hat{\sigma}_r$ quantifies how close an estimate (r) is likely to be to the parameter it is estimating (ρ).

Equation 8.7 will work poorly for estimating confidence intervals for ρ (see Section 8.2.1). This is because r is a biased estimator (biased low) of ρ (Snedecor and Cochran 1989), and the sampling distribution of r will be skewed when $\rho \neq 0$ (Quinn and Keough 2002; Figure 8.6). Nonetheless, we will use $\hat{\sigma}_r$ for hypothesis testing procedures concerning $\rho = 0$ (see below).

8.2.4 Hypothesis Tests for ρ

In general, we are interested in testing for the presence of a true linear association between Y_1 and Y_2. As a result, we use the two-tailed hypotheses:

$$H_0: \rho = 0,$$

versus

$$H_A: \rho \neq 0.$$

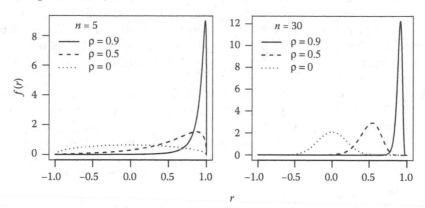

example(r.dist)

FIGURE 8.6

Assuming a bivariate normal parent distribution, the sampling distribution of *r* can be mathematically derived (Kenney and Keeping 1951, pp. 217–221). The function `r.dist` provides these derivations. We note that the sampling distribution of *r* is positively skewed if ρ is less than zero, and negatively skewed if ρ is greater than zero. Increasing sample size improves normality, particularly if ρ is close to 0 or 1. The function `see.r.dist.tck` can be used to interactively explore the sampling distribution of *r*.

Assuming bivariate normality (see below), this will be a test for the independence of Y_1 and Y_2.

Alternatively, we can specify the lower- and upper-tailed alternative hypotheses:

$$H_A: \rho < 0,$$

and

$$H_A: \rho > 0.$$

To test these sets of hypotheses, we first calculate the test statistic

$$t^* = \frac{r}{\hat{\sigma}_r}. \tag{8.8}$$

If the assumptions for the test hold (see below), and if H_0 is true, then t^* will be a random outcome from a *t*-distribution with $n - 2$ degrees of freedom. The sampling distribution of t^* has $n - 2$ degrees of freedom (not $n-1$) because the estimation of two parameters (μ_{Y1} and μ_{Y2}) is required to estimate ρ. Let $T \sim t(n-2)$, then a two-tailed *P*-value is calculated as $2P(T \geq |t^*|)$, while the lower- and upper-tailed *P*-values are calculated as $P(T \leq t^*)$ and $P(T \geq t^*)$, respectively.

For interval estimation (Section 8.2.1) and hypothesis testing, we have the following assumptions:

1. The underlying joint distribution Y_1, Y_2 is bivariate normal.
2. As a result, the association between Y_1 and Y_2 is linear.
3. Data constitute independent samples from the joint distribution, Y_1, Y_2^*.

---

[*] For all correlation techniques in this chapter, it is important to note that while observations within variables (among experimental units) are assumed to be independent, observations within experimental units (bivariate observations) will not be independent if $Cov(Y_1, Y_2) \neq 0$.

The null distribution will hold, approximately, for nonnormal bivariate distributions given large sample sizes. This is because the sampling distribution of r will be approximately normal, given both large sample sizes and $\rho = 0$ (Figure 8.6). In this case, however, independence cannot be verified by ρ, and hypothesis tests using r will only provide inference concerning the presence of dependence, and the form of linear dependence.

The library *asbio* contains several functions that are useful for bivariate diagnostics. The function bv.boxplot creates bivariate boxplots, useful for identifying outliers (Goldberg and Ingelwicz 1992), while the function DH.test performs *Doornik–Hansen* hypothesis tests for multivariate normality.[*] Note, however, that hypothesis testing procedures for multivariate normality, including the Doornik–Hansen test, tend to be extremely stringent; that is, it is very difficult not to reject the null hypothesis of multivariate normality.

Simple scatterplots (Y_1 vs. Y_2) can also be used to diagnose bivariate normality. The cloud of points should be approximately elliptical, and should not contain outliers. Normal quantile plots, histograms, and diagnostic hypothesis tests can also be used to check for univariate normality, which is a characteristic of bivariate normality (although the converse is not true). That is, Y_1 and Y_2 can be marginally normal, but jointly nonnormal. Thus, diagnostics for univariate normality are insufficient for insuring bivariate normality.

Nonlinear transformations can often be made to Y_1 and Y_2 to strengthen the linearity of their relationship and to improve bivariate normality. Common transformations, to one or both variables, include log (or $\log(Y + 1)$ if there are zeroes), square root transformations (often helpful for count data), or arcsine transformations, if data are proportions (see Table 7.5). If the relationship between Y_1 and Y_2 is nonlinear, and/or the joint distribution of Y_1, Y_2 is nonnormal, and transformations do not solve the problem, one should consider a robust correlation measure (Section 8.3) for inferences concerning independence and association.

EXAMPLE 8.4

Sokal and Rohlf (2012) described the mechanism of correlation analysis using gill weight and body weight data for 12 striped shore crabs (*Pachygrapsus crassipes*). We will calculate Pearson's correlation coefficient, and test the null hypothesis of independence using these data.

```
data(crab.weight)
```

A scatterplot shows a linear association between crab gill weight and body weight, and a roughly elliptical distribution of points, suggesting bivariate normality, and prompting the use of Pearson's correlation (Figure 8.7).

The sample covariance is easily calculated in **R**

```
cov(gill.wt, body.wt)
[1] 596.5107
```

---

[*] The Doornik–Hansen test is a test for multivariate normality based on the skewness and kurtosis of multivariate data that is transformed to ensure independence (Doornik and Hansen 2008). The test is more powerful than the classic Shapiro–Wilk test (Chapter 6) for many multivariate distributions.

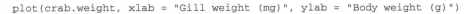

```
plot(crab.weight, xlab = "Gill weight (mg)", ylab = "Body weight (g)")
```

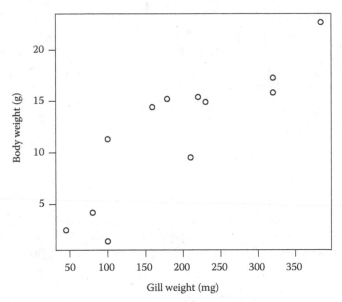

FIGURE 8.7
A scatterplot showing the relationship between body weight and gill weight for striped shore crabs (*Pachygrapsus crassipes*). Each point represents a pair of measurements for a single crab, $n = 12$.

Here is the sample variance covariance matrix:

```
var(crab.weight)
            gill.wt     body.wt
gill.wt 11306.2652 596.51068
body.wt   596.5107  42.04348
```

Here is the correlation of gill weight and body weight:

```
cor(gill.wt,body.wt)
[1] 0.8651857
```

The correlation coefficient $r = 0.87$ indicates a strong positive association between gill weight and body weight. Next, we calculate $\hat{\sigma}_r$:

$$\hat{\sigma}_r = \sqrt{\frac{1 - 0.865^2}{12 - 2}} = 0.1586.$$

The resulting test statistic is

$$t^* = \frac{0.865}{0.1586} = 5.4561,$$

and the P-value $= 2[P(T > |5.4561|)] = 0.0003$.

```
2*pt(5.456075,10,lower = F)
[1] 0.0002784622
```

The function cor.test calculates the correlation coefficient r, and conducts the associated hypothesis test.

```
with(crab.weight, cor.test(gill.wt, body.wt, method = "pearson")
Pearson's product-moment correlation
data: gill.wt and body.wt
t = 5.4561, df = 10, p-value = 0.0002785
alternative hypothesis: true correlation is not equal to 0
95 percent confidence interval:
0.5785547 0.9616151
sample estimates:
     cor
0.8651857
```

8.2.5 Confidence Interval for ρ

Earlier, I noted that the standard error of r works poorly in the creation of confidence intervals for ρ. An alternative estimation method was proposed by Fisher (1915, 1921). Under this approach, we convert r to a statistic, z', whose sampling distribution is approximately standard normal. Calculation of z' is based on the formula

$$z' = 0.5\ln\left(\frac{1+r}{1-r}\right) = \tanh^{-1}(r). \tag{8.9}$$

This transformation is called the *inverse hyperbolic tangent* and is often (as shown above) denoted $\tanh^{-1}$. Solving for r in Equation 8.9, we have

$$r = \frac{e^{2z'} - 1}{e^{2z'} + 1}. \tag{8.10}$$

In the z-transformed scale, we calculate $(1 - \alpha)100\%$ upper and lower confidence limits for ρ using

$$L_z = z' - z_{(1-(\alpha/2))}\sigma_{z'}, \tag{8.11}$$

$$U_z = z' + z_{(1-(\alpha/2))}\sigma_{z'}, \tag{8.12}$$

where $\sigma_{z'}$ is the standard error of z', and is defined to be $\sqrt{1/n-3}$. Converting back to r-scale units in the interval [–1, 1] using Equation 8.10, we have

$$C_L = \frac{e^{2L_z} - 1}{e^{2L_z} + 1}, \tag{8.13}$$

$$C_U = \frac{e^{2U_z} - 1}{e^{2U_z} + 1}. \tag{8.14}$$

EXAMPLE 8.5

For the striped crab data, we have

```
r <- with(crab.weight, cor(gill.wt, body.wt))
n <- with(crab.weight, length(gill.wt))
z <- atanh(r)
L.z <- z - (qnorm(0.975)/sqrt(n - 3))
U.z <- z + (qnorm(0.975)/sqrt(n - 3))
lower <- (exp(2 * L.z) - 1)/(exp(2 * L.z) + 1)
upper <- (exp(2 * U.z) - 1)/(exp(2 * U.z) + 1)
c(lower, upper)
[1]  0.5785547 0.9616151
```

Thus, we are 95% confident that the parameter ρ is contained in the interval (0.579, 0.961). This is in agreement with the `cor.test` output from Example 8.4.

Sokal and Rohlf (2012) recommend $n > 50$ for use of the z-transform confidence interval. However, given bivariate nonnormality, confidence intervals may not even be asymptotically correct (Duncan and Layard 1973). Bootstrapping has been recommended for these cases, although even this method will break down as ρ gets close to –1 or 1 (Wilcox and Muska 2001).

8.2.6 Power, Sample Size, and Effect Size

The parameter ρ provides the clearest definition of effect size in correlation analysis. Thus, r can be given as an estimate of the true effect size. To calculate the sample size sufficient to allow rejection of H_0 at significance level α, power $1 - \beta$, and effect size $|\rho|$, we use the formula

$$n = \left(\frac{z_{1-\alpha} + z_{1-\beta}}{\zeta} \right)^2 + 3, \qquad (8.15)$$

where $z_{1-\alpha}$ and $z_{1-\beta}$ denote the standard normal quantile function at probabilities $1 - \alpha$ and $1 - \beta$, respectively, and

$$\zeta = \tanh^{-1}(\rho) \frac{\rho}{2(n-1)}. \qquad (8.16)$$

The function `pwr.r.test` from library *pwr* uses an iterative root-finding algorithm to simultaneously solve for n, power, and α in Equations 8.15 and 8.16.

EXAMPLE 8.6

We assume that $\rho = 0.5$. What is the sample size required to reject H_0: $\rho = 0$ with probability 0.8, given that $\alpha = 0.05$?

```
install.packages("pwr"); library(pwr)
pwr.r.test(r = 0.5, sig.level = 0.05, power = 0.8)
       n = 28.87376
       r = 0.5
```

```
      sig.level = 0.05
         power = 0.8
   alternative = two.sided
```

A sample size of 29 paired observations is required.

8.3 Robust Correlation

Many robust methods for measuring correlation have been proposed (see Wilcox 2005). Two general approaches, rank-based permutation and nonrank robust estimation, are described in Sections 8.3.1 and 8.3.2, respectively.

8.3.1 Rank-Based Permutation Approaches

Spearman's rank correlation coefficient (also called ρ_s) and *Kendall's rank correlation coefficient* (also called Kendall's τ) utilize separate rank transformations of Y_1 and Y_2 with the pairing of observations preserved. Neither approach assumes bivariate normality. In addition, both methods will detect any monotonic association, including nonlinear associations. Note, however, that because this definition of correlation differs from that used by ρ, these methods should be seen as a measure of a different kind of association, rather than as alternative estimators for ρ.

8.3.1.1 *Spearman's* ρ_s

Spearman's estimator, r_s, is simply Pearson's correlation coefficient after Y_1 and Y_2 have been separately ranked-transformed, with the pairing of observations (y_{1i}, y_{2i}) preserved. This can be expressed as

$$r_s = \frac{\sum_{i=1}^{n}\left(r(Y_{1i}) - \frac{n+1}{2}\right)\left(r(Y_{2i}) - \frac{n+1}{2}\right)}{\sqrt{\sum_{i=1}^{n}\left(r(Y_{1i}) - \frac{n+1}{2}\right)^2 \sum_{i=1}^{n}\left(r(Y_{2i}) - \frac{n+1}{2}\right)^2}} = 1 - \frac{6 \times \sum_{i=1}^{n} D_i^2}{n^3 - n}, \qquad (8.17)$$

where $r(Y_{1i})$ is the rank of the ith observation from the Y_1 variable and $\Sigma_{i=1}^{n}D_i^2$ is sum of squared differences between the paired Y_1 and Y_2 ranks. A more complex derivation is necessary in the presence of ties (Hollander and Wolfe 1999, Siegel 1956):

$$r_s = \frac{\sum_{i=1}^{n} Y_{1i}^2 + \sum_{i=1}^{n} Y_{2i}^2 - \sum_{i=1}^{n} D_i^2}{2\sqrt{\sum_{i=1}^{n} Y_{1i}^2 \sum_{i=1}^{n} Y_{2i}^2}}. \qquad (8.18)$$

Here, the term $\Sigma_{i=1}^{n}Y_{1i}^2$ equates to $n^3 - n/12 - \Sigma_{i=1}^{n}T(Y_{1i})$, where $T(Y_{1i}) = t(Y_{1i})^3 - t(Y_{1i})/12$, and $t(Y_{1i})$ is the number of tied observations in Y_1 at the ith rank. In parallel, the term

$\sum_{i=1}^{n} Y_{2i}^2$ equates to $n^3 - n/12 - \sum_{i=1}^{n} T(Y_{2i})$, where $T(Y_{2i}) = t(Y_{2i})^3 - t(Y_{2i})/12$ and $t(Y_{2i})$ is the number of tied observations in Y_2 at the ith rank. For example, a single pair of tied observations for would result in $\frac{1}{2}[2(2-1)] = 1$. An untied observation is considered to be a tied group of size 1, resulting in $\frac{1}{2}[1(1-1)] = 0$.

As with other rank-based procedures, ties in data will result in approximate P-values unless the permutation distribution including ties is created and used as the distribution under H_0 (see Hollander and Wolfe, p. 403). If there are no ties, then Equation 8.18 reduces to Equation 8.17.

Our two-tailed hypotheses are

$$H_0: Y_1 \text{ and } Y_2 \text{ are independent,}$$

versus

$$H_A: Y_1 \text{ and } Y_2 \text{ are not independent.}$$

Hollander and Wolfe (1999) caution that hypothesis tests based on r_s should not be stated in terms of the parameter ρ_S. This is because the underlying measure of association linked to independence tests based on r_s is asymmetric and dependent on sample size (Hollander and Wolfe 1999, p. 405). As a result, the lower tailed alternative should be denoted as

$$H_A : Y_1 \text{ and } Y_2 \text{ are negatively associated,}$$

while the upper-tailed alternative should be

$$H_A : Y_1 \text{ and } Y_2 \text{ are positively associated.}$$

We have the following assumptions:

1. The association between Y_1 and Y_2 (if any) is monotonic (linear after rank transformation).
2. The bivariate observations are independent.

When n is small, the empirical distribution for r_s can be derived using the rank-based permutation methods discussed in Chapter 6. When n is large (>50), and H_0 is true, we can assume that

$$r_s \mathbin{\dot\sim} N\left(0, \ \frac{1}{n-1}\right), \tag{8.19}$$

where the dotted tilde signifies "follows approximately." The test statistic can be calculated as

$$z^* = \frac{r_s}{\sqrt{\dfrac{1}{n-1}}} = \sqrt{n-1} \times r_s. \tag{8.20}$$

The P-value for a two-tailed hypothesis test is calculated as $2P(Z \geq |z^*|)$, while lower- and upper-tailed P-values are calculated as $P(Z \leq z^*)$ and $P(Z \geq z^*)$,[*] respectively.

---

[*] Spiegel (1998) recommended using $t(n-2)$ as the null distribution for this test.

Because Spearman correlation estimates are not directly interpretable in terms of the parameter ρ_S, confidence interval estimates for the true correlation based on r_s are not useful (Hollander and Wolfe 1999).

EXAMPLE 8.7

Krochmal and Sparks (2007) measured the age and forelimb length of northern myotis bats (*Myotis septentrionalis*) captured in the field in Vermilion County, Indiana. Data for these variables are in the *asbio* file bats.

```
data(bats)
```

A scatterplot shows a monotonic relationship between bat age and forearm length (Figure 8.8). Thus, rank transformation will result in a linear association for the Y_1, Y_2. To calculate $\sum_{i=1}^{n} D_i^2$, we find the squared differences in the ranks.

```
Di.sq <- (rank(bats$forearm.length)-rank(bats$days))^2
sum(Di.sq)
[1] 307.5
```

Calculation of $\sum_{i=1}^{n} Y_{1i}^2$ and $\sum_{i=1}^{n} Y_{2i}^2$ is somewhat involved. We have

```
sum.T.Y.1 <- print((6^2 - 6)/12 + (2^3 - 2)/12 + (3^3 - 3)/12 + (9^3 -
9)/12+ (2^3 - 2)/12 + (2^3 - 2)/12 + (2^3 - 2)/12 + (2^3
- 2)/12)
[1] 67.5
sum.T.Y.2 <- print((2^3 - 2)/12 + (2^3 - 2)/12 + (2^3 - 2)/12 + (2^3 - 2)/12)
[1] 2
```

```
plot(bats, xlab = "Age (days)", ylab = "Forearm length (mm)")
```

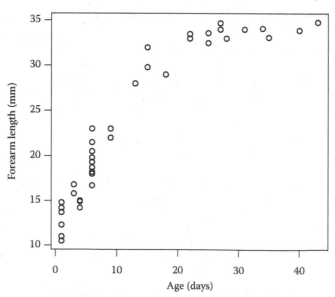

FIGURE 8.8

A scatterplot showing the relationship between age and forelimb length for northern myotis bats (*Myotis septentrionalis*). Each point represents a pair of measurements for a single bat, *n* = 38.

This results in

$$\sum_{i=1}^{n} Y_{1i}^2 = \frac{38^3 - 38}{12} - 67.5 = 4502,$$

$$\sum_{i=1}^{n} Y_{2i}^2 = \frac{38^3 - 38}{12} - 2 = 4567.5.$$

and

$$r_s = \frac{\sum_{i=1}^{n} Y_{1i}^2 + \sum_{i=1}^{n} Y_{2i}^2 - \sum_{i=1}^{n} D_i^2}{2\sqrt{\sum_{i=1}^{n} Y_{1i}^2 \sum_{i=1}^{n} Y_{2i}^2}} = \frac{4502 + 4567.5 - 307.5}{2\sqrt{4502 \times 4567.5}} = 0.966$$

The result is easily calculated in **R**.

```
cor(bats$forearm.length, bats$days, method = "spearman")
0.9660772
```

For a two-tailed test using the normal approximation, we have

$$2 \times P\left(Z \geq \left|r_s\sqrt{n-1}\right|\right) = 2 \times P(Z \geq 0.9660772\sqrt{37}) = 2 \times P(Z \geq 5.876418) = 4.2 \times 10^{-9}$$

Again, `cor.test` allows computation of the correlation coefficient and the corresponding *P*-value.

```
with(bats, cor.test(forearm.length, days, method = "spearman"))

        Spearman's rank correlation rho

data: forearm.length and days
S = 310.0208, p-value < 2.2e-16
alternative hypothesis: true rho is not equal to 0
sample estimates:
     rho
0.9660772
Warning message:
In cor.test.default(forearm.length, days, method = "spearman"):
   Cannot compute exact p-values with ties
```

The function `cor.test` does not use the normal approximation. Instead, it attempts to compute the exact Spearman's *P*-value using the AS-83 algorithm (Best and Roberts 1975). We are warned that *P*-values will not be exact in the case of ties.

To get the asymptotic results we derived by hand, we can use the function `spearman_test` in library *coin*.

```
install.packages("coin")
library(coin)
spearman_test(forearm.length ~ days, data = bats,
distribution = "asymptotic")

    Asymptotic Spearman Correlation Test
data: forearm.length by days
Z = 5.8764, p-value = 4.192e-09
alternative hypothesis: true mu is not equal to 0
```

8.3.1.2 Kendall's τ

Kendall's τ is a parameter defined as

$$\tau = 2P[(Y_{1j} - Y_{1i})(Y_{2j} - Y_{2i}) > 0)] - 1, \tag{8.21}$$

where Y_{1i} is the ith outcome from variable Y_1, Y_{1j} is the jth outcome from variable Y_1, Y_{2i} is the ith outcome from variable Y_2, and Y_{2j} is the jth outcome from variable Y_2, for $1 \le i < j \le n$.

Estimates of Kendall's τ can be computed using signed ranks for Y_1 and Y_2. In this way Kendall's τ can be considered a natural counterpart to the Wilcoxon signed-rank test described in Chapter 6. Kendall's τ can also be estimated by counting entities called concordant and discordant pairs. This approach is described below.[*]

Consider all possible combinations of (Y_{1i}, Y_{1j}) and (Y_{2i}, Y_{2j}), where $1 \le i < j \le n$. A pair is *concordant* if $Y_{1i} > Y_{1j}$ and $Y_{2i} > Y_{2j}$ or if $Y_{1i} < Y_{1j}$ and $Y_{2i} < Y_{2j}$. Conversely, a pair is *discordant* if $Y_{1i} < Y_{1j}$ and $Y_{2i} > Y_{2j}$ or if $Y_{1i} > Y_{1j}$ and $Y_{2i} < Y_{2j}$. For example, consider the following dataset:

Y_1	Y_2
1	5
2	6
4	4

There will be $(n^2 - n)/2$ pairs for concordant/discordant consideration, where n is the number of paired observations. Thus, we have $(3^2 - 3)/2 = 3$ possible concordant or discordant pairs. These are indicated with shading below:

In the first pairing, both Y_1 and Y_2 are increasing from the ith to the jth observation. As a result, these pairs from these variables are concordant. In the second pairing, Y_1 is increasing from the ith to the jth observation while Y_2 is decreasing. Thus, these pairs are discordant. In the final pairing, Y_1 and Y_2 are again discordant. Therefore, there are two discordant pairs and one concordant pair in the dataset.

We estimate Kendall's τ with Kendall's sample correlation coefficient r_k.

$$r_k = \frac{K}{\sqrt{\dfrac{n^2 - n}{2} - T(Y_1)}\sqrt{\dfrac{n^2 - n}{2} - T(Y_2)}} \tag{8.22}$$

where the number of concordant pairs is denoted as K', the number of discordant pairs equals K'', and K equals $K = K' - K''$. The term $T(Y_1)$ signifies $\frac{1}{2}\sum_{i=1}^{g} t(Y_{2i})[t(Y_{2i}) - 1]$, where $t(Y_{1i})$ is the number of tied observations in the ith group of ties ($i = 1, 2, \dots, g$) in Y_1. In parallel, the term $T(Y_2)$ equates to $\frac{1}{2}\sum_{i=1}^{g} t(Y_{2i})[t(Y_{2i}) - 1]$, where $t(Y_{2i})$ is the number of tied observations in

---

[*] **R** estimates a version of Kendall's τ called Kendall's $\tau - b$. This is also the procedure demonstrated here. Another version of Kendall's τ called Kendall's $\tau - a$ does not explicitly account for ties in ranks of Y_1 or Y_2. Kendall's $\tau - a$ is calculated as $R_k = K/((n^2 - n)/2) = 2K/n^2 - n$, where $K = K' - K''$ (Hollander and Wolfe 1999, pp. 366, 382).

the *i*th group of ties in Y_2. If there are no ties, then Equation 8.22 reduces to Kendall's $\tau - a$ (see footnote).

The two-tailed hypotheses are

$$H_0: Y_1 \text{ and } Y_2 \text{ are independent,}$$

versus

$$H_A: \tau \neq 0.$$

This comprises a more specific statement than $H_0: \tau = 0$ versus $H_A: \tau \neq 0$. This is because, although the marginal variables from a bivariate distribution will only be *independent* when $\tau = 0$ (see Hollander and Wolfe 1999 for derivation), they may be *dependent* if $\tau = 0$.

The lower- and upper-tailed alternative hypotheses are

$$H_A: \tau < 0,$$

and

$$H_A: \tau > 0.$$

We have the following assumptions for tests concerning Kendall's τ:

1. The association between Y_1 and Y_2 (if any) is monotonic.
2. The bivariate observations are independent.

When n is small, the empirical distribution for K can be derived using the rank-based permutation methods discussed in Chapter 6, and K can be used directly as a test statistic. When n is large (>50), and H_0 is true, we can assume that

$$K \sim N\left(0, \ \frac{n(n-1)(2n+5)}{18}\right). \tag{8.23}$$

As a result, we use the test statistic:

$$z^* = \frac{K}{\sqrt{\mathrm{Var}_k}}, \tag{8.24}$$

where, excluding ties, $\mathrm{Var}_k = n(n-1)(2n+5)/18$.

The formula for Var_k becomes complex with ties (see Hollander and Wolfe 1999, p. 366). A two-tailed *P*-value is calculated as $2P(Z \geq |z^*|)$, while lower- and upper-tailed *P*-values are calculated as $P(Z \leq z^*)$ and $P(Z \geq z^*)$, respectively. As with Spearman's correlation, ties will prevent the computation of exact *P*-values.

The percentile bootstrap method (Chapter 5) can be used to calculate confidence intervals for τ. Hollander and Wolfe (1999, p. 383) describe other asymptotic methods.

EXAMPLE 8.8

Here, we reexamine the bat forearm and age data from Example 8.7 using Kendall's τ.

We can identify concordant and discordant pairings with the function ConDis.matrix from *asbio*. The function creates a matrix of pairwise comparisons with concordant pairs labeled as ones, discordant pairs labeled as minus ones, and ties labeled as zeroes.

```
batm <- with(bats, ConDis.matrix(days,forearm.length))
```

The numbers of concordant and discordant pairs and the number of ties are

```
con <- print(sum(ifelse(batm[lower.tri(batm)] == 1, 1, 0)))
[1] 610
dis <- print(sum(ifelse(batm[lower.tri(batm)] == -1, 1, 0)))
[1] 29
tie <- print(sum(ifelse(batm[lower.tri(batm)] == 0, 1, 0)))
[1] 64
```

To account for the ties, we have

```
T.Y.1 <- print(1/2 * ((6 * 5) + (2 * 1) + (3 * 2) + (9 * 8) + (2 * 1) + (2 *
1) + (2 * 1) + (2 * 1) + (2 * 1)))
[1] 60
T.Y.2 <- print(1/2 * ((2 * 1) + (2 * 1) + (2 * 1) + (2 * 1)))
[1] 4
```

As a result, we have

$$r_k = \frac{K}{\sqrt{\frac{n^2 - n}{2} - T(Y_1)}\sqrt{\frac{n^2 - n}{2} - T(Y_2)}} = \frac{610 - 29}{\sqrt{\frac{38^2 - 38}{2} - 60}\sqrt{\frac{38^2 - 38}{2} - 4}} = 0.867.$$

The coefficient indicates a strong positive correlation between the forelimb length and age. This result is easily calculated in **R**.

```
cor(bats$days, bats$forearm.length, method = "kendall")
[1] 0.8666267
```

We use the function `vark` from library *agricolae* to calculate Var_k.

```
install.packages("agricolae")
library(agricolae)
vark(bats$days, bats$forearm.length)
[1] 6193.341
```

For our *P*-value, we have

$$2 \times P\left(Z \geq \left|\frac{K}{\sqrt{\text{Var}_k}}\right|\right) = 2 \times P\left(Z \geq \left|\frac{610 - 29}{\sqrt{6193.341}}\right|\right) = 2 \times P(Z \geq 7.382673) = 1.552 \times 10^{-13}.$$

Estimates of Kendall's τ and associated hypothesis tests can also be computed simultaneously in **R**.

```
with(bats, cor.test(days, forearm.length, method = "kendall"))
Kendall's rank correlation tau

data: bats[, 1] and bats[, 2]
z = 7.3827, p-value = 1.552e-13
alternative hypothesis: true tau is not equal to 0
sample estimates:
     tau
0.8666267
```

```
Warning message:
In cor.test.default(bats[, 1], bats[, 2], method = "kendall"):
Cannot compute exact p-value with ties
```

The function `cor.test` again warns that exact *P*-values cannot be computed in the presence of ties. The function `cor.test` will use the rank-based permutation distribution of *K* directly if sample sizes are less than 50 and no ties are present.

8.3.2 Robust Estimator Approaches

Many robust correlation analyses are possible that do not involve rank transformation. These include the Winsorized Pearson's correlation (r_w), the percentage bend correlation (r_{bp}), and the biweight midcorrelation (r_{bw}).

The Winsorized correlation can be considered as a robust replacement for *r* because it estimates the same parameter ρ. Conversely, the statistics r_{bp} and r_{bw} should not be interpreted as substitutes for *r*. This is because they estimate correlation parameters different than ρ that are not overly sensitive to distributional changes. An extensive treatment of these methods is given in Wilcox (2005).

8.3.2.1 Winsorized Correlation

The Winsorized correlation is straightforward to calculate following Winsorization of Y_1 and Y_2 using the function `win` from *asbio*. Significance tests for the null hypothesis of no association utilize the test statistic

$$t^* = r_w \sqrt{\frac{n-2}{1-r_w^2}}, \tag{8.25}$$

where r_w is the Pearson's correlation coefficient calculated after Winsorization.

Under H_0, the sampling distribution of t^* can be represented by a random variable, $T \sim t(h-2)$, where *h* is effective sample size (the number of non-Winsorized observations). The two-tailed *P*-value is calculated as $2P(T \geq |t^*|)$.

For hypothesis tests using Winsorized correlation, we assume

1. The relationship between Y_1 and Y_2 is linear following Winsorization.
2. The bivariate observations are independent.

8.3.2.2 Percentage Bend Criterion, and Sample Biweight Midvariance

The *midvariance* is the asymptotic variance of an *M*-estimator of location. Shoemaker and Hettmansperger (1982) distinguished two types of midvariance, the percentage bend midvariance and the biweight midvariance. Wilcox (1994a) adapted these definitions into measure of correlation he called the percentage bend criterion, r_{pb}, and the biweight midcorrelation, r_{mc}.

The percentage bend criterion produces very similar values to Pearson's correlation given bivariate normality, but will produce substantially different values in other situations (Wilcox 2005). The measure requires specification of a value for the bend argument β, which is usually given to be 0.2 (Wilcox 2005). The estimator can be run using the function r.pb in *asbio*, which also provides results for a null hypothesis test of independence based on the asymptotically normal sampling distribution of r_{pb} (see Wilcox 2005).

For hypothesis tests using the percentage bend criterion, we assume

1. The relationship between Y_1 and Y_2 is linear, after the application of the M-estimator weighting function.
2. The bivariate observations are independent.

The sample biweight midcorrelation (along with the biweight midvariance and biweight midcovariance) can be found using the function r.bw in *asbio*.

EXAMPLE 8.9

For the bat dataset, we have

```
cor(win(forearm.length), win(days))
[1] 0.9610091
r.pb(forearm.length, days)
       r.bp          t*        P(t > t*)
  0.9472897  17.74071  2.222623e-19
r.bw(forearm.length, days)
       s.xx       s.yy      s.xy        r.xy
  79.65606  160.3358  103.6698  0.9173346
```

8.4 Comparisons of Correlation Procedures

Pearson's correlation coefficient does not require bivariate normality for point estimation, although this is often stated in textbooks (cf. Nefzger and Drasgow 1957). However, interval estimation and hypothesis testing for independence *do* assume bivariate normality. Pearson's correlation is strongly affected by outliers. Given large sample sizes, Pearson's correlation can be used in hypothesis testing with distributions that are not bivariate normal, although (1) outliers (from the perspective of bivariate normality) will be more likely, undermining inferences concerning linear association, and (2) independence cannot be verified.

The statistic r will be more powerful than r_s or r_k for detecting independence in bivariate distributions with only moderate positive skew ($\gamma_1 < 1.75$) and moderate excess kurtosis ($\gamma_1 < 3$) (Chok 2010). However, r_s and r_k will always be less affected than r by outliers (Abdullah 1990).

Two important distinctions can be made with respect to r_s and r_k. First, except in the trivial case of perfect monotonicity, r_s will have a larger magnitude than r_k (it will be more negative or more positive) for a given dataset. Second, as noted in Section 8.3.1.1, hypothesis tests using r_s should not be stated in terms of the parameter ρ_s, and are therefore less specifically interpretable than those from Kendall's τ.

For bivariate distributions with nonnormality, bootstrapped confidence intervals for ρ using r_{pb} will be closer to nominal values than confidence intervals using r for practically all conditions and sample sizes (Wilcox 1997). This is despite the fact that the percentage bend parameter does not exactly equal ρ under dependence (Wilcox 1997). The sample biweight midcorrelation has a high breakdown point (approximately 0.5), and was found to be more efficient than 150 other correlation estimators for platykurtic parent distributions (Lax 1985). Because they do not estimate ρ, the estimators r_k, r_s, r_{pb}, and r_{mc} should not be considered as direct substitutes for r. Instead, they should be seen as robust approaches with a different conceptualization of correlation.

Five generalities can be made with respect to all the correlation measures described in this chapter.

- First, all estimators discussed here are unaffected by linear transformations (although the sign of the coefficient will be reversed if the sign for one of the variables is changed as a result of multiplication).

- Second, none of the correlation estimators described here will detect nonmonotonic association. Thus, modal, circular, and all other forms of dependence will not be quantified.

- Third, all estimators described here are sensitive to the range of values measured. That is, given a wider range of sampled values, the correlation estimates will tend to increase, even if a best fit line is identical for narrow and wide-range datasets (Figure 8.9). The exception will be a dataset with $n = 2$. Such a situation will always result in meaningless correlation of 1 or –1.

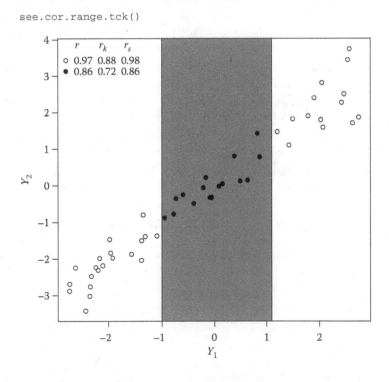

FIGURE 8.9
A demonstration of the effect of range on correlation. Correlations for all pairs of measurements (upper correlations in figure), and correlations for the subset black points (lower correlations). Apparent dependencies tend to be stronger if viewed over a wider range of values, except for the case when $n = 2$. In this case, a meaningless correlation of 1 or –1 will always result.

- Fourth, it should be noted that all conventional robust correlation approaches (including those described in this chapter) belong to a class called type-M correlations (not to be confused with M-estimators), which protect only against marginal distribution outliers (Wilcox 2005). As a result, strategically placed bivariate outliers can still have a substantial effect on estimators (see problem 9 from the exercises at the end of the chapter). This will not be true of the so-called type-O correlations, which consider the bivariate structure of data (see Wilcox 2005). From this perspective, an outlier will have shallow depth, that is, it will not be deeply subsumed in the bivariate cloud of data (see Wilcox 2005).

- Fifth, all hypothesis testing methods discussed in this chapter (including robust measures) will be adversely affected by heteroscedasticity (Wilcox and Muska 2001). In the context of correlation analysis, heteroscedasticity is the propensity of the variance in Y_1 to increase as Y_2 increases, or vice versa. In the presence of heteroscedasticity, all confidence intervals for the true correlation will be suspect.

8.5 Summary

- Covariance and correlation measure the strength and direction of association between two quantitative variables, Y_1 and Y_2. Unlike variances, a covariance can be negative. Conceptually, a covariance will have no upper and lower bound (although its absolute value must be less than $\sigma_1\sigma_2$).

- Correlation measures are unitless and range from –1 to 1. A correlation near –1 or 1 indicates a particularly strong linear association between two variables, while a correlation near 0 indicates a lack of association. Population covariances and correlations are denoted as $Cov(Y_1,Y_2)$ and ρ, respectively.

- If $\rho \neq 0$, this indicates bivariate dependence. However, the outcome $\rho = 0$ only indicates independence in the case of bivariate normal distributions.

- The population correlation is estimated with Pearson's correlation coefficient, r. This measure is sensitive to outliers, and assumes bivariate normality for inferences concerning independence.

- In the absence of bivariate normality, or in the presence of outliers, inferences concerning independence can be made using robust measures. Spearman's ρ_S is estimated with the statistic r_s. It is equivalent to Pearson's correlation using ranks instead of raw data.

- Kendall's τ is estimated with the statistic r_k. It is based on counts of concordant and discordant pairs.

EXERCISES

1. Calculate $f(0, 1)$ and graph the distribution using `bvn.plot` for

    a. $X \sim MVN\left(\begin{bmatrix} 0 \\ 0 \end{bmatrix}, \begin{bmatrix} 1 & 0 \\ 0 & 1 \end{bmatrix}\right)$

b. $X \sim MVN \left(\begin{bmatrix} 0 \\ 0 \end{bmatrix}, \begin{bmatrix} 1 & 0.5 \\ 0.5 & 1 \end{bmatrix} \right)$

c. $X \sim MVN \left(\begin{bmatrix} 0 \\ 0 \end{bmatrix}, \begin{bmatrix} 1 & -2.5 \\ -2.5 & 1 \end{bmatrix} \right)$. Is this even calculable? Why?

2. Distinguish the terms "independence," "dependence," and "linear correlation."

3. Soil nitrogen is only available to plants in certain inorganic forms, particularly NH_4 and NO_3. Ungulates often increase the available forms of soil N by speeding up decomposition rates. Decomposition is the conversion of organic forms of materials to inorganic forms (e.g., the conversion of DNA to NH_4), accompanied by the release of CO_2. The dataset goats in *asbio* compares mountain goat (*Oreomnos americanus*) fecal concentrations to total soil organic matter and concentrations of soil NO_3 for alpine locations on eight mountains in Yellowstone National Park (Aho 2012).

   a. Calculate Pearson's product moment correlation for feces versus NO_3 and feces versus organic matter, and calculate the two-tailed P-values "by hand" (i.e., without using cor or cor.test), but otherwise using **R** to help.

   b. Calculate a 90% confidence interval for ρ "by hand" (i.e., without using cor or cor.test), but otherwise using **R** to help.

   c. Confirm your results in (a) and (b) using cor or cor.test.

   d. What do these data indicate with respect to goat influences on nitrogen availability?

4. Linearly transform the variables in Question 3 by adding or multiplying by some constant. Does this transformation affect the correlation for any of the measures described in this chapter?

5. The following data document changes in a hypothetical bacterial population over time:

Time (Days)	Population Size (Thousands)
1	0.078
2	0.327
3	0.415
4	0.416
5	0.521
6	0.656
7	0.821
8	1.334
9	1.907
10	2.315

   a. Plot the data in **R**. Is the Pearson's product moment correlation appropriate here? Why or why not?

   b. Calculate Kendall's and Spearman's rank coefficients "by hand" (i.e., without using cor or cor.test), but otherwise using **R** to help. This will require use of the function ConDis.matrix in *asbio*. Confirm your results using cor or cor.test or spearman_test. Explain the output of the matrix from ConDis.matrix.

   c. Summarize your findings.

6. Consider bivariate normal distributions with

a. $\mu = \begin{bmatrix} 3 \\ 2 \end{bmatrix}, \sigma^2 = \begin{bmatrix} 1 & 0.9 \\ 0.9 & 1 \end{bmatrix},$

b. $\mu = \begin{bmatrix} 2.5 \\ 2.5 \end{bmatrix}, \sigma^2 = \begin{bmatrix} 9 & -8.1 \\ -8.1 & 9 \end{bmatrix},$ and

c. $\mu = \begin{bmatrix} 3 \\ 2 \end{bmatrix}, \sigma^2 = \begin{bmatrix} 2 & -0.2 \\ -0.2 & 4 \end{bmatrix}.$

Calculate ρ for (a), (b), and (c).

7. Calculate the number of discordant and concordant pairs by hand for the following data:

Y_1	Y_2
2	4
1	3
3	1

8. Calculate the Winsorized correlation of the data in Question 5 using 20% Winsorization with the function win. Calculate the P-value for the null hypothesis of no association.

9. Using **R**, create 20 random observations from $X \sim N(0,1)$, and 20 random observations from $Z \sim N(0,1)$. Let $y = x + z$ and find r, r_k, r_s, the biweight midcorrelation, and the percent bend correlation for x and y. Add an experimental unit with the outcomes (2.3, −2.5). Now, recalculate the correlation measures. Are any of these measures unaffected by the bivariate outlier? Are some measures more affected than others?

9

Regression

Deep snow in winter; tall grain in summer.

Estonian proverb

9.1 Introduction

If we are simply interested in the nature of association between variables (positive, negative, or none), then correlation analysis should be used (Chapter 8). Conversely, if we hope to show that one variable or group of variables explains variation in another variable, or to simply model this relationship, we would use *regression analysis*. Regression analyses require the designation of one response (Y) variable and one or more quantitative explanatory (X) variables (i.e., predictors).[*] For instance, to model the heritability of plant height, we could measure seedling height at maturity (Y), and mean parental height (X), and then "regress" Y on X.

> **EXAMPLE 9.1**
>
> Biologists have repeatedly observed that diversity for most groups of organisms increases as one nears the equator. Thus, it is reasonable to try to model diversity as a function of latitude (Figure 9.1a). Standard curves are another common application of simple linear regression. In biochemical assays, light absorbance at a particular wavelength is often modeled to be a function of the product from a reaction in which known amounts of a substance of interest (e.g., DNA, phenolics, or proteins) are added to a reagent (Figure 9.1b).

9.1.1 How to Read This Chapter

Most of this chapter is devoted to two topics: simple linear regression (with a single X variable, e.g., Figure 9.1) and multiple regression (with multiple X variables). These foci comprise Sections 9.1 through 9.12 and are appropriate for introductory statistics classes after a brief overview of linear algebra included in the Appendix.

Later sections of the chapter constitute advanced topics and can be skipped with little loss in continuity if related topics are avoided in Chapters 10 and 11. These include weighted least squares (WLS) (Section 9.13), polynomial regression (Section 9.14), comparison of regression parameters among different models (Section 9.15), likelihood estimation for linear models (Section 9.16), methods for multimodel inference (Section 9.17), robust regression (Section 9.18) (for situations with outliers), model II regression (Section 9.19)

---

[*] This is the model framework. Evidence of causality can, however, only come from a properly designed experiment, see Chapter 7.

```
Fig.9.1() ## See Ch. 9 code
```

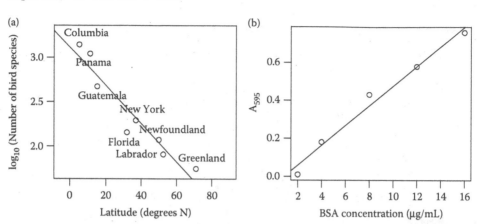

FIGURE 9.1

Applications for simple linear regression. (a) Log$_{10}$ numbers of breeding species of birds as a function of latitude. The data are based on the meta-analysis of Dobzhansky (1950). Note that transformation of Y is required to linearize the association between Y and X. (b) An example of a Bradford colorimetric protein assay (Bradford 1976). Absorbance units at 595 nm are plotted as a function of the concentration of bovine serum albumin (BSA), which has been mixed with the Bradford reagent. Note that in this case, we are interested in a model for predicting X from Y. While this is rather unusual (we are generally interested in the converse), regressions can also be used in this capacity (see Section 9.6).

(where X is not fixed, or is measured with error), GLMs (Section 9.20) (which require link functions for linearity, and do not require normal errors), nonlinear regression (Section 9.21) (for modeling relationships between Y and X that cannot be expressed as linear equations) smoother approaches to regression (Section 9.22) (which use local, not global models for predictive fitting), and Bayesian methods (Section 9.23).

9.2 Linear Regression Model

A general model for linear regression can be expressed as

$$Y_i = \beta_0 + \beta_1 X_{1i} + \beta_2 X_{2i} + \cdots + \beta_{p-1} X_{p-1i} + \varepsilon_i, \tag{9.1}$$

where $\beta_0, \beta_1, \ldots, \beta_{p-1}$ are parameters that define Y as a function of X, $X_{1i}, \ldots, X_{p-1}$, i denotes the ith observation from each of the $p-1$ predictors, $i = \{1,2,\ldots,n\}$, and ε_i is a random variable describing the variability of Y given the ith value of each of the predictors. For inferential purposes, we assume $\varepsilon_i \sim N(0, \sigma^2)$, where σ^2 (the error term variance) is the true variance of the difference of the observed and fitted values.

Because ε_is are normal with a mean zero (Chapter 7, Section 9.10), Y_is will be normal with mean

$$E(Y_i) = \beta_0 + \beta_1 X_{1i} + \beta_2 X_{2i} + \cdots + \beta_{p-1} X_{p-1i}. \tag{9.2}$$

That is, Y_i is a random variable, whose mean $E(Y_i)$ will be obtained from the true regression model.

There will be p parameters requiring estimation in the general regression model. These are β_0, the Y-intercept for the true regression line, and the coefficients associated with each explanatory variable, $\beta_1, \ldots, \beta_{p-1}$.

In simple linear regression, we have just one explanatory variable, and Equation 9.2 reduces to

$$E(Y_i) = \beta_0 + \beta_1 X_{1i}. \tag{9.3}$$

Because there are only two parameters requiring estimation, $p = 2$.

9.3 General Linear Models

In Equation 9.1, the response variable is expressed as a linear function of the explanatory variable(s). As a result, the general linear regression model is a type of *general linear model*.

It is important to note that a general linear model does not require that the "real" relationship between Y and X be a straight line. Consider *polynomial regression models* (Section 9.14) that are often used to fit curvilinear response surfaces and have higher-order predictor variables. A second-order polynomial regression with one predictor has the form

$$Y_i = \beta_0 + \beta_1 X_{1i} + \beta_2 X_{1i}^2 + \varepsilon_i. \tag{9.4}$$

Equation 9.4 is a linear model because it can be expressed in the form of Equation 9.1. For instance, let $Z_{1i} = X_{1i}$ and let $Z_{2i} = X_{1i}^2$. Then, we can rewrite Equation 9.4 as

$$Y_i = \beta_0 + \beta_1 Z_{1i} + \beta_2 Z_{2i} + \varepsilon_i,$$

which is in the form of Equation 9.1.

Other important general linear regression models include those with interactions. A two predictor regression with a term defining the interaction between predictors has the form

$$Y_i = \beta_0 + \beta_1 X_{1i} + \beta_2 X_{2i} + \beta_3 X_{1i} X_{2i} + \varepsilon_i. \tag{9.5}$$

Here an additional interaction variable is included in the model that is a product of the two predictors. By letting $X_{3i} = X_{1i} X_{2i}$, Equation 9.5 is in the form of Equation 9.1.

Linear models with nonlinear transformations (Table 7.5) may also be linear models. Consider

$$\ln(Y_i) = \beta_0 + \beta_1 X_i + \varepsilon_i. \tag{9.6}$$

By letting $Z_i = \ln(Y_i)$, we again have the form of Equation 9.1.

In nonlinear models, Y cannot be expressed as a linear function of X. For instance, the Michaelis–Menten kinetics equation is often used by biologists for defining asymptotic associations between Y and X:

$$E(Y_i) = \frac{\beta_0 X_i}{\beta_1 + X_i}.$$

(9.7)

However, this function cannot be mathematically re-expressed in the form of Equation 9.1. Thus, it is a *nonlinear model* (see Section 9.21).

9.3.1 `lm`

The function `lm` provides fitting of general linear models in **R**. Important arguments include `formula`, `weights`, `contrasts`, and `data`.

- `formula` (the only argument required by `lm`) defines the model form. A tilde (~) metacharacter indicates "function of," and is used to distinguish the response and predictor variables. For instance, `formula = Y ~ X1 + X2` indicates that `Y` should be modeled as an additive function of the predictors `X1` and `X2`. Interactions in the model are specified with the colon character (:). For example, `formula = Y ~ X1 : X2`, indicates a model consisting of the interaction of `X1` and `X2`. For quantitative predictors (required for regression) an interaction equates to the product of the crossed predictors. That is, `X1 : X2` equals `X1 × X2`. For categorical predictors (Chapters 10 and 11), interactions result in a new categorical variable resulting from the cross-classified factor levels.

  The function `I` can be included in formula specification to indicate that a predictor should be treated "as is" with respect to indicated arithmetic. For instance, while `~ X1 + X2` specifies a model with two predictors, `X1` and `X2`, `~ I(X1 + X2)` indicates that a single predictor should be constructed, consisting of the sum of `X1` and `X2`. Additional specifications for `formula` are addressed in the context of ANOVA (see Section 10.5).

- The argument `weights` allows postulation of a weighting vector of length n to be applied to model observations in WLS (see Section 9.13).

- The argument `contrasts` is used to customize the terms in `lm` hypothesis tests. It will have no effect on regression models. In this case, tests for each of the p predictors, $k = \{1, 2, \ldots, p\}$, will always use the same null hypothesis: $\beta_k = 0$. However, changing this argument *will* have dramatic effects on linear models with categorical predictors, for example, ANOVA and ANCOVA (see Chapter 10).

- The `data` argument specifies a dataframe, list, or environment (coercible with `as.data.frame`) that contains variables used in `formula`. That is, `lm` contains its own `with()` function.

9.4 Simple Linear Regression

With simple linear regression, we consider a Y variable as a linear function of a single X variable. The best possible linear fit of Y as a function of X is the *regression line*.

9.4.1 Parameter Estimation

We use the method of OLS (Chapter 4) to derive estimators for regression model parameters. The estimated regression model has the form

$$\hat{Y}_i = \hat{\beta}_0 + \hat{\beta}_1 X_{1i}. \tag{9.8}$$

Note that we use a "hat" symbol to distinguish parameters and estimates, or to distinguish observations from predictions (see $\hat{Y}$ description below). It can be shown that the OLS estimators for β_0 and β_1 are

$$\hat{\beta}_0 = \overline{Y} - \hat{\beta}_1 \overline{X}, \tag{9.9}$$

$$\hat{\beta}_1 = \frac{\sum_{i=1}^{n}(Y_i - \overline{Y})(X_i - \overline{X})}{\sum_{i=1}^{n}(X_i - \overline{X})^2}. \tag{9.10}$$

The rearrangement of Equation 9.9 into the form $\overline{Y} = \hat{\beta}_0 + \hat{\beta}_1 \overline{X}$ reveals that a regression line must always run through the points $(\overline{x}, \overline{y})$. OLS estimators for regression parameters will be identical to their ML estimators (Chapter 4) if ε_i is assumed to be normal. In addition, these estimators are unbiased, consistent, and have minimum variance (maximum efficiency) among all linear estimators (Plackett 1950).[*]

In simple linear regression $\hat{\beta}_0$ estimates the *Y-intercept*, the true mean value of Y when $X = 0$ (Figure 9.2). As a result, the units for $\hat{\beta}_0$ will be the same as the units for the Y variable. When the scope of the regression model includes $X = 0$, then the intercept term has an interpretable meaning. However, when the value of Y given $X = 0$ is extrapolated; see Section 9.6, then $\hat{\beta}_0$ will not be interpretable. $\hat{\beta}_1$ estimates the true *slope*, that is, the average linear change in the response variable as the result of a one unit increase in the explanatory variable (Figure 9.2). Its units are the ratio of the units of Y and X.

By substituting estimates for parameters in the regression using Equation 9.8, we obtain fitted and predicted values on the regression line. The ith fitted value, $\hat{Y}_i$, represents an estimate for the true mean of Y given X_i. The differences between the observed and fitted values are called *residuals* (Figure 9.2).

$$\hat{\varepsilon}_i = Y_i - \hat{Y}_i. \tag{9.11}$$

The ith residual estimates the error term ε_i in Equation 9.1.

---

[*] The character of OLS estimators as best linear unbiased estimators (BLUEs) was famously demonstrated by the *Gauss–Markov theorem*, which assumes only that observations are uncorrelated and error variances are equal (homoscedastic) though not necessarily normal.

```
Fig.9.2() # See Ch. 9 code
```

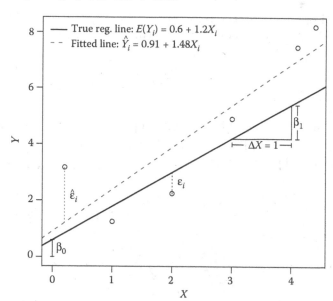

FIGURE 9.2

Graphical representation of parameters from simple linear regression. The fitted (dashed) line is the best possible linear fit for the six random points generated from $0.6 + 1.2X_i + \varepsilon_i$, where $\varepsilon_i \sim N(0, 1)$. The solid line is the true (population) regression line.

In a scatterplot with an overlaid fitted regression, the residuals are the vertical distances of the observed points to the regression line (Figure 9.2). They will be in a vertical direction because the prediction error will be with respect to response variable, which is plotted on the ordinate (Y-axis).

Recall that with OLS estimation we minimize the quantity Q in

$$Q = \sum_{i=1}^{n} (Y_i - f_i(\theta))^2 \tag{4.33}$$

This equates to minimizing the squared residuals, $\sum_{i=1}^{n} \hat{\varepsilon}_i^2$ because

$$\sum_{i=1}^{n} \hat{\varepsilon}_i^2 = \sum_{i=1}^{n} [Y_i - (\hat{\beta}_1 X_i + \hat{\beta}_0)]^2 = \sum_{i=1}^{n} (Y_i - \hat{Y}_i)^2, \tag{9.12}$$

where Y_i is the ith observed response and $\hat{Y}_i$ is the ith fitted value. $\Sigma \hat{\varepsilon}_i^2$ is called the *residual sum of squares* or the *sum of squares error* (SSE).

The *mean squared error* or *MSE* is an unbiased, consistent, and, assuming normal errors, maximally efficient estimator for the error term variance, σ^2 (Graybill 1976, p. 632). MSE is obtained by dividing Equation 9.12 by $n - p$, where n is the number of observations and p is the number of parameters estimated in the model.

$$\text{MSE} = \frac{\sum_{i=1}^{n}(Y_i - \hat{Y})^2}{n-p}. \tag{9.13}$$

The quantity $n - p$ represents the degrees of freedom required for estimation of σ^2. In simple linear regression, MSE has $n - 2$ degrees of freedom (not $n - 1$) because there are two dimensions in the subspace of the linear model. Adding more explanatory variables, *á la* multiple regression, further decreases the degrees of freedom for MSE.

Note that Equation 9.13 is simpler than the version of MSE introduced with *t*-tests in Chapter 6. This is because regression is only concerned with the variability of Y given X (see Section 9.10).

The standard errors of $\hat{\beta}_0$ and $\hat{\beta}_1$ are calculated using MSE:

$$\hat{\sigma}_{\hat{\beta}_0} = \sqrt{\text{MSE}\left[\frac{1}{n} + \bar{X}^2 \middle/ \sum_{i=1}^{n}(X_i - \bar{X})^2\right]}, \tag{9.14}$$

$$\hat{\sigma}_{\hat{\beta}_1} = \sqrt{\text{MSE} \middle/ \sum_{i=1}^{n}(X_i - \bar{X})^2}. \tag{9.15}$$

EXAMPLE 9.2

Male magnificent frigate birds (*Fregata magnificens*), a relative of pelicans, have an enlarged red gular pouch (a prominent area of featherless skin that joins the lower mandible of the beak to the neck) that has probably evolved as the result of sexual selection. During courtship displays, males display the pouch and use it to make a drumming sound. Madsen et al. (2004) noted that certain conditions (e.g., oblique viewing angles) may prevent a female from accurately appraising pouch size. Because females choose mates based on pouch size, a question of interest is whether pouch size determines sonic frequencies produced by courtship drumming. This would allow females to consider a gular pouch using both sight and sound.

Madsen et al. (2004) estimated the pouch volume and fundamental frequency (the lowest frequency of a periodic waveform) produced by drumming for 43 male frigate birds at Isla Isabel in Nayarit Mexico. The observations yielded 18 distinct pouch volumes. The mean fundamental frequencies at these volumes are contained in the *asbio* dataset Fbird. While the authors used only correlation analysis, it is natural to view sonic frequency as a response variable and pouch volume as an explanatory variable.

```
data(Fbird)
attach(Fbird)
```

Below, we obtain estimates for β_0 and β_1 (see Equations 9.9 and 9.10) using **R**.

```
XminusXbar <-(vol - mean(vol))
YminusYbar <-(freq - mean(freq))
beta.hat1 <- sum(XminusXbar * YminusYbar)/sum(XminusXbar^2)
beta.hat0 <- mean(freq) - mean(vol) * beta.hat1
c(beta.hat0, beta.hat1)
[1] 557.10215950 -0.02037952
```

```
plot(vol, freq, xlab = expression(paste("Estimated gular pouch size
(",cm^3,")")),ylab = "Fundamental frequency (Hz)")
abline(beta.hat0, beta.hat1)
```

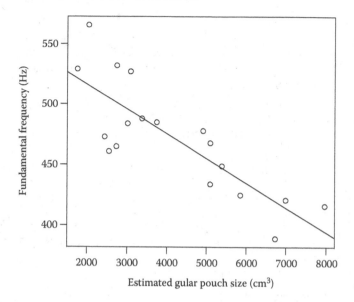

FIGURE 9.3
Frigate bird drumming frequency shown as a function of male pouch volume using linear regression. (Data from Madsen, V. et al. 2004. *Condor* 106: 156–160.)

Thus, the regression model is $\hat{Y}_i = 557.1022 - 0.0203X_i$. The Y-intercept (557.1 Hz) lies in the range of radical extrapolation, and is therefore not interpretable. The slope estimate indicates that for every 1 cm^3 increase in pouch volume we expect an approximate 0.02 Hz decrease in fundamental frequency. A plot of the regression is shown in Figure 9.3.

The fitted values, $\hat{Y}_i$, are

```
yhat <- beta.hat0 + beta.hat1 * vol
```

and the residuals are

```
ei <- freq - yhat
```

Given the residuals, $\hat{\sigma}_{\hat{\beta}_0}$ and $\hat{\sigma}_{\hat{\beta}_1}$ can be calculated as

```
n <- length(vol)
MSE <- print(sum(ei^2)/(n - 2))
[1] 742.2165

sb0 <- sqrt(MSE*(1/n + mean(vol)^2/sum(XminusXbar^2)))
sb1 <- sqrt(MSE/sum(XminusXbar^2))
c(sb0, sb1)
[1] 16.216397288 0.003552017
```

Thus, we have $\hat{\sigma}_{\hat{\beta}_0} = 16.22$ and $\hat{\sigma}_{\hat{\beta}_1} = 0.0036$.

```
anm.ls.reg(vol, freq)
```

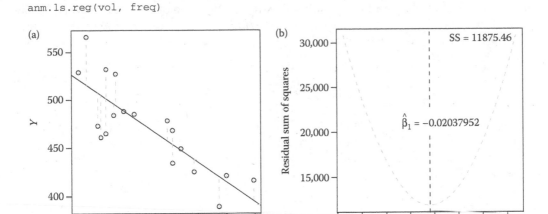

FIGURE 9.4

Animated depiction of OLS regression parameter estimation for the frigate bird dataset. (a) The best linear fit (OLS) regression line with the residuals shown as dashed lines. (b) The model SSE for different potential values of $\hat{\beta}_1$. The best fit line (shown in (a)) has the slope -0.0204. At this value, SSE (in (b)) is minimized.

By definition, the OLS parameter estimates for linear regression will minimize the residual sum of squares. This is demonstrated in Figure 9.4. Note that as the animation progresses and different regression lines (in gray) are examined, based on different slopes (the intercept is fixed at $\hat{\beta}_0$ if optimization for $\hat{\beta}_1$ is specified in anm.ls.reg) the sum of the vertical distances of the observations to the lines is minimized at the black line in Figure 9.4a, which results from OLS estimation of β_0 and β_1. Figure 9.4b shows that the minimum SSE (11,875.5) occurs at the OLS estimates for β_0 and β_1.

9.4.2 Hypothesis Testing

Hypothesis testing in simple linear regression is primarily used to evaluate whether or not β_1 is equal to some null value, usually zero. Thus, in the typical case, we use the two-tailed hypotheses:

$$H_0: \beta_1 = 0$$

versus

$$H_A: \beta_1 \neq 0.$$

If H_0 is true, and β_1 does in fact equal zero, then X will have no linear effect on Y. Lower and upper tailed alternative hypotheses are

$$H_A: \beta_1 < 0$$

and

$$H_A: \beta_1 > 0.$$

For inferential purposes, we assume that the underlying sampling distribution of $\hat{\beta}_1$ is normal. This follows directly from the assumption that ε_is are assumed to be normally distributed, that is, $\varepsilon_i \sim N(0, \sigma^2)$ (see Section 9.10). The population residuals, ε_i, describe the random variability of Y_i, and as a result $Y_i \sim N(\mu_i, \sigma^2)$. Because of this, and because we are required to estimate σ^2 with MSE, we quantify evidence concerning H_0 using probabilities from the t distribution. The test statistic is

$$t* = \frac{\hat{\beta}_1}{\hat{\sigma}_{\hat{\beta}_1}}.$$
(9.16)*

If H_0 holds, and the assumptions for the regression model are valid (see Section 9.10), then $t*$ will be a random outcome from a t distribution with $n - p$ degrees of freedom. Thus, the two-tailed P-value is calculated as $2P(T \geq |t*|)$, where $T \sim t(n-2)$, and lower and upper tailed P-values are calculated as $P(T \leq t*)$ and $P(T \geq t*)$, respectively.

EXAMPLE 9.3

To demonstrate hypothesis testing, we again use the frigate bird data. We use the conventional hypotheses:

$$H_0: \beta_1 = 0$$

versus

$$H_A: \beta_1 \neq 0.$$

The test statistic is

```
tstar <- beta.hat1/sb1
```

and the P-value is

```
2 * pt(abs(tstar), df = 16, lower.tail = FALSE)
[1] 3.056371e-05
```

The entire analysis can be run in **R** using lm:

```
Fbird.lm <- lm(freq ~ vol, data = Fbird)
summary(Fbird.lm)

             Estimate Std. Error t value Pr(>|t|)
(Intercept) 557.102159  16.216397  34.354  < 2e-16 ***
vol          -0.020380   0.003552  -5.737 3.06e-05 ***
---
Signif. codes:  0 '***' 0.001 '**' 0.01 '*' 0.05 '.' 0.1 ' ' 1

Residual standard error: 27.24 on 16 degrees of freedom
Multiple R-squared: 0.6729,     Adjusted R-squared: 0.6525
F-statistic: 32.92 on 1 and 16 DF, p-value: 3.056e-05
```

In the lm output, we are given estimates for each parameter in the model, along with associated standard errors, and P-values for hypotheses that the parameter values equal zero. We have very little evidence supporting the null hypothesis $\beta_1 = 0$ because $P = 3.1 \times 10^{-5}$. Indeed, the negative parameter estimate for β_1 indicates a negative linear effect of volume on frequency (Figure 9.3).

---

* In the case that the hypothesized value for β_1 is not equal to zero, Equation 9.16 becomes $t* = (\beta_1 - \beta_{1,H_0})/\hat{\sigma}_{\hat{\beta}1}$ where β_{1,H_0} is the hypothesized value for β_1.

The residual standard error is simply the square root of MSE. Multiple R^2 and adjusted R^2 are addressed in Section 9.8.

9.4.3 An ANOVA Approach

In the last line of the output from lm in Example 9.3 is the result from a test of H_0: $\beta_1 = 0$ that uses an F statistic rather than a t statistic. This represents an ANOVA approach to regression (Chapter 10). In this case, instead of considering the probability of the observed slope under H_0, the variance of the fitted values with respect to the conditional mean of Y, and the variance of observed values with respect to fitted values (i.e., MSE) are compared under the null hypothesis that the true variances corresponding to these estimates are equal. The interchangeability of these techniques illustrates the fundamental similarity of all general linear models.

An ANOVA approach requires the partitioning of a quantity called the *total sum of squares* or *SSTO*. SSTO is the sum of the squared deviations between each Y-value and $\bar{Y}$. That is

$$SSTO = \sum_{i=1}^{n} (Y_i - \bar{Y})^2. \tag{9.17}$$

We can break SSTO into two components:

$$SSTO = SSR + SSE \tag{9.18}$$

where SSR is the regression sum of squares, and SSE is the sum of squares error. We are already acquainted with SSE. SSR is the sum of the squared deviations between each fitted Y value and $\bar{Y}$. That is

$$SSR = \sum_{i=1}^{n} (\hat{Y}_i - \bar{Y})^2. \tag{9.19}$$

SSR is analogous to SSTR (the sum of squares treatment) in ANOVA (Chapter 10).

Mean square terms are variance estimates for components of the model, and are calculated by dividing the sums of squares by their corresponding degrees of freedom (effective sample size). MSR estimates the variance of the fitted values with respect to the true mean of Y, and is calculated as

$$MSR = \frac{SSR}{p-1}. \tag{9.20}$$

MSE describes the variability between the fitted and observed values, and has the form

$$MSE = \frac{SSE}{n-p}. \tag{9.21}$$

Regression sums of squares, associated degrees of freedom, and mean squares can be summarized in an ANOVA table (Table 9.1).

Note that if the null hypothesis $\beta_1 = 0$ is true, then $E(MSE) = E(MSR) = \sigma^2$, and the ratio $E(MSR)/E(MSE)$ will equal one (Table 9.1). In all other cases, however, the ratio will be

TABLE 9.1

ANOVA Table for Simple Linear Regression

Source	Sum of Squares	df	Mean Squares	E(MS)	F^*
Regression model	$SSR = \sum_{i=1}^{n}(\hat{Y}_i - \bar{Y})^2$	$p - 1$	$MSR = \dfrac{SSR}{p-1}$	$\sigma^2 + \beta_1^2 \sum_{i=1}^{n}(X_i - \bar{X})^2$	$\dfrac{MSR}{MSE}$
Error	$SSE = \sum_{i=1}^{n}(Y_i - \hat{Y})^2$	$n - p$	$MSE = \dfrac{SSE}{n-p}$	σ^2	
Total	$SSTO = \sum_{i=1}^{n}(Y_i - \bar{Y})^2$	$n - 1$			

greater than one because the quantity $\beta_1^2 \sum_{i=1}^{n}(X_i - \bar{X})^2$ will be greater than zero. Recall that we can use an F-test statistic to test the null hypothesis that variances for two normally distributed random variables are equal (Chapter 5).[*]

$$F* = \frac{MSR}{MSE}.$$ (9.22)

If H_0 is true and regression model assumptions are valid (see Section 9.10), then F^* will be a random outcome from an F distribution with $p - 1$ degrees of freedom in the numerator and $n - p$ degrees of freedom in the denominator. The P-value is $P[F(p - 1, n - p) \geq F^*]$.

EXAMPLE 9.4

For the frigate bird analysis in Examples 9.2 and 9.3, we have

```
MSR <- sum((yhat - mean(freq))^2)
```

The test statistic is

```
F.star <- MSR/MSE
```

and the P-value is

```
pf(F.star, 1, n -2, lower.tail = FALSE)
[1] 3.056371e-05
```

In a simple linear regression, the correct answer is provided by the function anova.

```
anova(Fbird.lm)
Analysis of Variance Table

Response: freq
          Df Sum Sq Mean Sq F value    Pr(>F)
vol        1  24433 24432.5  32.918 3.056e-05 ***
Residuals 16  11876   742.2
---
Signif. codes: 0 '***' 0.001 '**' 0.01 '*' 0.05 '.' 0.1 ' ' 1
```

---

[*] Unlike the parameter it estimates, F^* may occasionally be less than one due to sampling error.

The *P*-value obtained from the *F* test is identical to the *P*-value from Example 9.2. The test statistic, F^*, will be the *t* statistic from Equation 9.16, squared. For instance, we have $t^* = -5.737$, and $-5.737^2 \approx 32.92$. This is the value for F^* in the output above. The area above F^* in the distribution $F(1, 16)$ will be equal to two times the area above $|t^*|$ in $t(16)$. Thus, the *F* and *t*-tests are interchangeable.

9.5 Multiple Regression

Recall that in a multiple regression we have more than one quantitative predictors corresponding to a single *Y* variable (Section 7.6). In this approach, *Y* is modeled as a simultaneous linear function of all the *X* variables.

EXAMPLE 9.5

Aho (2006) considered alpine plant responses in Yellowstone National Park as a function of multiple predictors. Among the studied species were *Carex paysonis*, a sedge (i.e., a graminoid with triangular cross sections) associated with late snowmelt areas (Figure 9.5). The model shown considers the percent cover of *C. paysonis* as a simultaneous function of both the number of days with soil above ("wetter" than) the permanent wilt point of plants ($\psi_{soil} = -1.5$ MPa) and the length of the growing season (the number of days in which soils are above 10°C). Note that the regression model provides a fitted plane instead of a fitted line.[*]

9.5.1 Parameter Estimation

A method of OLS parameter estimation, applicable to all general linear models, is provided by matrix algebra. To use this approach, we solve for β in the matrix format for general linear models introduced in Chapter 7:

$$\mathbf{Y} = \beta \mathbf{X} + \varepsilon. \tag{7.13}$$

In regression, $\mathbf{Y}$ is a $n \times 1$ (i.e., *n* row and one column) vector of response variable outcomes, $\mathbf{X}$ is an $n \times p$ matrix with ones in the first column (representing the intercept)[†] and explanatory variable outcomes in the remaining $p - 1$ columns, β is a $p \times 1$ vector of parameters, and ε is an $n \times 1$ vector of random error terms.

$$\underset{n \times 1}{\mathbf{Y}} = \begin{bmatrix} Y_1 \\ Y_2 \\ \vdots \\ Y_n \end{bmatrix} \quad \underset{n \times p}{\mathbf{X}} = \begin{bmatrix} 1 & X_{11} & \cdots & X_{p-11} \\ 1 & X_{12} & \cdots & X_{p-12} \\ \vdots & \vdots & \ddots & \vdots \\ 1 & X_{1n} & \cdots & X_{p-1n} \end{bmatrix} \quad \underset{p \times 1}{\beta} = \begin{bmatrix} \beta_0 \\ \beta_1 \\ \vdots \\ \beta_{p-1} \end{bmatrix} \quad \underset{n \times 1}{\varepsilon} = \begin{bmatrix} \varepsilon_1 \\ \varepsilon_2 \\ \vdots \\ \varepsilon_n \end{bmatrix}$$

---

[*] In multiple regression models with more than two explanatory variables, the response surface is a $p - 1$-dimensional *hyperplane* that cannot be graphically depicted for all predictors simultaneously.
[†] This is because $Y_i = \beta_0 + \beta_1 X_{1i} + \beta_2 X_{2i} + \cdots + \beta_{p-1} X_{p-1i} + \varepsilon_i$ is equivalent to $Y_i = \beta_0 X_{0i} + \beta_1 X_{1i} + \beta_2 X_{2i} + \cdots + \beta_{p-1} X_{p-1i} + \varepsilon_i$ when $X_{0i} \equiv 1$.

```
Fig.9.5()## see Ch.9 code
```

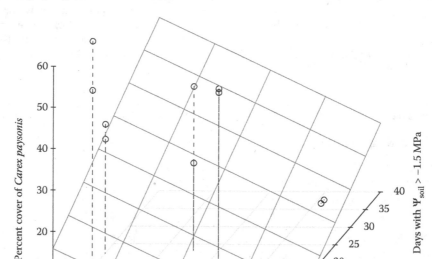

FIGURE 9.5
Representation of a multiple regression. The plane in the figure shows the fitted response (i.e., % cover).

The vector of regression parameters, β, can be estimated using*

$$\hat{\beta} = (X'X)^{-1}X'Y, \tag{9.23}$$

where $\hat{\beta}$ is a $p \times 1$ vector of parameter estimates. That is

$$\hat{\beta}_{p \times 1} = \begin{bmatrix} \hat{\beta}_0 \\ \hat{\beta}_1 \\ \vdots \\ \hat{\beta}_{p-1} \end{bmatrix}.$$

In multiple regression, $\hat{\beta}_0$ is the predicted mean value of Y when all X variables equal zero. Its units are the same as the Y variable. As with simple linear regression, interpretability of

---

* The method of parameter estimation used by most modern statistical software (including lm) is called *QR decomposition*. QR decomposition is based on the idea that any $n \times p$ matrix X can be written as $X = QR$, where Q is an $n \times n$ orthogonal matrix created from the multiplication of a series of matrices based on rotations of X, and R is an $n \times p$ matrix with zero elements except for the first p rows. For the plant richness example (Example 9.6), we have

```
X <- with(wash.rich, cbind(rep(1, 19), X1, X2, X3, X4, X5))
Y <- wash.rich[,2]
solve(qr(X, LAPACK = TRUE), Y) #solves the equation QR %*% beta.hat = Y for beta.hat
[1] 43.96128332 −3.75040840 0.02957584 −0.02423404 −0.15364577 −2.49368175
```

$\hat{\beta}_0$ will depend on the coverage of the model. Coefficients other than the intercept are often called *partial regression coefficients* because they give the partial linear effect of one predictor on the response when the other predictors are held constant. As a result, the estimated coefficient for a predictor in simple linear regression, that is, $\hat{\beta}_1$ in $\hat{Y}_i = \hat{\beta}_0 + \hat{\beta}_1 X_{1i}$, will *not* be the same as the regression coefficient of the same predictor in multiple regression, that is, $\hat{\beta}_1$ in $\hat{Y}_i = \hat{\beta}_0 + \hat{\beta}_1 X_{1i} + \hat{\beta}_2 X_{2i}$, except in the case that the predictors have a correlation of zero.

The partial regression coefficients allow assessment of the effect of the each of the predictors on the response, after adjustment for the remaining predictors. For example, for the kth predictor, Y is expected to change $\hat{\beta}_k$ units for a one unit increase in X_k, given that all other variables are held constant. The units for $\hat{\beta}_k$ are the ratio of the units of Y and X_k.

After finding the parameter estimates: $\beta_0, \beta_1, \ldots, \beta_{p-1}$, the fitted values for the response variable can be obtained using

$$\hat{Y} = X\hat{\beta}, \tag{9.24}$$

where $\hat{Y}$ is an $n \times 1$ vector of fitted responses

$$\underset{n\times1}{\hat{Y}} = \begin{bmatrix} \hat{Y}_1 \\ \hat{Y}_2 \\ \vdots \\ \hat{Y}_n \end{bmatrix}.$$

$\hat{Y}$ can also be obtained from the *hat matrix* (**H**), so named because it puts a hat on **Y**. In this case

$$\hat{Y} = HY, \tag{9.25}$$

where

$$H = X(X'X)^{-1}X'. \tag{9.26}$$

Matrix algebra can also be used to efficiently derive MSR and MSE:

$$MSR = \hat{\beta}' X' Y - \left(\frac{1}{n}\right) Y'1Y/(p-1), \tag{9.27}$$

$$MSE = Y' Y - \hat{\beta}' X' Y/(n-p), \tag{9.28}$$

where **1** is an $(n \times n)$ matrix of ones.

To get the standard errors for $\hat{\beta}$, we first calculate

$$\hat{\sigma}_{\hat{\beta}}^2 = MSE(X'X)^{-1}. \tag{9.29}$$

The resulting variance covariance matrix will have the form

$$\hat{\sigma}_{\hat{\beta}}^2 = \begin{bmatrix} \hat{\sigma}_{\hat{\beta}_0}^2 & \hat{\sigma}_{\hat{\beta}_0,\hat{\beta}_1} & \cdots & \hat{\sigma}_{\hat{\beta}_0,\hat{\beta}_{p-1}} \\ \hat{\sigma}_{\hat{\beta}_1,\hat{\beta}_{p-1}} & \hat{\sigma}_{\hat{\beta}_1}^2 & \cdots & \hat{\sigma}_{\hat{\beta}_1,\hat{\beta}_{p-1}} \\ \vdots & \vdots & \vdots & \vdots \\ \hat{\sigma}_{\hat{\beta}_{p-1},\hat{\beta}_0} & \hat{\sigma}_{\hat{\beta}_{p-1},\hat{\beta}_1}^2 & \cdots & \hat{\sigma}_{\hat{\beta}_{p-1}}^2 \end{bmatrix}. \tag{9.30}$$

Thus, the square root of the diagonal elements of $\hat{\sigma}_{\hat{\beta}}^2$ will contain the standard errors of $\hat{\beta}$.

EXAMPLE 9.6

Aho and Weaver (2010) examined the effect of environmental characteristics on alpine vascular plant species richness on Mount Washburn (3124 m), a volcanic peak in north-central Yellowstone National Park. A subset of these data, detailing 40 10-m² sites, is contained in wash.rich in *asbio*. The dataset has the following variables:

Y = number of vascular plant species
X_1 = percent Kjeldahl (total) soil nitrogen
X_2 = slope, in degrees
X_3 = aspect from true north, in degrees
X_4 = percent cover of surface rock
X_5 = soil pH

We note that while Y consists of discrete counts (and thus cannot be normal), it has a broad range and a large mean. This results in approximate Poisson convergence to normality, and allows conventional multiple regression approaches. Specific methods for handling discrete response variables are presented in Section 9.20.

To estimate β, we first define the *design matrix*, **X**. For regression applications, this will be a column of ones, followed by $p - 1$ additional columns containing the values of each predictor.

```
data(wash.rich)
X <- with(wash.rich, cbind(rep(1, 40), X1, X2, X3, X4, X5))
Y <- wash.rich[,2]
beta.hat <-print(solve(t(X) %*% X) %*% t(X) %*% Y)
         [,1]
    43.96128332
X1 -3.75040840
X2  0.02957584
X3 -0.02423404
X4 -0.15364577
X5 -2.49368175
```

Thus, $\hat{\beta}_0 = 43.96$, $\hat{\beta}_1 = -3.75$, $\hat{\beta}_2 = 0.03$, $\hat{\beta}_3 = -0.024$, $\hat{\beta}_4 = -0.15$, and $\hat{\beta}_5 = -2.49$. Note that the relative importance of the predictors cannot be ascertained from the regression coefficients alone because these are constrained by the units of measurement. For instance, the largest magnitude coefficient is associated with the first predictor, percent Kjeldahl nitrogen. Because $\hat{\beta}_1 = -3.75$, we expect to lose 3.75 species for every 1% increase in soil N, holding all other predictors constant. However, this would constitute a huge increase in soil N as all measured values are well below 1%.

9.5.2 Hypothesis Testing

In multiple regression, we use hypothesis testing for two purposes. First, we can test whether individual parameters equal some hypothesized value (usually zero). Second, we can evaluate whether the combined effect of the X variables produces a linear change in Y.

9.5.2.1 Tests Concerning Regression Parameters

The first type of test conventionally evaluates whether the kth parameter (of p) equals zero. The associated hypotheses are

$$H_0: \beta_k = 0$$

versus

$$H_A: \beta_k \neq 0.$$

The test statistic is calculated as

$$t* = \frac{\hat{\beta}_k}{\hat{\sigma}_{\hat{\beta}_k}}. \tag{9.31}$$

The hypotheses concerning β_k and the calculation of the test statistic follow the exact form described for making inferences about β_1 in Section 9.4.2. The distribution of the test statistic under H_0, described in Section 9.4.2, is also valid for the general test of parameters described here.

Frequently, an investigator will want assurance about the reliability of the results for an entire set (family) of tests. Thus, we can distinguish between a *statement* confidence interval and hypothesis test (those discussed so far), which provide a statement assuming an infinite number of repeated samples from the same population(s) and reapplication of the same analytical method, and a *family* confidence interval or test that indicates the probability, in advance of analysis, that an entire family of statements is correct. For instance, by running a set of hypothesis tests using a family confidence of 0.95, we are assured that the entire set of tests will be correct (will correctly reject H_0) 95% of the time. While seldom applied in practice, one can make a family-wise statement about multiple regression parameter tests using the *Bonferroni method*. This is accomplished by simply multiplying P-values in individual tests by the number of parameters in the model. Note, however, that because the Bonferroni inequality provides the lower bound to the intersection of a set of probabilities (Chapter 2), the method is conservative and prone to type II error. Other methods for simultaneous inference are discussed in the context of ANOVA (Section 10.3).

EXAMPLE 9.7

Using the Washburn data, we will address the first type of hypothesis test (a test whether predictor coefficients equal zero) first. We first calculate MSE:

```
n <- 40; p <- 6
MSE <- (t(Y) %*% (Y) - t(beta.hat) %*% t(X) %*% Y)/(n - p)
```

The estimated variance covariance matrix, $\hat{\sigma}_{\hat{\beta}}^2$, is

```
sigma.sq.beta.hat <- MSE[1] * solve(t(X) %*% X)
```

By taking the square root of the diagonal of $\hat{\sigma}_{\hat{\beta}}^2$, we obtain the standard errors for the parameter estimates.

```
s.beta.hat <- print(sqrt(diag(sigma.sq.beta.hat)))
            X1         X2         X3         X4         X5
9.49283800 5.00803800 0.11063296 0.01881505 0.03947785 1.39578160
```

The test statistics can now be calculated:

```
t.star <- beta.hat/s.beta.hat
```

And the resulting *P*-values are

```
2 * pt(abs(t.star), n - p, lower.tail = FALSE)
          [,1]
   5.142154e-05
X1 4.590807e-01
X2 7.908278e-01
X3 2.064387e-01
X4 4.408796e-04
X5 8.292472e-02
```

This entire operation is handled efficiently by lm.

```
wash.lm <- lm(Y ~ X1 + X2 + X3 + X4 + X5, data = wash.rich)
Coefficients:
            Estimate Std. Error t value Pr(>|t|)
(Intercept) 43.96128    9.49284   4.631 5.14e-05 ***
X1          -3.75041    5.00804  -0.749 0.459081
X2           0.02958    0.11063   0.267 0.790828
X3          -0.02423    0.01882  -1.288 0.206439
X4          -0.15365    0.03948  -3.892 0.000441 ***
X5          -2.49368    1.39578  -1.787 0.082925 .
---

Signif. codes: 0 '***' 0.001 '**' 0.01 '*' 0.05 '.' 0.1 ' ' 1

Residual standard error: 4.347 on 34 degrees of freedom
Multiple R-squared: 0.679, Adjusted R-squared: 0.6318
F-statistic: 14.38 on 5 and 34 DF, p-value: 1.394e-07
```

The null hypothesis of no linear effect on the response is rejected at $\alpha = 0.05$ only for the fourth predictor, soil rockiness. The first and second predictors appear to have very little effect on richness.

9.5.2.2 Combined Effect of X on Y

The second type of hypothesis test evaluates the simultaneous effect of all predictor variables on the response. It uses the following hypotheses:

$$H_0: \beta_1 = \beta_2 = \cdots = \beta_{p-1} = 0$$

versus

$$H_A: \text{not all } \beta_k, k \in \{1,\ldots,p-1\} \text{ equal } 0.$$

For this test, an ANOVA approach is used. That is, the ratio MSR/MSE is used as the test statistic. The distribution of the F^* under H_0, and the testing procedure is described in Section 9.5.3.

9.5.3 An ANOVA Approach

While the ANOVA approach does not provide any new information in simple linear regression, it will be very useful in multiple regression for testing the combined effect of all predictors simultaneously on the response. That is, it addresses the second type of multiple regression hypothesis test described in Section 9.5.2.2.

EXAMPLE 9.8

For the plant richness example, we have

$$H_0: \beta_1 = \beta_2 = \beta_3 = \beta_4 = \beta_5 = 0$$

versus

$$H_A: \text{not all } \beta_k \text{ equal } 0.$$

The MSR, test statistic for this test, and resulting *P*-value, respectively, are

```
MSR <- ((t(beta.hat)%*%t (X)%*%Y) - (1/n) * t(Y)%*% matrix(1, nrow=n,
ncol =n)%*%Y)/(p - 1)

F.star <- MSR/MSE
pf(F.star, p - 1, n - p, lower.tail =FALSE)
[1,] 1.393899e-07
```

The results of this test ($F_{5,34} = 14.38$; $P = 0.0000001$) are also in the last line of summary output from lm in Example 9.7.

```
F-statistic: 14.38 on 5 and 34 DF, p-value: 1.394e-07
```

Because the *P*-value is extremely small, we reject H_0 and conclude that richness is linearly dependent on the predictors.

F-tests can also be conducted to determine whether $\beta_k = 0$ (the first type of null hypothesis test described in Section 9.5.2). However, this approach will now be problematic. This is because, when multiple explanatory variables are used in a linear model, and those variables are quantitative, or categorical and unbalanced, or both, then the results from anova(lm) will be untrustworthy.

For models with multiple predictors, sums of squares can be calculated separately for each explanatory variable in an analysis of variance. Conventional ANOVAs use *type I sum of squares* to partition SSTO. In this process, the sums of squares are calculated by adding explanatory variables sequentially to the model. Methods for apportioning SSR using type I SS are described by Kutner et al. (2005). When a type I SS approach is used with multiple regression, only the *F* test for the last added explanatory variable will be equivalent to the *t* test for the same variable. An implication is that *P*-values and sums of squares for variables will change depending on the order that they are specified in the model.

One solution to this is to use *type II sums of squares*, also called *extra sums of squares* (Kutner et al. 2005).[*] To calculate type II SS, we must first specify reference models and nested models. The *reference model* will contain all of the predictors, while in the *nested model*, the predictor of interest is removed.

---

[*] Type III sums of squares (described in Chapter 10) will be identical to type II SS if no interactions are prescribed. As a result, they would be identical to each other in Example 9.9.

For the kth predictor (which is included in the reference model, but not the nested model), we are testing the hypotheses:

$$H_0: \beta_k = 0$$

versus

$$H_A: \beta_k \neq 0.$$

The test statistic is

$$F* = \frac{SSE_N - SSE_R}{dfE_N - dfE_R} \Big/ \frac{SSE_R}{dfE_R} \tag{9.32}$$

where SSE_N is the sum of squares error for the nested model, SSE_R is the sum of squares error for the reference model, dfE_N is the degrees of freedom error for the nested model, and dfE_R is the degrees of freedom error for the reference model. The numerator of the test statistic will be the type II SS for the kth predictor (see Section 10.14).

If H_0 is true, and model assumptions are valid (see Section 9.10), then $F*$ will follow an F distribution with $dfE_N - dfE_R$ degrees of freedom in the numerator and dfE_R degrees of freedom in the denominator. The P-value will be $P[F(dfE_N - dfE_R, dfE_R) \geq F*]$.

EXAMPLE 9.9

In a multiple regression, the results from anova(lm) will be in disagreement with t-test results from lm because the function anova uses type I sums of squares. For instance, compare the P-values below to the results in Example 9.7.

```
anova(wash.lm)
Response: Y
          Df Sum Sq Mean Sq F value     Pr(>F)
X1         1 663.74  663.74 35.1310 1.071e-06 ***
X2         1  10.67   10.67  0.5645 0.4576093
X3         1 306.33  306.33 16.2138 0.0003000 ***
X4         1 317.55  317.55 16.8073 0.0002431 ***
X5         1  60.31   60.31  3.1919 0.0829247 .
Residuals 34 642.38   18.89
```

Only the P-value for the last predictor, X_5, agrees with the t-test results from lm.

Disturbingly, reordering the predictors in the model gives completely different results. For instance, try

```
anova(lm(Y ~ X2 + X1 + X4 + X5 + X3, data = wash.rich))
```

Once again, only the P-value for the last predictor (now X_3) agrees with the original lm results. To resolve this problem, we can calculate type II sums of squares. I will demonstrate this process using the first predictor, Kjeldahl N.

We first require a reference model (with all of the predictors) and a nested model (with X_1 left out).

```
ref <- anova(wash.lm)
nested.lm <- lm(Y ~ X2 + X3 + X4 + X5, data = wash.rich)
nested <- anova(nested.lm)
```

We then extract the sum of squares error and degrees of freedom error from both models.

```
SSE.ref <- tail(ref$"Sum Sq", 1);
SSE.nested <- tail(nested$"Sum Sq", 1)
dfE.ref <- tail(ref$"Df", 1)
dfE.nested <- tail(nested$"Df", 1)
```

The type II mean square error for X_1 is

```
MSX1.II <- print((SSE.nested - SSE.ref)/(dfE.nested - dfE.ref))
[1] 10.59577
```

This number divided by the MSE from the reference (original) model provides the F-test statistic for the null hypothesis that the linear effect of X_1 is zero.

```
pf(MSX1.II/MSE, dfE.nested - dfE.ref, dfE.ref, lower.tail = FALSE)
      [,1]
[1,] 0.4590807
```

The type II SS and associated null hypothesis test for a single variable can be obtained from the function anova by using the reference and nested models as arguments.

```
anova(wash.lm, nested.lm)
Analysis of Variance Table
Model 1: Y ~ X1 + X2 + X3 + X4 + X5
Model 2: Y ~ X2 + X3 + X4 + X5
  Res.Df    RSS Df Sum of Sq      F Pr(>F)
1     34 642.38
2     35 652.97 1    10.596 0.5608 0.4591
```

Type II SS and associated null hypothesis tests can be obtained simultaneously for all predictors with the function Anova from *car*.

```
library(car)
Anova(wash.lm)
Anova Table (Type II tests)
Response: Y
          Sum Sq Df  F value     Pr(>F)
X1         10.60  1   0.5608  0.4590807
X2          1.35  1   0.0715  0.7908278
X3         31.34  1   1.6590  0.2064387
X4        286.18  1  15.1473  0.0004409 ***
X5         60.31  1   3.1919  0.0829247 .
Residuals 642.38 34
```

The P-values are now identical to those from lm, and the F statistics are the t statistics from lm squared.

The take-home message here is that anova should not be used to directly evaluate the results from a multiple regression. Instead, one should evaluate the results from lm directly, or use a type II SS approach, as described above.

9.6 Fitted and Predicted Values

With a regression model, inferences can be made for Y given X. For instance, from the frigate bird data in Example 9.2, we obtained the model

$$\hat{Y}_i = 557.1022 - 0.0203X_i.$$

The first observed X value was a volume of 1760 cm³. Thus, the first *fitted value* is 577.1022 − 0.0203(1760) = 521.2 Hz. Fitted Y values can be found for any general linear model using the function `fitted`. For instance

```
fitted(Fbird.lm)
        1           2...              18
521.2342 515.5279...  394.8811
```

Using parameter estimates, we can predict the fundamental frequency produced by gular pouch volumes that were not directly observed. For example, given a volume of 3000 cm³, our *predicted value* is 577.1022 − 0.0203(3000) = 495.97 Hz. Predicted values for any combination of X values can be found using the function `predict`. For instance

```
predict(Fbird.lm, newdata = data.frame(vol = 3000))
495.9636
```

The prediction of the fundamental frequency for a volume of 3000 cm³ is valid because this value is between the smallest and the largest values of X in the dataset. Conversely, predicting Y for a value of X that lies beyond the range of X used to create the regression line is known as *extrapolation*. Predictions based on extrapolation are often inaccurate or meaningless. For instance, the prediction of fundamental frequency at a pouch volume of 30,000 cm³ is 557.1022 − 0.0203(30000) = −51.9 Hz. This prediction does not make sense. Not only is the pouch volume absurdly large (capable of holding 30 L), we cannot have negative sonic frequencies.

X values can be predicted from responses given simple algebraic rearrangement of the regression model, for instance

$$\hat{X}_i = \frac{Y_i - 557.1022}{-0.0203}.$$

Note, however, that (1) this will not (excluding the possibility that the correlation of X and Y is 1 or −1) be equal to the fitted value for X given Y_h for a model that fits X as a function of Y, and (2) confidence limits for X given Y_h will be complex to derive (see Zar 1999, p. 342, Snedecor and Cochran 1967, pp. 159–160).

Calculating fitted and predicted values for multiple regression is done in exactly the same way as for simple linear regression. Now, however, explanatory values will be required for all $p - 1$ predictors. For instance, for the plant richness analysis in Example 9.5, we obtained the model

$$\hat{Y} = 43.96 - 3.75X_1 + 0.03X_2 - 0.024X_3 - 0.15X_4 - 2.49X_5.$$

Given $X_1 = 0.2$, $X_2 = 10$, $X_3 = 30$ $X_4 = 20$, and $X_5 = 7$, we have

$$\hat{Y} = 43.96 - 3.75(0.2) + 0.03(10) - 0.024(30) - 0.15(20) - 2.49(7) = 22.3 \text{ species}.$$

In R, we have

```
predict(wash.lm, newdata = data.frame(X1 = 0.2, X2 = 10, X3 = 30, X4 = 20,
X5 = 7))
```

22.25125

One should be particularly careful when estimating a mean response or predicting a new observation with multiple regression. This is because the range of the X observations will generally exceed the scope of the model that will now be defined by a $p - 1$ dimensional hull. As a result, one cannot merely look at the ranges of the X variables to determine if a combination of X values will result in extrapolation.

The hat matrix, H, can be used to identify hidden extrapolation in multiple regression. This involves defining

$$h_{new} = \mathbf{X}'_{new} (\mathbf{X}'\mathbf{X})^{-1} \mathbf{X}_{new} , \qquad (9.33)$$

where $\mathbf{X}_{new}$ is a vector of length $p - 1$ containing X values for which inference about a fitted value or a new observation is to be made, and $\mathbf{X}$ is the original design matrix. If the scalar h_{new} is well within the range of the original leverages (i.e., the diagonal elements of the hat matrix), then no extrapolation is taking place (leverage is formally defined in Section 9.10.5). On the other hand, if h_{new} is much larger than the diagonal elements of $\mathbf{H}$, then the use of $\mathbf{X}_{new}$ will result in extrapolation.

9.7 Confidence and Prediction Intervals

Confidence intervals are calculable for all linear regression parameters, and for fitted and predicted values described above.

9.7.1 Confidence Interval for β_k

We address confidence intervals for model parameters first. Let β_k be the kth model parameter $k = \{0, 1, \ldots, p - 1\}$. We calculate the $(1 - \alpha)100\%$ confidence interval for β_k using

$$\hat{\beta}_k \pm t_{[1-(\alpha/2), df=n-p]} \times \hat{\sigma}_{\hat{\beta}_k} , \qquad (9.34)$$

where $t_{[1-(\alpha/2), df=n-2]}$ is the quantile function outcome for a t-distribution with $n - p$ degrees of freedom at probability $1 - (\alpha/2)$. The standard errors for all parameter estimates, $\hat{\sigma}_{\hat{\beta}_k}$ can be obtained simultaneously as the square root of the diagonal of the variance covariance matrix given in Equation 9.30.[*]

EXAMPLE 9.10

For the frigate bird model, the 95% confidence interval for β_1 is

$$\hat{\beta}_1 \pm t_{(0.975, 16)} \times \hat{\sigma}_{\hat{\beta}_1} = -0.0203 \pm 2.12 \times 0.0036 = (-0.0279, -0.0128)$$

Thus, we are 95% confident that the true regression slope, β_1, lies in the interval $(-0.0279, -0.0128)$.

---

[*] Note that $(1 - \alpha)100\%$ Bonferroni family-wise intervals for β_k, and all other regression parameters, can be calculated by simply replacing α with the product αp in confidence interval formulae, where p is the number of parameters about which one wishes to make simultaneous inference. Bonferroni confidence intervals can be directly obtained from the function `joint.ci.bonf`.

Recall that a $(1 - \alpha)100\%$ confidence interval provides analogous information to a two-tailed hypothesis test using a significance level of α (Section 6.2.6). Given this, note that zero is not in the 95% confidence interval for β_1. In fact, the confidence interval contains only negative values. This indicates a negative association between volume and frequency, and is in accordance with the hypothesis test conclusions from Example 9.3.

Statement confidence intervals for linear model parameters can be obtained using the function `confint`. The default confidence level is 95%.

```
confint(Fbird.lm)
                  2.5 %         97.5 %
(Intercept) 522.72493295  591.47938604
vol          -0.02790946   -0.01284958
```

9.7.2 Confidence Intervals for True Fitted Values and Prediction Intervals for Predicted Values

Two subtly different approaches are used in interval estimation for fitted and predicted values in regression. The first creates an interval estimate for the true fitted value of Y for the hth explanatory observation(s). That is, it provides a confidence interval for the population mean, $E(Y_h)$. The second approach provides an interval estimate for a *future value* of Y given particular X_hs. This is denoted as $Y_{h(new)}$, and its interval estimate is called a *prediction interval*. The first type of interval will have greater precision than the second. That is, a future response can be predicted less precisely then a conditional expectations using the current dataset. This is demonstrated with the standard error formulae below:

$$\hat{\sigma}_{\hat{Y}_h} = \sqrt{MSE\left[\frac{1}{n} + (X_h - \bar{X})^2 \middle/ \sum_{i=1}^{n}(X_i - \bar{X})^2\right]}, \qquad (9.35)$$

$$\hat{\sigma}_{\hat{Y}_{h(new)}} = \sqrt{MSE\left[1 + \frac{1}{n} + (X_h - \bar{X})^2 \middle/ \sum_{i=1}^{n}(X_i - \bar{X})^2\right]}. \qquad (9.36)$$

We note two things. First, owing to the squared difference in the numerator, standard errors will increase exponentially for predicted values corresponding to values of X_h further away from $\bar{X}$ (as we approach extrapolation). Second, the standard error associated with a prediction interval for $Y_{h(new)}$ will always be greater than the standard error associated with a confidence interval for $E(Y_h)$. Intuitively, this is because two types of variability need to be considered in a prediction interval. These are (1) the variability associated with estimating the new location of $E(Y_h)$, and (2) the variability around $E(Y_h)$ implicit to every distribution of Y given X.

A $(1 - \alpha)100\%$ confidence interval for $E(Y_h)$ is constructed using

$$\hat{Y}_h \pm t_{[1-(\alpha/2), df=n-2]} \times \hat{\sigma}_{\hat{Y}_h}, \qquad (9.37)$$

while a $(1 - \alpha)$ 100% prediction interval for $Y_{h(\text{new})}$ is constructed using

$$\hat{Y}_h \pm t_{[1-(\alpha/2),df=n-2]} \times \hat{\sigma}_{\hat{Y}_{h(\text{new})}}. \tag{9.38}$$

Note that both intervals are centered at the fitted value of Y given X_h.

EXAMPLE 9.11

For the frigate bird regression model, we will calculate a 95% confidence interval and prediction interval for a frequency corresponding to a pouch volume of 3080 cm³.

$$\hat{\sigma}_{\hat{Y}_h} = \sqrt{742.2 \left[\frac{1}{18} + \frac{1237038.3}{58827511} \right]} = 7.54,$$

$$\hat{\sigma}_{Y_{h(\text{new})}} = \sqrt{742.2 \left[\frac{19}{18} + \frac{1237038.3}{58827511} \right]} = 28.27.$$

The fitted value for Y given $X = 3080$ is

```
fitted(Fbird.lm)[8]
494.3332
```

The resulting interval estimates are

$$\hat{Y}_h \pm t_{(0.975,16)} \times \hat{\sigma}_{\hat{Y}_h} = 494.33 \pm 2.12 \times 7.54 = (478.35,\ 510.32),$$

$$\hat{Y}_{h(\text{new})} \pm t_{(0.975,16)} \times \hat{\sigma}_{\hat{Y}_{h(\text{new})}} = 494.33 \pm 2.12 \times 28.27 = (434.40, 554.26).$$

Thus, we are 95% confident that the mean of Y given $X = 3080$ will lie in the interval (541.12, 573.08), and we are 95% confident that a future observation of Y given $X = 3080$ will lie in the interval (497.17, 617.03).

Both confidence and prediction intervals can be obtained from the function predict.

```
predict(Fbird.lm, newdata = data.frame(vol = 3080), interval = "confidence")
       fit      lwr      upr
1 494.3332 478.3505 510.3159
predict(Fbird.lm, newdata = data.frame(vol = 3080), interval =
"prediction")
       fit      lwr      upr
1 494.3332 434.4086 554.2579
```

Confidence and prediction bands can be overlaid on a regression plot (Figure 9.6). As expected, the prediction intervals are much wider than the confidence intervals, and both confidence and prediction bands are wider as distance from the predictor mean increases.

```
with(Fbird, plotCI.reg(vol, freq, xlab = expression(paste("Gular pouch
size (",cm^3,")")),ylab = "Fundamental frequency (Hz)"))
```

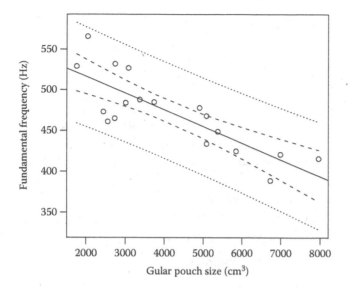

FIGURE 9.6

Frigate bird regression line along with 95% confidence intervals for $E(Y_h)$ (inner bands) and 95% prediction intervals for $Y_{h(new)}$ (outer bands).

9.8 Coefficient of Determination and Important Variants

From Chapter 8, we know that correlation coefficient, r, ranges from –1 to 1, and describes the strength and direction of the linear association between two variables. In the case of simple linear regression, the square of the correlation coefficient (r^2) will be equivalent to a statistic called the *coefficient of determination*, also called *multiple R^2*, or simply R^2 for multiple regression or r^2 for simple linear regression. For all regression approaches, R^2 will be the Pearson correlation of the fitted values and the observed responses, squared. R^2 can also be calculated using sums of squares:

$$R^2 = 1 - \frac{SSE}{TSS} = \frac{SSR}{TSS}. \tag{9.39}$$

R^2 can be interpreted as the proportion of variation in Y that is explained by the predictors. This is demonstrated in the form of Equation 9.39. SSE measures the variation in Y with respect to the regression line/plane/hyperplane. SSE is divided by SSTO, which measures the total variation in Y. As the fit of the line improves, the sum of squared residuals will decrease, and SSE/SSTO will approach 0, making R^2 approach 1. Thus, R^2 will have the range [0, 1]. These extremes represent a complete absence of explanatory power and perfect (100%) explanatory power.

It is important to emphasize that we cannot infer a causal association between X and Y just because $R^2 = 1$. While a high R^2 indicates that predictions for Y can be made effectively with X for the current dataset, it does not necessarily mean that the relationship is causal. Evidence of cause and effect can only come from effective experimental design: for instance, by measuring Y after randomly assigning experimental units to levels in X (Chapter 7).

A modified form of R^2 can be used that adjusts for the number of explanatory variables in a model. R^2_{adj}, or *adjusted* R^2, is calculated as

$$R^2_{adj} = 1 - \frac{SSE}{SSTO} \times \frac{n-1}{n-p}. \tag{9.40}$$

Unlike R^2, the adjusted coefficient of determination may become smaller when another X variable is added. This is because a decrease in SSE may be offset by the decrease in $n - p$. R^2 and R^2_{adj} are both reported in the summary output from lm (see Examples 9.3 and 9.7).

The *coefficient of partial determination* measures the reduction in sum of squares error after a variable of interest, X_k, is introduced into a multiple regression. As with extra sums of squares calculations, this statistic requires designation of nested and reference models:

$$R^2_{partial} = \frac{SSE_N - SSE_R}{SSE_N}, \tag{9.41}$$

where SSE_N is the sum of squares error for the nested model (that excludes the kth predictor), and SSE_R is the sum of squares error for the reference model (that includes the kth predictor).

Consider a model with three predictors X_1, X_2, and X_3. The partial effect of X_1 on Y, holding X_2 and X_3 constant, can be estimated using

$$R^2_{1|2,3} = \frac{SSE(X_2 + X_3) - SSE(X_1 + X_2 + X_3)}{SSE(X_2 + X_3)}.$$

where $SSE(X_2 + X_3)$ is the sum of squares error for a model with just predictors X_2 and X_3, and $SSE(X_1 + X_2 + X_3)$ is the sum of squares error for the reference model. A *partial correlation coefficient* is simply the square root of coefficient of partial determination with the sign taken from the parameter estimate in the reference model.

EXAMPLE 9.12

Consider the plant richness analysis. Assume that we are interested in the variability in richness explained by soil rockiness (X_4) holding other variables constant.

The function partial.R2 calculates coefficients of partial determination by serving as a wrapper for anova.

```
ref.lm <-lm(Y ~ X1 + X2 + X3 + X4 + X5, data = wash.rich)
nested.lm <- update(ref.lm, ~. - X4)
partial.R2(nested.lm, ref.lm)
[1] 0.3082016
```

When X_4 is added to a model with the other predictors, SSE is reduced by 31% (i.e., explanatory power is increased by 31% by adding X_4).

Nested models can be created efficiently from reference models by using the function update. The argument ~. - X4 tells **R** to recreate the entire reference model, while leaving X_4 out.

9.9 Power, Sample Size, and Effect Size

R^2 is a useful estimator of effect size in simple linear regression while R^2_{adj} is similarly useful for multiple regression models (Thompson 1998). Thus, these measures should be reported with significance test results from these analyses.

Cohen (1988) recommended that, in the context of power analyses, regression effect size should be defined as a function of the true proportion of variability in Y that is explained by the model, that is, the population multiple R^2, denoted here as ρ^2. In particular

$$f^2 = \frac{\rho^2}{1 - \rho^2}.\tag{9.42}$$

Cohen (1992) suggested that f^2 values of 0.02, 0.15, and 0.35 could be used to define the lower bound of small, medium, and large effect sizes. Such cutoffs, however, have been criticized as arbitrary, since the conceptualization of effect size magnitude is likely to vary with the study system (Quinn and Keough 2002). The function `pwr.f2.test` from library *pwr* uses an iterative root-finding procedure with respect to an F-distribution with non-centrality parameter $f^2(v_1 + v_2 + 1)$ to obtain power, the significance level, or the required sample size (see Johnson et al. 1995).

EXAMPLE 9.13

For a simple linear regression model, we assume that $\rho^2 = 0.3$. What is the sample size required to reject H_0: $\beta_1 = 0$ with probability 0.8, given $\alpha = 0.05$?
  We have

```
library(pwr)
pwr.f2.test(u = 1, v = NULL, power = 0.8, f2 = 0.3/(1 - 0.3), sig.
level = 0.05)
    u = 1
    v = 18.42547
    f2 = 0.4285714
sig.level = 0.05
    power = 0.8
```

The first argument u refers to the degrees of freedom in the numerator of the regression F-test statistic. Because this is a simple linear regression, u $= p-1 = 2-1 = 1$. The argument v signifies the denominator degrees of freedom for F^*. Since I have defined it to be NULL, the function will estimate this value. We can use the algorithm's solution for v to find n since v $= n - p = n - 2$. From the output above it appears that a sample size of approximately $19 + 2 = 21$ is required.

9.10 Assumptions and Diagnostics for Linear Regression

Hypothesis tests for Pearson's correlation coefficient and simple linear regression will produce identical P-values given the same data. That is, the P-values from summary(lm(Y ~ X)),

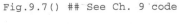

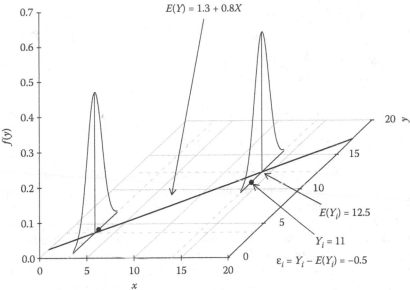

FIGURE 9.7

Example of a population regression line for simple linear regression. The mean response is a straight-line function of the explanatory variable. The true model is $E(Y) = 1.3 + 0.8X$. We have two random observations of Y given X. These are indicated as Y_i, $i = \{1,2,\ldots,n\}$. The values are 4.2 and 11, corresponding to the X values 3 and 14. The expectation of Y given $X = 14$ is 12.5. The error, ε_i, associated with $Y_i = 12.5$, is -0.5. The figure can be interactively depicted by typing `see.regression.tck()`.

`summary(lm(X ~ Y))`, `cor.test(Y, X)`, and `cor.test(X, Y)` will all be the same. These approaches, however, have different assumptions.

With Pearson's correlation coefficient, we assume an underlying bivariate normal distribution for two variables of interest. This means that we do not define explanatory and response variables since the variables are considered jointly. Conversely, with regression, we are not concerned with the distribution of X *per se*. Values of X are considered to be fixed and measured without error. Our only concern is that for each X_i, we have normal distributions of Y_i with the same variance. In particular

$$Y_i \sim N(\beta_0 + \beta_1 X_{1i} + \beta_2 X_{2i} + \cdots + \beta_{p-1} X_{p-1i}, \sigma^2) \qquad (9.43)$$

$$\sim N(\mu_i, \sigma^2).$$

Because variances are assumed to be constant for all distributions of Y_i, subtracting μ_i from every possible outcome in the random variable Y_i will result in identical normal distributions of the random error term ε_i^* (Figure 9.7).

---

* If $Y_i \sim N(\mu_i, \sigma^2)$, then $Y_i - \mu_i \sim N(0, \sigma^2)$. That is, $E(Y_i)$ is decreased by μ_i, but the variance of the distribution is unchanged by subtraction of this constant (see Chapter 4).

The general linear regression model (Equation 9.1) has four assumptions:

1. The error terms are independent.
2. The error terms are normally distributed.
3. The error terms have the same variance, σ^2.
4. The relationship between X and Y is linear.

The first three assumptions can be succinctly stated as *iid*, $\varepsilon_i \sim N(0, \sigma^2)$, where *iid* denotes independent and identically distributed. The fourth assumption emphasizes that linear regression only quantifies the linear association between X and Y.

While not explicitly stated as a model assumption, outliers may significantly affect all linear models, including regression analyses. Indeed, if n is small, a fitted regression line can be so distorted by an outlier that it may appear that there are problems with independence or nonconstant variance.

Linear model assumptions are generally checked using both diagnostic plots and hypothesis tests. There are three reasons that hypothesis tests should not be alone in this capacity. First, some hypothesis testing procedures for comparing error variances (particularly the F-test) are sensitive to departures from normality, particularly positive skewness (Conover et al. 1981). Thus, nonnormal errors will lead to false inferences concerning homoscedasticity. Note that this is *not* a concern for the variance constancy tests used in this text. Both the Fligner–Killeen test (Section 6.7.1) and the modified Levene test (introduced in Chapter 10) are robust to large departures from normality (Conover et al. 1981). Second, frequentist null hypothesis tests will produce smaller P-values given larger sample sizes (Chapter 6). Thus, larger sample sizes will result in a propensity to reject the H_0 even when the linear model is reliable. Conversely, given small sizes, tests may fail to reject H_0 even when the linear model is unreliable. Finally, hypothesis tests provide little information about the underlying causes and corrective procedures that will be required. Diagnostic plots (e.g., residual plots) and other procedures will generally be required to decide on the appropriate corrective actions (e.g., the type of transformation required, Section 9.11).

To demonstrate regression diagnostics, we will use an extended exploration of the frigate bird regression model. The function `plot.lm()` provides diagnostic plots for objects from class `lm` (Figure 9.8).

9.10.1 Independence of Error Terms

Lack of independence will cause regression estimators to be inefficient (although they will remain unbiased) and will result in the underestimation of error term and predictor variances. As a result, regression confidence intervals and hypothesis tests using the t- and F-distributions will no longer be strictly applicable.

If the error terms are independent, then there should be no pattern to error term estimates (residuals) plotted against fitted values. In particular, adjacent residuals should not be grouped above or below the residual mean (0), and should not increase or decrease in magnitude with increasing fitted values. The frigate bird data evince no obvious problems with respect to nonindependence (Figure 9.8a).

If data are collected in a time series or spatial sequence, then *sequence plots* should be created, in which the sequence order is plotted against the corresponding residuals. Lack of independence will be evident as a pattern because observations closer together in the sequence will tend to have similar errors.

```
par(mfrow = c(2, 2), mar = c(4, 4, 2, 1.5))
plot(Fbird.lm)
```

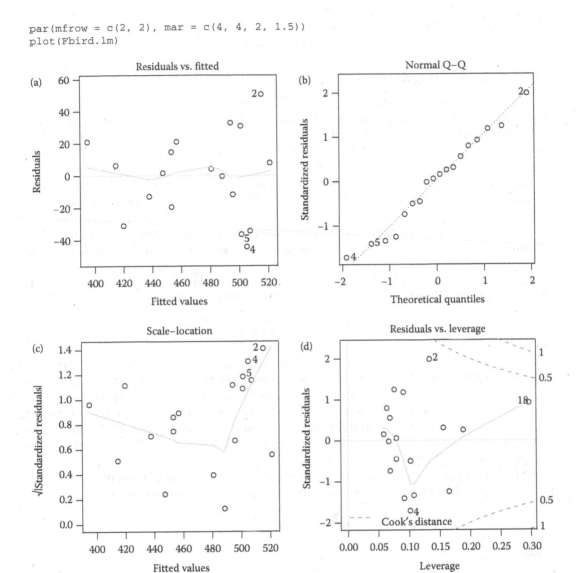

FIGURE 9.8
Default diagnostic plots produced for the frigate bird linear model: (a) fitted values versus residuals, (b) normal quantile plot, (c) scale–location plot, and (d) leverage plot with Cook's distances.

Independence can also be diagnosed with hypothesis tests. For instance, we learned in Section 7.5.5 about the Durbin–Watson hypothesis test for temporal autocorrelation. This approach is not used here because the frigate bird data are not the result of a serial process.

When sample sizes are large compared to the number of parameters in the model, the dependency of residuals is relatively unimportant, and can be ignored in most situations (Kutner et al. 2005). In other cases, one can use a formal autoregressive approach (e.g., time series analysis) or use remedial measures to account for nonindependence. These include adding predictors that account autoregressive trend, and time series *differencing*, in which

Y' is modeled as a function of X', where $Y_t' = Y_t - Y_{t-1}$ and $X_t' = X_t - X_{t-1}$, and specialized transformations (see Kutner et al. 2005, pp. 480–500).

9.10.2 Normality of Error Terms

We assume that the populations of Y given X_i are each normally distributed, and consequentially, the error terms in each population (which are the result of variable centering) are also normally distributed. Because we expect Y to vary randomly in the same way at each value of X, residuals can be pooled to test both this assumption and the assumption of constant variance (see below).

The most common diagnostic of normality is the normal quantile plot.[*] In Section 6.2.7, quantile plots of raw data were used to check for the appropriateness of t-procedures. For general linear models, quantile plots are made using residuals. In Figure 9.8b, the relationship between the sample and theoretical quantiles (the z-scores corresponding to residual percentiles, see Example 6.6) is linear. This supports the assumption that the error terms are normally distributed.

Figure 9.8b–d each use *standardized residuals* that are calculated by dividing the residuals by their standard deviations. The ith standardized residual is obtained from

$$\hat{\varepsilon}_i / \sqrt{MSE}\sqrt{1 - h_{ii}}, \tag{9.44}$$

where $\hat{\varepsilon}_i$ is the ith residual and h_{ii} is the ith diagonal element in the hat matrix, $\mathbf{H}$ (Equation 9.26). That is, h_{ii} is the ith diagonal element of $\mathbf{H}$. Standardization is used because standard deviations of residuals may vary greatly from one observation to the next, even when the variances of the errors terms are constant. This occurs because the regression line is drawn most strongly to outlying points. Standardization will give the residuals the same variance (one) for any fitted value, facilitating the correct identification of outliers and patterns among residuals.

We can also perform hypothesis tests for normality using the Shapiro–Wilk test (Section 6.2.7). In the context of general linear models, the appropriate hypotheses are

$$H_0: \varepsilon_i \text{s are normal,}$$

versus

$$H_A: \varepsilon_i \text{s are not normal.}$$

```
shapiro.test(resid(Fbird.lm))
    Shapiro-Wilk normality test

data: resid(Fbird.lm)
W = 0.9746, p-value = 0.8785
```

As indicated in the normal quantile plot, there is essentially no evidence against H_0.

While small departures from error term normality will not affect valid inference, major departures will often lead to spurious models (Kutner et al. 2005). Nonnormality can often be effectively addressed with data transformations (see Section 9.11).

---

[*] Empirical distribution plots, for example, histograms, ecdfs, and stem and leaf plots (type `example(stem)`) can also be used as diagnostics for normality.

9.10.3 Constancy of Error Variance

As with independence and normality, constancy of variance can be assessed using both diagnostic plots and hypothesis tests. Figures 9.8a and 9.8c address error variance constancy. If the assumption of homoscedasticity is correct, there should be no pattern to the points in either plot. Figure 9.8c is called a *scale–location plot*. It plots the square roots of the absolute values of the standardized residuals versus the fitted values. Heteroscedasticity will occur when the variance of the response is a function of the mean. This will often result in a funnel shape in the scatter of residuals. Figures 9.8a and 9.7b identify three large residual observations (2, 4, and 5) and demonstrate a slight funnel shape.

In the context of general linear models, the Fligner–Killeen test for homoscedasticity (Section 6.2.7) will have the following hypotheses:

$$H_0: \varepsilon_i\text{s are homoscedastic, } \sigma_1^2 = \sigma_2^2 = \cdots = \sigma_n^2,$$

versus

$$H_0: \varepsilon_i\text{s are not homoscedastic.}$$

The Fligner–Killeen test requires multiple Ys from each X_i to allow between group variance comparisons. While this is generally not realistic in regression analysis, datasets can be subset into groups and the error constancy among these groups can be estimated. Below, I divide the frigate bird dataset into two groups, birds with pouch volumes ≤ 3370 cm^3 and those with pouch volumes > 3370 cm^3. Because the data are ordered by pouch volume, we have

```
fligner.test(resid(Fbird.lm),factor(c(rep(1,9),rep(2,9))))

        Fligner-Killeen test of homogeneity of variances

data: resid(Fbird.lm) and factor(c(rep(1, 9), rep(2, 9)))
Fligner-Killeen:med chi-squared = 3.056, df = 1, p-value = 0.08044
```

In agreement with the diagnostic plot, the Fligner–Killeen test indicates possible issues with nonconstant variance.

Violation of homoscedasticity will generally be more detrimental to efficient parameter estimation than violation of normality (Quinn and Keough 2002). Remedial measures for nonconstant variance include data transformation (Section 9.11), and an approach called weighted least squares (WLS) (Section 9.13).

9.10.4 The Relationship between *X* and *Y* Is Linear

Curvilinear relationships between the predictor and response variables will not be recognized by general linear models unless particular adjustments are made. Model linearity in simple linear regression can be verified by examining raw data scatterplots, which may or may not include regression lines (i.e., Figures 9.3, 9.4, and 9.6), and plots of residuals versus fitted values (i.e., Figure 9.8a and c). Unaccounted for nonlinear associations will result in residual being grouped above or below the residual mean although this will not signify error term dependence (Cook 1998).

If *Y* appears to change linearly with *X*, and residuals show no pattern above or below the mean residual value, then the relationship between *X* and *Y* can be assumed to be linear. The frigate bird data satisfies this assumption.

Nonlinearity in multiple regression is difficult to ascertain because of the potential high dimensionality of the response surface. The partial relationships between predictors and response can be examined using both partial residual plots and matrix plots (see Section 9.10.6). Nonlinearity in simple linear and multiple regression can be addressed with transformations (Section 9.11), the addition of quadratic or other polynomial terms to the X variable (Section 9.14), or by explicitly defining a nonlinear model (Section 9.21).

9.10.5 Outliers

In the presence of outlying observations, the use of conventional regression estimators may lead to false inferences, particularly if the sample size is small. When outliers exist in data, and do not represent data entry or recording errors, and thus, cannot be discarded, one approach is to create two models, one with and one without the outliers. If there is little difference in interpretation, then a single model (with outliers) can be presented. If the outliers strongly affect the model, then both models should be described along with a biological explanation for the outliers (Quinn and Keough 2002). An alternative approach is robust regression (Section 9.18).

The existence and importance of outliers can be deduced by indices, including leverage, Cook's distance, and DFFITS (the difference between fitted values and leave one out fitted values) (see below). The final diagnostic plot (Figure 9.8d) shows leverage (on the X-axis) and Cook's distances of 0.5 and 1.0, as dashed contour lines. Points with large leverages (outliers) and large Cook's distances may distort a regression model, leading to false inferences.

The *leverage* of a particular point quantifies how unusual it is in predictor space. The leverage of the ith observation is its corresponding diagonal element in the hat matrix, $\mathbf{H}$. The sum of leverages will always equal p, the number of model parameters. Observations with leverage $>2p/n$ can generally be considered X outliers because their leverages are twice the average leverage (Kutner et al. 2005). For the frigate bird model, $2p/n = 2(2)/18 = 0.222$. Only one point (observation 18) has a leverage above this cutoff (Figure 9.8d). High leverage points may not be influential if they reflect trends demonstrated by other points. The *influence* of a point can be deduced from its Cook's distance and DFFITS value.

Cook's distance is calculated as

$$D_i = \frac{\sum_{j=1}^{n} (\hat{Y}_j - \hat{Y}_{j(i)})^2}{p \times \text{MSE}},$$

(9.45)

where $\hat{Y}_j - \hat{Y}_{j(i)}$ is the difference in fitted values when including the ith point, and when discarding the ith point. If a point has a large Cook's distance, this means that deleting the observation will substantially affect the regression. Such a point is said to be *influential*. Cooks distance can be related to lower tailed probabilities from the F-distribution by assuming $D_i \sim F(p, n - p)$. If the lower tailed F probability for the point is between 0.1 and 0.2, then the point has little influence on the regression. Conversely, if the probability is > 0.5, then the point has a substantial effect on the regression (Kutner et al. 2005).

```
cd <- round(cooks.distance(Fbird.lm),4)
```

Here are the *F* probabilities of these distances:

```
round(pf(cd, 2, 16),3)
    1      2      3      4      5      6      7      8      9     10     11     12
0.009  0.260  0.101  0.151  0.094  0.068  0.009  0.062  0.000  0.001  0.021  0.020
   13     14     15     16     17     18
0.011  0.000  0.014  0.140  0.008  0.163
```

The small probabilities indicate that none of the distances are particularly large.

Cooks distance considers the influence of the *i*th case on all fitted values. This is in contrast to measures that consider the influence of the *i*th case on the *i*th fitted value. A useful index for measuring the second type of influence is DFFITS. DFFITS is computed as

$$\text{DFFITS}_i = \hat{\varepsilon}_i \left[\frac{n-p-1}{\text{SSE}(1-h_{ii})-\varepsilon_i^2} \right]^{1/2} \left(\frac{h_{ii}}{1-h_{ii}} \right). \tag{9.16}$$

where h_{ii} indicates the *i*th diagonal element of the hat matrix.

As a guideline for identifying influential cases, Kutner et al. (2005) suggest that for small to medium datasets ($n < 30$), DFFITS absolute values >1 can be considered influential, while for large datasets, absolute values greater than $2\sqrt{p/n}$ can be considered influential.

```
round(dffits(Fbird.lm), 3)
     1       2       3       4       5       6       7       8       9      10
 0.130   0.876  -0.477  -0.615  -0.462   0.381  -0.129   0.366  -0.004   0.038
    11      12      13      14      15      16      17      18
 0.206  -0.198   0.148   0.017  -0.164  -0.563   0.121   0.598
```

No points in the frigate bird model are strongly influential using these criteria.

9.10.6 Multicollinearity

In addition to other regression assumptions and concerns, multiple regressions have the potential for *multicollinearity*, that is, correlation among predictors. Multicollinearity decreases the precision and reliability of regression estimators, by increasing sampling distribution variances. A model with perfectly correlated predictors is said to be *rank deficient*. In this case, an infinite number of coefficient estimates can be used to predict *Y* with equal effectiveness. Rank deficiency will also prevent parameter estimation using QR decomposition.

EXAMPLE 9.14

Consider the example below:

```
set.seed(1)
Y <- rnorm(10); X1 <- rnorm(10); X2 <- X1 * 3
lm(Y ~ X1 + X2)
Coefficients:
(Intercept)            X1            X2
     0.2006       -0.2749            NA
```

Because the model is rank deficient, and because X_1 was included in the model before X_2, X_2 is dropped. Rank deficiency can also occur with categorical variables that are linear combinations of each other in ANOVA (see Chapter 10).

Recall that for multiple regression models, a partial regression coefficient can be interpreted as a measurement of the expected change in the response given a one unit change in the predictor, holding all other predictors constant. As a practical matter, multicollinearity prevents this interpretation because it is no longer possible to hold one predictor constant while varying the other correlated predictor. Consider a model of crop yield as a function of the days with precipitation, X_1, and the number of days without precipitation, X_2. Clearly, it is impossible to vary X_1 while holding X_2 constant.

Potential consequences of serious multicollinearity include

1. Large changes in estimated regression coefficients when correlated variables are added or deleted.
2. Nonsignificant effects for predictors that are known to be important.
3. Irrational algebraic signs in the regression coefficients, that is, an association between predictor and response that is known to be positive is shown to be negative.

It should be noted that even a high degree of correlation between explanatory variables will affect neither the precision of fitted values nor the inferences about mean responses or the prediction of a new response (provided these are in the range of observed data). It may, however, dramatically affect estimates of regression coefficients, and test results based on those estimates. Techniques for addressing multicollinearity in the context of polynomial regression are described in Section 9.14.

Partial residual plots describe the marginal role an independent variable plays in explaining Y when adding it to a regression with other independent variables already in the model. Plots showing small residual effects indicate that either the predictor is (1) poor in explaining Y, or (2) is strongly correlated with other predictors, and as a result adds little additional information to the model.

Consider a dataset with two explanatory variables X_1 and X_2. We are interested in the effect of X_1 given that X_2 is already in the model. We first regress Y on X_2 and get the fitted values $\hat{Y}_i| X_2$, and residuals $\hat{\varepsilon}_i(Y| X_2)$

$$\hat{Y}_i| X_2 = \hat{\beta}_0 + \hat{\beta}_2 X_{2i}, \tag{9.47}$$

$$\hat{\varepsilon}_i(Y| X_2) = Y_i - \hat{Y}_i| X_{2i}, \tag{9.48}$$

where $\hat{\varepsilon}_i(Y| X_2)$ is the residual effect of X_1 with X_2 already in the model. We then regress X_1 on X_2 to obtain the fitted values $\hat{X}_{1i}| X_2$, and residuals $\hat{\varepsilon}_i(X_1| X_2)$:

$$\hat{X}_{1i}| X_2 = \hat{\beta}_0 + \hat{\beta}_2 X_{2i}, \tag{9.49}$$

$$\hat{\varepsilon}_i(X_1| X_2) = X_{1i} - \hat{X}_{1i}| X_{2i}. \tag{9.50}$$

In the partial residual plot, we would show $\hat{\varepsilon}_i(Y| X_2)$ versus $\hat{\varepsilon}_i(X_1| X_2)$.

EXAMPLE 9.15

Figure 9.9 shows partial residual plots for variables X_1, X_2, X_4, and X_5 in the plant richness analysis.

After accounting for other variables, X_1 and X_2 add little to the regression model (Figure 9.9a and b). Conversely, X_4 (soil rockiness) has a pronounced negative effect on richness after accounting for other variables (Figure 9.9c). Richness also decreases marginally with soil acidity (X_5) (Figure 9.9d).

```
lm1245 <-lm(Y ~ X1 + X2 + X4 + X5, data = wash.rich)
par(mfrow = c(2, 2), mar = c(4, 4.5, 2, 1))
partial.resid.plot(lm1245)
```

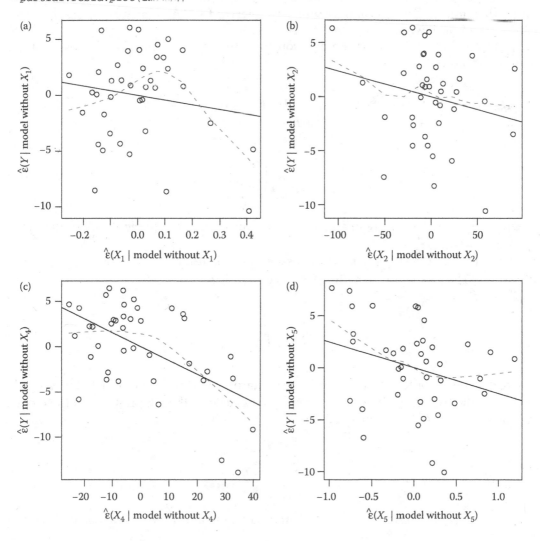

FIGURE 9.9

Partial residual plots for (a) X_1, (b) X_2, (c) X_4, and (d) X_5. The sold line in each plot is a linear regression; the dashed line is a lowess smoother (see Section 9.22).

Pairwise linear relationships among predictors can also be examined using matrix plots (scatterplots of all possible variable combinations). The code below customizes the *base* `pairs` function to provide a matrix plot with accompanying correlation tests.

```
x <- with(wash.rich, cbind(X1, X2, X3, X4, X5))
pairs(x, cex.labels = 1.3, cex = 1, gap = 0.1, lower.panel = panel.cor.res,
upper.panel = panel.lm)
```

Variance inflation factors (VIFs) constitute a widely accepted method for quantifying multicollinearity. VIFs measure the degree that variances of regression coefficients are inflated relative to when predictors are not linearly related. Specifically, the square root of the kth VIF is a scalar of proportionality, quantifying how much the confidence interval for β_k is expanded compared to a situation with orthogonal predictors (Fox 2002). Thus, a VIF larger than four indicates that the confidence interval for the predictor is more than two times as wide as it would be given orthogonality.

Let Y_i' and X_{ik}' (ik is the ith observation from the kth predictor) be the result of the standardizations:

$$Y_i' = \frac{1}{\sqrt{n-1}}\left(\frac{Y_i - \bar{Y}}{S_Y}\right), \tag{9.51}$$

$$X_{ik}' = \frac{1}{\sqrt{n-1}}\left(\frac{X_{ik} - \bar{X}}{S_k}\right), \tag{9.52}$$

where S_Y is the sample standard deviation of Y and S_X is the sample standard deviation of X.

Now, let $\hat{\beta}_k'$ be the kth estimated regression coefficient for a regression of Y' on the X' predictors. The population sampling variance of $\hat{\beta}_k'$ has the form

$$\sigma^2_{\hat{\beta}_k'} = \frac{(\sigma')^2}{1 - R_k^2}, \tag{9.53}$$

where σ' is the error term variance for the standardized regression (estimated with MSE') and R_k^2 is the coefficient of multiple determination *without* the kth predictor in the model. The denominator of Equation 9.53 is the kth VIF. That is

$$\text{VIF}_k = \frac{1}{1 - R_k^2}. \tag{9.54}$$

Note that as R_k^2 increases, $1/(1 - R_k^2)$ approaches 0, and the variance of $\hat{\beta}_k'$ becomes inflated.

The largest VIF among predictors is often used as an indication of the severity of multicollinearity. The value 5 is generally used as a cutoff for designating when multicollinearity is excessively influencing the efficiency and reliability of parameter estimation.

EXAMPLE 9.16

Here are the VIFs for the plant richness model:

```
library(car)
vif(wash.lm)
      X1        X2        X3        X4        X5
3.052300  1.640259  1.928907  3.571260  1.209886
```

While these VIFs do not appear to indicate excessive problems, values for X_1 and X_4, (soil N and soil rockiness) are relatively high because of their collinearity.

As a first response to multicollinearity, one or several predictors may be dropped from the model. There are two problems to this approach. First, no information will be acquired about the dropped predictors. Second, the magnitudes of the regression coefficients for the predictors remaining in the model remain affected by correlated predictors not included in the model.

Two other analytical approaches are frequently used to account for multicollinearity. The first is to use *eigenvectors* from a *principal component analysis (PCA)* as explanatory variables in multiple regression instead of raw data responses. By definition, these will be noncorrelated. The second is to use *ridge regression*. Under this (somewhat controversial) approach, a biasing constant is included in estimator algorithms. The resulting regression coefficients are biased for the true model parameters, but have reduced standard errors due to decreased predictor multicollinearity. An optimal value for the biasing constant will be close to 0 (equivalent to OLS regression) while reducing *VIFs* to acceptable levels. Techniques for estimating the biasing constant including the graphical *ridge trace* are discussed by Kutner et al. (2005).

9.11 Transformation in the Context of Linear Models

If a regression is in violation of model assumptions (Section 9.10), then the investigator has two options:

1. Abandon the general regression model (Equation 9.1) and use a more appropriate model (e.g., WLS, robust regression, nonlinear regression, polynomial regression, and GLM).
2. Transform the data so that the regression model is appropriate for the data.

There are two general transformation approaches in the context of linear models. First, when a nonlinear association is present, but error terms are normal and constant, then a transformation of the predictor(s) may be helpful for meeting model assumptions. Second, if there is nonconstant variance and/or nonnormality, with or without linearity, then the transformation of the response variable is likely to be helpful. Occasionally, transformation of both the predictors and the response may be necessary. For instance, if both explanatory and response variables are measured at exponential scales, then a common approach for biologists is to log transform both X and Y (see Exercise 8 at the end of the chapter). This is called a *power transformation* because it converts a power association of Y as a function

```
Fig.9.10()## See Ch. 9 code
```

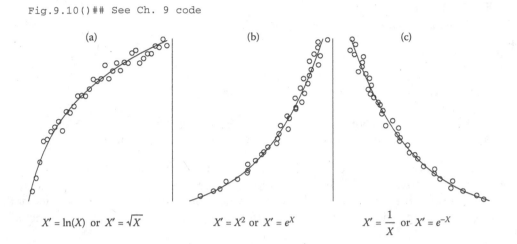

$$X' = \ln(X) \text{ or } X' = \sqrt{X} \qquad X' = X^2 \text{ or } X' = e^X \qquad X' = \frac{1}{X} \text{ or } X' = e^{-X}$$

FIGURE 9.10
Simple transformations to the X variable (X′) helpful in accounting for nonlinearity when variances are constant (plot follows Kutner et al. 2005). Three common nonlinear associations are shown: (a) logarithmic increase, (b) exponential increase, and (c) exponential decline.

of X, $Y = aX^b$, into a linear relationship, $\log(Y) = \log(a) + b\log(X)$. The general guidelines for transformations are shown in Figures 9.10 and 9.11.

Figure 9.10 depicts nonlinearity with constant variance, prompting predictor transformations. In Figure 9.10a, Y is increasing logarithmically with X, suggesting the usefulness of a log or square root transformation. In Figure 9.10b, Y is increasing exponentially with X, suggesting a square or exponential transformation. Finally, in Figure 9.10c, Y is decreasing exponentially with X, suggesting an inverse or inverse exponential transformation.

Figure 9.11 shows a nonlinear association with nonconstant variance, prompting transformation of the response. All three depicted scenarios may be helped by square root,

```
Fig.9.11()## See Ch. 9 code
```

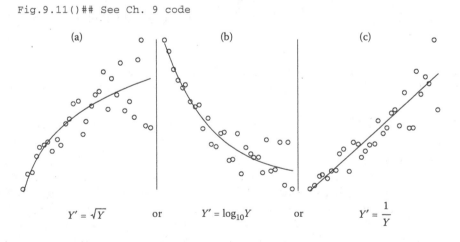

$$Y' = \sqrt{Y} \qquad \text{or} \qquad Y' = \log_{10}Y \qquad \text{or} \qquad Y' = \frac{1}{Y}$$

FIGURE 9.11
Simple transformations of the Y variable (Y′) helpful in simultaneously accounting for nonlinearity and nonconstant variance. Plots show (a) nonconstant variance with a logarithmic increase, (b) nonconstant variance with exponential decrease, and (c) nonconstant variance with linear increase.

logarithmic, and inverse transformations. Simultaneous transformation of X may also be helpful or necessary.

9.11.1 Optimal Transformation

Box and Tidwell (1962) and Box and Cox (1964) suggested methods for obtaining optimal predictor and response transformations, respectively. Both approaches maximize normal likelihood using iterative least squares procedures.

9.11.1.1 Box–Tidwell

The *Box–Tidwell* procedure obtains optimal power transformations of predictor variables for linearizing the relationship of the response and predictors under homoscedasticity. This is done by finding ML estimate for η_1 through η_{p-1} in

$$Y_i = \beta_0 + \beta_1 X_{1i}^{\eta_1} + \beta_2 X_{2i}^{\eta_2} + \cdots + \beta_{p-1} X_{p-1i}^{\eta_{p-1}} + \varepsilon_i, \tag{9.55}$$

where $\varepsilon_i \sim N(0, \sigma^2)$. The Box–Tidwell procedure constitutes a nonlinear model, and hypothetically can be fit using techniques described in Section 9.21. Box and Tidwell (1962), however, describe a more efficient computational approach that is utilized in the function `boxTidwell` from *car*.

9.11.1.2 Box–Cox

The *Box–Cox* procedure finds the optimal power transformation of the response variable for correcting nonlinearity and heteroscedasticity. This is done by finding the ML estimate for λ in

$$Y^\lambda = \beta_0 + \beta_1 X_{1i} + \beta_2 X_{2i} + \cdots + \beta_{p-1} X_{p-1i} + \varepsilon_i, \tag{9.56}$$

assuming $\varepsilon_i \sim N(0, \sigma^2)$. The function `box.cox` creates a profile log-likelihood plot (Box and Cox 1964, Venables and Ripley 2002, p. 170) with respect to the parameter λ showing the neighborhood of the ML estimate.

By the definition of the power transformation (Box and Cox 1964), $\lambda = 0$ and $\eta_k = 0$, denote $Y' = \ln(Y)$ and $X_k' = \ln(X_k)$, respectively. Otherwise, λ specifies the transformation $Y' = Y^\lambda$, while η_k designates $X_k' = X_k^{\eta_k}$.

EXAMPLE 9.17

Allelopathy is a phenomenon in which an organism produces chemicals that appears to influence the development, survival, and fecundity of other potentially competing organisms. Seal et al. (2004) examined the allelopathic potential of rice (*Oryza* sp.) with respect to an Australian rice invader called arrowroot (*Sagittaria montevidensis*). Data for the proportional growth of arrowhead in the presence of different densities of a rice variety called ET1444 were

```
arrow <- c(100, 54, 26, 12, 4, 1)
rice <- c(1, 2, 3, 4, 5, 6)
```

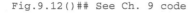

```
Fig.9.12()## See Ch. 9 code
```

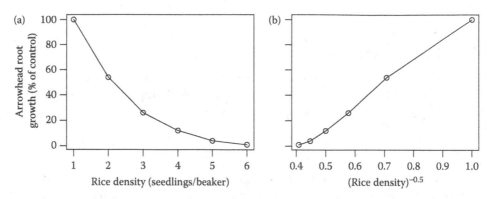

FIGURE 9.12

Rice (*Oryza* sp.) allelopathic dose–response curve for arrowhead (*Sagittaria montevidensis*) (a) before and (b) after inverse square root transformation of the predictor variable.

Note that while variability in Y does not appear to change as X changes, the association itself is negatively nonlinear (Figure 9.12a). For this sort of situation, an inverse transformation to the X variable may be helpful (Figure 9.10c). We can verify this using the Box–Tidwell procedure.

```
library(car)
boxTidwell(arrow ~ rice)
 Score Statistic p-value MLE of lambda
       25.87896        0      -0.4771498
```

The maximum likelihood estimate for η_1 is −0.477. This number is very close to −0.5, indicating that an inverse square root transformation would be helpful for improving linearity. The large magnitude of the score statistic and significant P-value for the null hypothesis of no increase in linearity indicates that the rice density variable *should* be transformed. The transformation dramatically improves the linearity of X and Y (Figure 9.12b).

In this scenario one should not forget that the real relationship between Y and X is nonlinear, although the relationship between Y and $X^{-0.5}$ is linear and is now appropriate for analysis with simple linear regression. The slope estimate for this model would now be interpreted as the change in the percentage of arrowroot growth, compared to the control, given a one unit increase in the inverse square root of rice density.

EXAMPLE 9.18

Polyamines are a class of organic compounds having two or more primary amino groups. These substances appear to have important functions in many organisms, including regulation of protein synthesis, cell proliferation, cell differentiation, and cell death. Polyamine plasma levels taken for healthy children of different ages were summarized by Hollander and Wolfe (1999). These data are contained in the **R** object polyamine.

A plot of the raw data indicates problems with both nonlinearity and nonconstant variance (Figure 9.13a). To use the Box–Cox procedure to obtain on an optimal transformation for Y, we create Figure 9.13b.

```
Fig.9.13() ## See Ch. 9 code
```

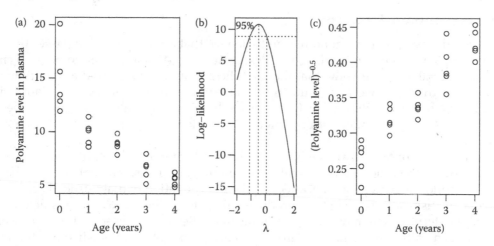

FIGURE 9.13
Transformation of the Y variable to account for nonconstant variance and nonlinearity. (a) Relationship between polyamine levels in plasma and the age of children before transformation. (b) Box–Cox results for optimal power transformation of the Y variable. (c) Plot of age versus (polyamine level)$^{-0.5}$.

```
data(polyamine)
with(polyamine, boxcox(p.amine ~ age))
```

The 95% confidence interval for λ includes both –1 and 0. However, the optimal value of λ is around –0.5, indicating that an inverse square root transformation would be helpful for improving linearity and nonconstant variance. The relationship between (polyamine concentration)$^{-0.5}$ and age is shown in Figure 9.13c.

The regression model using this transformation is

```
tran.lm<-lm(1/sqrt(p.amine) ~ age, data = polyamine)
summary(tran.lm)
Coefficients:
             Estimate Std. Error t value Pr(>|t|)
(Intercept) 0.268026    0.008033   33.36  < 2e-16 ***
Age         0.040062    0.003280   12.22 1.55e-11 ***
---
Signif. codes: 0 '***' 0.001 '**' 0.01 '*' 0.05 '.' 0.1 ' ' 1
```

The model units are now $Y' = \dfrac{1}{Y^{0.5}}$. To get back to the original units, we have

$$\frac{1}{Y^{0.5}} = Y'; \quad Y^{0.5} = \frac{1}{Y'}; \quad Y = \frac{1}{(Y')^2}.$$

Because $\hat{Y}' = 0.268 + 0.04X$, fitted values in the original units are obtained using

$$\hat{Y} = \frac{1}{(0.268 + 0.04X)^2}.$$

9.12 Fixing the Y-Intercept

Occasionally, we may have good reason to believe that a regression line passes through the origin. For instance, if we denote the exact time we plant a grass seed below 5 cm of soil at time = 0, then we know that the plant height above the ground at this time will be zero. In this case, we may want to fix the Y-intercept for a model for plant height as a function of time at zero. In **R**, we place the Y-intercept at the origin by specifying lm(Y ~ X - 1). A model with no intercept can also be specified as lm(Y ~ 0 + X) or lm(Y ~ X + 0).

Several words of caution are advisable for this kind of model. First, unlike other general linear models, the residuals will usually not sum to zero. Thus, residuals will not be balanced around the regression line, affecting the interpretation of diagnostic plots. Second, SSE may occasionally exceed *SSTO*, producing negative values of R^2. As a result, the coefficient of determination may have no clear meaning. Because of these and other concerns, one will generally want to fit an intercept to a model except in cases when $\beta_0 = 0$ is absolutely assured.

★ 9.13 Weighted Least Squares

Heteroscedasticity in linear models can be addressed with an approach called *WLS*. This allows the direct interpretation of regression coefficients in original units.

In WLS, an observation will generally be weighted by the inverse of its sample variance.

$$w_i = \frac{1}{S_i^2}. \tag{9.57}$$

A least squares model is then fit by minimizing Q in

$$Q = \sum_{i=1}^{n} w_i(y_i - (\hat{\beta}_0 + \hat{\beta}_1 X_{1i} + \hat{\beta}_2 X_{2i} + \cdots + \hat{\beta}_{p-1} X_{p-1i}))^2 = \sum_{i=1}^{n} w_i(Y_i - \hat{Y}_i)^2. \tag{9.58}$$

The WLS maximum likelihood estimator of linear model coefficients is

$$\hat{\beta} = (\mathbf{X'WX})^{-1}\mathbf{X'WY}, \tag{9.59}$$

where **W** is a *diagonal matrix* of weights, that is, w_is are on the diagonal of the matrix and all off-diagonal elements are zeroes. The variance–covariance matrix for WLS is found using

$$\hat{\sigma}_{\hat{\beta}}^2 = (\mathbf{X'WX})^{-1}. \tag{9.60}$$

A problem with this framework is that sample sizes are rarely large enough to effectively estimate σ_i^2 using s_i^2. One approach is to simply use the squared residuals from an OLS regression as estimates of $\sigma_i^2 s$ (Quinn and Keough 2002). Another approach is to regress $|\hat{\varepsilon}_i|$ (an estimator for σ_i) as a function of X_i, and let w_i be the inverse of the ith predicted value, squared (Kutner et al. 2005).

The method used to derive the variance function or standard deviation function should reflect the type of heteroscedasticity in the model (see Davidian and Carroll 1987). For instance, if a plot of residuals against the predictor is funnel shaped (the polyamine model has this pattern), then one should obtain the standard deviation function by regressing the absolute value of the residuals against the predictor (the second method discussed above).

Confidence intervals and prediction intervals for response expected values will be only approximate when using WLS, although this approximation will often be quite good when sample sizes are large (Kutner et al. 2005). Straightforward interpretation of R^2 will also be hindered because the estimation of variances, σ_i^2, introduces another source of variability to the model.

EXAMPLE 9.19

In **R**, one can call a vector of weights by using the `weights` argument in `lm`.
For the polyamine example, we have

```
ei <- resid(lm(p.amine ~ age, data = polyamine))
wi <- 1/fitted(lm(abs(ei) ~ polyamine$age))^2
wls.lm <- lm(p.amine ~ age, weights = wi, data = polyamine)
```

★ 9.14 Polynomial Regression

Nonlinear associations can often be handled effectively with additional predictor terms raised to a higher power, that is, *polynomial regression*. Polynomial regressions are general linear models (see Section 9.3). Thus, assumptions described in Section 9.10 also hold for these procedures.

Quadratic (*second-order*), cubic (*third-order*), and higher-order models can be constructed using one or several predictors. A second-order polynomial regression with a single predictor has the form

$$Y_i = \beta_0 + \beta_1 X_{1i} + \beta_2 X_{1i}^2 + \varepsilon_i. \tag{9.61}$$

Second-order polynomials are used to fit simple parabolic curves. As a result, they are often used for modeling differentiable associations with a single mode or global minimum. Third-order polynomials can be used to fit parabolic rates of curvature with the possibility of a single *inflection point*, (a point where the curve switches from concave to convex, or vice versa). Each additional higher order allows for another change in the rate of curvature, and an additional inflection point.

Polynomial regressions have two advantages over other approaches for modeling curvilinear association. First, lower-order terms allow direct interpretation of the effect of X

on Y, although this will not be as straightforward as for a nonpolynomial regression. For instance, in a quadratic model with a single predictor, positive values of β_2 indicate convex associations between Y and X, while negative values for β_2 indicate concave associations. The parameter β_1 gives the instantaneous slope when X_1 equals zero. This can be seen by taking the first derivative of Equation 9.61 with respect to X:

$$\frac{d}{dX}\beta_0 + \beta_1 X_1 + \beta_2 X_1^2 = \beta_1 + 2\beta_2 X_1.$$

Thus, when $X_1 = 0$, the rate of change in Y with respect to X equals β_1. Straightforward interpretation of higher-order terms in polynomial regression is generally extremely difficult. A second advantage is that polynomial regressions provide a single predictive *global model*, defined for all values of X. This will not be true for smoother approaches discussed in Section 9.22, which are based on local models.

Polynomial regressions have three disadvantages. First, because a polynomial model of sufficiently high order can always be created that will fit data—with no repeated observations—perfectly, polynomial regressions may prompt overfitting (see Section 9.17). An overfit model will generally be erratic, and useless for interpolation. Second, the inclusion of unnecessary higher-order terms will decrease power, because adding terms will decrease the degrees of freedom error. Third, polynomial terms will be highly correlated, affecting efficiency in estimation and inferences. This will be particularly true for higher-order models ($\geq$ third order) (Kutner et al. 2005). One solution to this final problem is to center predictor variables used in polynomial terms with

$$X_i' = X_i - \bar{X}. \tag{9.62}$$

This transformation will generally reduce the correlation of even- and odd-powered terms, but will not affect the correlation *among* even-powered terms (Sokal and Rohlf 2012). Other approaches for handling polynomial multicollinearity, including orthogonal polynomial functions, are discussed in Kutner et al. (2005) and Sokal and Rohlf (2012).

EXAMPLE 9.20

The dataset `SM.temp.moist` contains 30 growing-season soil temperatures taken from a depth of 5 cm at an alpine late snowmelt site on Mount Washburn in Yellowstone National Park. Seasonal temperatures from temperate latitudes have a unimodal association with time (Walter and Leith 1967). Thus, we will fit soil temperature as a quadratic function of day of the year (whereby Jan 1 = day 1 and Dec 31 = Day 365, on non-leap years).

```
data(SM.temp.moist)
```

We first center the predictor

```
cent.day <- with(SM.temp.moist, (day - mean(day)))
```

Here is the model:

```
poly.lm <- with(SM.temp.moist, lm(Temp_C ~ cent.day + I(cent.day^2)))
summary(poly.lm)
```

```
Coefficients:
                Estimate Std. Error t value Pr(>|t|)
(Intercept)     8.4372493  0.5271550  16.005 2.65e-15 ***
cent.day        0.0176484  0.0155425   1.135    0.266
I(cent.day^2)  -0.0022785  0.0004336  -5.255 1.54e-05 ***
```

Note that for the quadratic term, I specify I(cent.day^2). Without the function I, the carat (^) metacharacter is assumed by lm to represent higher-order interactions among predictors (see Chapter 10).

Here is code for a plot of the regression (Figure 9.14):

```
xv <- seq(-70, 70,.1)
new = data.frame(cent.day = xv)
yv <- predict(poly.lm, newdata = new)
plot(cent.day, SM.temp.moist$Temp_C, xlab = "Centered day of year",
ylab = "Soil temperature (\u00B0C)")
lines(xv, yv)
```

The fitted responses do not change as a result of centering time (recall that precision in estimates of $E(Y)$ is not affected by collinearity). However, the *P*-value for day of the year is much higher for the centered model (0.266 vs. 0.0003) because the standard error for the centered coefficient is higher. More importantly, collinearity is dramatically reduced by centering.

```
poly.nc <- with(SM.temp.moist, lm(Temp_C ~ day + I(day^2)))
vif(poly.nc)
      julian.day I(julian.day^2)
        304.2975         304.2975
vif(poly.lm)
      cent.day I(cent.day^2)
      1.908109      1.908109
```

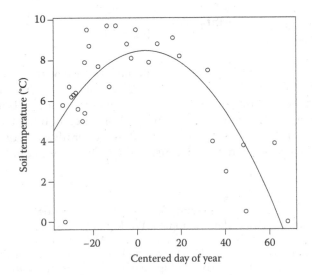

FIGURE 9.14
Second-order polynomial regression for Mount Washburn summer soil temperature data.

★   **9.15 Comparing Model Slopes**

Biologists are often interested in comparing parameter estimates from different regression models. For instance, we might be interested in comparing the growth rates (slopes) or the elevations (Y-intercepts) of two varieties of domestic wheat over a range of soil water levels.

A test for the equality of regression slopes from two models, A and B, will have the following two-tailed hypotheses:

$$H_0: \beta_{1(A)} = \beta_{1(B)},$$

$$H_A: \beta_{1(A)} \neq \beta_{1(B)}.$$

where $\beta_{1(A)}$ is the slope parameter for model A, and $\beta_{1(B)}$ is the slope parameter for model B.

Evidence concerning the equality of slopes can be quantified using a t-test. Assuming $\sigma^2_{\hat{\beta}_{1(A)}} = \sigma^2_{\hat{\beta}_{1(B)}}$, the test statistic is calculated as

$$t^* = \frac{\hat{\beta}_{1(A)} - \hat{\beta}_{1(B)}}{\hat{\sigma}_{\hat{\beta}_{1(A)} - \hat{\beta}_{1(B)}}}, \tag{9.63}$$

where

$$\hat{\sigma}_{\hat{\beta}_{1(A)} - \hat{\beta}_{1(B)}} = \sqrt{\frac{MSE_{pooled}}{\sum_{i=1}^{n}(X_{1i} - \bar{X}_1)^2} + \frac{MSE_{pooled}}{\sum_{i=1}^{n}(X_{2i} - \bar{X}_2)^2}},$$

and

$$MSE_{pooled} = \frac{SSE_A + SSE_B}{dfE_A + dfE_B},$$

where SSE_A and dfE_A are the residual sum of squares and degrees of freedom for model A, respectively, and SSE_B and dfE_B are the residual sum of squares and degrees of freedom for model B, respectively. If linear model assumptions are valid for both regressions, the sampling distribution variances are equal, and if H_0 is true, then the two-tailed P-value is calculated as $2P(T \geq t^*)$, where $T \sim t(n_A + n_B - 4)$. More generally, the equality of slopes and Y-intercepts for ≥ 2 categories can be tested using ANCOVA (see Section 10.13).

> **EXAMPLE 9.21**
>
> Wright et al. (2000) examined the foraging behavior of red wood ants (*Formica rufus*), a species that harvests honeydew in aphids. Worker ants traveled from their nests to nearby trees to forage honeydew from homopterans. Ants descending trees were laden with food and, for particular size classes, weighed more than unladen, ascending ants. The authors were interested in comparing regression parameters, for ant mass as a function of head width, for the ascending and descending ants. Relevant data are in the **R** object ant.dew. The problem is depicted in Figure 9.15.

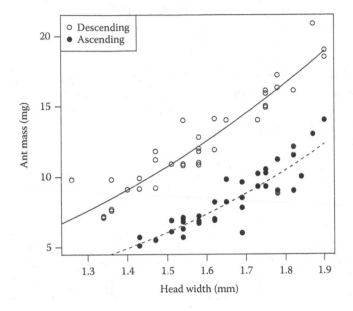

FIGURE 9.15

Comparison of regression slopes and intercepts for models of honeydew harvesting. *Formica rufus* mass as a function of ant head width, for both ascending (honeydew-unladen) and descending (honeydew-laden) ants.

After log transforming both the predictor and response variables to account for non-constant variance and nonlinearity, the authors found evidence for differences between the slopes for descending and ascending ants, $t_{68} = 2.08$, $P = 0.0415$, and a great deal of evidence for differences in the regression Y-intercepts, $t_{68} = 19.6$, $P < 1 \times 10^{-16}$. These results can be obtained using an ANCOVA approach by specifying

```
library(car)
Anova(lm(log(ant.mass) ~ log(head.width) + direction + log(head.width)
:direction, data = ant.dew)) # type II SS
```

In the model above, I specify the predictors `log(head.width)`, `direction`, and the interaction `log(head.width):direction`. Tests that the associated parameters for these predictors equal zero provide evidence concerning the effectiveness of the linear fit, the equality of Y-intercepts, and equality of the slopes, respectively.

★ ## 9.16 Likelihood and General Linear Models

From Chapter 4, recall that maximum likelihood is the product of densities for an assumed probability distribution with ML estimates given for pdf parameters. Thus, the likelihood of a general linear model is simply the product of the densities of the residuals assuming $\varepsilon_i \sim N(0, \sigma^2)$, and using the ML estimator for σ^2. Recall that this estimator is equivalent to the second sample moment (Chapter 4):

$$\hat{\sigma}^2_{ML} = \frac{\sum\limits_{i=1}^{n}(X_i - \bar{X})^2}{n}. \tag{4.37}$$

Because S^2 is unbiased for all Var(X) (Chapter 4), and because $S^2 \neq \hat{\sigma}^2_{ML}$, $\hat{\sigma}^2_{ML}$ must be biased for σ^2. Indeed, since the denominator of $\hat{\sigma}^2_{ML}$ is always larger than the denominator of S^2 (the numerators are identical, and are equivalent to SSE), $\hat{\sigma}^2_{ML}$ underestimates σ^2, although the bias will be minimal for large values of n.

In **R**, we have

```
sigma.hat.ML <-function(x){# x = residuals from linear model
SSE <- sum((x-mean(x))^2)
SSE/length(x)
}
```

Better-fitting models will produce smaller residuals, closer to $E(X)$ in the assumed error distrubution, and as a result they will have larger average densities, and higher likelihoods. Thus, likelihood can be used as a measure of model efficacy assuming *any* distribution of error terms (including nonnormal distributions). This is an important consideration for generalized linear models (Section 9.20) and generalized additive models (GAMs) (Section 9.22.1).

EXAMPLE 9.22

Consider the Washburn plant richness model

```
wash.lm <- lm(Y ~ X1 + X2 + X3 + X4 + X5, data = wash.rich)
```

Here is the maximum likelihood estimate for σ^2:

```
s.ML <- print(sqrt(sigma.hat.ML(resid(wash.lm))))
[1] 4.007419
```

The likelihood is

```
lik.wash <- print(prod(dnorm(resid(wash.lm), mean = 0, sd = s.ML)))
[1] 1.721555e-49
```

And the log-likelihood is

```
logLik.wash <- print(log(lik.wash))
[1] -112.2834
```

Log-likelihoods can be calculated directly using the function `logLik`.

```
logLik(wash.lm)
'log Lik.' -112.2834 (df = 7)
```

Biologists using likelihood for model evaluation should be aware of two issues. First, likelihood will only be useful for comparing models with the same response variable. This is true because (1) likelihood will largely be a function of the number of residuals, and (2) the reference criteria for models with respect to residual density will be completely altered given a different response variable. Second, like R^2, likelihood will almost always increase with the addition of predictors because SSE (and other measures of model deviance) will almost always decrease. Thus, likelihood alone is a poor criterion for model comparisons because, holding other factors constant, the model with the most parameters will always have the highest likelihood. A number of likelihood-based evaluators have been developed that penalize excessively complex models. These are addressed next.

★ ## 9.17 Model Selection

A fundamental concern with many analyses is the selection of predictors that provide useful models. One approach is to identify models that are *parsimonious*, that is, simple but effective. Such an approach utilizes *Occam's razor*[*] to "shave" away predictors from models whose complexity adds little to their explanatory power (Box and Jenkins 1970, Bozdogan 1987). A parsimonious model should (Burnham and Anderson 2002, Crawley 2007)

1. Be based on (be subset from) a set of parameters identified by the investigator as biologically important, including, if necessary, covariates, interactions, and higher-order terms.
2. Have as few parameters as possible (be as simple as possible, but no simpler).

Aside from being philosophically objectionable and difficult to interpret, an overly complex model may simply provide a poor representation of reality due to *overfitting*. For instance, assume that the "true" relationship between Y and X is $Y_i = e^{(X_i - 0.5)} - 1 + \varepsilon_i$, where $\varepsilon_i \sim N(0, 0.01)$. We can randomly create 10 datasets applying these constraints, and fit them with two different models: a simple linear regression, $E(Y_i) = \beta_0 + \beta_1 X_i$, and a fifth-order polynomial $E(Y_i) = \beta_1 X_{1i} + E(Y_i) = \beta_0 + \beta_2 X_{1i}^2 + \beta_3 X_{1i}^3 + \beta_4 X_{1i}^4 + \beta_5 X_{1i}^5$. The two-parameter model *underfits* the data and misses the nonlinear association of Y and X (Figure 9.16a). However, the six-parameter model is overfit and introduces erratic variability (Figure 9.16b). Thus, both overfitting and underfitting negatively affect valid inferences.

Some texts encourage the use of R_{adj}^2 as a parsimony criterion. This is because, while R^2 will never decrease with the addition of variables, R_{adj}^2 *can* decrease in this framework. Specifically, R_{adj}^2 will only be higher in a model B compared to a simpler model A if $\text{MSE}_A/\text{SSTO}_A > \text{MSE}_B/\text{SSTO}_B$. Nonetheless, R_{adj}^2 is generally a poor tool for model selection (Burnham and Anderson 2002). In simulations comparing 18 model selection algorithms, R_{adj}^2 consistently had the lowest efficiency for situations in which the "true" model belonged to a set of candidate models, and for situations when it did not (McQuarrie and Tsai 1998). Even worse are sources that encourage the use of P-values as a model selection criterion (i.e., choose the simplest model in which $P < \alpha$). This is problematic since inclusion of more parameters will generally lead to a model with better

---

[*] William of Occam was a 14th-century British philosopher who argued that, given a set of equally good explanations for a problem, the simplest explanation is the correct one. To quote Occam: "It is vain to do with more what can be done with fewer."

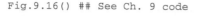

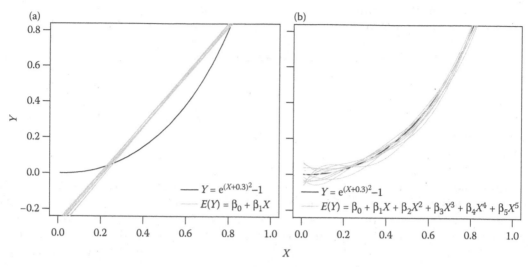

FIGURE 9.16

Ten randomly created datasets, each with $n = 100$, for $Y_i = e^{(X_i - 0.5)} - 1 + \varepsilon_i$, where $\varepsilon_i \sim N(0, 0.01)$ fit to two different linear models, (a) a two-parameter model and (b) a six-parameter model. The black thickened line indicates the "true" relationship between Y and X, while the gray lines are fitted lines. (Adapted from Sakamoto, Y., Ishiguro, M., and Kitagawa, G. 1986. Aka*ike Information Criterion Statistics*. KTK Scientific Publishers, Tokyo, Japan.)

fit and more power, making the choice of α difficult. In addition, there is no underlying theory to suggest that such an approach will lead to inferentially valid models (Burnham and Anderson 2002).

9.17.1 Model Selection Approaches

9.17.1.1 AIC

A large number of methods for identifying parsimonious models utilize a likelihood-based approach (e.g., Akaike 1973). Log-likelihood is strongly tied to the concepts of *information* from information theory and *entropy** from physics (Rényi 1961, Edwards 1972). Kullback and Leibler (1951) related information to the log-likelihood ratio of two hypotheses or models (also see Kullback 1959, p. 2). This term is now called *Kullback–Leibler information* or *KL-information*.[†] Given discrete probability models, KL-information is

$$I(f, g) = \sum_x f(x) \ln \left[\frac{f(x)}{g(x)} \right].$$

(9.64)

Because the log of a quotient is the difference of logs, Equation 9.64 can be separated into two summations. The first is equivalent to the biological information-based

---

* It can be shown that minimizing KL-information is equivalent to maximizing Boltzman entropy from physics (Burnham and Anderson 2002).

† Fisher (1925) provided a different definition for information.

criterion, Shannon Weiner diversity (Equation 5.31). For continuous probability models, KL-information is

$$I(f,g) = \int_{-\infty}^{\infty} f(x)\ln\left[\frac{f(x)}{g(x)}\right]dx. \tag{9.65}$$

For the purposes of model selection, the numerator of the KL-information, $f(x)$, defines the full reality or truth, while the denominator $g(x)$ represents an approximating model with known parameters. The term $I(f, g)$ represents the information lost when a candidate model is used to represent truth. Because both the true model and the parameter values for the approximating model will generally be unknown, an estimator is required for KL-information. Akaike (1973) proposed a biased estimator based on a maximized log-likelihood function. An adjusted estimator that accounts for this bias in the form of a *penalty term* is called the *Akaike information criterion* or AIC:

$$\text{AIC} = -2\ln \mathcal{L}(\hat{\theta}\,|\,data) + 2q, \tag{9.66}$$

where $\mathcal{L}(\hat{\theta}\,|\,data)$ is the likelihood of the estimated model, and q is the total number of parameters that are estimated in the model, including σ. Thus, for the purposes of general linear models, $q = p + 1$. The quantity $2q$ is the penalty term. The use of two as a multiplier is not arbitrary. It formally ties AIC to KL-information (Burnham and Anderson 2002). If model complexity does not increase likelihood appreciably, then a complex model will have a larger AIC than a simpler model. Thus, lower values of AIC indicate greater parsimony.

Model comparisons using AIC are often summarized by subtracting the AIC score of the optimal model from the scores of the other candidate models. This process results in a vector of differences with the ith difference given by

$$\Delta_i = \text{AIC}_i - \text{AIC}_{\min}, \tag{9.67}$$

where $\text{AIC}_{\min}$ is the AIC score from the best candidate model.

From the perspective of *multimodel inference*, the minimum AIC model may not be the "true" best model from a set of candidate models. Table 9.2 lists Δ_i cutoffs defining the support for the ith model.

TABLE 9.2

Support for the ith Model from a Set of Candidate Models

Δ_i	Empirical Support for the ith Model
0–2	Substantial
4–7	Little
>10	Essentially none

Source: Adapted from Burnham, K. P., and Anderson, D. R. 2002. *Model Selection and Multimodel Inference, A Practical Information-Theoretic Approach*, 2nd edition. Springer, New York.

Δ_i can also be used in more quantitative summaries concerning multimodel inference. For instance, the ith *Akaike weight* is defined as

$$w_i = \frac{\exp\left[-\frac{1}{2}\Delta_i\right]}{\sum_{i=1}^{m}\exp\left[-\frac{1}{2}\Delta_i\right]}, \tag{9.68}$$

where m is the number of candidate models. The statistic can be interpreted as the proportion of times under repeated sampling that the ith model would be the best (i.e., low AIC) candidate model (Burnham and Anderson 2002).

9.17.1.2 AICc

A number of authors have noted that AIC tends to overfit models when sample sizes are small (e.g., Hurvich and Tsai 1980). The *second-order "corrected" Akaike information criterion*, AIC_c, is robust to this problem and converges to AIC when n is large.

$$AIC_c = AIC + \frac{2q(q+1)}{n-q-1}, \tag{9.69}$$

where n is the number of observations.

AIC_c outperforms other parsimony criteria in a wide variety of model selection situations (McQuarrie and Tsai 1998). Burnham and Anderson (2002) recommend using AIC_c whenever $n/p < 40$.

9.17.1.3 BIC

Another frequently used information-theoretic measure is the *Bayesian information criterion (BIC)*, also called the Schwarz or SIC criterion:

$$BIC = -2\ln\mathcal{L}(\hat{\theta}\,|\,data) + q\ln(n). \tag{9.70}$$

BIC is an asymptotic approximation to the natural log of the Bayes factor (Equations 6.25 and 6.26; Schwarz 1978, Robert 2007). As with AIC, lower values of BIC indicate better models.

9.17.1.4 Mallows' C_p

A noninformation-theoretic criterion known as *Mallows' Cp* provides the same rankings of candidate models as AIC given identical normal errors (Burnham and Anderson 2002). It has the form

$$C_p = \frac{SSE_k}{MSE_R} - (n-2p), \tag{9.71}$$

where SSE_k is the sum of squares error for the fitted subset regression model with k parameters (out of p potential parameters), and MSE_R is the mean squared error for the saturated reference model with all p predictors. From the perspective of Mallows' C_p, better models will have a small C_p value, and/or a value of C_p near p.

9.17.1.5 PRESS

A final sum of squares error criterion for model evaluation/selection is the *prediction sum of squares* or PRESS:

$$\text{PRESS} = \sum_{i=1}^{n} \left(Y_i - \hat{Y}_{i(-i)} \right)^2, \tag{9.72}$$

where $\hat{Y}_{i(-i)}$ is the fitted value for the ith case when the ith case is left out of the linear model. PRESS values can be calculated without n separate regressions, with a single case deleted in each run. This is because

$$Y_i - \hat{Y}_{i(-i)} = \frac{\hat{\varepsilon}_i}{1 - h_{ii}},$$

where h_{ii} is the ith diagonal element from the hat matrix. Smaller PRESS values result from smaller prediction errors, and indicate better models.

9.17.1.6 Comparison of Selection Criteria

With respect to the criteria described above, BIC is consistent for the true model, while AIC, AIC_c, and C_p are asymptotically efficient (see below). AIC, AIC_c, BIC, PRESS, and R^2_{adj} allow the comparison of models with nonnested predictors. However, the same response variable *is* required. That is, models using different transformations of Y are not comparable. Nonnested comparisons are not possible when using C_p, or an extra sum of squares approach (Section 9.5.3).

Several authors have found that sums of squares model selection approaches, for example, PRESS, C_p, and R^2_{adj}, tend to select overfit models compared to information-theoretic criteria AIC, AIC_c, and BIC (Hurvich and Tsai 1989). AIC may also overfit data for small datasets (Hurvich and Tsai 1989).

BIC and AIC are often used interchangeably by researchers; however, they represent divergent views with respect to processes underlying models (Aho et al. 2014). Some authors favor BIC because it is consistent (Kass and Raftery 1995). That is, in simulations that include the true model in a set of candidate models, BIC tends to pick the correct model, while AIC picks a more complex model. This reflects a "worldview" in which hypotheses/models are being compared, and one of these is correct. Conversely, if one views the true model as unknown because it has essentially infinite dimensions (a realistic view for many biological phenomena), then AIC provides an asymptotically efficient selection of a finite dimension approximating model (Shibata 1976, Hurvich and Tsai 1989).

It should be emphasized that information-theoretic criteria will only measure the relative optimality of models. Thus, for a set of bad models, the algorithms will merely select the most parsimonious example from a poor batch. Other statistical measures are more useful in gauging model stand-alone efficacy (e.g., predictive usefulness), including R^2 and R^2_{adj}.

EXAMPLE 9.23

Given likelihood, information-theoretic summaries can be easily made. Consider a plant richness model using only the first four predictors:

```
wash.lm4 <- lm(Y ~ X1 + X2 + X3 + X4, data = wash.rich)
n <- 40; p <- 5
AIC <- print(-2 * logLik(wash.lm4) + (p + 1) * 2)
[1] 240.1561
```

We can calculate AIC and BIC directly in **R** using the function AIC.

```
AIC(wash.lm4)
[1] 240.1561
AIC(wash.lm4, k = log(n)) #BIC
[1] 250.2894
```

EXAMPLE 9.24

Consider the following candidate models:

```
lms <- list(
model.1 = lm(Y ~ X1 + X2 + X3 + X4 + X5 + X2:X3, data = wash.rich),
model.2 = lm(Y ~ X2 + X3 + X4 + X5, data = wash.rich),
model.3 = lm(Y ~ X2 + X3 + X4 + X5 + X2:X3, data = wash.rich),
model.4 = lm(Y ~ X1 + X4 + X5, data = wash.rich),
model.5 = lm(Y ~ X4 + X5, data = wash.rich),
model.6 = lm(Y ~ X4, data = wash.rich)
)
```

Model 1 includes predictors X_1 through X_4, along with the interaction of slope and aspect (X_2 and X_3) as a surrogate for incident radiation. To allow the calculation of C_p, all other models will be subsets from this model. Model 2 is the original model from Example 9.7. Model 3 eliminates soil N, which is strongly correlated with soil rockiness. Model 4 emphasizes surface and subsurface soil characteristics: rockiness (X_4), nitrogen (X_1), and pH (X_5). Model 5 examines the effect of subsurface soil characteristics nitrogen (X_1) and pH (X_5). Finally, model 6 considers plant richness as a function of only the strongest predictor, soil rockiness (X_4).

The function lm.select from *asbio* provides comparative summaries for a list of candidate models.[*]

```
lm.select(lms)
                                      Model   AIC  AICc   BIC   Cp PRESS
1 Y ~ X1 + X2 + X3 + X4 + X5 + X2:X3 222.2 226.9 235.7  9.0 557.4
2                 Y ~ X2 + X3 + X4 + X5 237.2 239.8 247.4 25.1 898.9
3       Y ~ X2 + X3 + X4 + X5 + X2:X3 221.3 224.8 233.2  7.9 540.9
4                 Y ~ X1 + X4 + X5 236.5 238.3 244.9 24.8 871.7
5                 Y ~ X4 + X5 235.4 236.5 242.1 24.0 826.0
6                 Y ~ X4 237.4 238.0 242.4 27.9 847.4
```

In this case, all five criteria favor model 3. Note, however, that the introduction of the interaction term results in multicollinearity [try vif(model.3)]. As a result, it is

---

[*] If C_p is to be used as a selection criterion, then the saturated model should be placed first in this list.

reasonable to favor model 5, which has the smallest AIC_c of all noninteraction models. Below is a ΔAIC comparison of the noninteraction models.

```
lm.select(lms.new, deltaAIC = TRUE)
                   Model deltaAIC Rel.likelihood Akaike.weight
1 Y ~ X2 + X3 + X4 + X5 1.830410      0.4004345     0.1704669
2      Y ~ X1 + X4 + X5 1.094709      0.5784783     0.2462610
3           Y ~ X4 + X5 0.000000      1.0000000     0.4257048
4                Y ~ X4 1.987787      0.3701327     0.1575673
```

Under repeated random sampling, we would expect that, given this set of candidate models, the model with predictors X_4 and X_5 would be optimal 43% of the time.

9.17.2 Stepwise and All Possible Subset Procedures

Algorithms exist to automate the process of model selection. In *stepwise regression*, variables are added or subtracted one at a time, and the best model, given a particular number of variables, is used as the starting point in the next stepwise evaluation. Another conventional method of variable selection called *all possible subsets* compares models from an exhaustive list of all possible sets of predictors for a given number of predictors. It then applies this approach for all models sizes, from 1 to $p - 1$ predictors. Stepwise regression is inadequate for comparing the efficacy of all possible models because the algorithm builds only on the optimal model from previous step, preventing the evaluation of all possible models. Nonetheless, because of computational run-time issues, these procedures have traditionally been recommended over all possible subsets for handling large numbers of predictors (Kutner et al. 2005). With the introduction of increasingly powerful computers, this advantage has become less important.

EXAMPLE 9.25

The simplest way to run a stepwise regression in **R** is to create a model containing all the variables of interest and then use this model as the first argument in step or stepAIC (from *MASS*). Both functions use AIC as the optimality criterion.

```
step(lm(Y ~ X1 + X2 + X3 + X4 + X5, data = wash.rich), trace = FALSE)
Call:
lm(formula = Y ~ X4 + X5, data = wash.rich)
```

This approach favors the two-parameter (rockiness and pH) model.

An all possible subset approach can be implemented using the function regsubsets from library leaps. The function uses BIC and Mallows' C_p as optimality criteria.

```
install.packages("leaps");library(leaps)
subs <- regsubsets(Y ~ X1 + X2 + X3 + X4 + X5, data = wash.rich, nbest = 2)
summary(subs)
2 subsets of each size up to 5
Selection Algorithm: exhaustive
         X1    X2    X3    X4    X5
1  (1)  " "   " "   " "   " "   "*"   " "   " "
1  (2)  " "   " "   " "   "*"   " "   " "   " "
2  (1)  " "   " "   " "   " "   "*"   "*"
2  (2)  " "   " "   " "   "*"   "*"   " "   " "
3  (1)  " "   " "   " "   "*"   "*"   "*"
3  (2)  "*"   " "   " "   " "   "*"   "*"
4  (1)  "*"   " "   " "   "*"   "*"   "*"
4  (2)  " "   " "   "*"   "*"   "*"   "*"
5  (1)  "*"   "*"   "*"   "*"   "*"
```

The argument nbest = 2 indicates that the function should identify the best two models with one, two, three, four, and five parameters. The best one-parameter model uses x4 (soil rockiness) only, while the second best one-parameter model uses X3 (slope). The best two-parameter model is in agreement with the best stepwise model, Y ~ X4 + X5.

9.17.2.1 Data Dredging

Automated model selection should not be used as a substitute for critical thinking. *Data dredging* refers to the careless use of algorithms to explore and compare a large number of models (Burnham and Anderson 2002). This approach is problematic for four reasons. First, these processes will not consider polynomial terms and interactions unless they are established in the *maximal model* (a model that includes all predictors of interest). Second, comparisons of nonnested sets of predictors are not possible. Third, the theory associated with hypothesis testing and parameter estimation for a data dredged model is poorly developed (Goutis and Casella 1995, Ye 1998). Fourth, and most importantly, data dredging can reflect a dangerous lack of consideration of important biological mechanisms. As a result, the use of stepwise and all possible subset approaches potentially *decreases* the usefulness of parsimony criteria for model selection. Building on unexpected negative results has often led to profound scientific insights, for instance, the development of a polio vaccine, and the discovery of missing planets (Section 1.3.1). Such examples suggest that hypotheses (and models) should be inductively constructed, and that successful research can often take on a meandering exploratory form (Kuhn 1963, Austin 1978). Nevertheless, if final models are the product of data dredging, then experts recommend that this should be reported in summaries and publications of results (e.g., Burnham and Anderson 2002).

★ ## 9.18 Robust Regression

In Section 9.10, we noted that OLS estimators will be strongly affected by outliers. In Section 9.11, we discussed transformation methods for dealing with nonlinearity, nonindependence, nonnormality, and heteroscedasticity. It is unlikely however that any of these methods will allow useful analyses in the presence of strong outliers. In lieu of deleting outliers, which may have biological importance (Chapter 7), one can use parameter estimators and inferential procedures that are resistant to outliers.

9.18.1 Bootstrapping Methods

Bootstrapping can be used to create an empirical distribution of $\hat{\beta}_1$s, allowing derivation of confidence intervals (Chapter 6). This can be done by randomly rearranging Y, X pairs with replacement in the sample (Gotelli and Ellison 2004). For instance, for the frigate bird analysis, we could use the following code:

```
library(boot)
reg.boot <-function(data,i){
cor(data$vol[i],data$freq[i])*(sd(data$freq[i])/sd(data$vol[i]))
}
boot(Fbird, reg.boot, R=1000, stype="i")
```

Venables and Ripley (2002, p. 164) caution that this method may not effectively mimic random variation and may result in singular fits with high probability. Methods discussed below in Section 9.18.2, and applied in Example 9.26, provide a more theoretically sound approach to outliers.

9.18.2 Robust Estimators

M estimators (introduced in Section 4.3.6) minimize the importance of outliers with weighting functions. Huber (2007) estimates of β_0 and β_1 minimize the function:

$$\sum_{i=1}^{n} \rho\left(\frac{Y_i - X_i \hat{\beta}}{s}\right) = n \ln s, \tag{9.73}$$

where ρ, is the negative log of a scaled pdf describing the distribution of model errors, n is the number of observations, and s is a scale estimator. Assuming we know s, the ML estimate $\hat{\beta}$ solves:

$$\sum_{i=1}^{n} X_i \psi\left(\frac{Y_i - X_i \hat{\beta}}{s}\right) = 0, \tag{9.74}$$

where ψ is the first derivative of ρ from Equation 9.73. The term s is typically MAD, although see Huber (2007) for other options. Equation 9.74 is solved by a process of iteratively reweighted least squares, with weights

$$w_i = \psi\left(\frac{Y_i - X_i \hat{\beta}}{s}\right) \bigg/ \left(\frac{Y_i - X_i \hat{\beta}}{s}\right). \tag{9.75}$$

The function `rlm` in *MASS* uses this overall approach.[*]

M estimator functions down-weight, but do not completely eliminate outliers (Chapter 4). As a consequence, they remain efficient (given normality), but lack resistance for extremely contaminated data. Because of this, several regression methods highly resistant to outliers have been proposed. Among these, a procedure called *S estimation* has the highest efficiency, although it is still low, 28.7%, given normality (Rousseeuw and Yohai 1984, Venables and Ripley 2002). *S* estimation can be specified using function `lqs` in MASS. A procedure called *MM estimation* combines the resistance of an *S* estimator with the efficiency of Huber's *M* estimator (Yohai et al. 1991), and can be specified in `rlm` using `method = "MM"`.

---

[*] While Huber's estimator (`psi = psi.Huber`) will result in a single set of estimates for parameters, redescending weighting functions such as Hampel's and Tukey's bisquare methods will have multiple roots (local minima). Thus a user should define valid initial estimates for regression coefficients. Default starting points in `rlm` are least squares estimates (`init = "ls"`); however, for situations with highly contaminated data, Venables and Ripley (2002) recommend least trimmed squares estimates, that is, `init = "lts"`.

EXAMPLE 9.26

To demonstrate robust regression, we use the frigate bird data from Example 9.2 with the addition of two artificial outliers.

```
Fbird1 <- rbind(Fbird, matrix(nrow = 2, ncol = 2,data = c(7700, 8200, 560,
555), dimnames = list(c(19, 20), c("vol", "freq"))))
```

Here is the OLS regression model:

```
Fbird1.lm <- lm(freq ~ vol, data = Fbird1)
```

Here are the coefficients using *M* estimation:

```
M <- coef(rlm(freq ~ vol, data = Fbird1))
```

Here are the coefficients using *S* estimation.

```
S <- coef(lqs(freq ~ vol, data = Fbird1, method = "S"))
```

Finally, here are the coefficients using *MM* estimation:

```
MM <- coef(rlm(freq ~ vol, data = Fbird1, method = "MM"))
```

A plot compares the approaches. We note, first of all, that the robust techniques are less affected by the outliers than OLS regression. *MM* estimation provides results closest to the original uncontaminated model (Figure 9.17).

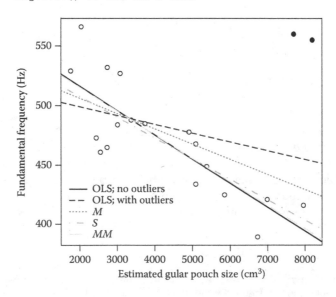

FIGURE 9.17
Conventional and robust regression for the contaminated frigate bird dataset. Black filled points are manufactured outliers.

★

9.19 Model II Regression (*X* Not Fixed)

I noted in Section 9.10 that in linear regression *Y* is assumed to be a random variable, while values within *X* are assumed to be fixed (i.e., Figure 9.7). As a result, we are only concerned with distributions of *Y* given *X*. The fixed *X* assumption is likely to be invalid in many applications associated with observational data where particular levels of *X* are not of interest, or are potentially measured with error.[*] Regardless of this, regressions that assume that *X* is random (i.e., *model II regressions*[†]) are seldom used by biologists. This is true for three reasons. First, fixed *X* regressions are generally the only type discussed in statistical textbooks and are therefore unknown to biologists (it should be noted that these books often use examples from business or industry where the assumption of fixed *X* may be valid). Second, biologists are generally interested in the test of H_0: $\beta_1 = 0$. The *P*-value from this test will be identical whether we regress *Y* on *X* or *X* on *Y*, and will be equivalent to the Pearson correlation *P*-value from the test H_0: $\rho = 0$. As a result, this test will be valid whether *X* is random or fixed. Finally, commercially available software seldom contains algorithms for model II regression. This is not an issue in **R**, which has libraries specifically designed for this purpose.

If the goal is: (1) testing H_0: $\beta_1 = 0$, or (2) obtaining predictive model of *Y* as a function *X*, then a conventional OLS model is reasonable even when *X* is random, although we must limit inferences about *Y* to the measured values of *X* (Kutner et al. 2005, p. 83). However, if the goal is (1) accurate estimation of β_1, or (2) inferential comparison of β_1 to a nonzero value, and *X* is random, then an OLS approach may be inappropriate.

If *X* is random, then the simple linear regression model is still $Y_i = \beta_0 + \beta_1 X_i + \varepsilon_i$, but *X* is now more complex. Specifically, a measurement of *X* is now $X_i' = X_i + \omega_i$, where ω_i is the *i*th "measurement error" associated with the *X* variable (Snedecor and Cochran 1989).

If *X*, ε_i, and ω_i are normally distributed and independent, then *X′* and the value of *Y* given *X* will follow a joint bivariate normal distribution (Snedecor and Cochran 1989). However, the OLS estimator for β_1 will now be biased toward zero. In particular, it will now provide an unbiased estimate for β_1', not β_1, where

$$\beta_1' = \beta_1/(1 + \lambda), \tag{9.76}$$

$$\lambda = \frac{\sigma_\omega^2}{\sigma_X^2}, \tag{9.77}$$

where σ_ω^2 is the variance of ω_i and σ_X^2 is the variance of *X*. If *X* is fixed, then λ will equal zero because $\sigma_\omega^2 = 0$, and β_1' will be equivalent to β_1. The bias in the OLS estimator will depend on the ratio of the "measurement error" variability. The greater the magnitude of σ_ω^2 compared to σ_X^2, the greater the downward bias in the OLS estimate.

Legendre and Legendre (1998) list a number of methods to address the problem of parameter estimation in random *X* regression. I describe three here: major axis regression (MA),

---

[*] Measurement error in *X* may be more dangerous to inference than random variability in *X*. This is because levels in *X* that are assumed to be fixed are incorrectly matched with responses (Kutner et al. 2005).

[†] The term "model II regression" is in parallel to model II ANOVAs (i.e., random effect ANOVAs), which have random predictors (Chapter 10). Gelman et al. (2003) consider random regression from a Bayesian perspective.

standardized (or reduced) major axis regression (SMA), and ranged major axis regression (RMA). Other alternatives include random X models derived from maximum likelihood estimation (see Madansky 1959, Kendall and Stuart 1966). McArdle (1988) presented methods for multiple regression in the context random predictors not considered here.

9.19.1 MA Regression

In *MA* (Pearson 1901, Jolicoeur 1973), the slope of the regression line is estimated by minimizing the sums of squared perpendicular distances to the MA line (Figure 9.18). The MA regression line will be the first principal component from PCA, a multivariate technique not discussed here, given standardized data. In MA, the error variance of Y given X, σ^2, is assumed to be equal to the error variance of X given Y, σ_ω^2 (Legendre and Legendre 1998). We estimate the MA slope with

$$\hat{\beta}_{1(MA)} = \frac{d \pm \sqrt{d^2 + 4}}{2}, \tag{9.78}$$

where

$$d = \frac{\hat{\beta}_{1(OLS)}^2 - r^2}{r^2 \hat{\beta}_{1(OLS)}}, \tag{9.79}$$

Fig.9.18()## See Ch. 9 code

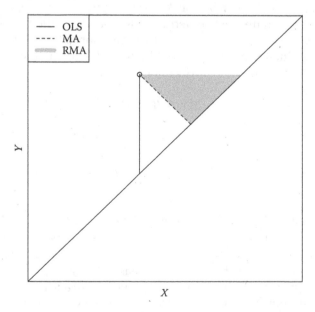

FIGURE 9.18
Heuristic comparison of OLS and random regression parameter estimation. A regression line is shown (line bisecting figure) along with a single observed value (point). Distances or areas that are minimized for parameter estimation in OLS regression (fixed X) versus two forms of model II regression are shown. (Adapted from Quinn, G. P., and Keough, M. J. 2002. *Experimental Design and Data Analysis for Biologists.* Cambridge University Press.)

and r is the Pearson correlation coefficient. In Equation 9.78, we use the positive square root of $d^2 + 4$ if r is positive and the negative square root if r is negative. Standard errors and confidence intervals can be calculated by bootstrapping, while the hypothesis H_0: $\beta_1 = 0$ can be tested using permutation approaches.

9.19.2 SMA Regression

MA regression is only appropriate if X and Y are both measured on the same scale, using the same units. A modification to MA regression called *standardized MA regression* or *SMA regression* (Kermack and Haldane 1950) addresses this problem. Confusingly, this method has also been called *reduced major axis regression*, and denoted RMA (McArdle 1988, see Sokal and Rohlf 2012, p. 539). The acronym RMA is reserved here for ranged major axis regression. In SMA regression, MA regression is calculated using standardized variables (Equation 7.12) and the slope is back-transformed to the original scale by multiplying it by S_Y/S_x. In SMA regression, it is assumed that σ^2 and σ_ω^2 are proportional to σ_Y^2 and σ_X^2, respectively. This condition is often met with count data or with log-transformed data (McArdle 1988). The SMA estimate for β_1 is simply the sample standard deviation of Y divided by the sample standard deviation of X.

$$\hat{\beta}_{1(SMA)} = \pm \frac{S_Y}{S_X}. \tag{9.80}$$

The sign of the slope is determined by the sign of r. Equation 9.80 is equivalent to be the geometric mean of the OLS slope of Y on X and the reciprocal of the slope of X on Y (Legendre and Legendre 1998). For example, using the frigate bird data, we have

```
YonX.b1 <- lm(freq ~ vol, data = Fbird)$coefficients[2]
XonY.b1 <- lm(vol ~ freq, data = Fbird)$coefficients[2]
G.mean(c(YonX.b1, 1/XonY.b1))
[1] 0.02484339
with(Fbird, sd(freq)/sd(vol))
[1] 0.02484339
```

The sign for the slope estimate should be negative in this instance because the sign of Pearson's correlation is negative.

A $(1 - \alpha)100\%$ confidence interval for $\beta_{1(SMA)}$ can be calculated using a method proposed by Jolicoeur and Mosimann (1968).

$$\hat{\beta}_{1(SMA)}(\sqrt{B+1} \pm \sqrt{B}), \tag{9.81}$$

where

$$B = [t_{(1-(\alpha/2);n-2)}]^2 (1 - r^2)/(n - 2). \tag{9.82}$$

A formal hypothesis test of H_0: $\beta_1 = \beta_{1(H_0)}$ is also possible using the test statistic (Clarke 1980, McArdle 1988):

$$t^* = \frac{|\log(\hat{\beta}_{1(SMA)}) - \log(\beta_{1(H_0)})|}{\sqrt{(1 - r^2)(n - 2)}} \tag{9.83}$$

If H_0 is true, and model assumptions are valid (see comments at the end of this section), then t^* will be a random outcome from a t-distribution with $n - 2$ degrees of freedom, and the two-tailed P-value will be $2P(T \geq |t^*|)$, where $T \sim t(n-2)$.

In this case, we note that a hypothesis test of H_0: $\beta_1 = 0$ is not possible because $\log(0)$ is undefined. In addition, using SMA confidence intervals (Equation 9.81) to test this hypothesis is inappropriate because S_Y/S_x (the estimator for β_1, divided by r) cannot be zero except in the trivial case, $S_Y = 0$. This does not constitute a problem, however, because as we noted earlier, OLS methods can used if the null hypothesis is $\beta_1 = 0$.

9.19.3 RMA regression

As an alternative to SMA, Legendre and Legendre (1998) recommend *RMA*. In RMA, regression X and Y are standardized by their ranges. An MA regression is then calculated using the standardized responses, and the slope is back-transformed to the original scale by multiplying it by a ratio of the ranges of Y and X ($Y_{max} - Y_{min})/(X_{max} - X_{min})$. Parameter estimation using RMA regression is equivalent to minimizing the area of a triangle formed by the regression line and the vertical and horizontal distances to the regression line from observations (Figure 9.18).

> **EXAMPLE 9.27**
>
> Hines et al. (1997) studied the effect of the eagle ray (*Myliobatis tenuicaudatus*), a marine predator, on wedge shell bivalves (*Macomona liliana*) larger than 15 mm in size on a New Zealand sandflat. The authors randomly established 20 0.25 m² quadrats and counted bivalve individuals. They then counted the number of eagle rays within 15 m radius circles that enclosed the quadrats. Since the number of predators was not fixed by the investigators, it was considered a random variable. The library *lmodel2* creates model II regressions, and provides associated confidence interval estimation and hypothesis testing. The bivalve data are also accessible in the *lmodel2* library.
>
> ```
> install.packages("lmodel2")
> library(lmodel2)
> data(mod2ex2)
> ```
>
> All model II regression methods discussed in this section can be derived using the function lmodel2:
>
> ```
> bivalve.reg <- lmodel2(Prey ~ Predators, data = mod2ex2, range.y =
> "relative", range.x = "relative", nperm = 1000)
> ```
>
> ```
> n = 20  r = 0.8600787  r-square = 0.7397354
> Parametric P-values: 2-tailed = 1.161748e-06 1-tailed = 5.808741e-07
> Angle between the two OLS regression lines = 5.106227 degrees
> ```
>
> ```
> Regression results
>   Method Intercept     Slope  Angle (degrees)  P-perm (1-tailed)
> 1    OLS  20.02675  2.631527         69.19283         0.000999001
> 2     MA  13.05968  3.465907         73.90584         0.000999001
> 3    SMA  16.45205  3.059635         71.90073                  NA
> 4    RMA  17.25651  2.963292         71.35239         0.000999001
> ```

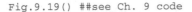

Fig.9.19() ##see Ch. 9 code

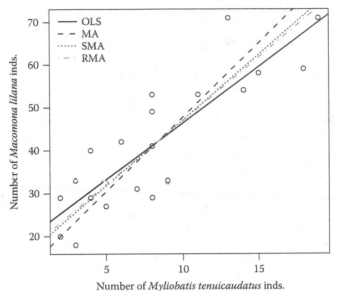

FIGURE 9.19
The number of prey (*Macomona lilana*) shown as a function of the number of predators (*Myliobatis tenuicaudatus*) with OLS (fixed *X*) and model II regressions (MA, SMA, and RMA) overlaid. The point that the lines cross is the estimated centroid of the joint bivariate distribution of *Y* and *X*.

```
Confidence intervals
  Method  2.5%-Intercept 97.5%-Intercept  2.5%-Slope 97.5%-Slope
1    OLS        12.490993        27.56251    1.858578    3.404476
2     MA         1.347422        19.76310    2.663101    4.868572
3    SMA         9.195287        22.10353    2.382810    3.928708
4    RMA         8.962997        23.84493    2.174260    3.956527
```

A two-tailed *P*-value is included for the OLS estimator of β_1, $P = 1.16 \times 10^{-6}$. Note that this is identical to the *P*-value one would obtain from lm for the same model. One-tailed permutation *P*-values are also included for the OLS, MA, and RMA regressions (the tail of the test is determined by the sign of Pearson's *r*). Finally, 95% bootstrap confidence intervals for β_1 are calculated for all four regression approaches. CI calculation methods vary with the model II approach. For details, see Legendre (2008). Zero is not contained in any of these intervals (although we note that the OLS estimator is biased toward zero). Regression lines for these models are shown in Figure 9.19.

9.19.4 Assumptions and Additional Comments

If *X* is random, and a predictive model for *Y* given *X* is desired, then OLS methods are sufficient, so long as the scope of inference is understood to be limited to the measured values of X_i (Kutner et al. 2005). If one is simply interested in a test of the hypothesis H_0: $\beta_1 = 0$, then an OLS approach is appropriate if general linear model assumptions are valid (Section 9.10). In addition, McArdle (1988) and Legendre (2008) recommend the use of OLS over model II regression if the random variation of *Y* given *X*, σ^2 is more than three times that of *X*, σ_ω^2.

All model II regression techniques discussed here assume

1. Independent observations.
2. Bivariate normality for the joint distribution of Y and X.

These assumptions can be checked with both diagnostic plots and formal hypothesis tests (see Chapter 8). An MA regression should only be used when the explanatory and response units of measurement are identical. McArdle (1988) found that the SMA method produced less biased estimate for β_1 than MA regression, but did not consider ranged major axis regression. Legendre (2008) recommended the use of RMA over SMA regression because the hypothesis $H_0: \beta_1 = 0$ is directly testable. This concern is trivial, however, since OLS methods can be used to test this hypothesis.

★ 9.20 Generalized Linear Models

The general linear regression model (Equation 9.1) assumes a linear association between Y and X and, for inference, identical normal errors. However, these assumptions will generally be invalid when Y is not continuous (e.g., dichotomous or discrete) or strictly bounded. *Generalized linear models (GLMs)* (Nelder and Wedderburn 1972), which include all general linear models (including linear regression) and which can incorporate categorical predictors (*à la* ANOVA and ANCOVA), can be used to fit models and make inferences given noncontinuous and bounded response variables.

A GLM will have two components that distinguish it from the more specialized general linear model:

1. A user-defined link function that transforms the response expectation and allows the expression of the mean function in linear terms.
2. A user-defined error term distribution that provides ML optimization criteria and allows inferences concerning model parameters.

A large number of error term distributions within the exponential family (Appendix) can be specified for GLMs. The three most common are normal (resulting in general linear models if an identity link is used), Poisson, and binomial.* Link functions linearize the mathematical form of these models. *Canonical links* arise naturally for particular error distributions (see canonical link derivations for normal, Poisson, and binomial errors in the Appendix). GLM users, however, are not limited to the specification of these relationships in models.

Binomial errors are used when the response variable is categorical or dichotomous. In this case, the canonical link is a logit function, although other reasonable choices include the probit function (Chapter 3, see Appendix for derivations). Poisson GLMs are often appropriate when the response variable is composed of counts. In this case, a Poisson error distribution is generally assumed for ML estimation and inference, and the canonical link is a $\log_e$ transformation of the response mean.

---

* Beta distributed errors frequently occur with proportion data, which are characteristic of many types of biological phenomena. Beta regressions can be fitted in a similar fashion to GLMs (see Smithson and Verkuilen 2006). For implementation, see the *betareg* package in **R**.

9.20.1 Model Estimation

GLMs use the method of maximum likelihood to estimate model parameters, $\beta_0, \beta_1, \ldots, \beta_{p-1}$ (Chapter 4). Recall that OLS and ML estimates for general linear models will be identical (for parameters other than σ^2). Parameter estimates that maximize the Poisson and binomial log-likelihood functions, however, cannot be obtained analytically (see Section 9.20.5). As a result, estimates are found through the method of iteratively weighted least squares, generally using variants on the Raphson–Newton algorithm (Section 4.3.6). Estimated variances and covariances are obtained by finding the second-order partial derivatives of the log-likelihood function for each parameter (see Equations 9.92 and 9.93), and evaluating these results at the maximum likelihood estimates for the parameters (see Kutner et al. 2005, p. 577).

9.20.2 glm

The **R** function for obtaining GLMs is `glm`. Important arguments include `formula`, `family`, `offset`, and `data`.

- The `formula` argument is analogous to the `formula` argument in `lm`. That is, response variables are placed to the left of a tilde, and the response function is placed to the right.
- The `family` argument is used to simultaneously specify the link function and error term distribution. For instance, to create a logistic GLM, one would use `family = binomial(link = "logit")`. A Poisson error distribution and its canonical link can be specified with `family = poisson(link = "log")`. General linear models can be specified using `gaussian(link = "identity")`, although this procedure is handled much more efficiently by `lm`.
- The `offset` argument is a component of the model that is known in advance (e.g., from theory or mechanistic models) and does not require a parameter estimate. The term will be redundant in general linear models since one can simply add or subtract the offset values from the values of the response variable and work with the residuals instead of the response (Crawley 2007). For GLMs, however, specifying an offset is often necessary. Consider a Poisson regression where the response is a rate: $\ln(\lambda_t/t_i) = \beta_0 + \beta_1 X_i$, where λ_t is the true number of successes given X_i, and t is some unit of time. Then, $\ln(\lambda_t) - \ln(t_i) = \beta_0 + \beta_1 X_i$ and $\lambda_t = t_i \exp(\beta_0 + \beta_1 X_i)$. In this case, the term $-\ln(t)$ can be specified in the model as an offset.
- The `data` argument, analogous to `data` in `lm`, specifies a dataframe, list, or environment, coercible with `as.data.frame`, containing variables used in `formula`.

9.20.3 Binomial GLMs

In a binomial GLM, we model a dichotomous response as a function of quantitative explanatory variables. Consider the simple linear regression model (Equation 9.3).

$$Y_i = \beta_0 + \beta_1 X_i + \varepsilon_i,$$

$$E(Y_i) = \beta_0 + \beta_1 X_i.$$

If Y_i is a Bernoulli random variable, that is, a binomial random variable with one trial (Chapter 3), then we have

$$P(Y_i = 0) = 1 - \pi$$
$$P(Y_i = 1) = \pi,$$

where π is the probability of a success (i.e., the outcome $Y_i = 1$).

From the definition of the expected value of a discrete random variable (Equation 4.1), we obtain the expected probability of success for one binary trial as

$$E(Y_i) = 1(\pi) + 0(1 - \pi) = \pi = P(Y_i = 1).$$

Combining equations we have

$$E(Y_i) = \beta_0 + \beta_1 X_i = P(Y_i = 1).$$

We learned in Chapter 3 that the logistic distribution is very similar to the normal distribution, but with a closed-form cdf. If we let $Y \sim \text{LOGIS}(1, 0)$, then we have

$$P(Y \le y) = F_L(y) = \frac{1}{1 + e^{-y}} = \frac{1}{1 + 1/e^y}$$

$$= \frac{1}{(e^y + 1)/e^y} = \frac{e^y}{1 + e^y}, \qquad (9.84)$$

where F_L denotes the cdf for LOGIS(1, 0). The logistic cdf function is sigmoidal (i.e., it exists largely near 0 or 1) and, like all cdfs, will take on values between 0 and 1. If Y_i is a Bernoulli random variable, considered as a linear function of $X_{1i}, X_{2i}, \ldots, X_{p-1i}$, then it can be shown (Kutner et al. 2005, p. 563) that

$$E(Y_i) = P(Y_i = 1) = F_L(\beta_0 + \beta_1 X_{1i} + \cdots + \beta_{p-1} X_{p-1i}) = F_L(\mathbf{X}_i' \boldsymbol{\beta}).$$

In the expression above, the linear predictor is stated in simplified matrix terms. $\mathbf{X}_i'$ denotes the transposed ith explanatory vector and $\boldsymbol{\beta}$ is a vector of model parameters. Substituting $E(Y_i)$ for y in Equation 9.84, we have

$$E(Y_i) = \pi_i = \frac{e^{(\mathbf{X}_i'\boldsymbol{\beta})}}{1 + e^{(\mathbf{X}_i'\boldsymbol{\beta})}}. \qquad (9.85)$$

This is called the logistic *mean function* and will be in units of probability of success. To remind us that the mean function gives the Bernoulli probability for success, we specify $E(Y_i)$ as π_i. The canonical logit link function is generally used to give the model a linear form and allow parameter estimation. Recall that the logit function (Ex. 3.17) is equivalent to the log odds ratio and the logistic inverse cdf.

To linearize the model, we apply the logit function to each π_i:

$$F_L^{-1}(\pi_i) = \text{logit}(\pi_i) = \ln\left(\frac{\pi_i}{1-\pi_i}\right),$$

where π_i is the ith probability of success. We have

$$\text{logit}(\pi_i) = \ln\left[\frac{e^{(X_i'\beta)}}{1+e^{(X_i'\beta)}} \bigg/ 1 - \frac{e^{(X_i'\beta)}}{1+e^{(X_i'\beta)}}\right]$$

$$= \ln\left[\frac{e^{(X_i'\beta)}}{1+e^{(X_i'\beta)}} \bigg/ \frac{1}{1+e^{(X_i'\beta)}}\right]$$

$$= \ln e^{(X_i'\beta)} = X_i'\beta\ln(e) = X_i'\beta.$$

This results in the familiar linear regression form, which will now be in logit units.

To summarize, the logistic mean function, that is, the cdf for LOGIS(1, 0), provides conditional probabilities of success. The logit link function, that is, the logistic inverse cdf, gives the model a linear form, and provides quantiles (as log odds), given lower tailed probabilities.

9.20.4 Deviance

The efficacy of GLMs can be assessed with a generalized measure of model variability called *deviance*:

$$D(\text{fitted}) = 2[\lambda(\text{saturated}) - \lambda(\text{fitted})], \tag{9.86}$$

where the *saturated model* is a theoretical construct containing an explanatory variable for every observation, and the fitted model contains only the measured explanatory variables of interest. Note the strong resemblance of deviance to KL-information (and the first product term from AIC). Deviance is also identical to a likelihood ratio test statistic comparing saturated and fitted models (see Section 9.20.5, Equation 9.88).

If log-likelihood for the fitted model is high compared to the saturated model, then deviance will be low. Thus, deviance can be used to measure goodness of fit. Deviance estimators for particular error families and link functions are shown in Table 9.3. The deviance for a model with ≥ 1 predictor is often called the *residual deviance*. A measure called the *null deviance* is calculated using only an intercept term in the fitted model. For mathematical derivations of formulae, see Hardin and Hilbe (2007).

EXAMPLE 9.28

Saint-Germain et al. (2007) modeled the presence or absence of a saprophytic wood-boring beetle *Anthophylax attenuatus* as a function of the wood density of decaying aspen trees (*Populus tremuloides*) in Western Quebec, Canada. Their data are in the object `beetle`.

TABLE 9.3

Error Families, Canonical Links, and Formulae for Deriving Deviance for GLMs

Error Family	Canonical Link	Residual Deviance
Normal	Identity	$\sum_{i=1}^{n}(Y_i - \hat{Y}_i)^2$
Binomial	Logit	$-2\sum_{i=1}^{n}\{Y_i \ln(\hat{Y}_i) + (1 - Y_i)\ln(1 - \hat{Y}_i)\}$
Poisson	Log	$2\sum_{i=1}^{n}\{Y_i \ln(Y_i / \hat{Y}_i)\}$

Note: For all equations, Y_i is the ith observed response and $\hat{Y}_i$ is the ith fitted response. Note that for general linear models, deviance is equivalent to SSE.

```
data(beetle)
beetle.glm <- glm(ANAT ~ Wood.density, family = binomial(link = "logit"),
data = beetle)
summary(beetle.glm)

Coefficients:
            Estimate Std. Error z value Pr(>|z|)
(Intercept)   15.659     6.856    2.284   0.0224 *
Wood.density -52.632    23.838   -2.208   0.0272 *
(Dispersion parameter for binomial family taken to be 1)
Null deviance: 31.755  on 23   degrees of freedom
Residual deviance: 14.338  on 22   degrees of freedom
AIC: 18.338
```

The ratio of coefficients and their standard errors resulting from ML estimation are called *Wald statistics*, and are asymptotically standard normal under H_0: $\beta_k = 0$. A plot of the model is shown in Figure 9.20. Note that the relationship of the mean function to the predictor is sigmoidal, and in units of probability, while the relationship of the link function and the predictor is linear and in logit units. Parameter estimates in the model are given in logit units. Thus, for instance, we can say that log odds for beetle presence decrease by 52.6 given a 1 g cm^{-3} increase in wood density.

To get fitted or predicted values, I apply the logistic mean function (Equation 9.85) to the parameter estimates from the model. For instance, to find the model probability of a wood-boring beetle being present given a wood density of 0.35 g cm^{-3}, I have

$$\frac{e^{(15.659-(52.632\times0.35))}}{1 + e^{(15.659-(52.632\times0.35))}} = 0.059.$$

As with general linear models, fitted values for GLMs can be obtained using the function `fitted`, while predicted values can be obtained using the function `predict`.

```
predict(beetle.glm, newdata = data.frame(Wood.density = 0.35),
type = "response")
0.05938482
```

```
Fig.9.20() ## See Ch. 9 code
```

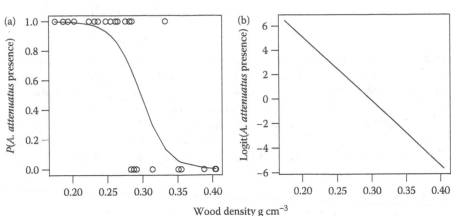

FIGURE 9.20
Logistic regression for the presence of wood-boring beetles (*Anthophylax attenuatus*) as a function of the wood density of decaying aspen trees. The model is shown both in (a) mean response units (probability) and (b) log odds (logit) units. Note that the Y observations shown in (a) are dichotomous. Note also that when logit($X = 1$) = 0, $P(X = 1) = 0.5$. Thus, positive log odds indicate probabilities of success greater than 0.5.

9.20.5 GLM Inferential Methods

9.20.5.1 Wald Test

For GLMs with nonnormal error families, inference procedures rely on the assumption of large sample sizes. Given this condition, the simplest method to test $\beta_k = 0$ is the *Wald test*. The test statistic concerning an arbitrary parameter β_k with an ML estimator $\hat{\beta}_k$ is

$$z^* = (\hat{\beta}_k - \beta_{H_0})/\hat{\sigma}_{\hat{\beta}_k ML}, \tag{9.87}$$

where $\hat{\sigma}_{\hat{\beta}_k ML}$ is the maximum likelihood estimate for the standard error of $\hat{\beta}_k$, and β_{H_0} is the value for β_k specified in H_0. Because ML estimators are asymptotically normal, the Wald test statistic will be asymptotically standard normal under H_0. Tests for H_0: $\beta_k = 0$ using the Wald procedure are included as a part of the standard glm output (see Example 9.28).

9.20.5.2 Likelihood Ratio Test

The Wald test assumption of normality is often problematic for GLMs, particularly those with binomial errors, and small to moderate sample sizes (Fox 2002, Agresti 2012). A method that generally comes much closer to matching actual type I error rates with nominal levels is the *likelihood ratio test*. In the context of GLMs (and GAMs, Section 9.22.4), the procedure is often referred to as a *drop in deviance test* or an *analysis of deviance* (Ramsey and Schafer 1997).

Let R be a reference model containing p parameters, (including the Y-intercept) and let N be a model nested in R with q parameters (i.e., $q < p$). The residual deviance of R subtracted from the residual deviance of N is the *drop in deviance* (Ramsey and Schafer 1997). This difference will be identical to the likelihood ratio test statistic, commonly denoted as G^2

$$G^2 = \hat{D}(N) - \hat{D}(R) = 2\ln\left[\frac{\hat{\mathcal{L}}(R)}{\hat{\mathcal{L}}(N)}\right] = 2[\hat{\ell}(R) - \hat{\ell}(N)]. \tag{9.88}$$

Generally (but not always), we let R and N differ only with respect to the fact that R contains a predictor interest, while N does not. If the null hypothesis $\beta_k = 0$ is true and n is reasonably large, then G^2 will be a random outcome from a χ^2 distribution with $p - q$ degrees of freedom. Note that the degrees of freedom correspond to the difference in the degrees of freedom error of the two fitted models, $(n - q) - (n - p) = p - q$. The P-value is computed as $P(X \geq G^2)$, where $X \sim \chi^2(p - q)$. The function Anova simultaneously provides drop in deviance tests for all predictors in a GLM. For Example 9.28, we have

```
library(car)
Anova(beetle.glm)
Analysis of Deviance Table (Type II tests)
            LR Chisq Df Pr(>Chisq)
Wood.density   17.417  1   3.002e-05 ***
```

The value 17.417 is the difference of the residual deviance and null deviance in Example 9.28. This will comprise the drop in deviance for models with a single predictor. The default approach in Anova is to create nested models without higher-order relatives of the predictor of interest X_k in a type II sum of squares approach (see Section 9.5.3). A type III sum of squares approach is also possible, wherein X_k is deleted but higher-order relatives of X_k are maintained in the nested model (see Section 10.14).

For logistic GLMs, one can also use an *error matrix* (also called a confusion matrix) to gauge the overall effectiveness of the model as a classifier of success and failure. Using the arbitrary probability of 0.5 as the cutoff for defining beetle presence (Peng et al. 2002), we have

```
table(ifelse(fitted(beetle.glm) > 0.5, 1, 0), beetle$ANAT)
      0  1
   0  6  1
   1  3 14
```

The table cross-classifies trees with predicted beetle presence/absence (in rows) versus the observed presence/absence (in columns). The diagonal contains predicted and observed agreements. Thus, the sum of the diagonal divided by the sum of all the matrix elements gives the proportion of correctly classified observations: 20/24 = 0.83. The *sensitivity* (i.e., *true-positive rate*) is the number of true positives divided by the number of observed positives and false negatives: 14/15 = 0.93. The *specificity* (i.e., *true-negative rate*) is the number of true negatives divided by the number of observed negatives and false positives: 6/9 = 0.67 (also see Example 2.11).

9.20.5.3 ROC and AUC

Criticisms of the previous approach include the fact that a single arbitrary value, denoted here as c, is used to define the classification of a success (the value 0.5 is used in the example above). The quantity given for c strongly affects measures of classification effectiveness. Extremely small values cause the rates of true positives and false negatives to increase, while extremely large values increase the rates of true negatives and false positives. A diagram called a *receiver operating characteristic (ROC) curve* plots sensitivity (i.e., true-positive rate) as a function of 1–specificity (i.e., false-positive rate) for all possible values of c (Figure 9.21). ROC curves are unaffected by changing values for c because they consider all possible values. Increasing the distinction between dichotomous groups with respect to predictors causes the curve to be pushed into top left-hand corner of the ROC plot, while increasing group similarity causes the curve to approach the dashed line indicating that the classification is no better than random. Because of these characteristics, the area under the ROC curve, denoted AUC, can be used as a measure of binomial GLM efficacy.

The set of binomial observations will contain n_0 zeroes and n_1 ones. Consider an arrangement of all possible pairs of observed zeroes and ones. Let Y_{0i} and Y_{1i} represent the ith pair, $i = \{1, 2, \ldots, n_0 n_1\}$, and denote the corresponding ith pair of fitted probabilities as $\hat{\pi}_{0i}$ and $\hat{\pi}_{1i}$. We then define the following *concordance index* (Section 8.3.1.1):

$$U_i = \begin{cases} 1, & \text{if } \hat{\pi}_{0i} < \hat{\pi}_{1i} \\ 0, & \text{otherwise} \end{cases}. \tag{9.89}$$

AUC can now be calculated as

$$\text{AUC} = \frac{\sum_{i=1}^{n_0 n_1} U_i}{n_0 n_1}. \tag{9.90}$$

```
see.roc.tck()
```

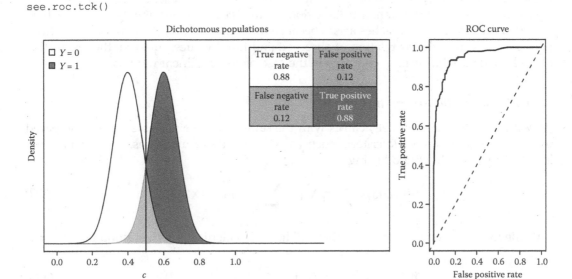

FIGURE 9.21
Demonstration of ROC curves. The classes from the error matrix in the left-hand figure are identical to those from a table describing type I and II hypothesis testing error (e.g., Table 6.1). The only difference is that the error matrix has correct decisions arranged on the diagonal.

A coarse scale, corresponding to academic letter grades, is often used to define model usefulness. Thus, AUC values >0.9 indicate excellent models, while AUC values ≤0.5 indicate that the classification is no better than a random assignment of classes (Figure 9.21). The function auc from *asbio* creates ROC curves, and calculates AUC from observed dichotomous responses and predicted probabilities.[*]

```
auc(beetle$ANAT, fitted(beetle.glm), plot = FALSE)
[1] 0.962963
```

This result indicates that the model effectively distinguishes between beetle presence and absence.

9.20.5.4 Measures of Explained Variance

Somewhat controversial are omnibus R^2 type measures for GLMs. The residual deviance is equivalent to SSE for general linear models (Table 9.2), while the null deviance is analogous to the general linear model *SSTO*. Given this, a *pseudo* R^2 statistic (Efron 1978) can be calculated using

$$\text{Pseudo } R^2 = 1 - \frac{\text{residual deviance}}{\text{null deviance}}. \tag{9.91}$$

Efron's pseudo R^2 statistic has been recommended over other similar measures because it allows "intuitively reasonable" interpretations and is insensitive to changes in the intercept, which can change the proportion of zeroes and ones observed in Y (Menard 2000). Agresti (2012) recommended using the correlation of the fitted and observed values to compare the effectiveness of GLMs with the same response variable. This method has the advantage of having a value proportional to the effect size. For instance, in a simple linear regression, the correlation coefficient is simply the slope (one measure of effect size) multiplied by the ratio of the standard deviations of the response and explanatory variables. Controversy for both approaches arises because the multiple conventional interpretations possible for R^2 (i.e., the correlation of observed and fitted values, squared, the proportion of explained variability) only converge in the case of general linear models.

9.20.5.5 Information-Theoretic Criteria

As with general linear models, GLMs with the same response variable can be compared using information-theoretic criteria such as AIC. For these measures, we calculate the logistic model log-likelihood using

$$\ell(\boldsymbol{\beta}|\text{data}) = \sum_{i=1}^{n} Y_i(\mathbf{X}_i'\boldsymbol{\beta}) - \sum_{i=1}^{n} \ln[1 + \exp(\mathbf{X}_i'\boldsymbol{\beta})]. \tag{9.92}$$

While for a Poisson GLM with a canonical link, we calculate log-likelihood using

$$\ell(\boldsymbol{\beta}|\text{data}) = \sum_{i=1}^{n} Y_i \ln(\mathbf{X}_i'\boldsymbol{\beta}) - \sum_{i=1}^{n} \mathbf{X}_i'\boldsymbol{\beta} - \sum_{i=1}^{n} \ln(Y!). \tag{9.93}$$

---

[*] The package *pROC* allows more sophisticated implementation of ROC curves, including confidence intervals for AUC.

These are also the functions that are iteratively optimized in the process of ML parameter estimation.

9.20.6 Poisson GLMs

Poisson GLMs are appropriate for modeling counts when large outcomes are rare. When counts are large and span a large range, then general linear models are generally appropriate due to the convergence of the Poisson to the normal distribution (Example 9.6).

Assuming that Y_i is a Poisson random variable (Chapter 3) with expectation λ_i, the most commonly used Poisson GLM mean function is

$$E(Y_i) = \lambda_i = e^{(X_i'\beta)}, \tag{9.94}$$

which provides the mean number of counts (successes) given X_i. We can linearize the response function using the canonical $\log_e$ link

$$\ln(\lambda_i) = X_i'\beta. \tag{9.95}$$

A model that assumes Poisson errors and uses a log link function is called a *Poisson loglinear model*.

EXAMPLE 9.29

Atlantic horseshoe crabs[*] (*Limulus polyphemus*) arrive at beach spawning sites on the Atlantic Coast of North America as attached male/female pairs. Additional males called satellites crowd a nesting couple and compete with the attached male for access to females. Satellite groups vary in size and appear to be unassociated with local environmental characteristics. Brockman (1996) examined satellite number as a function of female phenotypic characteristics, including carapace width. Because the satellite data are composed of small magnitude counts (median = 2), the association between the predictor and response is extremely nonlinear, and the distribution of responses is nonnormal. The situation calls strongly for the use of a Poisson GLM.

The horseshoe crabs' data are contained in the **R**-object `crabs`.

```
data(crabs)
crab.glm <- glm(satell ~ width,  family = poisson(link = "log"), data =
crabs)
summary(crab.glm)
Coefficients:
            Estimate Std. Error z value Pr(>|z|)
(Intercept) -3.30476    0.54224   -6.095  1.1e-09 ***
width        0.16405    0.01997    8.216  < 2e-16 ***
---
Signif. codes:  0 '***' 0.001 '**' 0.01 '*' 0.05 '.' 0.1 ' ' 1
(Dispersion parameter for poisson family taken to be 1)
Null deviance: 632.79  on 172  degrees of freedom
Residual deviance: 567.88  on 171  degrees of freedom
AIC: 927.18
```

---

[*] Not crabs at all (which have mandibles) but chelicerate arthropods in class Merostomata.

```
Fig.9.22()# See Ch. 9 code
```

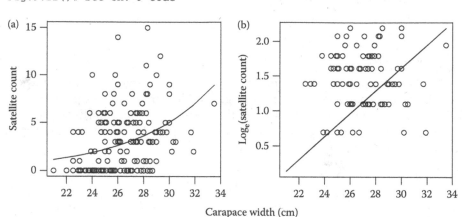

FIGURE 9.22

Poisson regression for counts of horseshoe crab (*Limulus polyphemus*) satellites as a function of female carapace width. The model is shown in both (a) mean response units (counts) and (b) log units.

A plot of the model is shown in Figure 9.22. Note that the mean function is nonlinear, but in the original (satellite count) units. The $\log_e$ link provides a linear fit with the response in $\log_e$ units. On an average, we would expect log satellite counts to increase by 0.16 for every 1 cm of increased carapace width.

9.20.7 Assumptions and Diagnostics

GLMs assume that

1. The true response surface is linear after the application of link function.
2. The true model error is correctly defined.
3. Observations are independent.

In addition, as with general linear models, inferences for GLMs will be hampered by multicollinearity and outlying, influential observations.

9.20.7.1 GLM Residuals

Diagnostics for GLMs without normal errors are problematic. For instance, in the case of binomial errors, residuals will have two possible values:

$$\hat{\varepsilon}_i = \begin{cases} 1 - \hat{\pi}_i & \text{if } Y_i = 1 \\ -\hat{\pi}_i & \text{if } Y_i = 0 \end{cases}.$$

These outcomes will be both heteroscedastic and nonnormal. In fact, the sampling distribution of $\hat{\varepsilon}_i$ under the assumption that the fitted model is correct will be unknown

(Kutner et al. 2005). Consequently, the analysis of conventional residuals will often be uninformative. To account for this, error estimators have been designed for particular GLM error term distributions. *Pearson residuals* are conventional residuals divided by the error term estimated standard deviation. Thus, for a logistic GLM (with Bernoulli errors), we have

$$\hat{\varepsilon}_{iP} = \frac{Y_i - \hat{\pi}_i}{\sqrt{\hat{\pi}_i(1 - \hat{\pi}_i)}}. \tag{9.96}$$

Deviance residuals are the square roots of casewise elements of the residual deviance, with the sign taken from $Y_i - \hat{\pi}_i$. For instance, in a logistic GLM, these would be calculated as

$$\hat{\varepsilon}_{iD} = \text{sign}(Y_i - \hat{\pi}_i)\sqrt{-2[Y_i \ln(\hat{\pi}_i) + (1 - Y_i)\ln(1 - \hat{\pi}_i)]}. \tag{9.97}$$

By default, four diagnostic plots are created by `plot(glm.object)`. Figure 9.23 shows these for the beetle example. Figure 9.23a shows residuals as a function of predicted values on the linear (link) scale. In the case of logistic GLMs, we assume, asymptotically, that $E(Y_i - \hat{\pi}_i) = 0$. Therefore, we would expect a smoother fit (Section 9.22) to these points to be approximately horizontal. Figure 9.23b shows the standardized deviance residuals as a function of standard normal theoretical quantiles. If the linearity assumption is appropriate, points should be near the fit line. We note that except for the outliers 1, 2, and 10, this assumption holds. Figure 9.23c, the scale–location plot, shows the square root of the absolute value of the standardized deviance residuals as a function of the linear predicted values, and re-emphasizes the presence of the three outliers. Figure 9.23d shows standardized Pearson residuals as a function of leverage, and can be used to identify outliers and influential points. A generalized analog of Cook's distance is overlaid (see Williams 1987) showing that only point 10 is strongly influential.

9.20.7.2 Model Goodness of Fit

Adequacy of a logistic model can also be assessed using the *Hosmer Lemeshow goodness of fit test*, which unlike most other assessment approaches does not require replication within predictor levels (Hosmer and Lemeshow 2002). The procedure first groups data into classes with similar fitted values, $\hat{\pi}_i$, with approximately the same number of observations in each class. The use of 10 classes (i.e., *deciles of risk*) is recommended by Hosmer et al. (1997). Observed and expected (fitted) frequencies in each class are then tabulated and compared using a two-way table (Kutner et al. 2005, p. 590; Chapter 11). This results in the test statistic

$$X^2 = \sum_{j=1}^{c} \sum_{k=0}^{1} \frac{\left(y_{jk} - E_{jk}\right)^2}{E_{jk}}, \tag{9.98}$$

where c is the number of classes $j = \{1, 2, \ldots, c\}$, k indexes successes and failures, that is, $k \in \{0, 1\}$, y_{j1} is the observed number of successes ($k = 1$) for the jth class, y_{j0} is the observed number of failures ($k = 0$) for the jth class, and E_{j1} and E_{j0} are the expected number of successes and failures under H_0, respectively. Under the null hypothesis of model appropriateness, $E_{j1} = n_j\hat{\pi}_j$ and $E_{j0} = n_j - E_{j1}$. If H_0 is true, then X^2 will approximately follow a χ^2 distribution with $c - 2$ degrees of freedom. The P-value is calculated as $P(\chi^2_{c-2} \geq X^2)$.

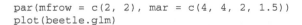

```
par(mfrow = c(2, 2), mar = c(4, 4, 2, 1.5))
plot(beetle.glm)
```

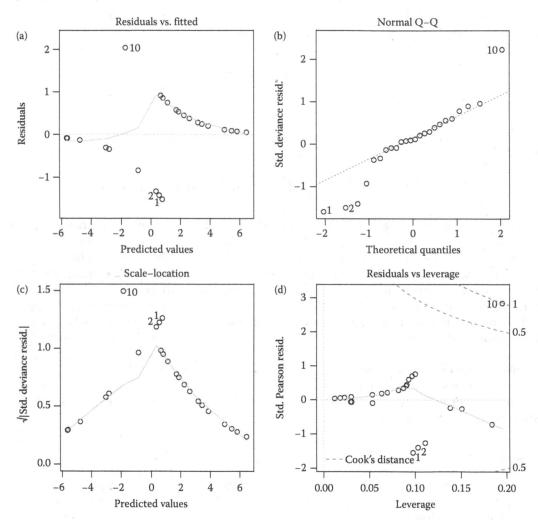

FIGURE 9.23

Default diagnostic plots produced for the beetle logistic general linear model: (a) fitted values versus residuals, (b) half-normal normal quantile plot, (c) scale–location plot, and (d) leverage plot with Cook's distances.

The function `HLgof.test` from *MKmisc* runs the HL test, as well as the more recently developed le Cessie–van Houwelingen–Copas–Hosmer goodness of fit test (Hosmer et al. 1997). For the beetle example, we have

```
install.packages("MKmisc"); library(MKmisc)
HLgof.test(fitted(beetle.glm), beetle$ANAT)
        Hosmer-Lemeshow C statistic
data:   fitted(beetle.glm) and beetle$ANAT
X-squared = 3.4537, df = 8, p-value = 0.9028
```

There is little evidence to refute the appropriateness of the model.

9.20.7.3 Dispersion, Overdispersion, and Quasi-Likelihood

For distributions in the exponential family, the variance is the product of the distributional *variance function* specified as $V[\theta]$ and a *dispersion* parameter denoted ϕ (see Appendix). That is, for an exponential family random variable Y, $\text{Var}(Y) = V[\theta]\phi$. For the normal distribution, $\phi = \sigma^2$, $V[\theta] = 1$, and as a result $\text{Var}(Y) = \sigma^2$. For the Poisson distribution, $\phi = 1$ and $V[\theta] = \lambda$, thus $\text{Var}(Y) = \lambda$. For the binomial distribution, $\phi = 1$ and $V[\theta] = \pi(1 - \pi)$, and as a result $\text{Var}(Y) = \pi(1 - \pi)$ (for derivations, see Appendix).

Overdispersion describes a situation when the observed variance exceeds the model (theoretical) variance. This will cause standard error estimates in a GLM to be artificially deflated and will increase the probability of type I error. Overdispersion will not be an issue for general linear models because, in the assumed normal distribution for error terms, the variance is independent of the mean. However, this is not true for most other exponential family distributions. For instance, the Poisson distribution variance is equal to its mean; while for the binomial distribution, the variance is a fixed proportion of the mean (it is the mean times 1 minus the probability of success). Because Poisson and binomial distributions have variances that are fixed functions of the mean, and because the dispersion parameter, ϕ, is taken to be 1 for both distributions, overdispersion requires ϕ to be *greater* than 1. The degree of overdispersion can be estimated as

$$\hat{c} = \text{residual deviance}/dfE, \tag{9.99}$$

where dfE indicates the residual degrees of freedom.

For Poisson and binomial GLMs, values of $\hat{c}$ greater than one indicate overdispersion; however, this is generally not a serious problem until $\hat{c} > 2$ (Hardin and Hilbe 2007). To address overdispersion, two steps can be taken. First, one can specify a more appropriate error distribution. For instance, we learned in Chapter 3 that the negative binomial distribution is appropriate for modeling overdispersed counts.[*] Second, one can use a *quasi-likelihood* approach that estimates ϕ (Wedderburn 1974). To account for error deflation, the standard errors in these models will be multiplied by the square root of the estimated dispersion parameter, $\hat{\phi}$ (Fox 2002). Poisson and binomial quasi-likelihood functions are specified in glm with the arguments family = quasipoisson and family = quasibinomial.

In retrospect, overdispersion is evident in the horseshoe crab analysis, since $\hat{c} = 567.88/171 = 3.32$. We note that a reanalysis (see below) using quasi-likelihood does not change parameter estimates, but increases P-values because of larger standard errors. This is because the standard errors from the quasi-likelihood model are the standard errors from the Poisson family model (Example 9.29) multiplied by the square root of the estimated dispersion parameter. That is, they are multiplied by $\sqrt{3.182205}$.

```
crab.glm1 <- glm(satell ~ width,  family = quasipoisson(link = "log"),
data = crabs)
summary(crab.glm1)
            Estimate Std. Error t value Pr(>|t|)
(Intercept) -3.30476    0.96729   -3.417 0.000793 ***
width        0.16405    0.03562    4.606 7.99e-06 ***

(Dispersion parameter for quasipoisson family taken to be 3.182205)
```

---

[*] Function glm.nb in *MASS* implements negative binomial GLMs.

9.20.7.4 *Fitted Probabilities Numerically 0 or 1*

A common problem in binomial GLMs occurs when the dichotomous responses are completely distinct with respect to X. That is, below a certain X threshold, Y is always 0, while above the value, it is always 1. In these cases, glm returns the message: fitted probabilities numerically 0 or 1 occurred. This scenario is mathematically problematic because coefficient values will be indeterminate between the largest X for Y = 0 and the smallest X for Y = 1, since, between these values, the slope estimate is infinite. This scenario has been called *perfect discrimination* because a user can perfectly predict sample outcomes by knowing predictor values (Agresti 2012). It has been suggested that maximum likelihood estimates do not exist when fitted probabilities are extremely close to zero or one (Santner and Duffy 1989), although it has been described as a desirable outcome in the pattern recognition literature (Ripley 1996). The lack of curvature due to perfect discrimination greatly inflates standard errors. When using such a model, one should be aware of these effects on estimators and their concomitant effects on inferences (Venables and Ripley 2002).

★   ## 9.21 Nonlinear Models

In earlier sections, we learned that (1) simple transformations may linearize relationships that seem hopelessly nonlinear, (2) polynomial regressions can be used to define many forms of nonlinearity, and (3) nonlinearity as a result of discrete, categorical, or bounded response variables can be addressed with GLMs. However, in many instances, conventional response variable transformations and higher-order terms provide nonintuitive models with many parameters. Furthermore, these approaches may often result in poorly generalizable (over- or under-fit) models. *Nonlinear regression models*, that is, models that cannot be expressed in the form of Equation 9.1, are often helpful in these situations. These models have the form

$$Y_i = f(\mathbf{X}_i, \gamma) + \varepsilon_i, \tag{9.100}$$

where $f(\mathbf{X}_i, \gamma)$ is a nonlinear response function, X_i is the vector of predictor values for the *i*th case, that is, $\mathbf{X}_i = \{1, X_{1i}, \ldots, X_{p-1i}\}$, and γ is a set of nonlinear parameters. Error terms are generally assumed to be independent and identically normal, $\varepsilon_i \sim N(0, \sigma^2)$. Parameter estimates are obtained using *nonlinear least squares*. This involves minimizing Q in

$$Q = \sum_{i=1}^{n} \left(y_i - f(\mathbf{X}_i, \gamma)\right)^2. \tag{9.101}$$

Optimization of Q requires an iterative search algorithm because analytical solutions cannot generally be obtained. The most common approach is the *Gauss–Newton method*, which uses a Taylor series approximation to express the nonlinear model in linear terms, and then uses OLS to estimate parameters iteratively (see Kutner et al. 2005, pp. 518–520).

9.21.1 nls

The function nls implements nonlinear models in **R**. Its major arguments include formula, start, data, and algorithm.

- formula is analogous to the formula argument in lm and glm. That is, a response variable is specified to the left of a tilde metacharacter, and a response function (containing all necessary variables and parameters) is specified to the right. A number of common nonlinear response functions have been precoded for formula, including asymptotic regression, two-parameter exponential (biexponential), three-parameter logistic, Gompertz, Weibull, and Michaelis Menten models (see ?selfStart and see.nlm()).

- The argument start is a named list or numeric vector containing starting estimates for parameters. Functions from selfStart (see above) do not require user-defined starting values.

- The data argument specifies a dataframe, list, or environment, coercible with as.data.frame, containing variables used in formula.

- The algorithm argument specifies the method used to minimize Equation 9.101. By default, the Gauss–Newton method is used.

9.21.2 Model Examples

Subject area expertise often guides the development of nonlinear models. For instance, the *Michaelis–Menten* model was initially developed to depict enzyme kinetics:

$$E(Y_i) = \frac{\gamma_1 X_i}{\gamma_2 + X_i}, \tag{9.102}$$

where X_i represents the ith substrate concentration and Y_i gives the resultant ith reaction rate. The term γ_1 represents the maximum reaction rate achievable by the system (the maximum possible value of Y), while γ_2 is the substrate concentration at which the reaction rate is half of γ_1. This algorithm and other nonlinear models can be explored using the function see.nlm (Figure 9.24).

> **EXAMPLE 9.30**
>
> The bacterium *Pseudomonas aeruginosa* causes disease in human hosts leading to sepsis and even death in part by secreting lipases (proteins that break down lipids) into cellular environments. The protein ExoU is a phospholipase produced by particularly virulent strains of *P. aeruginosa*. Benson et al. (2009) measured ExoU enzymatic activity at varying levels of the fluorescent phospholipase substrate PED6. The data are in the dataframe enzyme.
>
> ```
> data(enzyme)
> ```
>
> Under the Michaelis–Menten model, we have
>
> ```
> MM.nls <- nls(rate  ~ SSmicmen(substrate, Vm, K), data = enzyme)
> summary(MM.nls)
> Parameters:
> ```

```
see.nlm()
```

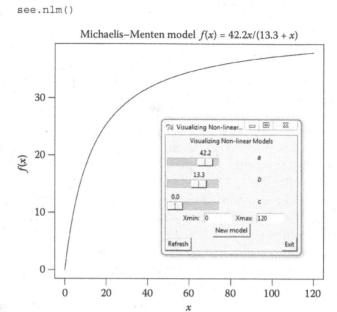

Michaelis–Menten model $f(x) = 42.2x/(13.3 + x)$

FIGURE 9.24

Interactive depiction of the Michaelis–Menten nonlinear model for Example 9.30. The X-axis represents concentration of the fluorescent phospholipase substrate PED6 in μM. The Y-axis represents the cleavage rate for ExoU in nM PED6 per mg ExoU.

```
     Estimate Std. Error t value Pr(>|t|)
Vm    42.169      1.493   28.243 2.67e-09 ***
K     13.283      1.709    7.774 5.37e-05 ***
```

The fitted model is shown in Figure 9.24.

Note that I used the `selfStart` function `SSmicmen`. I could have also defined something like `formula = rate ~ (Vm * substrate)/(K + substrate)`, although this would have required specification of starting values for `Vm` and `K` (γ_1 and γ_2 in Equation 9.102) in the `start` argument.

P-values in the `nls` output concern null hypothesis tests of $\gamma_k = 0$. General tests of $\gamma_k = \gamma_{k0}$ can be run for any γ_{k0} using the test statistic

$$t^* = \frac{\hat{\gamma}_k - \gamma_{k0}}{\hat{\sigma}_{\hat{\gamma}_k}},\tag{9.103}$$

where $\hat{\sigma}_{\hat{\gamma}_k}$s are obtained from an approximate covariance matrix made up of the outer product of partial derivatives from the regression model evaluated at the final least squares parameter estimates (see Kutner et al. 2005). These are given for γ_{k0} in `nls` output. The distribution under H_0 is $t(n - p)$.

9.21.3 Assumptions and Additional Comments

Inferences concerning nonlinear model parameters require valid assumptions of

1. Independent observations.
2. Identical, normally distributed errors.

Diagnostic plots are not a utility function of nls. However, one can get residuals using fitted values. For instance

```
e.i <- fitted(MM.nls) - enzyme$rate
```

allowing diagnostic plots and hypothesis tests, for example, qqnorm(e.i). Nonetheless, even given the validity of these assumptions, inferences concerning nonlinear models will only be approximate (see below).

Nonlinear regression models have two advantages over other nonlinear approaches presented in this chapter. First, unlike smoother approaches (discussed next), they provide a mechanistic global model of Y as a function of X. Second, although coefficients do not directly summarize how Y varies with X (they do not represent the change in Y given a one unit increase in X), they often represent important model characteristics that can be directly inferred to the phenomenon under study, for example, the maximum kinetics rate.

There are two disadvantages of nonlinear regression models. First, as an inconvenience, user-supplied initial guesses for parameters are required for many nonlinear least square algorithms. Second, inferences concerning nonlinear regression parameters are based on *large sample theory*. This theory asserts that, given independent and homoscedastic errors and large sample sizes, the sampling distributions of least squares estimators for nonlinear regression parameters will (1) be approximately normally distributed, (2) be almost unbiased, and (3) have approximate maximum efficiency. Unfortunately, no set rules exist for defining when a sample size large enough for large sample theory be applicable to all nonlinear models (Kutner et al. 2005). To aid in the diagnosis of sufficient sample size, Box (1971) and Hougaard (1985) have developed methods for determining bias and skewness in nonlinear regression estimates, respectively. Bootstrap methods can also be used to determine bias and to make inferences concerning estimates.

9.22 Smoother Approaches to Association and Regression

In all regression models discussed so far, we calculate estimates parameters for a *global model*. That is, parameter estimates are based on all X, Y pairs of observations, and can be used to mechanically interpolate predictions of Y within the hull of X. In contrast are approaches that fit *local models* based on response outcomes in adjacent predictor space. These approaches are called *smoothers* and make no *a priori* assumptions about the form of association of Y and X, for example, linear or some specific type of nonlinearity. Instead, these methods let the data "speak for themselves" and as a result can have response surfaces of virtually any shape.

Smoothers have two major drawbacks. First, they do not provide straightforward deterministic models, and as a result, they are somewhat difficult to evaluate except graphically. Second, smoothers may require subjective specification of smoothing parameters, and may provide no guidance as to when data are overfit or underfit. To address the second problem, methods have been developed for some smoother approaches that define the optimal size of the local neighborhood, and other parameters that affect smoothing. Despite these disadvantages, biologists often use smoothers as a means of determining the correct polynomial function or nonlinear model, to express associations bereft of assumptions, or to express relationships that are locally complex.

There have been a large number of proposed smoother approaches, including relatively simple methods such as local neighborhood medians and means (Tukey 1977). More recent and more useful approaches include *locally weighted scatter plot smoother (LOWESS)*, kernel-based methods, and splines.

9.22.1 LOWESS

A LOWESS approach obtains predicted values based on locally WLS regressions (Cleveland 1979). Regression weights will either be zeroes, if observations are outside the defined local neighborhood in predictor space, or will be the result from a *bicubic function* if points are within the neighborhood.

To find the LOWESS predicted value, $\hat{Y}_h$, we first calculate *Euclidean distances* for the predictor value(s) of interest, X_h, to all other values in predictor space. Euclidean distance constitutes a *p*-dimensional application of the Pythagorean theorem. For instance, if we have two predictors, then we would calculate the distance of the *i*th predictor values to X_{1h}, X_{2h} using

$$d_i = \left[\left(X_{1i} - X_{1h} \right)^2 + \left(X_{2i} - X_{2h} \right)^2 \right]^{1/2}, \tag{9.104}$$

where X_{1h} is the *h*th observation from predictor X_1, X_{2h} is the *h*th observation from predictor X_2, X_{1i} is the *i*th observation from predictor X_1, and X_{2i} is the *i*th observation from predictor X_2.

For models with multiple predictors, explanatory variables will generally be normalized preceding this step to preventing overweighting of large magnitude variables. This is typically accomplished by dividing observations by their standard deviation after 10% trimming (Cleveland et al. 1992).

The number of observations that contribute to a fitted value, and their degree of contribution, depends on the distance, d_i. Let d_q denote the boundary of the local neighborhood of interest around X_h; then the weights are

$$w_i = \begin{cases} \left[\left(1 - d_i/d_q \right)^3 \right]^3 & \text{when } d_i < d_q \\ 0 & \text{when } d_i \geq d_q \end{cases}. \tag{9.105}$$

The weights are used directly in WLS (Section 9.13) to estimate local regression parameters. The "smoothness" of the overall curve will depend on the size of the local neighborhood, that is, the *span*. A larger span will result in greater smoothness (less curve complexity). It follows that spans that are too large will underfit data while spans that are too small will result in overfitting.

9.22.2 Kernel-Based Approaches

Kernel-smoothers provide fitted/predicted values based on the weighted means of neighboring responses. With the Nadaraya–Watson approach (Nadaraya 1964, Watson 1964), weights are dependent upon the kernel function:

$$\hat{Y}_h = \sum_{i=1}^{n} Y_i \left[K\left(\frac{X_h - X_i}{b}\right) \middle/ \sum_{i=1}^{n} K\left(\frac{X_h - X_i}{b}\right) \right], \tag{9.106}$$

where b is a bandwidth parameter that defines the width of the probability mass around X_h, and K is a kernel function. The bandwidth creates kernel quantiles that are scaled differences from X_h. Y values that are adjacent to Y_h in predictor space will have higher corresponding kernel densities, and will be given greater weight when computing $\hat{Y}_h$. Thus, b is analogous to span for LOWESS. K is required to be symmetric, continuous, and bounded with the form $\int K(u)du = 1$. Of course, all continuous pdfs meet these criteria, including the normal distribution.

Optimal bandwidths can be determined using mechanistic approaches (Ruppert et al. 1995, Wand and Jones 1995). The "direct plug-in methodology" of Sheather and Jones (1991) is implemented in the function dpik of package *KernSmooth* (Wand 2012).

9.22.3 Splines

Splines are long flexible strips of wood that shipbuilders bend and hold at particular control points. The elasticity of the splines is employed to form the curved shape of a ship's hull. A *spline smoother* is analogous to this definition because it consists of linear fits bent at particular control points called *knots* (Schoenberg 1964).

The basis for a *cubic spline* can be tied to framework of general linear models established earlier in this chapter. Assuming a scaled [0, 1] range of predictor values using

$$X_{\text{scaled}} = (X - \min(X))/\max(X - \min(X)), \tag{9.107}$$

let

$$R(x,z) = [(z - 1/2)^2 - 1/12][(x - 1/2)^2 - 1/12]/4$$

$$- [(|x - z| - 1/2)^4 - 1/2(|x - z| - 1/2)^2 + 7/240]/24. \tag{9.108}$$

The degree of spline smoothness will be largely due to the size of its *basis dimension*, q, where $q =$ the number of knots $+ 2$.

Assume the form of a general linear model $\mathbf{Y} = \mathbf{X}\beta$, whose parameters, β, are estimated using OLS, and whose design matrix is

$$\mathbf{X} = \begin{bmatrix} 1 & X_{1,1} & R(X_{1,1}, X_1^*) & \cdots & R(X_{1,1}, X_{q-2}^*) \\ \vdots & \vdots & \vdots & \ddots & \vdots \\ 1 & X_{1,n} & R(X_{1,n}, X_1^*) & \cdots & R(X_{1,n}, X_{q-2}^*) \end{bmatrix}, \tag{9.109}$$

where $X_{1,1}$ is the first scaled observation from the first predictor variable (we have only one predictor in this example), and $R(X_{1,1}, X_1^*)$ is the function defined in Equation 9.108 applied to the first observation of the first predictor at the location of the first knot. Knots are generally evenly spaced or placed at unique X values.

A basic concern of splines is the specification of the knot number. One solution is to hold q constant at some level slightly larger that would be believed to be reasonably necessary and alter the "wiggliness" at the knots with a smoothing parameter, λ (Wood 2006). Given λ, spline coefficients are obtained using the method of *penalized least squares*, defined as

$$\hat{\beta} = X(X'X + \lambda S)^{-1}X',$$

(9.110)

where S is an $n \times q - 2$ matrix of coefficients resulting from the application of Equation 9.108. Unfortunately, this still leaves the problem of specifying λ. If λ is too large, then the data will be overfit; conversely, if it is too small, the data will be underfit. A method discussed by Wood (2006) uses cross-validation procedures to identify the optimal value of λ. Given outcomes from Equations 9.109 and 9.110, fitted values can be obtained using $\hat{Y} = X\hat{\beta}$.

EXAMPLE 9.31

Phosphorous levels in the Portneuf River near Pocatello Idaho are closely monitored by federal, state, and tribal (Shoshone-Bannock Tribe) environmental agencies. This is because an elemental phosphorous refinery has been located adjacent to the river since 1944. Storage wells and industrial processes at the refinery have resulted in significant discharge of P into the river, particularly during spring runoff. Phosphorous levels since January 15, 1998 have been continuously recorded, and are contained in the dataset portneuf.

```
data(portneuf)
```

The first column of portneuf is a time series object incorporating day, month, and year. We can convert this to a quantitative variable, the number of days from the beginning of monitoring, with the function difftime.

```
date <- as.POSIXlt(portneuf[,1])
ndays <- -1 * as.numeric(round(difftime("1998-01-15", date,
units = "days"), 0))
```

Assume that we are interested in predicting the level of phosphorous 500 days after the start of the monitoring without any assumptions concerning the form of association (linear or otherwise) between P and time.

Lowess Approach

We will (subjectively) use a local neighborhood whose boundary spans one-fifth the data. That is, we let span = 0.2. First, we calculate Euclidean distances, and rank the distances from the smallest to the largest.

```
e.dist <- sqrt((ndays - 500)^2)
s <- sort(e.dist)
```

Because the span is 0.2, and because there are 172 observations, the 34 responses associated with predictor observations closest to $X = 500$ will be used (weighted) to predict phosphorous when the number of days $X = 500$, that is, $172(0.2) \approx 34$. The 34th ranked distance is 436.

```
s[34]
[1] 436
```

Thus, $d_q = 436$.

Next, we calculate the weights using Equation 9.105:

```
w.i <- sapply(e.dist, function(x){ifelse(x<436, (1 - ((x/436)^3))^3,
0)})
```

Finally, we use WLS to calculate parameter estimates (Equation 9.59).

```
X <- as.matrix(cbind(rep(1, 172), ndays))
Y <- as.matrix(portneuf$total.P)
W <- diag(w.i)
beta.WLS <- solve(t(X) %*% W %*% X) %*% (t(X) %*% W %*% Y)
beta.WLS
             [,1]
       0.4780665059
ndays  0.0004371348
```

Thus, the LOWESS predicted value of phosphorous on day 500 is $0.00048 + 0.478$ $(500) = 0.696$ mg/L.

We can bypass the individual steps above by using the function `loess`.

```
P.lowess <- loess(Y ~ ndays, degree = 1, span = 0.2)
predict(P.lowess, 500)
[1] 0.6956891
```

The argument `degree` is used to specify first- or second-order polynomial WLS fits. A second-order model (`degree` = 2) can be used to detect or depict substantial curvature.

Kernel Approach

The use of the bandwidth selection function `dpik` indicates that $b \approx 465$ is optimal.

```
install.packages("KernelSmooth"); library(KernelSmooth)
dpik(ndays)
[1] 464.837
```

Given $a = \dfrac{X_h - X_i}{b}$, we have

```
a <- (500 - ndays)/465
```

For a standard normal kernel function, we have

```
K <- dnorm(a)
```

Under Equation 9.106, the predicted level of phosphorous on day 500 is 0.77 mg/L.

```
with(portneuf, sum((total.P * K)/sum(K)))
[1] 0.7734064
```

Spline Approach

We first we scale the predictor using Equation 9.107.

```
scale <- function(x) (x - min(x))/max(x - min(x)); x <- scale(ndays)
```

The cubic spline function from Equation 9.108 is

```
R <- function(x, z){
((z - 0.5)^2 - 1/12) * ((x - 0.5)^2 - 1/12)/4 - ((abs(x - z) - 0.5)^4 -
(abs(x - z) - 0.5)^2/2 + 7/240)/24}
```

Next, we set up the design matrix using a function (D below) whose arguments will be the scaled predictor values and the number of knots. Code here follows Wood (2006).

```
D <- function(x, k){
q <- length(k) +2
n <- length(x)
X <- matrix(1, n, q)
X[,2] <- x
X[,3 : q] <- outer(x, k, FUN = R)
X
}
```

We will (arbitrarily) use 30 evenly spaced knots.

```
k <- 1 : 29/30 # Define thirty evenly spaced knots in (0,1):
X <- D(x, k) # Create the design matrix
beta.hat <- solve(t(X)%*% X)%*% t(X)%*% Y
```

The scaled value for day 500 is

```
scale.500 <- (500 - min(ndays))/max(ndays - min(ndays))
```

A design matrix for $X_h = 500$ can be created using

```
Xh <- D(scale.500, k)
```

Thus, the predicted level of phosphorous on day 500 is 0.67 mg/L.

```
beta.hat%*% Xh
          [,1]
[1,] 0.6708901
```

The three smoother approaches can be viewed simultaneously and interactively with the function see.smooth.tck (Figure 9.25). Note that periodicity is essentially nonexistent in the scatterplot, and that predictor values are not even spaced, preventing the analysis with conventional time series approaches. Despite this, the smoothers capture the basic trends over time. Of particular interest is a dip in river phosphorous levels after approximately day 3500. This closely coincides with the time of repair and reinforcement of sump wells at the refinery (Idaho DEQ, personal communication).

9.22.4 Generalized Additive Models

GAMs extend GLMs by allowing specification of both smoother and conventional parametric terms for models. As with GLMs, nonlinearity and nonnormal errors can be accounted for by specifying particular link functions and error distributions. The name "additive model" comes from a smoother approach that fits the response as an additive function of penalized regression splines, for ≥ 1 predictors, while assuming that $\varepsilon_i \sim N(0, \sigma^2)$. Thus, just as GLMs follow from general linear models, GAMs follow from general additive models with normal errors. GAM coefficients are obtained by maximizing the likelihood function from the specified error distribution using penalized iteratively reweighted least squares (see Wood 2006).

Specification of error distributions allows hypothesis tests using likelihood ratio tests (in the case of normal errors, an *F*-test is used based on MSE). The effective degrees of

`see.smooth.tck()`

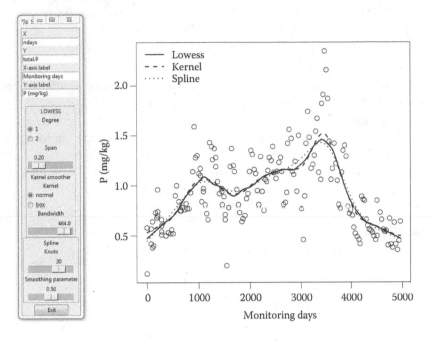

FIGURE 9.25

Smoother fits to the Portneuf phosphorous data. The spline parameter `spar` from `smooth.spline` is monotinically assocatied with the penalized spline wiggliness penalty criterion, λ. Note that the kernel fit does not exactly match the hand-calculated results. This is becasue kernels in `ksmooth` (used in the figure) are scaled so that the bandwidth $\times$ ±0.25 comprises the kernel interquartile range.

freedom for smoothers can be estimated using the GAM influence matrix (for details, see Wood 2006, p. 171). The relative parsimony of GAMs (which can be complex) can be compared to other GAMs (and to GLMs) using information-theoretic criteria such as AIC. The number of parameters here is defined using the effective degrees of freedom. As with other smoother approaches a straightforward mechanistic model will not be obtained as the result of spline, and mixed non-spline/spline fitting.

9.22.4.1 gam

The function `gam` from the library *mgcv* (mixed GAM computation vehicle) implements GAMs in **R** (Wood 2011). Its major arguments include `formula`, `family`, `knots`, and `data`.

- `formula` is analogous to the same argument in `lm`, `glm`, and `nls` with the difference that smoother terms can be specified in a GAM model. Smoother fits are defined within `formula` using the function `s`, which itself has several arguments. The first (triple dot) argument in `s` defines the predictors that the smooth is a function of, while the argument `bs` defines the smoothing basis to be used. Options for `bs` include `"tp"` for thin plate spline (the default) and `"cr"` for cubic spline. *Thin plate splines* eliminate the need for subjective assignment of knot location and spline bases (necessary for cubic splines), although they are computationally more costly than cubic splines (Wood 2006). Consider a situation where Y is to be considered as a function of two predictors X_1 and X_2. We would define the additive component due to X_2 to be a simple cubic smooth function with

```
formula = Y ~ X1 + s(X2, bs = "cr")
```

- `knots` is used to specify the number of knots to be used in smoothers (it can also be specified within individual smoothing terms, s, within `formula`).
- The `family` argument is analogous to the same argument in `glm` that is used to define the link function and error term distribution. By default, an identity link is used, and normal errors are assumed for estimation and inference.
- The `data` argument specifies a dataframe, list, or environment, coercible with `as.data.frame`, containing variables used in `formula`.

EXAMPLE 9.32

Application of `gam` to the Portneuf dataset reveals

```
summary(gam(total.P ~ s(ndays), data = portneuf))
Family: gaussian
Link function: identity

Parametric coefficients:
            Estimate Std. Error t value Pr(>|t|)
(Intercept)  0.93397    0.01916   48.74   <2e-16 ***
---
Approximate significance of smooth terms:
            edf Ref.df     F  p-value
s(ndays) 8.654  8.965 25.94   <2e-16 ***
---
R-sq.(adj) =  0.569   Deviance explained = 59.1%
GCV score = 0.066898  Scale est. = 0.063143  n = 172
```

The smooth term is highly significant assuming Gaussian errors. Thus, we reject the null hypothesis that the identified model is no better than the null model (containing just the intercept). Note that Efron's pseudo-R^2 (deviance explained) is included in the `summary.gam` output.

9.22.4.2 Assumptions and Additional Comments

As with GLMs, GAM assumptions will vary with error term specification. In general, GAMs assume that

1. The true model error and link function are correctly defined.
2. Observations are independent.

Scale term estimates are included with model output from `gam`, and can be used to assess the legitimacy of binomial and Poisson errors, which have dispersion parameters fixed at one. Models with Gaussian errors can be assayed using conventional general linear model approaches, although, of course, linearity is no longer assumed. Residuals, fitted and predicted values, and smoother plots for individual predictors can be obtained from `gam` utility functions.

★ ### 9.23 Bayesian Approaches to Regression

With a Bayesian approach to regression, we first assume that the parameters in β are not fixed entities, but are random variables. As a result, we do not evaluate whether an element of β equals a particular value. Instead, we are interested in the posterior distributions of β, σ^2, and $E(Y_i)$. The approach, shown below, is similar to the one described for normal hierarchical models in Example 5.21.

We start with a conventional regression model; one that assumes a linear relationship between Y and X, and independent identically normal errors (Section 9.10). If we assume *standard noninformative priors* that are uniform on β and $\log(\sigma^2)$, then the distribution of β conditional on σ^2 and Y will be multivariate normal (Gelman et al. 2003). In particular

$$\beta|\,\sigma^2, Y \sim \text{MVN}(\hat{\beta},\ \mathbf{V}_\beta \sigma^2),$$

where $\hat{\beta}$ is obtained using the OLS estimator (Equation 9.23), while

$$\mathbf{V}_\beta = (\mathbf{X'X})^{-1}.$$

The marginal posterior distribution for σ^2 follows the two-parameter scaled inverse χ^2 distribution*. In particular

$$\sigma^2\,|\,Y \sim \text{Inv}\chi^2(n-p, \text{MSE}),$$

where n is the number of observations, p is the number of model parameters, and MSE is calculated using Equation 9.28.

Derivation of the posterior distribution for β requires two simulation steps. First, one randomly draws from the inverse χ^2 distribution to obtain the posterior values for σ^2. Second, one uses these values to define the covariance structure of a multivariate normal pdf. One draws from this distribution to create a posterior distribution for β.

EXAMPLE 9.33

We will derive the posterior distribution of β using the frigate bird regression model from Example 9.2.

```
n <- 18; p <- 2
X <- with(Fbird, cbind(rep(1, 18), vol))
Y <- Fbird$freq
beta.hat <- solve(t(X)%*%X)%*%t(X)%*%Y
V.beta <- solve(t(X)%*%X)
MSE <-(t(Y)%*%(Y)-t(beta.hat)%*%t(X)%*%Y)/(n - p)
```

---

* The scaled inverse χ^2 pdf, parameterized as stated above, is

$$f(x) = \frac{[(n-p)/2]^{(n-p)/2}}{\Gamma[(n-p)/2]}\,\text{MSE}^{(n-p)}x^{-[(n-p)/2+1]}e^{-[(n-p)\text{MSE}/2x]}$$

```
#step 1

sigma.sq <- rinvchisq(1000, n - p, MSE) # Scaled inverse χ² deviates

#step 2
library(mvtnorm)
beta.f.x <- matrix(ncol = 2, nrow = 1000)
for(i in 1 :1000){
beta.f.x[i,] <- rmvnorm(1, beta.hat, V.beta *sigma.sq[i])}
```

Here is code to plot the result (see Figure 9.26).

```
plot(beta.f.x, xlab = expression(paste(beta[0], "|Y, ",
sigma^2)),ylab = expression(paste(beta[1], "|Y, ", sigma^2)))
```

The function bayes.lm calculates credibile intervals for Bayesian general linear models:

```
bayes.lm(Y, X, model = "reg")
Bayesian linear model with standard uniform priors
                2.5%        Median          97.5%
sigma.sq 433.09328317 806.35612506 1699.25626974
beta0    522.97970147 558.58278874  593.43752308
beta1     -0.02854152  -0.02068323   -0.01317231
```

The 95% credible interval for β_1 is (−0.0285, −0.0132). This is very similar (although slightly wider) than the conventional 95% confidence interval for β_1 (see Example 9.10).

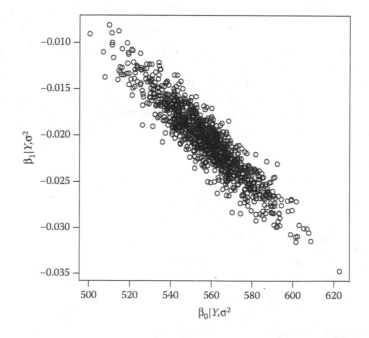

FIGURE 9.26
Joint conditional posterior distribution for parameters in the frigate bird example. Note that β_0 and β_1 are negatively correlated.

9.24 Summary

- Regression models are applied in situations where X is quantitative. Most of Chapter 9 deals with conventional regression models that have quantitative Y variables and one or more quantitative X variables. These characteristics describe simple and multiple linear regression models, which are a subclass of general linear models. Parameter estimates for these models are derived using the method of OLS, which minimizes a function that sums the squared differences of fitted values (based on parameter proposals) and observed responses.

- General linear regression models assume that Y can be expressed as a linear function of X, and that error terms are independent and are identically normally distributed. Outliers may compromise reliability for all general linear regression models. Multicollinearity (correlation of predictors) may prevent reliable parameter estimation and inference in multiple regression.

- Problems with model assumptions can often be addressed with transformations of the response and/or predictors. Transformations, however, will generally not solve problems with outliers. For these situations, robust regression techniques, including M estimation of parameters is often useful. WLS provides another method for addressing nonconstant variance by incorporating weights, which vary with residual variability, into parameter estimation.

- Likelihood is calculable for any linear model with an assumed error distribution, and is easily understood in the context of general linear models. In this case, likelihood is simply the product of the densities of residuals obtained from a normal distribution with a mean of 0 and a variance obtained from ML estimation. Modifications of likelihood, such as AIC and AICc, provide parsimony criteria, and the capacity to evaluate the relative optimality of regression models.

- Conventional regression models assume that X values are fixed and measured without error. An approach called model II regression can be used to address situations occurring frequently in biological research when X is a random variable.

- GLMs encompass general linear models, but allow nonnormal errors resulting from unconventional (discrete or dichotomous) response variables, and have link functions allowing models to be linearly expressed. GLMs use the method of maximum likelihood to estimate parameters and a generalized measure of error term variability called deviance.

- Nonlinear models can be used when Y cannot be expressed as a linear model. Nonlinear models are frequently driven by the knowledge of particular biological systems, and model parameters often represent important characteristics of the study system.

- Unlike other regression approaches, smoothers use local models to predict Y. As a result, they have no assumptions concerning the form of association of Y and X (linear or otherwise). Smoothers, however, do not provide straightforward mechanistic models.

- Bayesian approaches can be used with regression. In this context, parameters are not viewed as fixed, but as variables with conditional posterior distributions.

EXERCISES

1. In Example 2.1, data from Gosline et al. (1984) were used to heuristically illustrate the statistical analysis of random variables. Perform a regression analysis using these data, which are located in the dataframe webs.

   a. Correctly state your hypotheses.

   b. Calculate $\hat{\beta}_0$, and $\hat{\beta}_1$ and R^2 by hand (use **R** to help, but do not use lm).

   c. Correctly interpret the meaning of your results for $\hat{\beta}_1$ and R^2.

   d. Calculate t^* for $\hat{\beta}_1$ by hand.

   e. What P-value is obtained from (d)?

   f. Predict $\hat{Y}$ when the temperature = 15.2°C.

   g. What is the residual result from (f)?

   h. Verify your results using lm. Attach your results.

   i. Create diagnostic plots and interpret them.

   j. Reverse the explanatory variable and response variables. Is R^2 different from your answer in (b)? Is the P-value for the slope different from (e)?

★ k. Calculate the likelihood of the model by hand (without using logLik). Does reversing the order of the variables change the likelihood? Comment after rereading Sections 9.10 and 9.16.

★ l. Calculate AIC by hand (without using aic).

2. Ott and Longnecker (2004) describe a study that measured chicken weight gain (in grams) given a lysine additive (in grams).

```
weight.gain <- c(14.7, 17.8, 19.6, 18.4, 20.5, 21.1, 17.2, 18.7,
20.2, 16.0, 17.8, 19.4)
lysine.eaten <- c(0.09, 0.14, 0.18, 0.15, 0.16, 0.23, 0.11, 0.19,
0.23, 0.13, 0.17, 0.21)
```

   a. State the null and alternative hypotheses for the regression.

   b. Run the regression using lm in **R**. Attach output.

   c. State your conclusions. Can we reject H_0?

   d. Calculate the 90% CI for β_1. Correctly summarize your results.

   e. Calculate a 95% CI for $E(Y_h)$ by hand when $X_h = 0.14$. Correctly summarize your results.

   f. Calculate a 95% PI for $Y_h(\text{new})$ by hand when $X_h = 0.14$. Correctly summarize your results.

   g. Verify your results for (d), (e), and (f) using the functions confint and predict in **R**. Attach output.

   h. Create a plot of the regression. Overlay the confidence and prediction interval bands on the plot using plotCI.reg.

   i. Create diagnostic plots for the regression and interpret them.

3. Show that the following are linear models, that is, given p parameters, they have the general form: $Y = \beta_0 + X_1\beta_1 + \cdots + X_{p-1}\beta_{p-1}$.

a. $Y = \beta_0 + \beta_1 e^X$

b. $Y = \beta_1 X_i$

c. $Y = \beta_0 + \beta_1 X + \beta_2 X^2$

d. $Y = \exp(\beta_0 + \beta_1 X)$

4. Using the data from Question 2 above:

a. Complete an ANOVA table by calculating the sums of squares and mean squares by hand.

b. Verify your results using **R**; attach output.

c. Calculate R^2.

d. Correctly interpret the meaning of the R^2 value.

e. Calculate F^*.

f. Calculate the P-value.

g. If you have not done so already, complete Questions 2a, through 2c. Compare your conclusions in 2c to your conclusion in 4f.

5. Given that $\hat{\beta}_0 = \bar{Y} - \hat{\beta}_1 \bar{X}$ and $E(\bar{Y}) = \beta_0 + \beta_1 \bar{X}$, show that $\hat{\beta}_0$ is an unbiased estimator for β_0; that is, show $E(\hat{\beta}_0) = \beta_0$.

6. Using the data in Question 2 above, which points, if any, have high leverage according to the criteria of Kutner et al. (2005)? Which points, if any, have high Cook's distance?

7. Below are hypothetical counts of big horn sheep for a series of mountain aerial surveys on 10 consecutive years.

```
year <- 1:10
sheep <- c(1, 3, 11, 40, 133, 155, 335, 594, 588, 768)
```

a. Create a scatterplot. Does a linear regression appear to be valid idea?

b. Use the Box–Cox or Box–Tidwell method to find an appropriate transformation.

c. Create a linear regression using the transformed data.

   i. State hypotheses

   ii. Run the regression in **R**. Attach output

   iii. State conclusions

d. Calculate a 95% confidence interval for the slope.

e. Calculate confidence and prediction intervals for all of the fitted values using **R**.

f. Back-transform the regression equation to express variables in their original units.

g. Create diagnostic plots for the regression and interpret them.

8. Among other things the *theory of island biogeography*[*] predicts that species richness will increase with increasing island area. Islands, however, need not be a body

---

[*] The theory of island biogeography (Macarthur and Wilson 1963) remains one of the most important ideas of geographical ecology. The theory predicts that while particular types of species that inhabit an island will change (because of constant immigration and extinction), the number of species will be more or less constant after allowing for a certain length of time for equilibrium. The rates of immigration and extinction will vary depending on island size and island isolation. As a result, large islands (low extinction) that are near the mainland (high immigration) will have the highest number of species.

of land in water. Lomolino et al. (1989) investigated the relationship between the area of montane[†] forest patches (islands) and the richness of mammal fauna in the Southwestern United States. His data are contained in the *asbio* data object `mon-tane.island` Regress species richness (number of species) as a function of island.

a. Create a plot to examine the number of species as a function of island area.

b. A common transformation used when measuring the linear association of richness as a function of area is the log transformation of both X and Y variables (Begon et al. 1996). Perform this transformation and examine a scatterplot of the association.

c. Does the transformed data appear suitable for linear regression? Why or why not?

d. Perform a linear regression using the transformed data.

e. Preston (1962) described the following relationship:

$$S = CA^{\gamma}$$

where S is the number of species, A is the island area, C is a constant describing the number of species when $A = 1$, and γ is a biologically meaningful parameter describing a particular group of organisms, for example, flowering plants, birds, and zooplankton (Begon et al. 1996). In particular, the quantity γ represents the number of species that would be lost given the destruction of their habitat. Use the slope of the simple linear regression model $\log(S) = \log(A)\gamma$ to estimate γ (Ramsey and Schafer 1997). Assuming that the sampling distribution of γ is normally distributed calculate a 95% confidence interval for γ.

9. Of concern to public health officials and biologists is the safety and health of ecosystems in and around urban areas. Using a meta-analysis of government publications, Sokal and Rohlf (2012) compiled a dataset that described the quantity of air pollution (measured as annual mean SO_2 concentration per m^3) with respect to six environmental variables for 32 major cities in the United States. Whenever the data were available, variable outcomes are based on averages of 3 years 1969, 1970, and 1971. The dataset is included in *asbio* under the name `so2.us.cities`. Variables are described in `?so2.us.cities`.

a. Correctly state your hypotheses.

b. Create $\mathbf{X}$ and $\mathbf{Y}$ matrices and calculate $\hat{\boldsymbol{\beta}} = (\mathbf{X}'\mathbf{X})^{-1}\mathbf{X}'\mathbf{Y}$.

c. Calculate the fitted values using $\hat{\mathbf{Y}} = \mathbf{X}\hat{\boldsymbol{\beta}}$.

d. Calculate MSR $= \hat{\boldsymbol{\beta}}'\mathbf{X}'\mathbf{Y} - \left(\dfrac{1}{n}\right)\mathbf{Y}'\mathbf{1}\mathbf{Y}/(p-1)$.

e. Calculate MSE $= \mathbf{Y}'\mathbf{Y} - \hat{\boldsymbol{\beta}}'\mathbf{X}'\mathbf{Y}/(n-p)$.

f. Calculate F^*. Calculate the p-value for the F-statistic. What do you conclude?

g. Calculate $\boldsymbol{\varepsilon} = \mathbf{Y} - \hat{\mathbf{Y}}$

h. Calculate the hat matrix using $\mathbf{H} = \mathbf{X}(\mathbf{X}'\mathbf{X})^{-1}\mathbf{X}'$.

i. Calculate leverages by hand and interpret them.

[†] Areas associated with mountains.

     j. Calculate Cook's distances using the function `cooks.distance`. Interpret your results.

     k. Calculate a vector of t statistics, and calculate P-values for the t-statistics. What do you conclude?

     l. Run the multiple regression using `lm`.

     m. Using a type II sums of squares approach, find the P-value for annual precipitation by hand (using only `lm` and `anova`). Verify your result using `Anova`.

     n. Calculate 99% confidence intervals for all β_k using the function `confint`.

     o. Calculate 90% confidence and prediction intervals for all $E(Y_h)$ using the function `predict`.

     p. Perform diagnostics on the model using the functions `plot` and `shapiro.test`. Interpret the four diagnostic plots

     q. Create partial residual plots of the model using the function `partial.resid.plot`.

     r. Create a correlation graph with the function `pairs`.

     s. Calculate *VIFs* for the model using the function `vif` and interpret your results.

     t. Calculate the log-likelihood of the full (reference) model "by hand," using `dnorm`.

     u. Calculate AIC, AICc, and BIC of the full model "by hand" using **R** to help.

     v. Establish a set of candidate models. What is the optimal model according to AIC?

     w. Take at least two paragraphs and interpret your results as a biologist.

10. Type `see.regression.tck()` or click on the *Regression mechanics* tab from Chapter 9 in the book menu.

     a. Two normal distributions are overlaid on the population regression line for illustrative purposes. What variable is normally distributed?

     b. Verify that the values of $E(MSR)$ and $E(MSE)$ in the figure are correct.

     b. Adjust the slider widgets, or simply click *Refresh* repeatedly (>30 times). Are MSE and MSR consistently greatly than or less than $E(MSE)$ and $E(MSR)$. Why?

11. Fit the following data using weighted least squares.

```
X <- 1:25; Y <- 3 * X + 4 + rnorm(25, sd = 4*(0.1 * X))
```

12. To the Question 2 data, add the outlier: weight gain = 20, lysine = 0.1. Reanalyze using OLS, and robust M, S, and MM estimators. Which parameter estimates were closest and furthest away from the original estimates in Question 2?

13. The dataset `mod2ex3` in `lmodel2` contains data used by Sokal and Rohlf (2012) to illustrate model II regression. The data describe the mass (to nearest 100 g) of unspawned Californian cabezon females (*Scorpaenichthys marmoratus*) and the number of eggs (in thousands) that they produced. Since no fixed values of X (mass) were established by the researchers, X can be considered a random variable.

    a. Calculate the MA and SMA slopes by hand.

    b. Confirm your results by running the model II regression in **R** using the function `lmodel2`.

    c. Interpret your results.

★ 14. The presence of tropical trees with diameter at breast height equal or larger than 10 cm were recorded along with environmental factors at Barro Colorado Island by Pyke et al. (2001). A subset of the data is in the object `BCI.plant` in *asbio*. Using only the two quantitative explanatory variables (precipitation and elevation) as predictors, create a model for *Anacardium excelsum* presence.

    a. Create a GLM using the appropriate link function.

    b. Using the fitted values, calculate the residual deviance by hand.

    c. Use `stepAIC` for variable selection.

    d. Assess the classification efficacy of the optimal model by creating a confusion matrix (using $c = 0.5$) and AUC (using function auc). Interpret your results.

    e. Make and interpret diagnostic plots of the optimal model.

    f. Calculate Efron's pseudo-R^2.

    g. Calculate the significance of variables in the optimal model by hand using drop in deviance tests. Verify your results in **R** using `Anova`.

★ 15. Brown et al. (1996) showed that Australian women who live in rural areas tended to have fewer visits with general practitioners. It was not clear from this data, however, whether this was because they were healthier or because of other factors (e.g., shortage of doctors, higher costs of visits, and longer distances to travel for visits). To address this issue Dobson (2001) compiled data describing the number of chronic medical conditions for women visiting general practitioners in New South Wales. Women were divided into two groups; those from rural areas, and those from urban areas. All of the women were aged 70–75, had the same socioeconomic status, and reported to general practitioners three or fewer times in 1996. The question of central interest was: "do women who have the same level of general practitioner visits have the same medical need?" The data are in the **R** object `chronic`.

    a. Calculate the treatment means and variances. Does this appear to be a good candidate for analysis with a Poisson GLM? Why?

    b. Perform a GLM analysis on the data using `glm`, specify `family = "poisson"`. Is this a regression model? Why or why not?

    c. Does overdispersion appear to be a problem? Why or why not?

    d. Calculate the null and residual deviance by hand (without `glm`).

    e. Calculate the significance of the treatment variable by hand using a drop in deviance (likelihood ratio) test. Verify your results in **R** using `Anova`.

★ 16. Excess levels of a saccharide called glycosaminoglycans (GAGs) can be used as an indicator of genetic disorders associated with a deficiency of lysomal enzymes. GAG excretion in urine is largely age dependent (infants secrete more GAGs than adults). According to www.cambridgebiomedical.com (accessed December 9, 2012) normal ranges of GAGs (in mg/mmol creatinine) are 0–1 year: 0.0–36.0, 2–3 years: 0.0–15, 4–5 years: 0.0–10.5, 6–7 years: 0.0–10.3, 8–9 years: 0.0–9.6, 10–11 years:

0.0–8.2, 12–13 years: 0.0–6.4, >13 years: 0.0–5.5. GAGs for 314 children data are in the object GAGurine from the library *MASS*.

a. Create smooths of GAGs as a function of age using LOWESS, kernel, and spline approaches. This will require use of the functions loess, skernel, and smooth.spline.

b. Analyze the data using a GAM with a cubic regression spline. Interpret the results.

c. Plot the smooth result from (b). Does the optimal spline from (b) differ from the arbitrary spline in (a)?

★ 17. Reanalyze Question 2 using a Bayesian approach. Summarize your results.

10

___

ANOVA

___

It is better to have an approximate answer to the right question, than an exact answer to the wrong question.

John Tukey

___

10.1 Introduction

ANOVA is a statistical tool for estimating treatment effects and testing hypotheses when predictors are categorical and the response variable is quantitative. Unlike simpler analyses for comparing hypothetical population means (e.g., t-tests), ANOVA predictors (factors) may have more than two categories (factor levels). Furthermore, multiple factors and their interactions can be considered within a single model.

If you have not already done so, read pertinent sections concerning ANOVA in Chapters 7 and 9 before proceeding further in this chapter. Recall that Section 7.6.5 introduces ANOVA designs and distinguishes important terms such as factor, factor-level, and fixed and random effects. ANOVA approaches to regression are described in Sections 9.4.3 and 9.5.3. Important computational information, including the decomposition of the total sums of squares, is presented in both sections.

EXAMPLE 10.1

In his *Statistical Methods for Research Workers*, Fisher (1925) introduced the world to ANOVA using data from the Rothamsted Agricultural Experimental Station where he worked as a biometrist.[*][†] In one example, Fisher compared potato (*Solanum tuberosum*) yield per plant for 12 varieties and 3 fertilizer treatments. Three replicates were measured for each of the $12 \times 3 = 36$ factor-level combinations. Fisher quantified differences in varieties, fertilizer, and their interaction, by testing a null hypothesis of no difference in the true mean yields. His results can be summarized with an interaction plot (Figure 10.1). Fisher found ample evidence countering the hypotheses of no effect for variety (K of K, Ajax, and Up-to-Date tended to always have the highest yield, while Duke of York had the lowest), little evidence of fertilizer effects (fertilizer treatments had similar yields), and essentially no evidence of interactions (high yield varieties always had high yields regardless of fertilizer, while low varieties always had low yields). The experiment can be seen as a two-way factorial design (see Examples 10.8 and 10.9).

___

[*] Established in 1843, Rothamsted is one of the world's oldest agricultural research institutions.

[†] More recent applications of ANOVAs include comparisons of remote sensing algorithms for detecting forest disturbance (Martin 2009), analyses of the relationship between alpine plant species diversity, life history traits and phytogeographical history (Thiel-Egenter et al. 2009), and the analysis of DNA microarray data to find the genes controlling fruit development in apples (Janssen et al. 2008).

```
Fig.10.1() ## See Ch. 10 code
```

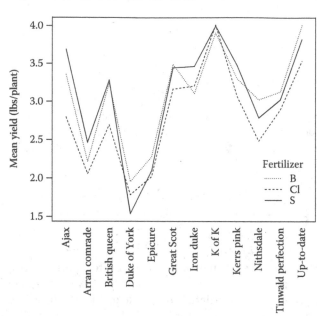

FIGURE 10.1

Potato yield experiment from Fisher (1925). ANOVA results indicated pronounced variety differences: $F_{11,72} = 11.9$, $P = 4.38 \times 10^{-12}$. Weak fertilizer effects: $F_{2,72} = 2.1$, $P = 0.13$, and no interaction between main effects: $F_{22,72} = 0.27$, $P = 0.999$.

10.1.1 How to Read This Chapter

The rationale for particular ANOVA layouts (completely randomized, randomized block, split plot, nested, repeated measures, etc.) was explained in Chapter 7. In this chapter, details for the analysis of these designs are presented, including methods for fixed, random, and mixed effect variants.

Because of their simplicity, a great deal of emphasis is given to one-way ANOVAs. This format is used to present ANOVAs as linear models, demonstrate contrast approaches, including pairwise tests, introduce random effects models, and to summarize general ANOVA assumptions. Coverage of sections given to these topics (Section 10.1 through 10.7) will be sufficient for most introductory classes.

Advanced ANOVA topics which are addressed in later sections of the chapter will be more appropriate for advanced students. These include multifactor analysis approaches (Section 10.8 through 10.12), ANCOVA (Section 10.13), approaches for unbalanced designs, including type II and III sums of squares (Section 10.14), robust methods (Section 10.15), and Bayesian approaches (Section 10.16).

10.1.2 Notation

ANOVA equations contain potentially confusing notation. I take a relatively simple approach.

- Factors will be indicated with capital letters, that is, A, B, and so on.
- The number of factor levels will be given the corresponding lowercase letters of the factor. Thus, A will have a factor levels, while B will have b factor levels.

- Individual factor levels within factors A, B, and C will be indexed using subscripts i, j, and k. For instance, factor A will have factor levels $i = 1,2,...,a$, and B will have factor levels: $j = 1,2,...,b$. The exception to this will be equations describing estimation at the level of model error. In this case, the use of the subscripts described above will be required to denote individual observations (see Equation 10.3 below).

- The total sample size across all factors and levels will be indicated with n, while the sample size for individual factor levels will be denoted with appropriate subscripts. For instance, the number of observations in the ith factor level in A will be n_i. For balanced designs, the level of replication will be denoted as n_0.

- The ith factor-level sample mean will be represented as $\bar{Y}_i$, while the sample mean of all n observations will be denoted as $\bar{Y}$.

- The constant μ in linear models will reflect the type of contrast (Section 10.3) and the purpose of the study; μ is estimated with $\hat{\mu}$. In some ANOVA frameworks with balanced designs, $\hat{\mu}$ will be equivalent to $\bar{Y}$.

- Factor-level population effects will be denoted with lowercase Greek letters corresponding to appropriate factors. For instance, the treatment effect of the ith factor level in A will be indicated α_i, while the treatment effect of the jth factor level in B will be denoted as β_j.

10.2 One-Way ANOVA

ANOVAs partition the total sums of squares (SSTO), that is, the sum of squared deviations from the sample grand mean, into variance components. This is expressed, for a one-way ANOVA (an ANOVA with a single factor), as

$$\text{SSTO} = \sum_{i=1}^{a} \sum_{j=1}^{n_i} (Y_{ij} - \bar{Y})^2, \tag{10.1}$$

where Y_{ij} represents the jth observation from the ith factor level.

In one-way ANOVAs, SSTO is divided into two components. These are the treatment sum of squares (SSTR), which is analogous to the regression sum of squares (SSR) in Chapter 9, and the sum of squares error (SSE). That is

$$\text{SSTO} = \text{SSTR} + \text{SSE}.$$

SSTR is the sum of squared deviations of the factor-level means and the mean of all n observations. In a one-way ANOVA, this quantity will be due to variation of factor-level means within a single factor, A. Thus, in this context, SSTR will only be composed of the sum of squares associated with factor A and can be called sum of squares A, or SS_A.

$$SS_A = \sum_{i=1}^{a} n_i (\bar{Y}_i - \bar{Y})^2. \tag{10.2}$$

Subscripts are used to emphasize that SSTR can be subset into more than one component. For example, in an ANOVA involving two factors A and B, and an interaction term $A{:}B$, SSTR will be split into components SS_A, SS_B, and $SS_{A:B}$, where $SS_{A:B}$ is the sum of squared deviations of individual observations from their respective means for their levels in A and B and the overall mean of all n observations (see Section 10.8).

As described for regression, SSE is the sum of squared deviations of the fitted values and the individual observations. Fitted values, however, are no longer values from the OLS regression line; they are factor-level means. Thus, for a one-way ANOVA

$$\text{SSE} = \sum_{i=1}^{a} \sum_{j=1}^{n_i} (Y_{ij} - \bar{Y}_i)^2 = \sum_{i=1}^{a} \sum_{j=1}^{n_i} \hat{\varepsilon}_{ij}^2, \tag{10.3}$$

where $\hat{\varepsilon}_{ij}$ is an estimate of the true model error for the jth observation from the ith factor level.

The degrees of freedom for SS_A and SSE are $a - 1$ and $n - a$, respectively. Dividing the sum of squares by their degrees of freedom provides estimates for the variance among factor levels, MS_A, and the variance of observations within factor levels, MSE.

$$MS_A = \frac{SS_A}{a - 1}, \tag{10.4}$$

$$MSE = \frac{SSE}{n - a}. \tag{10.5}$$

As noted in Chapter 9, MSE is unbiased, consistent, and assuming normality, maximally efficient for $E(\text{MSE})$ (Graybill 1976, p. 632). $E(\text{MSE})$ is equivalent to the factor-level population variance, σ^2. The parameter σ^2 is assumed to be identical across all factor-level populations (see Section 10.7).

For a one-way ANOVA, comparing a factor-level populations, we are concerned with the following hypotheses:

$$H_0: \mu_1 = \mu_2 = \cdots = \mu_a$$

versus

$$H_A: \text{at least one } \mu_i \text{ not equal to the others.}$$

This set of hypotheses is equivalent to:

$$H_0: \alpha_1 = \alpha_2 = \cdots = \alpha_a = 0$$

versus

$$H_A: \text{at least one } \alpha_i \text{ not equal to } 0,$$

where α_i is the true effect of the ith factor level in A, and is measured as

$$\alpha_i = \mu_i - \mu. \tag{10.6}$$

The natural estimator for α_i is

$$\hat{\alpha}_i = \bar{Y}_i - \hat{\mu}. \tag{10.7}$$

The ANOVA test statistic is the ratio of the estimates of variability among factor levels, and the variability within factor levels. Thus, unexpectedly, we test for the equality of population means using estimates of population variances. For a one-way ANOVA, the test statistic is

$$F^* = \frac{MS_A}{MSE}. \tag{10.8}$$

As a general rule, when specifying the denominator for F^*, we use variance estimators for parameters that contain error information describing all aspects of the model excepting the factor or interaction of interest. For instance, in a one-way ANOVA, MSE is an estimator for σ^2, which does not contain distinct information about the true variability among levels in A. Thus, MSE is used as the denominator in the test statistic for a hypothesis test concerning A (Table 10.1).

From Chapter 5, we know that if the underlying populations for MSE and MS_A are normal and homoscedastic, F^* will be a random outcome from an F-distribution with $a - 1$ numerator degrees of freedom and $n - a$ denominator degrees of freedom. P-values are calculated using the upper tail of the F-distribution. That is, $P[F(a - 1, n - a) \geq F^*]$.

The expectation of the sampling distribution of MS_A is determined by both the error variance and the sum of the weighted effect sizes divided by their degrees of freedom (Table 10.1):

$$E(MS_A) = \sigma^2 + \sum_{i=1}^{a} n_i \frac{\alpha_i^2}{a - 1}. \tag{10.9}$$

Assuming equal sample sizes, Equation 10.9 reduces to

$$E(MS_A) = \sigma^2 + n_0 \sum_{i=1}^{a} \frac{\alpha_i^2}{a - 1},$$

where n_0 is the sample size in each factor level.

TABLE 10.1

General Form of One-Way ANOVA; α_i Represents the True ith Factor-Level Effect in Factor A

Variation Source	df	SS	MS	E(MS)	F*
A (among groups)	$a - 1$	$SS_A = \sum_{i=1}^{a} n_i(\bar{Y}_i - \bar{Y})^2$	$MS_A = \frac{SS_A}{a - 1}$	$\sigma^2 + \sum_{i=1}^{a} n_i \frac{\alpha_i^2}{a - 1}$	$\frac{MS_A}{MSE}$
Error (within groups)	$n - a$	$SSE = \sum_{i=1}^{a} \sum_{j=1}^{n_i} (Y_{ij} - \bar{Y}_i)^2$	$MSE = \frac{SSE}{n - a}$	σ^2	
Total	$n - 1$	$SSTO = \sum_{i=1}^{a} \sum_{j=1}^{n_i} (Y_{ij} - \bar{Y})^2$			

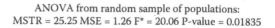

```
see.anova.tck()
```

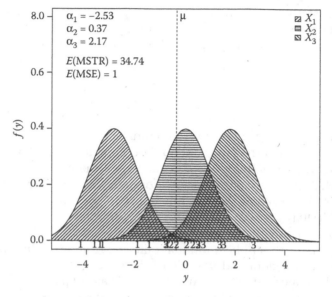

FIGURE 10.2

Mechanics of ANOVA. Differences in factor-level population means are ascertained from the variance among factor levels compared to the variance within factor levels. The variance of each group, σ^2, is defined to be 1, and the sample size from each group is 6. Numbers under the distributions indicate random sample outcomes from populations 1, 2, and 3.

If H_0 is true, then the true effects, α_is, will be 0 because the μ_is will not differ from the grand mean. In this case, $E(MS_A) = E(MSE) = \sigma^2$, resulting in a true test statistic of one. We would expect factor-level means of data randomly acquired from the populations to follow this pattern, with sampling error, resulting in $MS_A/MSE \approx 1$, and little evidence against H_0. Conversely, if the true variability among groups, $E(MS_A)$, is large, and the true variability within groups, $E(MSE)$, is small, then MS_A/MSE will also tend to be large, resulting in substantial evidence against H_0. This is illustrated in Figure 10.2. As treatment effects increase, due to increased factor-level differences, $E(MS_A)$ will also increase. On an average, this will increase the value of the test statistic, and increase the amount of evidence against H_0. Increasing sample sizes (balanced in this case) will also increase $E(MS_A)$ and tend to increase the test statistic. Increasing σ^2 [i.e., $E(MSE)$] will also increase $E(MS_A)$, viz., Equation 10.9; however, it will result in an overall decrease in $E(MS_A)/E(MSE)$, and therefore will tend to decrease the test statistic.

Unlike the true test statistic, it is conceivable that MS_A/MSE will be less than one as a result of sampling error. It has been argued that this outcome may indicate that an inappropriate model has been specified (for instance, pseudoreplication may artificially deflate MS_A) or the presence of serious outliers (Searle et al. 1992).

10.2.1 `lm` Revisited

The `formula` argument in `lm` can be defined using terms suited specifically for ANOVA (Table 10.2).

TABLE 10.2

Expressions for `formula` in `lm` and Their Meaning

Expression	Interpretation
A + B	Include both A and B in model
– B	Exclude B
A : B	Interaction of A and B
A * B	A + B + A:B
B %in% A	B nested in A
A/B	A + B %in% A
(A + B + C) ^k	All effects from A, B, and C crossed up to order k

As noted in Chapter 9, additive terms in `lm` are specified with a plus sign (+), while interactions are indicated with a colon (:). Main effects and interactions can be specified simultaneously using the asterisk character (*). That is, `A*B` is equivalent to `A + B + A:B`. The carat (^) sign defines crossed effects up to the order given in the exponent. Thus, `(A + B + C) ^2` is equivalent specifying to `A*B*C – A:B:C`. That is, the pairwise interactions, `A:B`, `B:C`, and `A:C`, would each be included in the model, but the three-way interaction `A:B:C` would not. The term `%in%` is used to define a nested experimental structure. For example, in split plot and nested designs with different sizes of experimental units we would specify B nested in A as `B %in% A`. The forward slash character (/) can be used a shortcut for specifying the main effects of both the nonnested term and the nested term. For example, `A/B` indicates `A + B %in% A`. Model definitions can also be combined. For instance, `B%in%A:C` indicates B nested in A crossed with C. These specifications are also valid for the `formula` argument in the functions `glm`, `nls`, and `gam` introduced in Chapter 9, and the functions `lme`, `lmer`, and `aov` introduced in this chapter.

EXAMPLE 10.2

Humans are born with rudimentary reflexes for walking, but these largely disappear by 8 weeks of age due to disuse. Accordingly, walking movements must be relearned by an infant following a significant passage of time, through a process of trial and error. Zelazo et al. (1972) performed a series of experiments to determine whether certain exercises could allow infants to maintain their walking reflexes, and allow them to walk at an earlier age. Study subjects were 24 white male infants from middle-class families. Infants were randomly assigned to one of four groups immediately following birth, and the age at the onset of unaided walking (in months) was verified for each infant. The groups were

- *Active exercise* (AE): parents were taught and were told to apply exercises that would strengthen the walking reflexes of their infant.
- *Passive exercise* (PE): parents were taught and told to apply exercises unrelated to walking.
- *Test-only* (TO): investigators did not specify any exercise, but visited and tested the walking reflexes of infants in weeks 1 through 8. This treatment was established to account for the potential effect of the walking reflex tests themselves.
- *Control* (C): no exercises were specified, and infants were only tested at weeks one and eight.

The data are contained in the dataframe `baby.walk`. Our hypotheses are

$$H_0\colon \mu_{AE} = \mu_{PE} = \mu_{TO} = \mu_C$$

versus

$$H_A\colon \text{at least one } \mu_i \text{ not equal to the others.}$$

In **R**, we have

```
data(baby.walk)
n.i <- print(with(baby.walk,tapply(date, treatment, length)))
AE  C PE TO
 6  5  5  6
Ybar.i <- print(with(baby.walk, tapply(date, treatment, mean)))
      AE        C       PE       TO
10.12500 12.35000 10.65000 11.70833
Ybar <- print(with(baby.walk, mean(date)))
[1] 11.18182
SSA <- print(sum(n.i * (Ybar.i - Ybar)^2))
[1] 16.60189
SSE <- print(with(baby.walk, sum((date - rep(Ybar.i[c(1, 3, 4, 2)],
n.i[c(1, 3, 4, 2)]))^2)))
[1] 27.92083
```

Thus

$$SS_A = 6(10.13 - 11.18)^2 + 5(12.35 - 11.18)^2 + \cdots + 6(11.71 - 11.18)^2 = 16.6$$

$$SSE = (9.0 - 10.12)^2 + (9.5 - 10.12)^2 + \cdots + (11.5 - 11.71)^2 = 27.92$$

and the completed ANOVA table has the form

Source of Variation	df	SS	MS	F*
Among groups	$a - 1 = 3$	16.6	5.53	3.568
Within groups	$n - a = 18$	27.92	1.55	
Total	$n - 1 = 21$			

The *P*-value, 0.035, allows rejection of H_0 at $\alpha = 0.05$.

```
pf(3.568, 3, 18, lower.tail = F)
[1] 0.03480957
```

This procedure can be efficiently handled by built-in **R** functions in a number of ways. One approach is to first create a linear model, as we did for regression.

```
lm.baby <- lm(date ~ treatment, data = baby.walk)
```

Next, we apply the function `anova` to partition the sums of squares of the linear model.

```
anova(lm.baby)
Analysis of Variance Table

Response: date
          Df Sum Sq Mean Sq F value  Pr(>F)
treatment  3 16.602  5.5340  3.5676 0.03482 *
Residuals 18 27.921  1.5512
```

These results agree with the hand-calculated values above.

10.3 Inferences for Factor Levels

The ANOVA procedure often serves as a precursor to a more detailed comparison of factor-level means. For instance, in Example 10.2, we conducted an ANOVA and rejected the hypothesis: $\mu_{AE} = \mu_{PE} = \mu_{TO} = \mu_C$. While this indicates that at least one μ_i is not equal to at least one of the others, we do not know which factor levels are significantly different. In addition, we do not know if there is evidence of a trend among μ_is, with some less than or greater than others. Factor-level comparisons to address these questions can be divided into three types: planned comparisons, unplanned comparisons, and data snooping.

In *planned comparisons* (often called *contrasts*), the investigator stipulates the null hypotheses for comparisons before obtaining ANOVA results or statistical summaries. For example, in the baby walk experiment, the investigators were particularly interested in comparing the AE and PE treatments. This is because a significant result would support the hypothesis that maintaining walking reflexes affects the inception of walking.

Planned comparisons are desirable because they generally involve a smaller number of potentially independent tests (see below). This increases statistical power when familywise error is being controlled (Sections 9.5.2 and 10.3.4). Planned comparisons can be stipulated in the contrast matrix of the linear model to prevent data snooping (see below and Section 10.3.3).

Unplanned comparisons are *post hoc*. That is, they are not prespecified by the investigator. A frequently used type of unplanned comparison is the set of all possible pairwise comparisons in which all individual factor-level means are compared to each other. *Data snooping* occurs when hypotheses for particular contrasts are specified after examining data summaries (e.g., boxplots). Both unplanned comparisons and data snooping can be used to *search* for significant results, and thus violate the familywise independence of individual tests. As a result, most experts recommend the application of methods that control familywise error when using these comparison approaches (see Section 10.3.4).

10.3.1 Introduction to Contrasts

We can contrast any two factor levels or groups of factor levels by defining a set of linear constraints on the factor-level means. This is called a *linear contrast*. Linear contrasts are made by (1) separating the factor levels into two groups, and (2) choosing two coefficients such that, after multiplicatively distributing the coefficients to factor levels within groups, the sum of the coefficients equals zero. Applied to the factor-level population means, this will result in the true difference, D, of the contrast. That is

$$D = \sum_{i=1}^{a} c_i \mu_i \quad \text{where} \sum_{i=1}^{a} c_i = 0, \tag{10.10}$$

where c_i is the ith linear contrast coefficient, and μ_i is the ith factor-level population mean.

For the baby walk analysis, suppose that we are interested in comparing the effect of AE treatment to the combined effect of all the other three treatments. We could use the following contrast:

$$D = 0.75(\mu_{AE}) - 0.25(\mu_{PE} + \mu_C + \mu_{TO}) = 0.75\mu_{AE} - 0.25\mu_{PE} - 0.25\mu_C - 0.25\mu_{TO}$$

We note that factor levels that are grouped are given the same sign, while the contrasted levels receive the opposite sign. The sum of the distributed coefficients equals zero.

Comparisons of individual factor-level means are called *pairwise comparisons*. For the pairwise comparison of a primary interest (AE vs. PE), we would use the contrast

$$D = 1(\mu_{AE}) - 1(\mu_{PE}) + (0)\mu_{C} + (0)\mu_{TO}.$$

Note again that the coefficients sum to 0.

We estimate D with $\hat{D}$

$$\hat{D} = \sum_{i=1}^{a} c_i \bar{Y}_i, \tag{10.11}$$

and estimate the standard error, $\sigma_{\hat{D}}$, with

$$\hat{\sigma}_{\hat{D}} = \sqrt{MSE \sum_{i=1}^{a} \frac{c_i^2}{n_i}}. \tag{10.12}$$

To test the hypothesis

$$H_0: D = 0$$

versus

$$H_A: D \neq 0,$$

we calculate the test statistic

$$t^* = \frac{\hat{D}}{\hat{\sigma}_{\hat{D}}}. \tag{10.13}$$

The two-tailed P-value is $2P(T \geq t^*)$, where $T \sim t(n - a)$. This procedure is identical to a t-test assuming equal variances, one of the assumptions for ANOVA (see Section 10.7).

EXAMPLE 10.3

For a contrast of the AE and PE treatments, we have the hypotheses

$$H_0: D = 0$$

versus

$$H_A: D \neq 0, \quad \text{where } D = \mu_{PE} - \mu_{AE}.$$

We estimate D with

$$\hat{D} = 1(10.65) - 1(10.125) = 0.525.$$

The test statistic is

$$t^* = \frac{0.525}{\sqrt{1.55\left(\frac{1}{6} + \frac{1}{5}\right)}} = 0.696,$$

and the P-value $= 2P(T \geq 0.696) = 0.495$, where $T \sim t(18)$. Thus, we fail to reject H_0, and conclude that there is little evidence supporting the hypothesis that reflex-specific exercises speed up the onset of walking. Note that this is identical to the third default contrast from lm.baby in Example 10.2 obtained by typing summary(lm.baby). This is because with default contrasts all factor levels are compared to the first alphanumeric factor level (thus, all treatments will be compared to AE). This process is explained in detail in Section 10.3.3.

10.3.2 Orthogonality

Given a large number of treatments, there may be a huge number of possible comparisons; however, there will be only $a - 1$ *orthogonal comparisons*, where a equals the number of distinct treatments (e.g., factor levels for a one-way ANOVA). Orthogonal contrasts will generally be planned, and will always be independent. As a result, they reduce type I errors that may result from correlated comparisons (Howell 2010), and allow the exact calculation of the rates of familywise error (see Section 10.3.4).

With orthogonal comparisons, things are only compared once. Consider a factor, A, with three factor levels, A_1, A_2, and A_3. Assume that we are primarily interested in comparing the combined effect of A_1 and A_2 with the individual effect of A_3. Because this contrast implicitly compares A_3 to both A_1 and A_2, there is only one orthogonal contrast remaining. This is A_1 versus A_2. It follows that orthogonal comparisons depend on the order that contrasts are specified by the investigator. For instance, another set of orthogonal contrasts would be the combined effect of A_1, and A_2 versus A_3, and A_2 versus A_3.

Linear contrasts, demonstrated in Example 10.3, can be summarized in the form of a *contrast matrix* in which rows represent factor levels, columns represent contrasts, and elements contain linear coefficients. In a one-way layout, a contrast matrix will have $a - 1$ nonredundant columns. In the case of orthogonal contrasts, each column will sum to zero, and the rowwise products of any of the two columns will sum to zero. For example, given a factor with three factor levels, we might have

Levels	Contrast 1	Contrast 2
A_1	0.5	-1
A_2	0.5	1
A_3	-1	0

In the first contrast, A_3 is compared to the combined effect of A_1 and A_2. As a result, we have $D = \mu_1(0.5) + \mu_2(0.5) + \mu_3(-1)$. For the second contrast, we compare A_1 and A_2, and as a result, $D = \mu_1(1) + \mu_2(-1) + \mu_3(0)$. The contrasts are orthogonal because the columns in the comparison matrix sum to zero, $0.5 + 0.5 - 1 = 0$, and $-1 + 1 + 0 = 0$, and because the sum of the product of the rows equals zero, $(0.5)(-1) + (0.5)(1) + (-1)(0) = 0$.

It has been argued that orthogonal contrasts have been overused by biologists at the cost of more meaningful comparisons (Quinn and Keough 2002). The desirable features of orthogonal contrasts (independence and calculability of familywise error) should not deter a scientist from specifying other more interesting comparisons. For instance, a series of nonorthogonal pairwise tests may be of greater interest than an awkward set of orthogonal factor-level groupings.

10.3.3 `lm` contrast

The `lm` argument `contrast`, while unimportant in regression, will have great importance given categorical predictors. Specifically, it will determine the form of the linear model design matrix, X (see Section 10.4), and define planned comparisons in `lm`. These specifications will not change the omnibus ANOVA results at the level of factors. Several contrast frameworks are commonly used by statistical software packages. These include *treatment, Helmert,* and *sum* contrasts. Custom contrast matrices can also be designed (with care) using the function `contrast`; see Crawley (2007) for examples. Note that while the baby walk dataset is analyzed in this section repeatedly for illustrative purposes, this would be inappropriate in an actual application of planned comparisons.

10.3.3.1 *Treatment Contrasts*

Treatment contrasts are the default for `lm` and other linear model functions in **R**. This is apparent by viewing the contrast options.

```
options("contrasts")
$contrasts
      unordered      ordered
"contr.treatment"  "contr.poly"
```

Note that there are two elements in the global defaults above, `contr.treatment` and `contr.poly`. These stipulate that treatment contrasts are to be used with unordered categorical explanatory variables and that a third-order polynomial model (Section 9.14) is to be applied in the presence of ordinal predictors. Recall that for the latter, we are not sure of the quantitative relationship of factor levels of each other, but only their order (e.g., $1 < 2 < 3 < 4$). The polynomial contrasts are forced (by `lm`) to be orthogonal to address multicollinearity among the predictors (Section 9.14). This is done by creating an orthogonal contrast matrix with the linear, quadratic, and cubic terms as columns and the ordinal factor levels as rows (see Crawley 2007, pp. 381–385).

With treatment contrasts, the first reported coefficient is the sample mean of the first alphanumeric factor level (say level 1 in A). Thus, the first factor-level sample mean is defined as the intercept $\hat{\mu}$ in the ANOVA linear model (Section 10.4). The associated null hypothesis for the first coefficient is that $\mu_1 = 0$. The remaining effects are estimated by subtracting $\bar{Y}_1$ from the remaining factor-level means, and the remaining null hypotheses are that the true difference of μ_1 and the other individual factor-level means is zero. The standard error of the first alphanumeric term will always be $\sqrt{MSE/n_1}$. Standard errors for other contrasts are found using Equation 10.12.

Here is the baby walk treatment contrast matrix:

```
contrasts(baby.walk$treatment)
     2 3 4
AE 0 0 0
C  1 0 0
PE 0 1 0
TO 0 0 1
```

Because effects are not computed with respect to the mean of the factor-level means (as other contrasts are), the treatment contrast matrix has an unusual form. Binary (0, 1) dummy codes are used, and the model contrasts are not explicitly shown. The matrix correctly shows, however, that the treatment contrasts are not orthogonal.

For the baby walk analysis, the following default null hypotheses will be tested by lm:

$$H_0: \mu_{AE} = 0,$$

$$H_0: \mu_C - \mu_{AE} = 0,$$

$$H_0: \mu_{PE} - \mu_{AE} = 0,$$

and

$$H_0: \mu_{TO} - \mu_{AE} = 0.$$

```
summary(lm.baby)
Coefficients:
            Estimate Std. Error t value Pr(>|t|)
(Intercept) 10.1250     0.5085  19.913 1.04e-13 ***
treatmentC   2.2250     0.7542   2.950  0.00856 **
treatmentPE  0.5250     0.7542   0.696  0.49523
treatmentTO  1.5833     0.7191   2.202  0.04095 *
---
Signif.codes: 0 '***' 0.001 '**' 0.01 '*' 0.05 '.' 0.1 ' ' 1

Residual standard error: 1.245 on 18 degrees of freedom
Multiple R-squared: 0.3729, Adjusted R-squared: 0.2684
F-statistic: 3.568 on 3 and 18 DF, p-value: 0.03482
```

The R^2 value indicates that 37% of the variability in the model is explained by differences in factor-level means. We can also see that the effect of AE is significantly different from zero, AE is very different from C (babies tend to walk more quickly under AE), and AE is also marginally different from TO and indistinguishable from PE. It is important to note that (1) these default comparisons are not orthogonal and (2) no contrasts specified from lm (including those that follow) are adjusted for simultaneous inference.

The model coefficients are obtained from

```
means <- with(baby.walk, tapply(date, treatment, mean))
c(means[1], (means - means[1])[-1])
       AE         C        PE        TO
10.125000  2.225000  0.525000  1.583333
```

10.3.3.2 Helmert Contrasts

Here is the Helmert contrast matrix for the baby walk example[*]:

```
contrasts(baby.walk$treatment) <- contr.helmert(4)
contrasts(baby.walk$treatment)
```

---

[*] A number of authors would refer to this as a set of *reverse Helmert contrasts*. For these individuals, Helmert contrasts would involve a reversal of the columns in the contrast matrix shown above. That is, in the first contrast, the first alphanumeric factor level would be compared to the average of the latter factor-level means; in the second contrast, the second alphanumeric factor levels would be compared to the average of the latter factor-level means, and so on.

```
     [,1] [,2] [,3]
AE    -1   -1   -1
C      1   -1   -1
PE     0    2   -1
TO     0    0    3
```

Note that the Helmert contrasts are orthogonal. That is, the columns in the contrast matrix sum to one, and all possible pairwise row products sum to zero. With Helmert contrasts, the linear model intercept term will be the mean of the factor-level means, and the first null hypothesis will be that $\mu = 0$ (the intercept term is never included in the contrast matrix). As indicated in the contrast matrix, the first contrast compares the population means of the first two alphanumeric factor levels. Note, however, that the coefficient given by the model will be the second factor-level mean subtracted from the average of the first two alphanumeric factor levels. These apparent discrepancies occur for computational reasons, and are "corrected" with adjustments to the standard errors. The second contrast compares the population means of the third factor level and the average of the first two factor levels. However, the coefficient reported is the average of the first two factor levels subtracted from the average of the first three factor levels. Finally, the third contrast compares the population means of the fourth factor level and the average of the first three factor levels. However, the coefficient reported is the average of the first three factor levels subtracted from mean of all the factor-level means.

Our null hypotheses are now

$$H_0: \mu = 0$$

$$H_0: \mu_C - \mu_{AE} = 0$$

$$H_0: \mu_{PE} - (\mu_{AE} + \mu_C)/2 = 0$$

$$H_0: \mu_{TO} - (\mu_{AE} + \mu_C + \mu_{PE})/3 = 0$$

By specifying Helmert contrasts above, we now have

```
summary(lm(date ~ treatment, data = baby.walk))
Coefficients:
              Estimate Std. Error t value Pr(>|t|)
(Intercept)   11.2083     0.2666   42.036< 2e-16   ***
treatment1     1.1125     0.3771    2.950  0.00856  **
treatment2    -0.1958     0.2242   -0.873  0.39392
treatment3     0.1667     0.1492    1.117  0.27867
Residual standard error: 1.245 on 18 degrees of freedom
Multiple R-squared: 0.3729,      Adjusted R-squared: 0.2684
F-statistic: 3.568 on 3 and 18 DF,   p-value: 0.03482
```

The overall average treatment effect is significant, and AE is significantly different than AE and C. Note that both the treatment contrasts (described earlier) and Helmert contrasts compare AE and C in their first contrast. The corresponding coefficient for the Helmert contrast is 1/2 the size of the treatment contrast. However, the standard error is also 1/2 as large. As a result, the test statistics and P-values for the two analyses are identical ($t_{18} = 2.950$, $P = 0.00856$). Also note that the residual standard errors (the square root of MSE), the R^2 measures, and the overall P-values are identical for the two approaches. This consistency will hold for any specified method of contrasts.

As tediously described above, the model coefficients can be obtained using

```
AE <- means[1]; C <- means[2]; PE <- means[3]; TO <- means[4]

c(mean(means), mean(c(AE, C)) - AE, mean(c(AE, C, PE)) - mean(c(AE, C)),
mean(c(AE, C, PE, TO)) - mean(c(AE, C, PE)))
11.2083333  1.1125000  -0.1958333  0.1666667
```

10.3.3.3 Sum Contrasts

Here is the sum contrast matrix for the baby walk example:

```
contrasts(baby.walk$treatment) <- contr.sum(4)
contrasts(baby.walk$treatment)
     [,1] [,2] [,3]
AE      1    0    0
C       0    1    0
PE      0    0    1
TO     -1   -1   -1
```

As with Helmert contrasts, the intercept term is set to the mean of the factor-level means, and the first null hypothesis is $\mu = 0$. However, the remaining contrasts concern the difference of the first $a - 1$ alphanumeric factor-level means and μ, and test the null hypothesis that this true difference is zero. Note that because the last factor level is not included in comparisons, it is used to represent the mean of the factor-level means in the contrast matrix above. The matrix shows that sum contrasts are not orthogonal.

The baby walk null hypotheses are now

H_0: $\mu = 0$

H_0: $\mu_{AE} - \mu = 0$ (i.e., the true effect of AE is not different from the overall effect)

H_0: $\mu_C - \mu = 0$

H_0: $\mu_{PE} - \mu = 0$

We have

```
summary(lm(date ~ treatment, data = baby.walk))
Coefficients:
            Estimate Std. Error t value Pr(>|t|)
(Intercept)  11.2083     0.2666  42.036 <2e-16    ***
treatment1   -1.0833     0.4476  -2.420  0.0263   *
treatment2    1.1417     0.4756   2.400  0.0274   *
treatment3   -0.5583     0.4756  -1.174  0.2557
```

These results indicate that AE and C treatment means each differ from the true mean of the factor-level means.

The coefficients can be obtained by taking factor-level means minus $\bar{Y}$, excluding the last alphanumeric factor level.

```
c(mean(means), means[1:3] - mean(means))
                     AE           C          PE
11.2083333  -1.0833333  1.1416667  -0.5583333
```

10.3.4 Issues with Multiple Comparisons

It is often desirable to have a measure of inferential correctness for an entire set of factor-level comparisons. One such measure is familywise type I error, that is, the probability that any one test in a family of tests incorrectly rejects H_0. Given m independent tests, this probability will be

$$1 - (1 - \alpha)^m,$$

where α is the significance level to be used for each test. Thus, five independent tests,[*] each using $\alpha = 0.05$, will have a familywise α of $1 - (1 - \alpha)^5 = 0.226$. That is, the probability of falsely rejecting H_0 across all five tests will be 0.226.

The problem of ballooning familywise type I error due to multiple comparisons has elicited a great deal of controversy and disparate opinions (see the reviews of Day and Quinn 1989 and Hancock and Klockars 1996). Many experts have argued that planned contrasts, particularly those that are orthogonal, require no familywise adjustments (Quinn and Keough 2002, Sokal and Rohlf 2012). Others have argued that all related nonorthogonal comparisons should be adjusted for familywise error (Keppel 1991, Todd and Keough 1994). Still others have maintained that type I error should be controlled in essentially any situation with related multiple comparisons (Ramsey 1993, Kutner et al. 2005). Essentially, all experts agree, however, that familywise error adjustments should be made in the case of unplanned comparisons or data snooping.

Two overall groups of approaches can be used for controlling familywise error in factor-level comparisons. The first uses MSE from an ANOVA as an estimate for the pooled variance in comparisons (Equations 10.11 through 10.13). P-values and confidence intervals are then adjusted using methods that vary in their definition of conservativeness (with respect to familywise error), and demarcation of the family of tests. These procedures are described in Section 10.3.5 and compared in Section 10.3.5.6. A second group of methods can be used to account for familywise error in any arbitrary set of tests (including non-general linear models) and are described in Section 10.3.6.

10.3.5 Simultaneous Inference Procedures

A large number of methods have been developed to address familywise type I error in the context of ANOVA. These procedures (1) address comparisons of population means, (2) utilize the ANOVA pooled variance estimator (MSE) based on all observations from all factor levels, and (3) assume normality and constant variance for the factor levels being compared (Shaffer 1995). Five important methods are briefly described here: Fisher's least significant difference (LSD), the Bonferroni correction, Scheffé's procedure, Tukey's honest significant difference (HSD), and Dunnett's procedure.

Hypothesis tests and confidence intervals can be considered simultaneously in these procedures. This is because a two-tailed hypothesis test for H_0: $D = 0$ and a confidence interval for D provide analogous information (Section 6.2.6). That is, if H_0: $D = 0$ is rejected at α, then the $(1 - \alpha)$ 100% confidence interval for D will not contain 0.

Of particular interest in *post hoc* analyses are comparisons of all possible pairs of individual factor levels. These will involve $(a^2 - a)/2 = (a - 1) + (a - 2) + \cdots + 1$ separate tests

---

[*] Familywise type I error rates cannot be calculated when comparisons are not independent/orthogonal (Quinn and Keough 2002). This is because the multiplication rule for the probability for independent events is no longer strictly applicable (Equation 2.11).

that will be nonorthogonal. For instance, in the baby walk example, we have $(16-4)/2 = 6$ possible pairwise comparisons, greatly exceeding the number of possible orthogonal comparisons.

10.3.5.1 Bonferroni Correction

The *Bonferroni correction* introduced in Section 9.5.2 is the simplest and the most conservative method for controlling familywise type I error. Given m comparisons, the Bonferroni correction uses a significance level of α/m instead of the level α used for individual tests. As a result, Bonferroni corrected P-values will be the original P-values multiplied by m, while corrected confidence intervals for the true mean difference will use a t-quantile function at a probability of $1 - (\alpha/2m)$ instead of the conventional $1 - (\alpha/2)$. A Bonferroni $(1 - \alpha)100\%$ confidence interval for D is obtained using

$$\hat{D} \pm t_{(1-(\alpha/2m),dfE)}\sqrt{MSE\sum_{i=1}^{a}\frac{c_i^2}{n_i}}, \tag{10.14}$$

where $t_{(1-(\alpha/2m),n-a)}$ is the t-quantile function with $n - a$ degrees of freedom for the probability $1 - (\alpha/2m)$, c_i is the ith contrast coefficient, and MSE is taken from the omnibus ANOVA.

10.3.5.2 Least Significant Difference

The *LSD* method (Fisher 1949) fixes the probability of type I error for each single pair of means being compared. It does not, however, control for the probability of false rejection of H_0 for other pairs of means. Thus, the procedure is equivalent to the method of contrasts described in Equations 10.11 through 10.13. Fisher (1949) recommended that the test be used only following the rejection of $H_0: \mu_1 = \mu_2 = \cdots = \mu_a$, allowing some protection against familywise type I error.

We define the *LSD* to be the difference between sample means necessary to reject any $H_0: \mu_i = \mu_i'$ (where $\mu_i' = not\,\mu_i$):

$$LSD_{i,i'} = t_{(1-(\alpha/2),dfE)} \times \sqrt{MSE(1/n_i + 1/n_i')}. \tag{10.15}$$

Using Equation 10.15, we calculate *LSD* for all pairs of sample means, or for selected planned comparisons. If $|\bar{Y}_i - \bar{Y}_i'| \geq LSD_{i,i'}$, then we reject $H_0: \mu_i = \mu_i'$. A $(1 - \alpha)100\%$ *LSD* confidence interval for the true difference $\mu_i - \mu_i'$ is calculated as

$$Y_i - Y_i' \pm t_{(1-(\alpha/2),dfE)} \times \sqrt{MSE(1/n_i + 1/n_i')}. \tag{10.16}$$

10.3.5.3 Scheffé's Procedure

Scheffé's method is recommended when an investigator wishes to address familywise type I error for a large number of tests possibly including, but not limited to, pairwise comparisons. Scheffé's method has the property that if $H_0: \mu_1 = \mu_2 = \cdots = \mu_a$, is rejected at α, then at least one comparison out of all possible comparisons (considered by this method) will also reject H_0 at α (Milliken and Johnson 2009). This is not assured with other methods.

For a set of comparisons, the procedure rejects H_0: $D = 0$ if $F^* \geq F_{(1-\alpha, a-1, dfE)}$, where

$$F^* = \frac{\hat{D}^2}{(a-1)\text{MSE}\sum_{i=1}^{a} c_i \bar{Y}_i},$$ (10.17)

and $F_{(1-\alpha, a-1, dfE)}$ is the F-quantile function with $a-1$ numerator degrees of freedom and $n-a$ denominator degrees of freedom, evaluated at the probability $1-\alpha$.

A Scheffe's $(1-\alpha)100\%$ confidence interval for D has the form

$$\hat{D} \pm C \times \sqrt{\text{MSE}\sum_{i=1}^{a} c_i Y_i},$$ (10.18)

where

$$C = \sqrt{(a-1)F_{(1-\alpha, a-1, dfE)}}.$$ (10.19)

10.3.5.4 Tukey–Kramer Method

The *Tukey's honest significant difference* and *Tukey–Kramer* methods control familywise type I error for all possible pairwise comparisons of factor-level means. This is in contrast to simultaneous inference methods discussed so far, which are applicable to any type of comparison. The original formula from an unpublished manuscript (Tukey 1953) did not allow unequal sample sizes. However, a simple modification from Kramer (1956) allowed imbalance. For balanced designs, the methods provide identical answers and a familywise significance level of exactly α (DeMuth 2006). When sample sizes are not equal, then the Tukey–Kramer method will be conservative with a familywise significance level less than α.

The method rejects H_0:$\mu_i = \mu_i'$ if $|q^*| \geq q_{(1-\alpha, a, dfE)}$, where $q_{(1-\alpha, a, dfE)}$ is the studentized range distribution quantile function[*] with parameters a and $n-a$, at probability $1-\alpha$, and

$$q^* = \frac{\sqrt{2}\hat{D}}{\sqrt{\text{MSE}(1/n_i + 1/n_i')}}.$$ (10.20)

A Tukey–Kramer $(1-\alpha)100\%$ confidence interval for D has the form

$$\hat{D} \pm T \times \sqrt{\text{MSE}(1/n_i + 1/n_i')},$$ (10.21)

where

$$T = \frac{1}{\sqrt{2}} q_{(1-\alpha, a, n-a)}.$$

---

[*] Assume a independent observations from $N(\mu, \sigma^2)$, and let w be the range of these observations. Now calculate an estimate, s^2, of σ^2 from these observations based on v degrees of freedom. Then the ratio w/s will follow the studentized range distribution $q(r, v)$. The studentized range distribution function in **R** is tukey, thus the quantile function is qtukey.

10.3.5.5 Dunnett's Method

Of interest in many biological studies is the comparison of a series of treatments to a control. This approach allows one to better estimate the true effect of a predictor by quantifying the background effect of unmeasured variables (see Section 7.6). *Dunnett's procedure* rejects the null hypothesis that a treatment mean, μ_i, equals the control mean, μ_0, whenever

$$|\mu_i - \mu_0| > d_{(\alpha, a, dfE)} \times \sqrt{MSE(1/n_i + 1/n_i')},$$

where $d_{(\alpha, a, dfE)}$ refers to the quantile function of the so-called *many-to-one-t-statistic* at probability α (Miller 1981).

Exact *P*-values and confidence interval calculations can be obtained using multivariate *t*-distributions. These methods are not described here (see Sokal and Rohlf 2012, p. 252). Like the Tukey–Kramer method, Dunnett's test is conservative for unequal sample sizes (Milliken and Johnson 2009).

> **EXAMPLE 10.4**
>
> The function `pairw.anova` from *asbio* can be used for adjusted simultaneous pairwise comparisons. Bonferroni, LSD, Scheffé, Tukey–Kramer, and Dunnett's procedures can be specified in the argument `method`. To make Tukey–Kramer adjusted comparisons for the baby walk data, I would specify:
>
> ```
> pairw.anova(baby.walk$date, baby.walk$treatment, method = "tukey")
> 95% Tukey-Kramer confidence intervals
> ```
>
> | | Diff | Lower | Upper | Decision | Adj. p-value |
> |-----------|----------|----------|----------|-----------|--------------|
> | muAE-muC | -2.225 | -4.35648 | -0.09352 | Reject H0 | 0.038997 |
> | muAE-muPE | -0.525 | -2.65648 | 1.60648 | FTR H0 | 0.897224 |
> | muC-muPE | 1.7 | -0.52625 | 3.92625 | FTR H0 | 0.172932 |
> | muAE-muTO | -1.58333 | -3.61562 | 0.44895 | FTR H0 | 0.160457 |
> | muC-muTO | 0.64167 | -1.48981 | 2.77314 | FTR H0 | 0.829542 |
> | muPE-muTO | -1.05833 | -3.18981 | 1.07314 | FTR H0 | 0.513366 |
>
> After adjustment for simultaneous inference, the only significant pairwise comparison is AE versus C.
>
> While we would only want to examine a single family of inferences, we could have also compared each treatment to the control. To do this, we specify `method = "dunnett"` and `control = "C"`.
>
> ```
> pairw.anova(baby.walk$date, baby.walk$treatment, method = "dunnett",
> control = "C")
> 95% Dunnett confidence intervals
> ```
>
> | | Diff | Lower | Upper | Decision |
> |----------|-----------|-----------|-----------|-----------|
> | muAE-muC | -2.226661 | -4.154739 | -0.298582 | Reject H0 |
> | muPE-muC | -1.701356 | -3.714497 | 0.311786 | FTR H0 |
> | muTO-muC | -0.643577 | -2.573064 | 1.28591 | FTR H0 |

Note that Dunnett's test provides more powerful pairwise comparisons of treatments to the control (i.e., it has narrower confidence bands) than the Tukey–Kramer procedure.

10.3.5.6 Comparing Simultaneous Inference Procedures

The Bonferroni, Scheffé, and Tukey–Kramer HSD procedures explicitly control familywise type I error. That is, for m factor-level comparisons, there will be at most a probability α

that one of the comparisons will falsely reject H_0. Surprisingly, by requiring the rejection of H_0: $\mu_1 = \mu_2 = \cdots = \mu_a$, Fisher's LSD has a familywise type I error rate similar to the significance level of the original ANOVA F-test (Carmer and Swanson 1973). Nonetheless, the protected LSD procedure does not explicitly control familywise type I error at the level of individual factor-level comparisons. As a result, LSD confidence intervals are not adjusted for simultaneous inference, and are not recommended for familywise summaries (Milliken and Johnson 2009).

Because it was designed specifically for the family of all possible pairwise comparisons, the Tukey–Kramer method will have greater power than the Bonferroni or Scheffé methods for this set of tests. That is, it will produce narrower confidence bands for the true comparison differences (Kutner et al. 2005).

Because the Bonferroni inequality (Chapter 2) gives the lower bound to the probability of the intersection of events, the Bonferroni correction will have low familywise power given more than a handful of tests, and a relatively high probability of familywise type II error. That is, an investigator will be assured that an entire family of tests will correctly fail to reject a true null hypotheses with a probability of at least $1 - \alpha$; however, the familywise probability of correctly rejecting a false null hypotheses will often be extremely low.

Scheffé's method controls familywise type I error for all possible linear comparisons (including all possible pairwise comparisons). Given this potentially immense set, Scheffé's method will provide a familywise error rate of α. For a smaller number of comparisons, however, the familywise significance level will generally be much smaller than α.

When the number of comparisons is approximately equal to the number of factor levels, then Bonferroni and Scheffé methods will provide confidence intervals of similar width. The number of comparisons generally must exceed the number of factor levels by a considerable amount before Scheffé's method becomes superior to the Bonferroni correction (Figure 10.3).

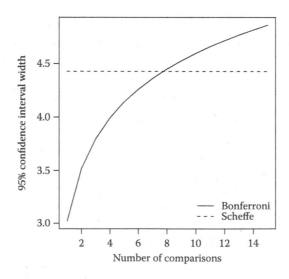

Fig.10.3()## See Ch. 10 code

FIGURE 10.3

Comparison of Bonferroni and Scheffé's 95% confidence bands for the baby walk example, with respect to AE versus TO factor-level comparisons. The Bonferroni confidence interval width increases as the number of comparisons (including AE vs. TO) increases; however, the Scheffé's interval width is unchanged. For eight or more comparisons, the Bonferroni confidence interval is wider.

Dunnett's test controls familywise type I error for a specialized family of inferences: the pairwise comparisons of the first $m - 1$ means with the mth mean. For instance, all noncontrol levels compared to the control. Dunnett's procedure will provide narrower confidence intervals than either the Scheffé or Bonferroni methods for this set of comparisons.

One can compare Scheffé and Tukey results and use the one that provides the narrower confidence bands. This sort of data snooping is allowed because the family of inferences for both methods includes all possible pairwise comparisons. For this same reason, Dunnett and Scheffé results can also be compared. Because it is not associated with any particular family of tests, the Bonferroni correction does not lend itself to post hoc methodological comparisons (Kutner et al. 2005).

10.3.6 General Methods for *P*-Value Adjustment

A number of procedures have been developed to adjust P-values for an arbitrary family set of tests (including non-general linear models) whose results are to be considered simultaneously. The Bonferroni correction can be used in this context. As before, the adjusted P-value will simply be the original P-value times the number of tests. That is, the significance level for each test will now be α/m. Recall, however, that this method strongly emphasizes the reduction of type I error over power. Dramatic improvement in Bonferroni familywise power can be obtained using methods that consider tests sequentially. These include the Bonferroni–Holm procedure and the Benjimani–Hochberg method for controlling false discovery rate (FDR).

The *Bonferroni–Holm procedure* (Holm 1979) consists of three steps:

1. P-values from the family of tests are first ranked from smallest to largest $(P_1, P_2, \ldots, P_m)$.
2. The conventional Bonferroni correction is applied to P_1. That is, if P_1 is less or equal to α/m, we reject the associated null hypothesis and go to P_2.
3. For P_2, use an adjusted significance level of $\alpha/(m - 1)$, while for P_3, use $\alpha/(m - 2)$, and so on. If at any point we fail to reject H_0, we conclude that all other lower-ranking tests in the family, corresponding to larger P-values, also fail to reject H_0.

Holm's procedure is more powerful than a strict Bonferroni correction because larger significance levels are used for all but the first comparison.

The *Benjamini–Hochberg method* (Benjamini and Hochberg 1995) controls the number of false-positive decisions by holding constant the number of false H_0 rejections (type I errors) divided by the total number of H_0 rejections. This is called the *false discovery rate*. Controlling the rate of false discovery for a family of tests is less conservative than controlling familywise type I error. That is, in some situations, FDR will be greater than α. In fact, FDR will equal one in the case that all null hypotheses from the family of tests are true (Shaffer 1995). As with the Holm–Bonferroni procedure, one would rank P-values from smallest to largest and find the largest (kth) rank corresponding to $P_k \leq \alpha(k/m)$. For all ranks $\leq k$, we would reject the associated H_0 hypotheses, by declaring that these are positive discoveries. For the remaining tests, we would fail to reject H_0.

The `p.adjust` function lets one adjust a vector of P-values for simultaneous inference using Holm's procedure, the Benjamini and Hochberg correction, and several other methods.

```
par(mfrow = c(2, 2), mar = c(4,4,2,1))
plot(pairw.anova(baby.walk$date,  baby.walk$treatment),  main = "",  las = 2)
bplot(baby.walk$date, baby.walk$treatment, simlett = TRUE,
lett = c("b","a","ab","ab"), ylab = "Onset of walking (months from birth)",
cex.lett = 0.8)
```

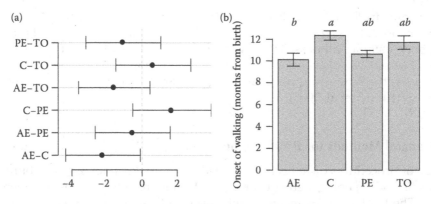

FIGURE 10.4

Methods for visually representing pairwise comparisons using the baby walk example. (a) 95% Tukey–Kramer confidence intervals for the true mean difference of each of the six pairwise comparisons. (b) Barplot of factor-level means with error bars representing ±1SE. Factor levels containing the same letter are not significantly different at $\alpha = 0.05$ using the Tukey–Kramer procedure.

10.3.7 Depicting Factor-Level Comparisons

Factor-level comparisons are usually represented with plots showing the confidence intervals for the comparison differences, or with a barplot showing factor-level means, overlaid with error bars (e.g., standard errors or confidence intervals for the true mean). For example, Figure 10.4a shows 95% Tukey–Kramer confidence intervals for true mean differences of all possible pairwise comparisons from the baby walk example. We note that the interval for the single significant comparison, AE versus C, does not contain zero. Figure 10.4b provides analogous information. In this case, factor-level means (not mean differences) are shown with standard errors. Letters are included above bars to summarize adjusted pairwise comparisons. Bars with the same letter are not significantly different after adjustment for familywise error. For instance, because AE is significantly different from C, and because C is greater than AE, AE is assigned the letter *b* while C is given the letter *a*. The remaining factor levels do not differ from each other or from AE or C, and consequently are assigned letters *a and b*.

For simple comparisons of two factor levels (e.g., *t*-tests) asterisks are often placed between factor-level bars to indicate the significance level at which a null hypothesis of no difference can be rejected. In the conventional system (used by many **R** functions), *** indicates $\alpha = 0.001$, ** indicates $\alpha = 0.01$, and * indicates $\alpha = 0.05$.

10.4 ANOVA as a General Linear Model

Like regression, ANOVA is a type of general linear model. That is, it quantifies the response variable as a linear combination of explanatory variables. A one-way ANOVA can be expressed as

$$Y_{ij} = \mu + \alpha_i + \varepsilon_{ij}, \tag{10.22}$$

where Y_{ij} represents the jth observation from the ith factor level. As we have already learned, the meaning of μ and α_i will vary with the contrasts specified in the linear model. However, the following will always be true: (1) μ will be a constant, (2) $\alpha_i = \mu_i - \mu$ (where μ_i is the true mean of the ith factor level), and (3) $\Sigma_{i=1}^{a} \alpha_i = 0$. In Equation 10.22, ε_{ij} is the jth random error term from the ith factor level. For inferential purposes, we assume that errors are independent, and that $\varepsilon_{ij} \sim N(0, \sigma^2)$ (see Section 10.7).

To demonstrate this generality, we use the matrix algebra form of general linear models introduced in Chapter 7, and used for regression in Chapter 9:

$$\mathbf{Y} = \beta\mathbf{X} + \varepsilon. \tag{7.13}$$

In the context of ANOVA, $\mathbf{Y}$ is an $n \times 1$ vector of responses, $\mathbf{X}$ is an $n \times a$ design matrix, β is an $a \times 1$ vector of parameters, and ε is an $n \times 1$ vector of random errors. In parallel to the general linear regression model, we estimate ANOVA parameters using

$$\hat{\beta} = (\mathbf{X}'\mathbf{X})^{-1}\mathbf{X}'\mathbf{Y}, \tag{10.23}$$

and find the fitted values (in this case, factor-level means) using

$$\hat{\mathbf{Y}} = \mathbf{X}\hat{\beta}. \tag{10.24}$$

Mean squares can also be derived using the same formulae that were used for regression. In particular, we find MSTR with

$$\text{MSTR} = \hat{\beta}'\mathbf{X}'\mathbf{Y} - \left(\frac{1}{n}\right)\mathbf{Y}'\mathbf{1}\mathbf{Y}/(a-1), \tag{10.25}$$

where $\mathbf{1}$ is an $n \times n$ matrix of ones, and calculate MSE using

$$\text{MSE} = \mathbf{Y}'\mathbf{Y} - \hat{\beta}'\mathbf{X}'\mathbf{Y}/(n-a). \tag{10.26}$$

The form of $\mathbf{X}$ and $\hat{\beta}$ will vary dramatically depending on the type of contrast specified by the user. The design matrices for treatment contrasts will use so-called *dummy coding*, composed of ones and zeroes. As with regression, the first column in $\mathbf{X}$ will define the intercept, and will be a vector of ones. Recall that the first alphanumeric factor-level mean is used for the intercept in treatment contrasts. The remaining columns in $\mathbf{X}$ will represent assignments of the remaining factor levels to observations. That is, elements in the second column will be assigned a 1 if they correspond to the second factor level and will be assigned a 0 otherwise. Likewise, elements in a third column will be given a 1 if they correspond to factor level 3 and a 0 otherwise, and so on. Other types of contrasts (e.g., Helmert and sum) will also define the intercept in the first column using n ones. Now, however, the intercept will be the mean of the factor-level means. The remaining columns in $\mathbf{X}$ will specify contrasts using linear constraints (Section 10.3). The vector $\hat{\beta}$ will contain estimated treatment effects, and consequently will change with the contrasts defined by the design matrix. Astonishingly, if the contrast matrix specifies $a - 1$ unique comparisons, then the form of the design matrix will have no effect whatsoever on the ANOVA fitted values, the resulting sums of squares, or the test statistic.

EXAMPLE 10.5

For the baby walk example, we have the following treatment design matrix:

```
contrasts(baby.walk$treatment) <- contr.treatment(4)
with(baby.walk, model.matrix(~treatment))
    (Intercept) treatmentC treatmentPE treatmentTO
```

	(Intercept)	treatmentC	treatmentPE	treatmentTO
1	1	0	0	0
2	1	0	0	0
3	1	0	0	0
4	1	0	0	0
5	1	0	0	0
6	1	0	0	0
7	1	0	1	0
8	1	0	1	0
9	1	0	1	0
10	1	0	1	0
11	1	0	1	0
12	1	0	0	1
13	1	0	0	1
14	1	0	0	1
15	1	0	0	1
16	1	0	0	1
17	1	0	0	1
18	1	1	0	0
19	1	1	0	0
20	1	1	0	0
21	1	1	0	0
22	1	1	0	0

From the coding, we see that observations 18–22 are in factor level C, observations 7–11 are factor level PE, and observations 12–17 are in factor level TO.

For comparison, here is the Helmert design matrix.

```
contrasts(baby.walk$treatment)<- contr.helmert(4)
with(baby.walk, model.matrix(~treatment))
    (Intercept) treatment1 treatment2 treatment3
```

	(Intercept)	treatment1	treatment2	treatment3
1	1	-1	-1	-1
2	1	-1	-1	-1
3	1	-1	-1	-1
4	1	-1	-1	-1
5	1	-1	-1	-1
6	1	-1	-1	-1
7	1	0	2	-1
8	1	0	2	-1
9	1	0	2	-1
10	1	0	2	-1
11	1	0	2	-1
12	1	0	0	3
13	1	0	0	3
14	1	0	0	3
15	1	0	0	3
16	1	0	0	3
17	1	0	0	3

18	1	1	-1	-1
19	1	1	-1	-1
20	1	1	-1	-1
21	1	1	-1	-1
22	1	1	-1	-1

The coding in the design matrix defines the contrasts. The first comparison, specified in the first column of **X**, indicates that H_0: $\mu = 0$ is to be tested. The first contrast, specified in the second column of **X**, compares the first and second alphanumeric factor levels, AE and C. The second contrast compares the average effect of AE and C to PE. The final contrast compares the average effect of AE, C, and PE to TO.

Now, if we define three different design matrices

```
contrasts(baby.walk$treatment) <- contr.treatment(4)
X1 <- with(baby.walk, model.matrix(~treatment))
contrasts(baby.walk$treatment) <- contr.helmert(4)
X2 <- with(baby.walk, model.matrix(~treatment))
contrasts(baby.walk$treatment) <- contr.sum(4)
X3 <- with(baby.walk, model.matrix(~treatment))
```

and derive three different sets of effect estimates

```
Y <- baby.walk$date
beta.hat1 <- solve(t(X1)%*% X1)%*% t(X1)%*% Y
beta.hat2 <- solve(t(X2)%*% X2)%*% t(X2)%*% Y
beta.hat3 <- solve(t(X3)%*% X3)%*% t(X3)%*% Y
```

the resulting fitted values, sums of squares, and mean squares (MSE shown below) and *P*-values will nonetheless be identical.

```
n <- 22; a <- 4
(t(Y)%*% (Y) - t(beta.hat1)%*% t(X1)%*% Y)/(n - a)
[1,] 1.551157
(t(Y)%*% (Y) - t(beta.hat2)%*% t(X2)%*% Y)/(n - a)
[1,] 1.551157
(t(Y)%*% (Y) - t(beta.hat3)%*% t(X3)%*% Y)/(n - a)
[1,] 1.551157
```

10.5 Random Effects

As noted in Chapter 7, *random factor levels* have two characteristics: (1) they can be considered a sample from a population of factor levels, and (2) they will not be informative. Because they are noninformative, comparisons among random factor level means are of little interest. Instead, investigators will generally be interested in making inferences concerning the true variance of the distribution of random factor-level means. For a random factor A, this would be σ_A^2 (see Figure 10.5). A single factor random effects model has the same form as Equation 10.22:

$$Y_{ij} = \mu + \alpha_i + \varepsilon_{ij},$$

(10.27)

`see.rEffect.tck()`

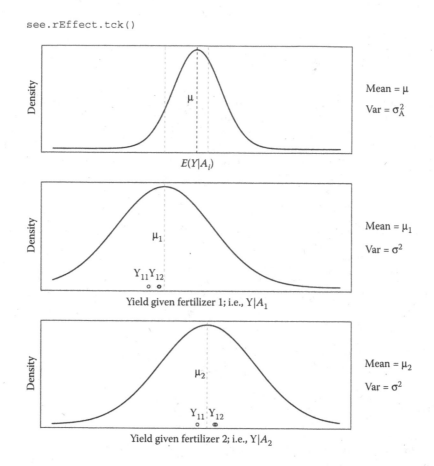

FIGURE 10.5
A random effects model created to ascertain the effect of randomly selected fertilizers on plant height. Except for the top graph, the model is identical to a conventional fixed effect linear model. Note that σ^2 (the factor-level variance) is larger than σ_A^2 (the variance for the population of factor-level means).

where Y_{ij} represents the jth observation from the ith factor level, μ is a constant (e.g., the mean of the random factor-level means, depending on contrasts), α_i is the ith random effect ($\alpha_i = \mu_i - \mu$), and ε_{ij} is a marginal error term. For purposes of inference and estimation, we assume that α_is are independent and follow $N(0, \sigma_A^2)$, ε_{ij}s are independent and follow $N(0, \sigma^2)$, and that α_is and ε_{ij}s are independent. Observations within random factors levels, however, are allowed to be dependent (see below and Section 10.7).

EXAMPLE 10.6

Figure 10.5 shows a hypothetical experiment to determine the effects of soil fertilizer on wheat yield, using two randomly selected brands. The upper figure shows the true mean yields produced by the entire population of fertilizers $E(Y|A_i)$. These means are normally distributed $\sim N(\mu, \sigma_A^2)$. An investigator randomly chooses two brands of fertilizer (1 and 2) from the population of factor levels (do this by clicking the *Sample* widget). These are indicated with gray lines in the upper figure. It is possible to select a factor level that will produce small average yields (compared to μ), or one that will produce large average yields, but it is more likely that a chosen factor level will produce a yield close to μ. The investigator proceeds by randomly assigning three experimental units (e.g., wheat fields) to both fertilizers. These are shown as points in the middle and bottom figures. We assume that the yield of the fields is normally distributed and homoscedastic for each fertilizer, $Y|X_i \sim N(\mu_i, \sigma^2)$. The fact that the true yield from fertilizer 2 is lower than for fertilizer 1 (or vice versa) is not of great interest. Of principal interest are inferences concerning σ_A^2.

10.5.1 Variance Components

Recall that for a one-way fixed effect ANOVA, $E(MS_A)$ will consist of the error variance, σ^2, plus a standardized measure of the true treatment effects. Specifically, $E(MS_A) = \sigma^2 + \sum n_i(\alpha_i^2/a-1)$. However, if A is a random factor, then

$$E(MS_A) = \sigma^2 + n - \frac{\sum_{l-1}^{a} n_i^2}{n(a-1)}\sigma_A^2. \tag{10.28}$$

Assuming a balanced design, this can be simplified to

$$E(MS_A) = \sigma^2 + n_0\sigma_A^2, \tag{10.29}$$

where n_0 is the sample size in each factor level (Table 10.3).

An unbiased and maximally efficient estimator for σ_A^2 (assuming normality for the population of factor-level means) can be obtained by solving for σ_A^2 in Equation 10.29 and substituting estimators for corresponding parameters:

$$\hat{\sigma}_{2A} = \frac{MS_A - MSE}{n_0}. \tag{10.30}$$

As with fixed effect models, an unbiased and (under normality) maximally efficient estimator of σ^2 is MSE. Equation 10.30 and MSE are both *method of moments estimators* (see Chapter 4; Searle 1971). Despite a lack of distributional assumptions, unbiased estimates, and high efficiency, there are two problems with these measures. First, MSE can potentially be greater than MS_A, making $\hat{\sigma}_A^2$ negative. However, variances cannot be negative. As a result, negative method of moments variance estimates are frequently given the value

TABLE 10.3

Form of a Balanced One-Way Random Effects ANOVA

Source	df	MS	E(MS)	Variance Component Estimator	Variance Component
A	$a-1$	$\dfrac{\sum_{i=1}^{a} n_i(\bar{Y}_i - \bar{Y})^2}{a-1}$	$\sigma^2 + n_0\sigma_A^2$	$\dfrac{MS_A - MSE}{n_0}$	σ_A^2
Error	$n-a$	$\dfrac{\sum_{i=1}^{a}\sum_{j=1}^{n_i}(Y_{ij} - \bar{Y}_i)^2}{n-a}$	σ^2	MSE	σ^2

zero, although this practice implies that the estimators are no longer unbiased (Kutner et al. 2005, Milliken and Johnson 2009). Second, method of moments variance estimates cannot be obtained if a design has multiple factors and is unbalanced.

As a result of these issues, either the method of maximum likelihood (ML) or *restricted maximum likelihood (REML)* is generally set as the default estimation procedure in random and mixed effect statistical software. With REML, variance–covariance components are estimated using ML, averaging over all possible values of the fixed effects (Kutner et al. 2005). As a result, REML adjusts for the degrees of freedom used in estimating the fixed effects, making variance component estimation invariant to fixed effect values (Crawley 2007). When a design is balanced, and method of moments estimates are nonnegative, then REML and method of moments estimates will be identical (Milliken and Johnson 2009). That is, REML can provide unbiased estimates for variance components. This is in contrast to ML variance estimates, which will, by definition, be biased low (West et al. 2008). ML and REML both preclude variance estimates being less than zero.

Because observations within random factor levels will be dependent (with covariance σ_A^2), the variance–covariance matrix of Y will have nonzero off-diagonal values, if H_0 is false. However, observations are still assumed to have the same variance, σ_Y^2, which will be the sum of the marginal error variance, σ^2, and the random effect variance(s), for example, σ_A^2. For instance, in a one-way random effects model, let there be two levels for factor A and two observations for each level. The variance–covariance matrix of Y can be written as

$$\sigma_Y^2 = \begin{bmatrix} \sigma_Y^2 & \sigma_A^2 & 0 & 0 \\ \sigma_A^2 & \sigma_Y^2 & 0 & 0 \\ 0 & 0 & \sigma_Y^2 & \sigma_A^2 \\ 0 & 0 & \sigma_A^2 & \sigma_Y^2 \end{bmatrix}.$$

Of particular interest in random and mixed effect models is the proportion of variance in σ_Y^2 attributable to random factors. The presentation and comparison of these ratios is often called *variance components analysis*. In a one-way random effect model, the true proportional variance component of A is

$$\frac{\sigma_A^2}{\sigma_Y^2} = \frac{\sigma_A^2}{\sigma_A^2 + \sigma^2} \tag{10.31}$$

This measure is called the *intraclass correlation coefficient* because it represents the correlation between any two observations in the same random factor level A_i. Because the components in Equation 10.31 must be greater than or equal to zero, the intraclass correlation coefficient must be in the range [0, 1]. We assume, prior to random sampling, that within-factor-level observations will be similar because they have the same effect size, α_i, and differ only with respect to random outcomes from ε_{ij}.

Equation 10.31 is generally estimated by directly substituting method of moments or REML variance estimates for parameters (Venables and Ripley 2002). The resulting proportions are generally reported as percentages.

10.5.2 Hypothesis Testing

Hypotheses for random effects concern inferences for the variance of the population of random factor-level means. In particular, we have

$$H_0: \sigma_A^2 = 0,$$

versus

$$H_A: \sigma_A^2 > 0.$$

As before, the significance of fixed factors is evaluated with respect to factor-level population means or treatment effects, that is

$$H_0: \alpha_1 = \alpha_2 = \cdots = \alpha_a = 0.$$

The approaches used for testing null hypotheses concerning parameters in random and mixed effect models is both a contentious issue and an active field of research. The significance of fixed and random effects is often evaluated using F-tests or t-tests (Fitzmaurice et al. 2004). A point of disputation with this approach, however, is that the t-test and F-test statistics will often not follow an exact t- or F-distribution (West and Galecki 2011). Approximate degrees of freedom are required because standard errors for fixed effects will be biased downwards because they do not take into consideration the uncertainty introduced by the process of covariance estimation (West et al. 2008). Furthermore, the computation of sums of squares and test statistics will be hindered by unbalanced designs. These issues have been addressed to some degree by procedures for degrees of freedom approximations, including the Satterthwaite and Kenward–Roger methods (see Kenward and Roger 1997), and by software allowing calculation of type II and III sums of squares (Section 10.14). These problems, however, have dissuaded creators of major packages for random and mixed effect models in **R** from providing F or t significance tests for random effects.[*]

10.5.2.1 Likelihood Ratio Test

A solution to this issue is provided by likelihood ratio tests (cf. Venables and Ripley 2002, Littell et al. 2006, West et al. 2008). Recall that the likelihood ratio test statistic has the form

$$G^2 = 2\ln\left[\frac{\mathcal{L}(R)}{\mathcal{L}(N)}\right] = 2[\ell(R) - \ell(N)], \tag{10.32}$$

where models R and N differ only with respect to the fact that the reference model R contains the factor of interest, X_k, while the nested model N does not. If the null hypothesis of no effect is true then G^2 will be a random outcome from a χ^2 distribution with degrees of freedom equivalent to the number of factors in the nested model, p, subtracted from

---

[*] The views of Douglas Bates (maintainer of *nlme*) with respect to hypothesis testing in mixed models are summarized in a 2006 blog communication available at my laboratory website https://sites.google.com/a/isu.edu/aho/

the number of parameters in the reference model, q. The P-value is computed as $P(X \geq G^2)$, where $X \sim \chi^2(p - q)$.

Because of their desirable characteristics, estimates of likelihood based on REML are generally used for hypothesis tests concerning the variance of the population of random factor-level means. This approach, however, is inappropriate if the fixed effects are not invariant (Venables and Ripley 2002, Pinheiro and Bates 2000). If likelihood ratio tests are to be used for assessing the significance of fixed effects, then ML parameter estimates must be used. That is, the likelihood for the reference model and the nested model (without the fixed effect) should be based on ML estimation.

For a test of $\sigma_A^2 = 0$, the hypothesized value of σ_A^2 lies on the edge of its parameter space. Consequently, the null distribution will be a mixture of χ_0^2 and χ_1^2 distributions with each given equal weight, 0.5 (Verbeke and Molenberghs 2000). Because the χ_0^2 distribution is concentrated entirely at zero, the P-value is calculated as $0.5 \times P(\chi_1^2 \geq G^2)$ (West et al. 2008). In the case that two random factors are used in a model and one wishes to consider one factor for deletion, in correspondence to $H_0 : \sigma_A^2 = 0$, the null distribution will be a mixture of χ_1^2 and χ_2^2 distributions with each given equal weight, and so on (see West et al. 2008).

It should be noted that the exact null distribution of G^2 under general conditions, including small samples, has recently been defined for random and mixed effect models (Crainiceanu and Ruppert 2004).[*] In the context of this distribution, classical likelihood ratio tests, described above, should be considered conservative (West and Galecki 2011).

10.5.3 `lme` and `lmer`

The most useful **R** functions for random and mixed models are `lme` in package *nlme* and `lmer` in package *lme4*.

Important arguments for `lme` include

- The argument `fixed` defines the fixed portion of the model. As in the function `lm`, a tilde (~) metacharacter indicates "function of," and is used to distinguish the response and predictor variables. To indicate that the intercept should be set as the only fixed factor, we would type: `fixed = Y ~ 1`. Complex relationships of fixed factors (crossed, nested, etc.) can be specified using the conventions of `formula` in `lm` (see Section 10.2.1).

- The argument `random` defines the random portion of the model. The statement `random = ~1|A` indicates that there is a single random effect for each level in factor A, and that the random effects will be evaluated with respect to the intercept. Nested and additive relationships can also be specified, although crossed random factors cannot. For nested structures, there must be replication for the nested factor. For example, a user cannot specify `random = ~1|block/treatment` unless the treatment is replicated at each block.

- The argument `correlation` allows the specification of an optional within-factor-level correlation structure (nonindependence) for random effects. See `?corClasses`.

- The argument `weights` allows modeling of heteroscedasticity of the within-group error. For example, by specifying `weights = varIdent(form = ~1|A)` separate variances for each level in A will be estimated and used in the model.

---

[*] Implementation for this distribution, allowing exact likelihood ratio tests, is provided by the function `exactLRT` in the package *RLRsim package2*.

- The argument method defines the likelihood optimization criterion used for parameter estimation. Choices are "REML" (the default) and "ML".
- The argument data is an optional dataframe containing variables from the formulae in fixed and random.

There are several characteristics that distinguish lmer.

- Fixed and random effects in lmer are specified together using a single formula argument, with random effects placed inside parentheses. As with lme, a vertical bar | is used to specify the grouping of random effects. For instance, to run a two-factor mixed effect model with fixed factor A, random factor B, assumed additivity (no interaction for A and B), and B evaluated with respect to the intercept, we would type: formula = Y ~ A + (1|B).
- Nesting must be specified using the forward slash character (/), where A/B indicates A + B%in%A.
- Crossed random factors are allowed.
- The likelihood estimation procedure is defined with the logical argument REML, which defaults to TRUE. REML = FALSE gives ML estimation.
- lmer reports variance components as both variances and standard deviations (lme reports only standard deviations).

EXAMPLE 10.7

Of particular interest to many biologists is the reliability and consistency of results from soil testing labs. With this in mind, Jacobsen et al. (2002) sent nine "identical" soil samples to eight randomly selected soil testing laboratories in the Great Plains region of the Central United States over 3 years. Among other characteristics, the labs were paid to measure soil potassium.

```
data(K)
library(nlme)
ref.K <-lme(K ~ 1,  random = ~1|lab,  data = K)
```

In the first argument above, I specify that I want the intercept to be a fixed effect, that is, K ~ 1. In the second argument, I specify that I want the soil testing labs to be considered as a random factor. Specifically, the statement ~1|lab indicates that there is single random effect for each level in the random factor lab, and that the random effects will be quantified with respect to the intercept term.

```
summary(ref.K)
      AIC      BIC     logLik
  745.0835 751.8715 -369.5417
Random effects:
 Formula: ~1 | lab
         (Intercept) Residual
StdDev:     30.41513 39.00521
Fixed effects: K ~ 1
                Value Std.Error DF  t-value p-value
(Intercept) 307.7917  11.69468 64 26.31894       0
```

The values 30.41513 and 39.00521 are REML estimates for σ_{lab} and σ. Because the design is balanced, we could have also obtained these values with methods of moments estimators using the output from a conventional one-way ANOVA.

```
aov.K<-print(anova(lm(K~lab,data=K)))
          Df Sum Sq Mean Sq F value    Pr(>F)
lab        7  68930  9847.1  6.4724 8.924e-06 ***
Residuals 64  97370  1521.4
```

Thus,

$$\hat{\sigma}_{lab} = \sqrt{(9847.125 - 1521.406)/9} = 30.41513,$$

and

$$\hat{\sigma} = \sqrt{1521.406} = 39.00521.$$

The estimated variance component of σ_{lab}^2 (expressed as a percentage) is

$$\frac{30.41513^2}{(30.41513^2 + 39.00521^2)} \times 100 = 37.8126.$$

Thus, we estimate that 38% of the magnitude of σ_Y^2 is due to σ_{lab}^2.

The fixed effects for this model consist of only the intercept term. In some settings, it might also be useful to know the effect size associated with a particular random factor level. Random effects are predicted with *empirical best linear unbiased predictors (EBLUPs)* (see West et al. 2008). These are known as *shrinkage estimators* because the correlation between observations within random factor levels causes the effect sizes to shrink compared to fixed effect estimates. The code below estimates effect sizes if the labs were fixed factors with treatment contrasts:

```
means <- tapply(K[,1],K[,2],mean); Yhat <- mean(means)
fixef <- round(means - Yhat, 2)
```

We can compare these to the EBLUPs for a demonstration of shrinkage:

```
cbind(fixef, round(ranef(ref.K), 2))
    fixef      ranef
B   18.32      15.49
D   13.32      11.26
E    8.76       7.41
F    7.88       6.66
G   -3.68      -3.11
H  -78.46     -66.34
I    5.43       4.59
J   28.43      24.04
```

We test the hypotheses: $H_0: \sigma_{lab}^2 = 0$ versus $H_A: \sigma_{lab}^2 > 0$, using a likelihood ratio test. To do this, we create a reference model including the random factor, and a nested model excluding it. In this case, the nested model will only contain an intercept, and this will be fixed. As a result, we use the function gls from *nlme* to create the nested model since it does not require random effects like lme.

```
nested.K <- gls(K ~ 1, data = K)
nested.K$logLik
[1] -378.3228
```

The function gls uses *generalized least squares (GLS)*. In this framework, the variance–covariance matrix describing the model error is allowed to have nonconstant diagonal elements and nonzero off-diagonal elements. As a result, GLS is appropriate for heteroscedastic and nonindependent data (see West et al. 2008). Weighted least squares (Section 9.13) is a special example of GLS in which the conditional variance–covariance matrix has nonconstant diagonal elements (heteroscedasticity) but off-diagonal elements that are all zeroes.

The likelihood ratio test statistic is

$$G^2 = 2(-369.5417 + 378.3228) = 17.562.$$

We obtain $P(X \geq G^2)$ and multiply this value by 0.5 to account for the fact that the null distribution is a mixture of χ_0^2 and χ_1^2 distributions:

```
pchisq(17.562, 1, lower.tail = FALSE) * 0.5
[1] 1.390305e-05
```

This answer can be obtained directly from anova.lme. However, here we would need to remember to multiply the *P*-value by 0.5.

```
anova(ref.K, nested.K)
          Model df     AIC      BIC    logLik    Test  L.Ratio p-value
ref.K         1  3 745.0835 751.8715 -369.5417
nested.K      2  2 760.6456 765.1709 -378.3228 1 vs 2 17.56210  <.0001
```

Our conclusion is that potassium measurements vary depending on the laboratory you send them to. We note that the information-theoretic criteria suggest retaining the more complex random effects model. That is, *AIC* and *BIC* are lower in the random effects model. Because the labs were randomly selected, inference extends to the population of factor levels. Specifically, we conclude that laboratories vary in their measures of potassium across the whole population of Great Plains soil testing laboratories.

10.6 Power, Sample Size, and Effect Size

Effect size estimators in ANOVA generally describe the *proportion of explained variation* (*PEV*). A natural candidate is R^2; however, this measure will tend to overestimate the true proportion of total variance explained by treatment groups for ANOVAs (Maxwell and Delaney 1990). An alternative PEV measure designed specifically for fixed factor ANOVAs is ω^2, pronounced "omega squared" (Hays 1994). We estimate ω^2 with

$$\hat{\omega}^2 = \frac{SS_A - (a-1)MSE}{SSTO + MSE}. \tag{10.33}$$

For random effects models, variance component estimates can be used to summarize factor effects.

Cohen (1988) recommended that the measure *f* be used to define the true effect size in a balanced one-way fixed effect ANOVA:

$$f = \sqrt{\sum_{i=1}^{a} \frac{\frac{n_i}{n}(\mu_i - \mu)^2}{\sigma^2}}. \tag{10.34}$$

Cohen (1992) suggested that the values 0.1, 0.25, and 0.4 could be used as lower-end cutoffs for small, medium, and large effect sizes, although see Section 9.9 for criticisms of this approach. The function pwr.anova.test from library *pwr* uses an iterative root-finding procedure to obtain power, the significance level, or the required sample size, based on the effect size definition given in Equation 10.34. The function power.anova.test from the base *stats* library uses the same computational approach but defines the effect size to be the true ratio of between to within group variance (i.e., the true F-statistic).

> **EXAMPLE 10.8**
>
> For a balanced one-way ANOVA layout with four groups, we assume that $f = 0.3$. What is the sample size required to reject H_0: $\mu_1 = \mu_2 = \cdots = \mu_a$ with probability 0.8, given $\alpha = 0.05$?
>   We have
>
> ```
> library(pwr)
> pwr.anova.test(k = 4, power = 0.8, f = 0.3, sig.level = 0.05)
>         k = 4
>         n = 31.27917
>         f = 0.3
>   sig.level = 0.05
>       power = 0.8
> ```
>
> A sample size of 32 for each group is required.

10.7 ANOVA Diagnostics and Assumptions

Because they are both general linear models, ANOVA and regression require similar assumptions (see Section 9.10). For ANOVA, we assume

1. The error terms are independent.
2. The factor-level populations are normally distributed.
3. The factor-level populations have the same variance, σ^2.

Thus, model errors are assumed to be independent and identically normally distributed, $\varepsilon_{ij} \sim N(0, \sigma^2)$. These assumptions hold for all ANOVAs. However, particular ANOVA designs may require still other assumptions (e.g., additivity), while particular analytical approaches allow assumption violations to be directly addressed, or even modeled. As always, we remain conscious of the detrimental effect of outliers on linear models.

Violations of ANOVA assumptions become more problematic when designs are unbalanced. This is true for four reasons. First, group means and treatment effects will be estimated with different levels of precision, resulting in possible experimental bias (Underwood 1997). Second, violations of homoscedasticity will have greater negative effects on valid inferences, particularly if the smaller group has the larger variance (Quinn and Keough 2002). Third, the lack of balance in multifactor models will make conventional sums of squares untrustworthy (see Section 10.14). Fourth, the estimation of variance components will be more complicated in random and mixed effect models.

To demonstrate applied diagnostics, we will explore the baby walk model with both residual plots and diagnostic hypothesis tests. Recall that hypothesis tests should not be used exclusively to check model assumptions (Section 9.10). The default diagnostic plots from `stats::plot.lm` are shown in Figure 10.6.

10.7.1 Independence of Error Terms

As with regression, nonindependence of error terms will result in estimation inefficiency, underestimation of σ^2, and false inferences. Unaccounted for positive association between

```
baby.lm <- lm(date ~ treatment, data = baby.walk)
par(mfrow = c(2, 2), mar = c(4, 4, 2, 1.5))
plot(baby.lm)
```

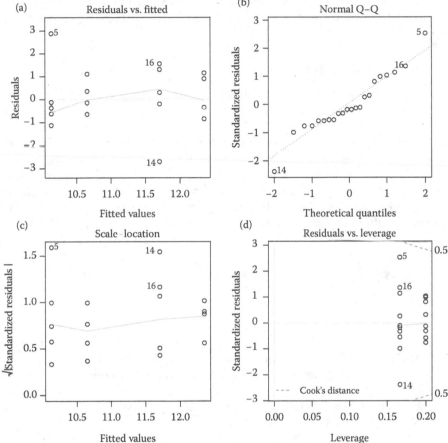

FIGURE 10.6

Default diagnostic plots produced for the baby walk ANOVA model: (a) fitted values versus residuals, (b) normal quantile plot, (c) scale–location plot, and (d) leverage plot with Cook's distances.

individuals within groups (due, for instance, to sequential measures) increases type I error, while negative association within groups increases type II error (Underwood 1997). Lack of independence will often be evident as a "pattern" of error term estimates ($\hat{\varepsilon}_i s$) when plotted against fitted values or the order of observations.

For the baby walk example, we do not know the order of recorded observations and therefore we cannot plot residuals against this variable. However, in this case, the observations should theoretically be independent. That is, it is unlikely that an observation of first-walking-time in one infant will influence the onset of walking in another infant from a different family. Plots of residuals versus fitted values do not exhibit trends, supporting the assumption of independence (Figure 10.6a and c).

Problems with independence should be considered at the design stage of the experiment (Chapter 7). For example, initial steps can often be taken to reduce sample dependency as a result of spatial and temporal autocorrelation (see Chapter 7). Transformations can be used to reduce some dependence effects (see Kutner et al. 2005). Lack of independence will not necessarily prevent effective analyses, but will prevent analyses with conventional linear models. Mixed effect repeated measures approaches can be used to formally address both a lack of temporal independence (see Section 10.12), and spatial autocorrelation (not addressed here).

10.7.2 Normality of Error Terms

With ANOVAs, we assume that the populations underlying the factor levels are normal, resulting in normal error terms. Because each factor level contains multiple observations, normality can be gauged within factors using diagnostic plots, hypothesis tests, and descriptive statistics. Significant correlations among factor-level means and variances indicate both nonconstant variances and nonnormality because the mean and variance are independent in a normal distribution. Because Y is expected to vary randomly in the same way at each factor level, residuals are generally pooled and summarized in a normal quantile plot to test the assumption of normality. For the baby walk example, the normal quantile plot is fairly linear, although outliers (points 5 and 10) are evident at either end of the fit line (Figure 10.6b). The assumption of normality is further supported by the Shapiro–Wilks test.

```
shapiro.test(resid(baby.lm))
 Shapiro-Wilk normality test

data:  resid(baby.lm)
W = 0.9424, p-value = 0.2222
```

Given the homoscedasticity and equal sample sizes, ANOVAs are robust to nonnormality (Quinn and Keough 2002). Nonnormality can often be effectively addressed with nonlinear data transformations (see Sections 7.5.6.5 and 9.11), or with robust procedures that do not assume normality (see Section 10.15). If the nonnormality and heteroscedasticity are a result of unconventional response variables units (e.g., counts), then the use of generalized linear models (Section 9.20) may be helpful.

10.7.3 Constancy of Error Variance

A central supposition of ANOVA is that factor-level populations have the same error variance, σ^2. Violation of this assumption is much more serious than the existence of nonnormal errors. That is, ANOVA is resistant to nonnormal homoscedastic populations, but not the converse (Coombs et al. 1996). For example, given normality, a balanced design, and moderately large sample sizes, but unequal variances, type I error can be as high as 0.09 when testing at $\alpha = 0.05$. With unequal sample sizes, the probability of type I error can exceed 0.3 (Wilcox 1986, Wilcox 2005).

Diagnostic plots can be used to identify nonconstant variance. These include side-by-side boxplots for individual factor levels and plots of residuals as a function of fitted values (i.e., factor-level means). In boxplots, the hinges in the plots (the first and third quartiles) should have a similar spread. As with regression, plots showing residuals as a function of fitted values should not demonstrate patterns of increased or decreased variability (i.e., a funnel shape) with increasing fitted values. For the baby walk example, some degree of heteroscedasticity is evident, but no association is apparent with respect to the fitted value magnitude and variability. This suggests that the constant variance assumption for the model is valid (Figure 10.6a and c).

The Fligner–Killeen test, introduced in Section 6.2.7, is a rank-based permutation alternative to Bartlett's test (which is a multilevel generalization of the F-test for equal variances). An alternative, used frequently in ANOVA, is the *modified Levene test* (Levene 1960). The procedure calculates $d_{ij} = |\hat{\varepsilon}_{ij} - \tilde{\varepsilon}_i|$, where $\hat{\varepsilon}_{ij}$ is the jth residual from the ith factor level, and $\tilde{\varepsilon}_i$ is the median of the residuals associated with ith factor level. An ANOVA is then run on the d_{ij}s. The modified Levene test (such as the Fligner–Killeen test) is robust to non-normality, and has the hypotheses

$$H_0 : \sigma_1^2 = \sigma_2^2 = \cdots = \sigma_a^2,$$

versus

$$H_A : \text{at least one of } \sigma_i^2\text{'s not equal to the others.}$$

```
modlevene.test(resid(baby.lm), baby.walk$treatment)
Modified Levene's test of homogeneity of variances
df1 = 3, df2 = 18, F = 0.26042, p-value = 0.85294
```

The test agrees with the diagnostic plots. There is little evidence of heteroscedasticity.

Nonconstant variance can often be addressed with nonlinear transformations (Section 7.5.6.5), particularly if the conditional response distribution is positively skewed, or with weighted least squares (Section 9.13). Alternative testing formats are also possible. For instance, a generalization of the Welch t-test can be run for heteroscedastic populations in one-way ANOVA formats (Welch 1951).[*] GLMs can also be used to address distributional situations in which variances are nonindependent of the mean (Section 9.20). Finally, several robust ANOVA variants do not require homoscedasticity (see Section 10.15).

10.7.4 Outliers

Outliers can have a strong detrimental effect on general linear models, including ANOVA. As noted in Section 9.10, one procedure for outliers is to create two linear models, one with outliers and one without, and report both models if they are dramatically different. Another approach is to use a robust ANOVA procedure (Section 10.15). Methods for outlier detection, including leverage, Cook's distance, and DFFITS, are described in Section 9.10. For the baby walk example, leverages are similar among observations, and appear noninfluential (Figure 10.6d).

10.7.5 Random and Mixed Effect Models

Violations of normality and constant variance are a more serious problem for random and mixed effect models (Kutner et al. 2005). Unfortunately, diagnostic tests are much more difficult to perform and interpret because of the additional presence of random effects and potentially complex covariance structures. For random and mixed effect models, we assume that

1. The marginal residuals, ε_{ij}, are normal, homoscedastic, and independent.
2. Random effects are normal, homoscedastic, and independent.
3. Observations are normally distributed because they are a linear combination of two normal random variables: the random effects and the marginal residuals.

---

[*] The appropriate R function here is `oneway.test`.

4. Marginal residuals and random effects are independent.

5. Observations from different random factor levels are independent from each other. However, observations from the same level will be dependent with covariance σ_A^2 (the variance of the population of random factor-level means).

Diagnostic plots for random and mixed effects can be obtained from the utility function nlme::plot.lme. The function can be customized to compare fitted values and residuals for any combination of fixed or random factors, including conditional relationships. By default, it displays standardized residuals as a function of fitted values for the innermost level of nesting in the model (see Equation 9.44).

Type I error in random and mixed effect models can be greatly inflated by the presence of outliers (Wilcox 2005), and influence metrics for their detection is an active area of

```
plot(ref.K, cex = 1.2, grid = F, id = .05, adj = -0.5)
qqnorm(ref.K, cex = 1.2,main = "", id = .05, adj = -0.5)
plot(ref.K, resid(., type = "p") ~ fitted(.) | lab, abline = 0)
qqnorm(ref.K, ~ ranef(.), cex = 1.2, grid = F, id = .05, adj = -0.5)
```

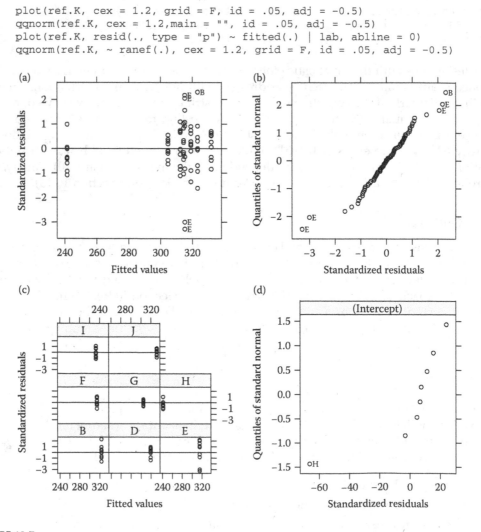

FIGURE 10.7
Diagnostic plots produced for the potassium random effects model: (a) fitted values versus residuals in the marginal model, (b) normal quantile plot for the marginal model, (c) residuals separated by level, and (d) normal quantile plot for the conditional model.

research (West et al. 2008). Outlying observations can be indicated in `nlme::plot.lme` by using the `id` argument to establish the probability contour cutoff for outliers at `1 - id/2` (see code for Figure 10.7).[*] Figure 10.7 shows diagnostic plots for the marginal and conditional portions of the soil potassium example (Example 10.7).

There are potential problems with heteroscedasticity. Specifically, level E is more variable than the other levels (Figure 10.7a and c). The marginal residual distribution is also heavy tailed, due to outlying observations from levels B and E (Figure 10.7b). The normal assumption of effect sizes also appears to be violated due to extremely low potassium measurements in the H level. Reanalysis of these data using a robust random effects model is left as an exercise at the end of the chapter.

It is important to note that within-factor heteroscedasticity and lack of marginal independence can be accounted for in random and mixed effect models (see Littell et al. 2006, Pinheiro and Bates 2009). For example, heteroscedasticity can be addressed by using variance weighting functions (Pinhiero and Bates 2000, p. 216), while temporal autocorrelation can be addressed with customized covariance structures that assume and quantify error covariance as a function of temporal proximity (Section 10.12).

★ 10.8 Two-Way Factorial Design

We begin our discussion of multifactor ANOVAs by considering a two-way *factorial design* in which two factors, A and B, and their interaction, A:B, are evaluated simultaneously. In a properly fashioned factorial design, factors are fully crossed. That is, every experimental unit receives a factor-level from each factor, and there will be an equal number of experimental units assigned to each factor-level combination (see Section 7.6.5.3). Lack of balance will hinder sum of squares calculations for all multifactor ANOVAs (see Section 10.14).

An important difference between one-way and two-way ANOVAs (and even more complex multifactor models) is the potential for interactions. Recall that factor-level effects from a single factor are called main effects, while interaction effects describe the nonadditive (synergistic or antagonistic) effect of factor-level combinations. Significant interaction effects indicate that the main effects cannot be considered independently because the factors are not independent. Thus, nonadditivity will complicate analyses by forcing the investigator to consider effects at the interaction level. On the other hand, significant interactions will often provide valuable insights into the complexities of the phenomena under study.

If interaction effects are absent and main effects are significant, then factor-level analyses can be conducted for the main effects using components of the two-way ANOVA (i.e., MSE and df_E), where these items are required in post hoc procedures. Figure 10.8 provides a flowchart for factorial analyses.

Because significant interactions that are important (see below) prevent consideration of main effects, emphasis is often given to "correcting" interactions with transformations. For instance, if an interaction is the result of the scale at which the response variable is measured, then interaction effects can often be decreased or eliminated with log or square root transformations (Kutner et al. 2005).

In an *interaction plot*, factor levels from one factor are placed along the X axis, and levels from a second factor are distinguished by varying plotting symbols or line types. Parallel

---

[*] The package *influence.ME* allows computation of Cook's distance for hierarchical groups and influence diagnostics for models created using the function `lmer` in *lmer4*.

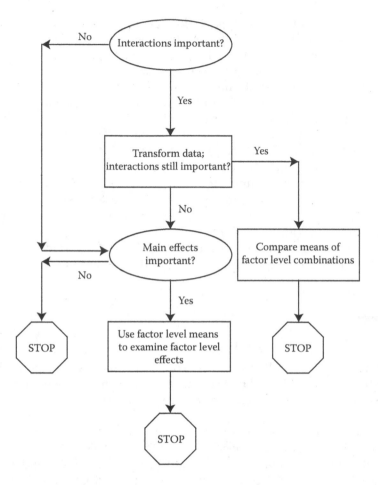

FIGURE 10.8
Order of analytical operations for a factorial design. (Adapted from Kutner, M. H. et al. 2005. *Applied Linear Statistical Models*, 5th ed. McGraw-Hill, Boston.)

lines indicate the absence of an interaction. Figure 10.9 shows interaction plots depicting the hypothetical effect of two factors, fertilizer and soil water, on plant biomass. Fertilizer, denoted *A*, has two levels (N+ = N addition and Co = control), while soil water availability, denoted *B*, has three levels (dry, mesic, and wet). In Figure 10.9a, both factors are significant, and there is no interaction effect. This is intimated by both the extremely high *P*-value for null hypothesis H_0: all $(\alpha\beta)_{ij} = 0$, and the perfectly parallel interaction plot lines. In Figure 10.9b, fertilizer (factor *A*) is significant, but soil water (factor *B*) is not. The approximately parallel lines and nonsignificant interaction *P*-value provide little evidence for an interaction. Figure 10.9c shows a significant interaction effect. In this case, factor-level combinations are no longer a simple additive combination of factors. Specifically, biomass is much lower in the high-water/N-addition treatment, obscuring interpretation of the significant main effects. Figure 10.9d also shows a significant interaction. In this case, however, the main effect of the fertilizer remains interpretable. This is because, regardless of water treatment, N addition always results in more biomass. This is a less severe form of interaction that will occur when interaction traces are not parallel but do not cross.

```
Fig.10.9() ## see Ch. 10 code
```

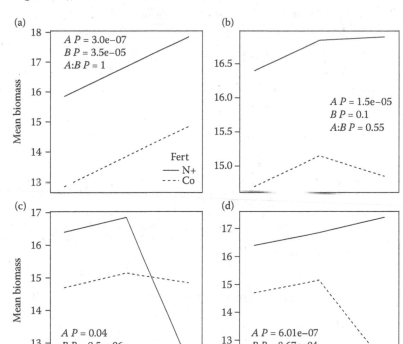

FIGURE 10.9
Four different interaction scenarios. Biomass is evaluated as a function of two fertilizer (A) and three water (B) treatments: (a) main effects are significant, and no interaction occurs, (b) one main effect is significant, and interaction effects are insignificant, (c) significant confounding interaction occurs (lines cross) obscuring the effect of both main effects, and (d) a significant interaction occurs, but the effect of fertilizer remains interpretable as a main effect.

In a two-way factorial design, we have the linear model

$$Y_{ijk} = \mu + \alpha_i + \beta_j + (\alpha\beta)_{jk} + \varepsilon_{ijk}, \tag{10.35}$$

where Y_{ijk} is the kth observation from the jth level in B and the ith level in A, μ is a constant (for instance, the true mean of all the factor-level combination means), α_i is the ith treatment effect from A ($\alpha_i = \mu_i - \mu$), β_j is the jth treatment effect from B ($\beta_j = \mu_j - \mu$), $(\alpha\beta)_{ij}$ is the interaction effect of ith level in A and the jth level in B ($(\alpha\beta)_{ij} = \mu_{ij} - \mu$), and ε_{ijk}s are independent random errors that follow $N(0, \sigma^2)$.

Three sets of hypotheses will be tested. For preliminary consideration are hypotheses concerning the presence of interaction between A and B:

$$H_0: \text{all } (\alpha\beta)_{ij} = 0; \quad \text{that is, there is no interaction,}$$

versus

$$H_A: \text{not all } (\alpha\beta)_{ij} = 0.$$

TABLE 10.4

ANOVA Table for a Balanced Two-Way Factorial Design with Fixed Factors

Source	df	SS	MS	F*
A	$a-1$	$n_0 b \sum_{i=1}^{a} (\bar{Y}_i - \bar{Y})^2$	$\dfrac{SS_A}{a-1}$	$\dfrac{MS_A}{MSE}$
B	$b-1$	$n_0 a \sum_{j=1}^{b} (\bar{Y}_j - \bar{Y})^2$	$\dfrac{SS_B}{b-1}$	$\dfrac{MS_B}{MSE}$
A:B	$(a-1)(b-1)$	$n_0 \sum_{i=1}^{a} \sum_{j=1}^{b} (\bar{Y}_{ij} - \bar{Y}_i - \bar{Y}_j + \bar{Y})^2$	$\dfrac{SS_{A:B}}{(a-1)(b-1)}$	$\dfrac{MS_{A:B}}{MSE}$
Error	$ab(n_0 - 1)$	$\sum_{i=1}^{a} \sum_{j=1}^{b} \sum_{k=1}^{n_0} (Y_{ijk} - \bar{Y}_{ij})^2$	$\dfrac{SSE}{ab(n_0 - 1)}$	

Note: A:B indicates the interaction of A with B, α_i represents the ith factor-level effect in factor A, β_j represents the jth factor-level effect in factor B. The term n_0 represents the sample size for each factor-level treatment combination.

The hypotheses concerning the main effects of A are

$$H_0: \mu_1 = \mu_2 = \cdots = \mu_a ,$$

versus

$$H_A: \text{at least one } \mu_i \text{ not equal to the others.}$$

The hypotheses concerning the main effects of B are

$$H_0: \mu_1 = \mu_2 = \cdots = \mu_b,$$

versus

$$H_A: \text{at least one } \mu_j \text{ not equal to the others.}$$

Table 10.4 shows the form of a balanced two-way factorial ANOVA with fixed factors.

We note that (1) the degrees of freedom for the interaction are the product of the individual degrees of freedom for the main effects, and (2) the denominator in the F statistic, for all three SSTR components, is MSE. As with a one-way fixed effect ANOVA, test statistics in factorial designs are the ratios of variance estimates. As before, if the null hypothesis is true, then the test statistic will be a random outcome from an F-distribution with the degrees of freedom taken from the numerator and denominator degrees of freedom in the test statistic. For example, for the factor A null hypothesis $\mu_1 = \mu_2 = \cdots = \mu_a$, the P-value is calculated as $P(X \geq F^*)$ where $X \sim F[a-1, ab(n-1)]$.

EXAMPLE 10.9

To demonstrate a two way factorial design, we revisit Fisher's potato yield experiment discussed in Example 10.1. We will first assume that levels in fertilizer and variety were of a particular interest, and are therefore of fixed effects.

```
data(potato)
potato.lm <- lm(formula = Yield ~ Variety * Fert, data = potato)
```

The linear model contains a long series of treatment contrasts. The ANOVA model is

```
anova(potato.lm)
Analysis of Variance Table

Response: Yield
              Df Sum Sq Mean Sq F value     Pr(>F)
Variety       11 43.638  3.9671 11.6000 4.379e-12 ***
Fert           2  1.434  0.7170  2.0965    0.1303
Variety:Fert  22  2.003  0.0910  0.2662    0.9995
Residuals     72 24.623  0.3420
---
Signif. codes:  0 '***' 0.001 '**' 0.01 '*' 0.05 '.' 0.1 ' ' 1
```

There is essentially no evidence to counter the null hypothesis of no interaction ($F_{22,72} = 0.2662$, $P = 0.9995$, Figure 10.1). This allows us to consider the main effects independently. Fertilizer is not significant ($F_{2,72} = 1.434$, $P = 0.1303$); however, variety is strongly significant ($F_{11,72} = 11.6$, $P = 4.4 \times 10^{-12}$), prompting factor-level comparisons.

Pairwise comparisons result in a cumbersome family of ($12^2 - 12)/2 = 66$ tests. For instance, try

```
pairw.anova(potato$Yield, potato$Variety, method="tukey", MSE=0.342,
df.err=72)
```

Note that I specified the MSE and df_E for the two-way ANOVA in the `pairw.anova` function. In this case, it is probably wise to consider a smaller subset of tests (not necessarily pairwise comparisons) using Scheffé's procedure to adjust for familywise type I error.

10.8.1 Random Effects

Factorial designs with random factors are also possible. In this case, inferences (e.g., hypothesis tests) are made with respect to the variance of the population of factor-level means. That is, we test $H_0 : \sigma_A^2 = 0$, instead of $H_0 : \mu_1 = \mu_2 = \cdots = \mu_a$.

We can express both a factorial mixed model (with A random and B fixed), and a factorial random effects model (with both A and B random) in the form of Equation 10.35.

Now, however, we are concerned with two types of random variability: the marginal errors and the random effects. Both are assumed to be normal, homoscedastic, and independent of each other.

Table 10.5 lists expected mean squares for fixed, random, and mixed two-way factorial models. By solving for the variance components in the E(MSE) column, and replacing parameters with estimators, we obtain the method of moments variance estimators (Table 10.6). We note, importantly, that the cross of a random and a fixed factor is itself a random factor (Littell et al. 2006).

It is interesting that the expected mean squares for the fixed factor require consideration of the random and fixed factors (Tables 10.5 and 10.6). In fact, the interaction mean square is used as the denominator of the F statistic to assess the significance of the fixed factor in conventional F-tests. The rationale for this is demonstrated in Figure 10.10.

The levels for a fixed factor are shown in the rows of the figure, while the columns distinguish levels for a random factor. Thus, the table depicts a mixed model. Assume that the values in the table are population means. For instance, the true mean of random level R1 for the fixed level F1 is 7. Using all possible information from all random factor levels, we see that the null hypotheses for the fixed factor and random factor are true. That is,

TABLE 10.5

Expected Mean Squares for Balanced Two-Way Factorial Models

Source	*A* and *B* Fixed	*A* and *B* Random	*A* Fixed, *B* Random
A	$\sigma^2 + bn_0 \dfrac{\sum \alpha_i^2}{a-1}$	$\sigma^2 + bn_0\sigma_A^2 + n_0\sigma_{A:B}^2$	$\sigma^2 + bn_0 \dfrac{\sum \alpha_i^2}{a-1} + n_0\sigma_{A:B}^2$
B	$\sigma^2 + an_0 \dfrac{\sum \beta_j^2}{b-1}$	$\sigma^2 + an_0\sigma_B^2 + n_0\sigma_{A:B}^2$	$\sigma^2 + an_0\sigma_B^2$
A:B	$\sigma^2 + n_0 \dfrac{\sum\sum (\alpha\beta)_{ij}^2}{(a-1)(b-1)}$	$\sigma^2 + n_0\sigma_{A:B}^2$	$\sigma^2 + n_0\sigma_{A:B}^2$
Error	σ^2	σ^2	σ^2

Note: The term n_0 represents the sample size for each factor-level treatment combination. α_i is the *i*th factor-level treatment effect from *A*. β_j is the *j*th factor-level true treatment effect from *B*.

$\mu_{F1} = \mu_{F2} = \mu_{F3} = 4$, and $\sigma_A^2 = 0$ (since all the random factor-level means equal 4). Thus, all the random variability in the table is coming from the nonzero $(\alpha\beta)_{ij}$s. However, when we select a subset of random levels, information concerning the fixed effect null hypothesis is obscured even though the means for the random effects are unchanged. In fact, for any subset of random factor levels, there appears to be evidence against H_0 because there is apparent variability among the fixed factor-level means (due to nonzero $(\alpha\beta)_{ij}$s). Thus, to avoid the inflation of type I error, we must consider the interaction of the random and fixed factors when making inferences to the fixed factor-level population.

EXAMPLE 10.10

Let us reconsider the potato experiment with variety as a random factor and fertilizer as a fixed factor. It is reasonable to define variety as a random factor. There are over 4000 potato varieties, and inference to such a population might be of interest. To examine random interactions, we will use `lmer` from package *lme4*. We have

TABLE 10.6

Variance Components and Method of Moments Estimators for Balanced Two-Way Factorial Designs

Source	Variance Component	Variance Component Estimator	
		A and *B* Random	*A* Fixed, *B* Random
A	σ_A^2	$\dfrac{MS_A - MS_{A:B}}{n_0 b}$	–
B	σ_B^2	$\dfrac{MS_B - MS_{A:B}}{n_0 a}$	$\dfrac{MS_B - MSE}{n_0 a}$
A:B	$\sigma_{A:B}^2$	$\dfrac{MS_{A:B} - MSE}{n_0}$	$\dfrac{MS_{A:B} - MSE}{n_0}$
Error	σ^2	MSE	MSE

Note: The term n_0 represents the sample size for each factor-level treatment combination.

```
see.mixedII()
```

Random

<table>
<tr><th></th><th></th><th>R1</th><th>R2</th><th>R3</th><th>R4</th><th>R5</th><th>R6</th><th>R7</th><th>R8</th><th>R9</th><th>R10</th><th>R11</th><th>R12</th><th>R13</th><th>R14</th><th>R15</th><th>R16</th><th>R17</th><th>R18</th><th>Mean
All levels</th><th>Mean
Selected levels</th></tr>
<tr><td rowspan="3">Fixed</td><td>F1</td><td>7</td><td>6</td><td>5</td><td>7</td><td>6</td><td>5</td><td>4</td><td>4</td><td>4</td><td>1</td><td>2</td><td>3</td><td>4</td><td>4</td><td>4</td><td>1</td><td>2</td><td>3</td><td>4</td><td>5</td></tr>
<tr><td>F2</td><td>4</td><td>4</td><td>4</td><td>1</td><td>2</td><td>3</td><td>7</td><td>6</td><td>5</td><td>7</td><td>6</td><td>5</td><td>1</td><td>2</td><td>3</td><td>4</td><td>4</td><td>4</td><td>4</td><td>5</td></tr>
<tr><td>F3</td><td>1</td><td>2</td><td>3</td><td>4</td><td>4</td><td>4</td><td>1</td><td>2</td><td>3</td><td>4</td><td>4</td><td>4</td><td>7</td><td>6</td><td>5</td><td>7</td><td>6</td><td>5</td><td>4</td><td>2</td></tr>
</table>

FIGURE 10.10
The effect of random factor-level selection on inferences concerning fixed effects. (Figure concept from Maxwell, S. E., and Delaney, H. D. 1990. *Designing Experiments and Analyzing Data: A Model Comparison Perspective.* Wadsworth Publishing, Belmont, CA.)

```
library(lme4)
potato.ref <- lmer(Yield ~ Fert + (1|Variety) + (1|Variety : Fert), data =
pot       ato)
potato.ref
   AIC   BIC logLik deviance REMLdev
 217.2 233.3 -102.6    198.9   205.2
Random effects:
 Groups        Name        Variance    Std.Dev.
 Variety:Fert (Intercept) 2.9231e-18 1.7097e-09
 Variety      (Intercept) 4.0932e-01 6.3978e-01
 Residual                 2.8326e-01 5.3222e-01
Number of obs: 108, groups: Variety:Fert, 36; Variety, 12

Fixed effects:
             Estimate Std. Error t value
(Intercept)  3.083958   0.200020  15.418
FertCl      -0.274375   0.133056  -2.062
FertS        0.006319   0.117344   0.054
```

In this case, the REML variance estimates are not equal to the method of moments estimates, which give negative results (see Exercise 14 at the end of the chapter).

Our estimates of $\sigma^2_{\text{Variety}}/\sigma^2_Y$ and $\sigma^2_{\text{Variety:Fert}}/\sigma^2_Y$ (expressed as percentages) are

```
4.0932e-01/(2.9231e-18 + 4.0932e-01 + 2.8326e-01)*100
[1] 59.10075
2.9231e-18/(2.9231e-18 + 4.0932e-01 + 2 8326e-01)*100
[1] 4.220595e-16
```

To determine the significance of the fixed fertilizer factor, we would perform a likelihood ratio test with model likelihoods based on ML estimates (see Section 10.5). This is most easily done with the function update.

```
pot.ref.ML <- update(potato.ref, REML = FALSE)
pot.nested.1 <- update(potato.ref, ~. - Fert, REML = FALSE)
anova(pot.ref.ML, pot.nested.1)
             Df    AIC    BIC  logLik Chisq Chi Df Pr(>Chisq)
pot.nested.1  4 211.91 222.63 -101.952
pot.ref.ML    6 210.87 226.96  -99.434 5.0356      2    0.08064 .
```

The test and information-theoretic criteria indicate that fertilizer has only marginal importance to yield.

Next, we test for the importance of variety, and the interaction of fertilizer and variety, both of which are random. To accomplish this, we test the hypotheses $H_0: \sigma^2_{\text{Variety:Fert}} = 0$ and $H_0: \sigma^2_{\text{Variety}} = 0$.

For the first hypothesis, $\sigma^2_{\text{Variety:Fert}} = 0$, we create a nested model without the variety:fertilizer interaction.

```
pot.nested2 <- update(potato.ref, ~. - (1| Variety:Fert))
```

We then calculate the likelihood ratio test statistic by finding 2 times the difference in REML likelihoods of the reference model and pot.nested2 model. We find that $X^2 = 2[\ell(R) - \ell(N)] \approx 0$.

If H_0 is true, then X^2 will be a random outcome from a mixture of χ^2_2 and χ^2_1 distributions. Thus, we calculate the P-value as $0.5[P(\chi^2_1 \geq X^2)] + 0.5[P(\chi^2_2 \geq X^2)]$. In this case, note that $X^2 \approx 0$, making the P-value approximately 1, and prompting both failure to reject H_0 and deletion of the interaction term.

To test $\sigma^2_{\text{Variety}} = 0$, we create a nested model (within pot.nested2) without the variety factor. We use gls since this model will be without random effects.

```
library(nlme)
pot.nested3 <- gls(Yield ~ Fert, data=potato)
```

We then find 2 times the difference in REML likelihood of pot.model2 and pot.model3.

This is $X^2 = 2[-102.6 + 133.22] = 61.24$.

The P-value allows clear-cut rejection of H_0 even without division by 2 to address the χ^2_1 and χ^2_0 mixture distribution under H_0.

```
pchisq(61.24,1,lower.tail=F)
[1] 5.052453e-15
```

★ 10.9 Randomized Block Design

In a *randomized block design*, experimental units are randomly assigned to all factor levels separately within blocks. The blocks can be used to increase precision by accounting for heterogeneity in a studied system (see Section 7.6.5.3).

Two factor designs without specified interactions, including randomized complete block designs, have the following linear model:

$$Y_{ijk} = \mu + \alpha_i + \beta_j + \varepsilon_{ijk}, \tag{10.36}$$

where Y_{ijk} is the kth observation from the jth level in blocking variable B and the ith level in A, μ is a constant (for instance, the true mean of the block and treatment factor-level combination means), α_i is the ith treatment effect from A ($\alpha_i = \mu_i - \mu$), β_j is the jth treatment effect from B ($\beta_j = \mu_j - \mu$), and ε_{ijk}s are independent random errors that follow $N(0, \sigma^2)$.

Lack of replication of treatments in blocks prevents assessment of interactions, and requires the assumption of additivity (Chapter 7). If this assumption is violated then the significance of neither the blocks nor the treatments is testable (Quinn and Keough 2002).

Note, however, that additivity is often improved with nonlinear transformations (Section 10.8). If an RBD has replication, then analysis as a factorial design is possible, interaction tests can be obtained, and the assumption of additivity is unnecessary (Section 10.8, Kutner et al. 2005, p. 908–909).

The assumption of additivity can be checked with interaction plots and with *Tukey's test for additivity* (Tukey 1949). This procedure can be viewed as a test for curvilinearity in the association of residuals and predicted values. The surface comprises the non-parallelism we are trying to detect in an interaction plot. The test has the following hypotheses:

$$H_0: \text{all } (\alpha\beta)_{ij} = 0, \text{ that is, main effects and blocks are truly additive,}$$

versus

$$H_A: \text{all } (\alpha\beta)_{ij} \neq 0.$$

Tukey's test for additivity is best for detecting simple block × treatment interactions: for instance, when lines in an interaction plot cross. As a result, interaction plots (not Tukey's test) should be used to check for other types of interactions. A high probability of type II error results from the inability of the test to detect complex interactions (Kirk 1995). As a result, a conservative value of α should be used, that is, 0.1–0.25 (Quinn and Keough 2002).

Missing data constitute a serious problem for randomized complete block designs (Patterson and Thompson 1971). This is because there are no replicates within blocks, making a missing observation equivalent to an empty cell (a treatment without an observation). Assuming additivity, a missing observation from the ith treatment and the jth block can be interpolated by finding[*]

$$\hat{Y}_{ij} = \bar{Y}_i + \bar{Y}_j - \bar{Y}. \tag{10.37}$$

For blocked designs without replication or with assumed additivity, two sets of hypotheses are tested. Of primary interest are hypotheses concerning the effects of the treatment factor, A:

$$H_0: \mu_1 = \mu_2 = \cdots = \mu_{a'}$$

versus

$$H_A: \text{at least one } \mu_i \text{ not equal to the others.}$$

Of less interest are hypotheses concerning the blocking factor, B:

$$H_0: \mu_1 = \mu_2 = \cdots = \mu_{b'}$$

versus

$$H_A: \text{at least one } \mu_j \text{ not equal to the others.}$$

Table 10.7 shows the form of a balanced randomized complete block design.

---

[*] Patterson and Thompson (1971), Snedecor and Cochran (1989), and Sokal and Rohlf (2012) describe more complex interpolation procedures for randomized block designs with missing data.

TABLE 10.7

ANOVA Table for a Randomized Block Design

Source	Df	SS	MS	F*
A	$a-1$	$b\sum_{j}(\bar{Y}_j - \bar{Y})^2$	$\dfrac{SS_A}{a-1}$	$\dfrac{MS_A}{MSE}$
Blocks	$b-1$	$a\sum_{i}(\bar{Y}_i - \bar{Y})^2$	$\dfrac{SS_B}{b-1}$	$\dfrac{MS_B}{MSE}$
Error	$(b-1)(a-1)$	$\sum_{i=1}^{a}\sum_{j=1}^{b}(Y_{ij} - \bar{Y}_j - \bar{Y}_i + \bar{Y})^2$	$\dfrac{SSE}{(b-1)(a-1)}$	

Note: For the equations, b is the number of blocks ($j = 1, ..., b$). Note that there is no multiplication by the number of replicates in the sums of squares. This is because there is no replication of A in the individual blocks. The error term, MSE, describes the interaction mean squares and thus is often denoted $MS_{A:B}$.

As before, if H_0 is true, a test statistic will be a random outcome from an F-distribution with the degrees of freedom taken from the degrees of freedom for the numerator and denominator error estimates in the test statistic. That is, for the null hypothesis concerning A, that is, $H_0: \mu_1 = \mu_2 = \cdots = \mu_a$, the P-value is $P(X \geq F^*)$, where $X \sim F[a - 1, (a - 1)(b - 1)]$. The denominator in the F statistics, MSE, represents the mean squares for the interaction of the treatment factor and the blocks. As a result, a number of texts define this term as $MS_{A:B}$ (e.g., Kutner et al. 2005).

If block effects are significant, then one can conclude that using blocks as treatment replicates was beneficial (i.e., the design addressed homogeneity among replicates). Conversely, large P-values indicate that the blocks are not different. Nonetheless, because the original design incorporated blocks, one should analyze the data as a randomized block design, and not as a CRD (Kutner et al. 2005).

EXAMPLE 10.11

An oft-used RBD example concerns an agricultural experiment to evaluate the effect of levels of soil K_2O (potash) on the Pressley strength index (breaking strength) of cotton fibers (Cochran and Cox 1957). In the experiment, five levels of K_2O were randomly assigned to soil plots (36, 54, 72, 108, and 144 lbs · acre^{-1}), which served as experimental units, and a single sample of cotton was taken from each unit. The experiment had three blocks (agricultural fields), and each of the K_2O treatments were randomly assigned to the five plots within each block. The data are in the dataframe potash.

To check for additivity, we would use interaction plots, and Tukey's test for additivity.

```
data(potash)
with(potash, tukey.add.test(strength, treatment, block))
Tukey's one df test for additivity
F = 0.2962553 Denom df = 7 p-value = 0.6031373
```

The additivity assumption appears valid. Running the ANOVA model, we have

```
K2O.lm <-lm(strength ~ treatment + block, data = potash)
anova(K2O.lm)
          Df  Sum Sq  Mean Sq F value  Pr(>F)
treatment  4 0.73244 0.183110  4.1916 0.04037 *
block      2 0.09712 0.048560  1.1116 0.37499
Residuals  8 0.34948 0.043685
```

The potash treatment is marginally significant, while the blocks are not significant at $\alpha = 0.05$.

The relative efficiency of a randomized block design compared to a completely random design can be estimated using the ratio of error variance of the CRD (without blocks) and the error variance of the randomized block design. The function `eff.rbd` calculates this for us:

```
eff.rbd(K2O.lm)
[1] 1.04765
```

If the relative efficiency ratio is greater than one, then a randomized block design is more efficient than a CRD in which the number of replicates per treatment is equal to the number of blocks. The result for the potassium experiment indicates that blocking improves efficiency to a small degree. Specifically, we would need to have 1.05 times as many replications per treatment with a CRD to achieve an identically powerful result.

10.9.1 Random Block Effects

As noted above, blocks are rarely of interest as a main effect. In addition, they may often constitute a real or conceptual sample from a population of levels. In these cases, blocking can be viewed as a random factor (Wilk 1955).

Consider an investigation to determine the effect of small input streams on the arthropod species richness in a larger river (Harris 2013). In the study, two diurnal periods (morning and afternoon) were defined for acquiring arthropod counts. These were the experimental units. Drift nets applications, preventing arthropod dispersal from the stream into the river, were randomly assigned to one of the periods, and counts were made downstream for both net and no net treatments. The experiment was replicated on four randomly selected days separated sufficiently in time so as to be independent. In this case, days can be seen as random blocks from a sample of all possible days that could have been chosen. That is, time is a random blocking factor. The net and no net treatments were fixed factor levels randomly assigned within each block.

The random block model can be expressed in the form of Equation 10.36. However, now we must assume that the block effects are normally distributed, independent, and homoscedastic, and that the block effects and marginal errors are independent. In addition, inferences (e.g., hypothesis tests) for the blocks are now made with respect to the variance of the population of random block means, σ_B^2.

Nonadditivity can be specified for RBDs with random blocks despite a lack of replication, although this requires a different $E(MS)$ for the treatment (see Kutner et al. 2005, p. 1063, Quinn and Keough 2002, p. 275). Such a framework allows testing for fixed treatment effects in the presence of additivity although a test for block significance is no longer possible. Sphericity (discussed in Chapter 7) is also assumed in mixed-model RBDs. In general, this assumption is likely to be valid. This is because treatments are randomly assigned within blocks, decreasing the likelihood that the variance of pairwise differences among the treatments will differ. Asphericity, however, is possible in repeated measures RBDs because of interference and carryover effects in time (Section 10.12). Asphericity will *not* be an issue for RBDs with fixed factors because the model being used implies that observations are uncorrelated in treatments and blocks (Quinn and Keough 2002, p. 282). Asphericity is also irrelevant for randomized block designs with only two treatments (i.e., paired *t*-tests with two-tailed hypotheses) because there is only one possible covariance value.

TABLE 10.8

Expected Mean Squares and Method of Moments Variance Estimators
for Randomized Block Designs with Assumed Additivity

Source	A, Blocks Fixed E(MS)	A Fixed, Blocks Random E(MS)	Variance Component Estimator
A	$\sigma^2 + b\dfrac{\sum \alpha_i^2}{a-1}$	$\sigma^2 + b\dfrac{\sum \alpha_i^2}{a-1}$	–
Blocks	$\sigma^2 + a\dfrac{\sum \beta_j^2}{b-1}$	$\sigma^2 + a\sigma_B^2$	$\dfrac{MS_B - MSE}{a}$
Error	σ^2	σ^2	MSE

Note: For the equations, b is the number of blocks ($j = 1,...,b$).

Expected mean squares for randomized block designs, and method of moments variance estimators for random blocks are shown in Table 10.8.

EXAMPLE 10.12

Allard (1966) sought to quantify variation in the yield of crosses of wheat grasses. Five genetic crosses of wheat (*Triticum* spp.) were selected from a breeding program and were grown at four randomly selected locations where the wheat would be planted commercially. At each location, crosses were randomly assigned to particular sections of blocks. Four blocks were established at each location. That is, at each location, a randomized block design with four blocks was conducted. Both location and block (in location) were considered random factors, while wheat crosses were fixed. Preliminary diagnostics indicated that the additivity of blocks and crosses was satisfied. The data are in the dataframe wheat.

To run the model in *nlme*, we have

```
data(wheat)
w.ref <- lmer(yield ~ 1 + cross + (1|cross : loc) + (1|loc/block),
data = wheat)
```

Here, for the first time, we have nested factors. This topic will be dealt with in greater detail in the next section. Note that I specify that block is nested in loc, and that loc, block%in%loc, and cross:loc are random effects.

To test for the significance of the crosses, we create reference and nested models with and without this (fixed) factor, and switch to ML parameter estimation. Again, this is easily done with the **R** update function.

```
w.ref.ML <- update(w.ref, REML = FALSE)
w.nested.ML <- update(w.ref, ~. - cross, REML = FALSE)
anova(w.ref.ML, w.nested.ML)
            Df   AIC    BIC   logLik  Chisq Chi Df Pr(>Chisq)
w.nested.ML  5 499.71 510.19 -244.86
w.ref.ML     9 485.32 504.17 -233.66 22.394      4  0.0001673 ***
---
```

We conclude that the crosses differ with respect to yield.

★ ## 10.10 Nested Design

Nested factors require hierarchical specification in models to allow proper inferences (see Section 7.6.5.3). For a two-factor nested design, let B be nested in A, the linear model can then be expressed as

$$Y_{ijk} = \mu + \alpha_i + \beta_{j(i)} + \varepsilon_{ijk}, \tag{10.38}$$

where Y_{ijk} is the kth observation from the jth level of B nested in the ith level in A, μ is a constant (for instance, the true mean of the factor-level combination means), α_i is the ith treatment effect from A ($\alpha_i = \mu_i - \mu$), and $\beta_{j(i)}$ is the jth treatment effect from B, nested in the ith level from A ($\beta_{j(i)} = \mu_{ij} - \mu_i - \mu$). As always, we assume that ε_{ijk}s are independent and follow $N(0, \sigma^2)$.

Table 10.9 shows an ANOVA table for a two-factor balanced nested design with fixed factors.

Nested factors are generally random in biology, while factors at the highest hierarchical level are either random or fixed (Quinn and Keough 2002). Expected mean squares for fixed, random, and mixed two-factor hierarchical designs are shown in Table 10.10.

Method of moments variance estimators for random and mixed hierarchical models are shown in Table 10.11. We note that to estimate variance components we subtract the mean square of the factor in the hierarchical rank immediately below the stratum of interest from the mean square of the factor at the level of interest and divide this difference by the product of the number of factor levels and/or pseudoreplicates in the level(s) below the stratum of interest.

EXAMPLE 10.13

Sokal and Rohlf (2012) measured the variability in wing length (in micrometers) for female mosquitos (*Aedes intrudens*). Four females were randomly selected from three cages and two measurements were made on the left wing of each. Both cage and female (in cage) can be seen as random effects. The data are in the dataframe mosquito.

TABLE 10.9

ANOVA Table for a Balanced Nested Design B (in A) with Fixed Factors

Source	Df	SS	MS	F^*
A	$a-1$	$n_0 b \sum\limits_{i=1}^{a} (\bar{Y}_i - \bar{Y})^2$	$\dfrac{SS_A}{a-1}$	$\dfrac{MS_A}{MSE}$
$B(A)$	$a(b-1)$	$n_0 \sum\limits_{i=1}^{a} \sum\limits_{j=1}^{b} (\bar{Y}_{j(i)} - \bar{Y})^2$	$\dfrac{SS_{B(A)}}{a(b-1)}$	$\dfrac{MS_{B(A)}}{MSE}$
Error	$ab(n-1)$	$\sum\limits_{i=1}^{a} \sum\limits_{j=1}^{b} \sum\limits_{k=1}^{n_0} (Y_{j(i)k} - \bar{Y}_{j(i)})$		

Note: There are a levels in A, b levels in $B(A)$, and n_0 replicates within each combination of A and $B(A)$.

TABLE 10.10

Expected Mean Squares for a Balanced Two-Factor Nested Model (B Nested in A)

Source	A and B Fixed	A and B Random	A Fixed, B Random
A	$\sigma^2 + bn_0 \dfrac{\sum_{i=1}^{a} \alpha_i^2}{a-1}$	$\sigma^2 + bn_0\sigma_A^2 + n_0\sigma_{A:B}^2$	–
$B(A)$	$\sigma^2 + n_0 \dfrac{\sum_{i=1}^{a}\sum_{j=1}^{b} \beta_{j(i)}^2}{a(b-1)}$	$\sigma^2 + an_0\sigma_B^2$	$\sigma^2 + an_0\sigma_B^2$
Error	σ^2	σ^2	σ^2

We have

```
data(mosquito)
aedes.ref <- lmer(length ~ (1|cage/female), data = mosquito)
aedes.ref
  AIC   BIC logLik deviance REMLdev
138.5 143.2 -65.25      135   130.5
Random effects:
 Groups      Name        Variance Std.Dev.
 female:cage (Intercept) 94.9424  9.7438
 cage        (Intercept) 17.7064  4.2079
 Residual                 1.3017  1.1409
```

Because the design is balanced, we can obtain the variance component estimates using mean squares from a conventional ANOVA.

```
anova(lm(length ~ cage/female, data = mosquito))
            Df  Sum Sq Mean Sq F value     Pr(>F)
cage         2  665.68  332.84  255.70 1.452e-10 ***
cage:female  9 1720.68  191.19  146.88 6.981e-11 ***
Residuals   12   15.62    1.30
```

TABLE 10.11

Method of Moments Variance Components Estimators for Balanced Hierarchical Designs with Three Levels of Nesting

Source	Variance Component	Variance Component Estimator	
		A, B, C Random	A Fixed, B, C Random
A	σ_A^2	$\dfrac{MS_A - MS_{B(A)}}{n_0 bc}$	–
$B(A)$	$\sigma_{B(A)}^2$	$\dfrac{MS_{B(A)} - MS_{C(B(A))}}{n_0 c}$	$\dfrac{MS_{B(A)} - MS_{C(B(A))}}{n_0 c}$
$C(B(A))$	$\sigma_{C(B(A))}^2$	$\dfrac{MS_{C(B(A))} - MSE}{n_0}$	$\dfrac{MS_{C(B(A))} - MSE}{n_0}$
Error	σ^2	MSE	MSE

Note: There are a levels in A, b levels of $B(A)$, c levels of $C(B(A))$, and n_0 replicates for each combination of $C(B(A))$.

We have

$\hat{\sigma}_{cage} = (332.8379 - 191.1864)/8 = 17.7064$ (4 females in each cage × 2 length measures for each female = 8), and $\hat{\sigma}_{female(cage)} = (191.1864 - 1.301667)/2 = 94.9424$ (2 length measures for each female).

Expressed as percentages (of the total variance in Y), these are

```
c(94.9424, 17.7064)/(94.9424 + 17.7064 + 1.3017) * 100
[1] 83.31898 15.53868
```

EXAMPLE 10.14

Sokal and Rohlf (2012) described a hierarchical experimental design to determine the effect of diet on glycogen levels in rat livers. Six rats were randomly assigned one of three diets: "control," "compound 217," and "compound 217 + sugar." After a short period of time, the rats were euthanized and glycogen levels in their livers were measured. Two glycogen measurements were made for each of three preparations of each liver from each rat. In this case, the diet is fixed, while liver preparation (in diet) and glycogen measurements (in liver in diet) are random.

```
data(rat)
```

We can use this example to show how pseudoreplication can result in incorrect analyses. Of primary interest is the detection of differences in the fixed diet factor. By treating each measure in each liver preparation in each rat as independent observations, we have

```
anova(lm(glycogen ~ diet, data = rat))
          Df Sum Sq Mean Sq F value    Pr(> F)
Diet       2 1557.6  778.78  14.498 3.031e-05 ***
Residuals 33 1772.7   53.72
```

The analysis tells us that $df_E = 33$; however, there are only six rats. Thus, df_E should be $n - a = 6 - 3 = 3$. Pseudoreplication has artificially inflated sample sizes, and increased the probability of type I error. We can average-out the pseudoreplication, and get the correct P-value, by using the means for the interaction of the pseudoreplicated levels as observations. Thus, creating new treatment and response vectors of length six (the number of true independent replicates)

```
resp <- with(rat, tapply(glycogen, interaction(diet, rat), mean))
trt <- factor(rep(seq(1,3),2))
anova(lm(resp ~ trt))
          Df Sum Sq Mean Sq F value Pr(>F)
trt        2 259.59 129.796   2.929 0.1971
Residuals  3 132.94  44.315
```

The fixed factor results are no longer significant.

Using a mixed-model approach, we can obtain variance components estimates for the model. Calculating these values by hand is left as an exercise.

```
rat.ref <- lmer(glycogen ~ diet+(1|diet:rat)+(1|diet:rat:liver), data = rat)
rat.ref
 AIC   BIC logLik deviance REMLdev
 233.6 244.7 -109.8    234.9   219.6
Random effects:
 Groups          Name        Variance Std.Dev.
 liver:(rat:diet) (Intercept) 14.1668  3.7639
```

```
rat:diet        (Intercept)   36.0651   6.0054
Residual                      21.1666   4.6007
```

★   ## 10.11 Split-Plot Design

In a split-plot design, we have two sizes of experimental units: larger units, to which whole-plot treatments are assigned, and smaller units, within whole plots, to which split-plot treatments are assigned (Chapter 7). While split-plot designs are hierarchical (the whole-plots treatments have pseudoreplicates), they can be distinguished from nested designs because split plots contain informative (generally fixed) factor levels that mean exactly the same thing in other whole plots. Split-plot designs can be distinguished from randomized block designs because each whole plot will generally be assigned to an informative whole-plot factor level (Chapter 7).

There are two types of split-plot designs: simple and usual (cf. Milliken and Johnson 2009). In the *simple whole-plot design*, the whole-plot factor A will have its levels randomly assigned to whole-plots experimental units in completely randomized (nonblocked) fashion, and split plots will be randomly assigned within whole plots. This approach is briefly addressed, in the context of repeated measures designs, in Section 10.12. In the *usual split-plot design*, we will have a blocking factor, B, in addition to a factor A, whose levels are assigned to whole plots, and a factor C whose levels are assigned to split plots. The blocks represent whole-plot replicates. That is, each block will contain a single iteration of all the whole-plot factor levels, randomly assigned to the whole plots in the block. The usual split-plot design can be expressed as

$$Y_{ijk} = \mu + \alpha_i + \beta_j + \gamma_k + (\alpha\gamma)_{ik} + \varepsilon_{ijk}, \tag{10.39}$$

where Y_{ijk} is the observation from kth level in C, the jth blocking level of B and the ith level in A, μ is a constant (for instance, the true mean of the factor-level combination means), α_i is the ith whole-plot treatment effect from A, β_j denotes the jth level of blocking factor B,[*] γ_k indicates the kth treatment effect from the split-plot factor C, and $(\alpha\gamma)_{ik}$ specifies the interaction effect of the ith level in A and the kth level in C. As with all general linear models, we assume that ε_{ijk}s are independent random errors that follow $N(0, \sigma^2)$.

The interaction between the whole-plot treatment A and the blocking factor B is not estimable (since, as with conventional RBDs, there is no replication for levels in A in the blocks). As a result, we assume that A and B are additive; that is, we assume that $(\alpha\beta)_{ij} = 0$. In addition, in conventional split-plot designs, the residual error term cannot be fully isolated, and as a result, we assume $(\beta\gamma)_{jk} = 0$ (see Underwood 1997).

Conventionally, factors A and C are fixed, while B is considered random. In this case, the hypotheses for the whole-plot factor A are

---

[*] The formulation $B(A)$, that is, B nested in A, is often used in textbooks. However, this can be confusing because the blocks, not the whole plots, constitute the largest hierarchical stratum of the design. As a result, this notation is not used in Equation 10.39 (cf. Littell et al. 2006, Milliken and Johnson 2009).

$$H_0: \mu_1 = \mu_2 = \cdots = \mu_a,$$

versus

$$H_A: \text{ at least one } \mu_i \text{ not equal to the others.}$$

Hypotheses for the random blocking factor B are

$$H_0: \sigma_B^2 = 0,$$

versus

$$H_A: \sigma_B^2 \neq 0 \text{ (although this effect is not tested in conventional split-plot ANOVAs).}$$

Hypotheses for the split-plot factor C are

$$H_0: \mu_1 = \mu_2 = \cdots = \mu_c,$$

versus

$$H_A: \text{ at least one } \mu_k \text{ not equal to the others.}$$

Finally, hypotheses for the interaction of A and C are

$$H_0: \text{ all } \alpha\gamma_{ik} = 0,$$

versus

$$H_A: \text{ all } \alpha\gamma_{ik} \neq 0.$$

Because of the lack of replication for whole plots within blocks, we use treatment interactions as error terms in both RBDs and split-plot designs. Note that in Table 10.7 the randomized block design SSE term is equivalent to the interaction sum of squares term from the factorial design in Table 10.4. As with RBDs, missing values constitute a significant problem for split-plot designs. Hoshmand (2006) discusses interpolation procedures for these scenarios.

In F-tests, we assay the significance of the whole-plot factor A using the interaction with the block; that is, $MS_{A:B}$ is the denominator of F^* (Table 10.12). This is because, as with RBDs, blocks function as replicates for the whole-plot treatments. While computational details for sums of squares are not included in Table 10.12, these procedures can be found in a number of texts, including Hoshmand (2006).

10.11.1 aov

The function `aov` can be used to analyze balanced *multistratum* (hierarchical) mixed-model designs using conventional F-test procedures. Like anova, aov is *not* appropriate for multifactor unbalanced designs (Section 10.14). However, these situations are capably handled by `lme` and `lmer`.

The aov function contains arguments analogous to those in `lm`, including `formula`, `data`, and `contrasts`. The `formula` argument, however, can now be used to specify multiple error terms. In a conventional split-plot design, this is done by defining a means formula, which includes all strata except for the largest factor, and an `Error` term. In the

TABLE 10.12

ANOVA Table for a Balanced Two-Factor Split-Plot Design

Source	df	MS	E(MS)	F^*
Whole plot				
A	$a-1$	$\dfrac{SS_A}{a-1}$	$\sigma^2 + c\sigma_W^2 + \phi_A$	$\dfrac{MS_A}{MS_{A:B}}$
Whole-plot error	$(a-1)(b-1)$	$\dfrac{SS_{A:B}}{(a-1)(b-1)}$	$\sigma^2 + c\sigma_{A:B}^2$	
Split plot				
C	$c-1$	$\dfrac{SS_C}{c-1}$	$\sigma^2 + \phi_C$	$\dfrac{MS_C}{MS_{SPE}}$
$A:C$	$(a-1)(c-1)$	$\dfrac{SS_{A:C}}{(a-1)(c-1)}$	$\sigma^2 + \phi_{A:C}$	$\dfrac{MS_{A:C}}{MS_{SPE}}$
Split-plot error	$a(b-1)(c-1)$	$\dfrac{SS_{SPE}}{a(b-1)(c-1)}$	σ^2	

Note: A (the whole-plot factor) and C (the split-plot factor) are fixed, and B (a blocking factor) is random. There are a levels in A, b levels of B, and c levels of C. The ϕ terms represent quadratic expressions of the fixed effects. They will have no effect on variance component estimation (see Milliken and Johnson 2009 for additional information). The term SPE denotes split-plot error. For details on the calculation of sums of squares, see Hoshmand (2006) or Snedecor and Cochran (1989).

`Error` term, all strata, except for the smallest factor, are listed from largest to smallest using forward slashes. For instance, in the *usual* split-plot design, one would specify

```
formula = Y ~ A + C + Error(B/A)
```

The term `Error(B/A)` defines the error strata. It identifies the largest stratum (the blocking factor B) and the next largest stratum (the whole-plot factor A). The smallest stratum (the split-plot factor C) is automatically included as output in the so-called within stratum level.

EXAMPLE 10.15

Snedecor and Cochran (1989) describe an experiment to test how varieties of alfalfa respond to the last cutting day of the previous year. In the fall, alfalfa plants stop adding above-ground biomass and store carbon resources (acquired from photosynthesis) in their roots for growth during the following year. Thus, we might expect that earlier cutting dates inhibit growth for the following year by decreasing available resources. On the other hand, if plants are cut after they have gone into senescence, there should be little effect on productivity during the following year. The agricultural whole plots in the experiment were grouped into six blocks, each with three plots. One of three alfalfa varieties (Cossack, Ranger, or Ladak) was randomly assigned to each whole plot (in each block). Each whole plot was then subdivided into four split plots. To each of these, one of the four cutting dates (A = none, B = September 1, C = September 20, or D = October 7) was randomly assigned (Figure 10.11).

We have

```
data(alfalfa.split.plot)
summary(aov(yield ~ variety * cut.time + Error(block/variety),
data = alfalfa.split.plot))
```

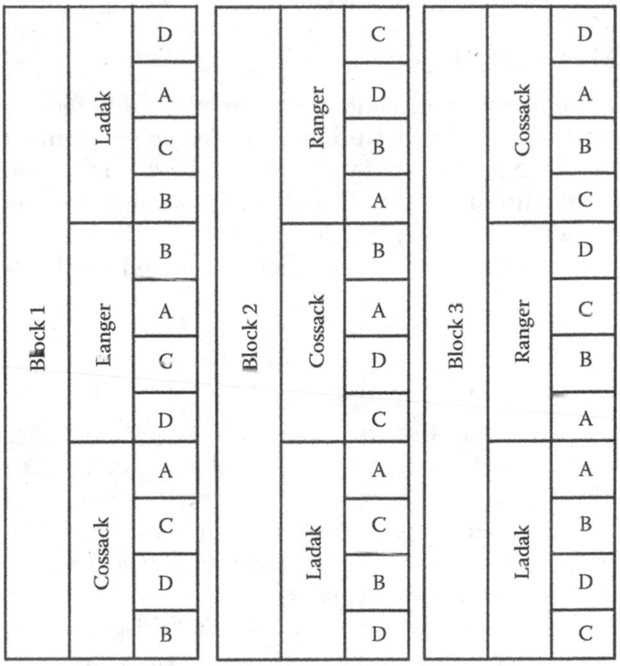

FIGURE 10.11
First three blocks of the alfalfa split-plot experiment showing the random assignment of whole- and split-plot treatments. (Adapted from Snedecor, G. W., and Cochran, W. G. 1989. *Statistical Methods*, 8th ed. Iowa State College Press, Ames, IA.)

```
Error: block:variety
         Df  Sum Sq Mean Sq F value Pr(>F)
variety    2 0.17802 0.08901  0.6534 0.5412
Residuals 10 1.36235 0.13623

Error: Within
                 Df  Sum Sq Mean Sq F value   Pr(>F)
cut.time          3 1.96247 0.65416 23.3897 2.826e-09 ***
variety:cut.time  6 0.21056 0.03509  1.2548   0.2973
Residuals        45 1.25855 0.02797
```

Recall that we have 3 levels in A, 6 levels in B, and 4 levels in C. Thus, $a = 3$, $b = 6$, and $c = 4$. The degrees of freedom in the whole-plot analysis occur because $a - 1 = 2$, and $(a - 1)(b - 1) = 10$. The split-plot residual degrees of freedom are calculated as $c - 1 = 3$, $(a - 1)(c - 1) = 6$, and $a(b - 1)(c - 1) = 45$.

We conclude that alfalfa yield differs greatly with respect to cut time, but not variety. Given the significance of the main effects and additivity, whole-plot and split-plot factor-level effects can be examined using their respective error term estimates as pooled variance values in simultaneous inference procedures.

Note that the procedure above could also be handled in lmer or lme. For instance,

```
lmer(yield ~ cut.time * variety + (1| block/variety), data = alfalfa.split.
plot)
```

This will allow inferential consideration of the random blocks.

★ 10.12 Repeated Measures Design

In repeated measures designs, experimental units are measured repeatedly. There are two general frameworks: (1) matched pairs studies, in which an experimental unit receives all of the treatments over time (often in a randomized order), and (2) longitudinal designs, in which an experimental unit is randomly assigned to a single treatment and is then followed over time (Chapter 7).

Matched pairs studies can often be viewed as randomized block designs. That is, they will have the form

$$Y_{ij} = \mu + \text{time}_i + \text{subject}_j + \varepsilon_{ij}, \tag{10.40}$$

where Y_{ij} is the ith observation of the jth subject, μ is a constant, time_i is the effect of the ith time frame, and subject_j is the effect of the jth subject. The level time_i will define the ith treatment while the subjects will be blocks. Here, the time frames are of interest themselves, or coincide with treatment applications.[*] The variation of subjects over time is often referred to as the *within-subject* effect. The tendency of subjects receiving different treatments to differ is called the *between-subject* effect.

Longitudinal studies can often be viewed as *simple* split-plot designs in which treatments are not randomly assigned within blocks. In this case, subjects define whole plots that receive a single whole-plot treatment and split-plots levels define the time series. That is

$$Y_{ijk} = \mu + \text{treatment}_i + \text{subject}_{j(i)} + \text{time}_k + (\text{treatment} : \text{time})_{ik} + \varepsilon_{ijk}, \tag{10.41}$$

where Y_{ijk} represents the application of the ith treatment to the jth subject at time k, μ is a constant, treatment_i defines the ith treatment effect, $\text{subject}_{j(i)}$ is the effect of the jth subject level (nested in the ith treatment), time_k is the effect of the kth time frame, and $(\text{treatment:time})_{ik}$ is the interaction effect of the ith treatment and the kth time. An analysis using this approach is called a *split plot in time*. The subjects recapitulate the treatment and should be specified as nested in the treatment factor in statistical models.[†] Note that, in contrast to conventional split-plot designs, the split-plot levels (time frames) cannot be randomly assigned. For both the RBD and split-plot formats, we assume that subject is a random factor (cf. Kutner et al. 2005).

As with their non-time series mixed-model counterparts, repeated measures RBDs and repeated measures split-plot designs with $n_0 = 1$ assume

1. Additivity between block and treatment factors (although see discussion of RBDs with random blocks in Section 10.9).

2. Normality and homoscedasticity of the marginal errors.

3. Normality and homoscedasticity of random effects.

4. Independence of outcomes from distinct random levels.

5. Independence of random effects and marginal residuals.

---

[*] See Hoshmand (2006) for a description of analyses in which randomized blocks contain treatments of interest followed over time.

[†] No worked examples are given here. For guidance type: `example(heart)`.

Additionally, for the F-distribution under H_0 to be correct, sphericity is necessary (Milliken and Johnson 2009). Recall from Chapter 7 that sphericity describes a situation in which the variances of treatment differences are the same for all pairs of treatments.

Asphericity will occur as a result of temporal autocorrelation. Examples include *order effects* in which the order that the treatment is received influences the response, or *carryover effects*, in which the effect of a previous treatment influences the response to the current treatment. Dependence from these effects can often be addressed by randomizing the order of treatment applications and/or by separating time frames by intervals that are sufficient to insure independence. However, such steps may not be possible. For instance, if time itself is a treatment (as in models (10.40) and (10.41)), then randomization is impossible, and asphericity, which will prevent analysis with conventional mixed (or fixed effect) models, is likely.

EXAMPLE 10.16

Driscoll and Roberts (1997) examined the impact of fire on the Walpole frog (*Geocrinia lutea*) in catchments in Western Australia by counting the number of calling males in six paired burn and control sites for 3 years following a prescribed fire. The experiment used a randomized block repeated measures design with catchments as a random blocking factor and year as a fixed factor.

Qualitative comparison of the variance of the differences in treatments (years) reveal potential problems with asphericity.

```
data(frog)
yrs <- as.matrix(unstack(frog, frogs ~ year))
```

Here are the differences among years:

```
yr.diff <- print(data.frame(yr1.minus.2 = yrs[,1] - yrs[,2], r1.minus.3
= yrs[,1] - yrs[,3], yr.2.minus.3 = yrs[,2] - yrs[,3]))
  yr1.minus.2 r1.minus.3 yr.2.minus.3
1         -13        -14           -1
2          -9        -18           -9
3          -5        -16          -11
4          -3        -12           -9
5         -10         -4            6
6          -5         -1            4
```

and here are the variances of the differences:

```
apply(yr.diff, 2, var)
 yr1.minus.2    r1.minus.3 yr.2.minus.3
    14.30000      46.56667     53.86667
```

10.12.1 Analysis of Contrasts

A conventional approach for analyzing repeated measures designs is to account for asphericity by decreasing the degrees of freedom, and the resultant power in the model. This procedure, called an *analysis of contrasts*, or a *repeated measures ANOVA*, computes the slopes of treatment on time regression lines for each subject. It then uses the slopes as data to assess treatment effects over time (Littell et al. 2006). The process requires a technique called *multivariate analysis of variance* (*MANOVA*), wherein the response variable is a matrix, with subjects as rows and points in the time series as columns. In this format, predictor columns (if any) will have a length equal to the number of subjects.*

The degree to which true experimental unit variance–covariance matrix of time frames departs from sphericity can be defined by the parameter epsilon, ε (Winer et al. 1991). The measure is defined in the range [0, 1]. When sphericity is met, then $\varepsilon = 1$, and as asphericity increases, ε approaches 0. In most statistical software, ε will be estimated using two methods: the *Greenhouse–Geisser epsilon* and the *Huynh–Felt epsilon*, both denoted $\hat{\varepsilon}$. For either approach, the degrees of freedom for sources of variation in the model will be reduced via multiplication by $\hat{\varepsilon}$. The Greenhouse–Geisser procedure is conservatively biased in its adjustment to degrees of freedom when ε is close to 0.75 (Winer et al. 1991). On the other hand, Huynh–Felt estimates for epsilon can exceed 1, making it "optimistically" biased. Principally for this reason, the Greenhouse–Geisser estimator is generally recommended over the Huynh–Felt estimator (Quinn and Keough 2002).

EXAMPLE 10.17

To conduct an analysis of contrasts for the frog dataset, we first specify a multivariate linear model. This will require that the "response" be a subject by year matrix. Because the block and treatment variables are defined by the response matrix rows and columns, respectively, we will make this matrix a function of the intercept in the model.

```
frog.mlm <- lm(yrs ~ 1) # yrs matrix from Exercise 10.15
```

The function Anova (from *car*) requires that we define the time frames in the time series as an ordinal variable. We will call this object idata.

```
idata <- data.frame(yrs = ordered(c(1992:1994)))
```

The function also requires specification of the structure of the repeated measures design using the argument idesign. Time frames specified in the argument idesign must be contained in the dataframe called by the argument idata.

```
frog.aoc <- Anova(frog.mlm, idata = idata, idesign = ~ yrs)
```

The summary output from the function is verbose. However, we can isolate the important aspects of the results, including a summary of the univariate model assuming sphericity:

```
summary(frog.aoc)
Univariate Type III Repeated-Measures ANOVA Assuming Sphericity

              SS num Df Error SS den Df      F   Pr(>F)
(Intercept)  2.72      1   955.61      5 0.0142 0.909649
yrs        369.44      2   191.22     10 9.6601 0.004615 **
---
Signif. codes:  0 '***' 0.001 '**' 0.01 '*' 0.05 '.' 0.1 ' ' 1
```

Note that the degrees of freedom error are calculated as if this were a randomized block design, that is, $df_E = (a - 1)(b - 1) = (3 - 1)(6 - 1) = 10$.

Of primary interest are the asphericity-corrected results for the random within-subject component, year.

---

* In a MANOVA, we test the null hypothesis that vectors of population means (each representing a treatment) are equal.

```
summary(frog.aoc)
Greenhouse-Geisser and Huynh-Feldt Corrections
 for Departure from Sphericity
     GG eps Pr(>F[GG])
yrs 0.71218    0.01252 *
     HF eps Pr(>F[HF])
yrs 0.91539    0.006175 **
```

Both the Greenhouse–Geisser or Huynh–Felt procedures report asphericity-corrected results for year that are significant.

An analysis of contrasts approach can also be used in longitudinal designs. Consider the example below.

EXAMPLE 10.18

Littell et al. (2002) described an experiment in which the effect of three asthma drugs were measured on 24 asthmatic patients. Each patient was randomly given a single drug (A, C, or P). Drug A was a standard asthma medication, drug C was a potential competitor for A, and drug P was a placebo consisting of only the gelatin capsule used to deliver drugs A and C. Forced expiratory volume in one second (FEV1) was measured at eight consecutive hours starting 11 h after drug ingestion. The data, contained in the dataframe asthma, are formatted (with subjects in rows) for an analysis of contrasts.

A baseline measure of FEV1 was also taken 11 h before the application of the treatment, but will be ignored because it was spaced unequally in time with respect to the other measurements.

```
data(asthma)
FEV1.data <- as.matrix(asthma[,3:10])
asthma.mlm <- lm(FEV1.data ~ DRUG, data = asthma)
idata <- data.frame(hr = ordered(11:18))
summary(Anova(asthma.mlm, idata = idata, idesign = ~ hr))

Univariate Type II Repeated-Measures ANOVA Assuming Sphericity
              SS num Df Error SS den Df         F    Pr(>F)
(Intercept) 5489.2     1   247.41     69 1530.8650 < 2.2e-16 ***
DRUG          25.8     2   247.41     69    3.5952  0.03271 *
hr            17.2     7    30.49    483   38.8569 < 2.2e-16 ***
DRUG:hr        6.3    14    30.49    483    7.1060 1.923e-13 ***

Greenhouse-Geisser and Huynh-Feldt Corrections
 for Departure from Sphericity
         GG eps Pr(>F[GG])
hr       0.49706  < 2.2e-16 ***
DRUG:hr 0.49706  1.087e-07 ***
         HF eps Pr(>F[HF])
hr       0.52668  < 2.2e-16 ***
DRUG:hr 0.52668  4.949e-08 ***
```

The between-subject factor, drug, can be assessed using a test that is unadjusted for asphericity (i.e., it is the result from a conventional ANOVA in which subject (in drug) is random).

The within-subject components hour, and the drug:hour require asphericity corrections. Corrected P-values are calculated by multiplying the degrees of freedom in the unadjusted model by $\hat{\varepsilon}$. For instance, the Greenhouse–Geisser P-value for drug:hour is calculated as

```
epsilon <- 0.49706
pf(7.106, 14*epsilon, 483*epsilon, lower.tail = FALSE)
[1] 1.086602e-07
```

10.12.2 Modern Mixed-Model Approaches

As a result of recent developments in software, researchers can interactively modify and evaluate the covariance structure of a mixed model. This approach is useful for four reasons. First, it allows missing observations that constitute a serious problem for traditional repeated measures designs. Second, unlike an analysis of contrasts, the approach directly addresses the problem of complex covariance structures resulting from nonindependent observations. This is done by allowing the marginal residuals to have nonconstant variances and nonzero covariances. Third, because the covariance structure can incorporate autocorrelation, sphericity is not required. Fourth, it allows outcomes whose patterns vary both over time and among experimental units.

As with RBDs with random blocks and split-plot mixed models, the approach here assumes that the random between-block (subject) effects are independent, normal, and homoscedastic. The covariance structure for random within-subject effects (repeated measurements) is potentially very complex given nonindependence, and is defined by a series of within-subject $k \times k$ covariance matrices, denoted Σ. In Σ, k is the number of time frames that the subject is observed. Note that because missing observations are allowed, k need not be equal across all subjects. The covariance structure will encompass the error structure of the marginal residuals because they are defined at the level of individual observations. The diagonal of Σ will contain variances for each time frame, while the off-diagonal elements will contain the covariances among time frames. Taken together, the Σ matrices constitute the diagonal of a *block matrix* denoted $\mathbf{V}$, that defines the marginal error covariance structure (cf. Littell et al. 2006). Below is an example in which two experimental units are each measured at two time frames.

$$\mathbf{V} = \begin{bmatrix} \Sigma & 0 \\ 0 & \Sigma \end{bmatrix}, \quad \text{where } \Sigma = \begin{bmatrix} \sigma_1^2 & \sigma_{12} \\ \sigma_{12} & \sigma_2^2 \end{bmatrix}.$$

Note that (1) the off-diagonal elements (blocks) of the matrix constitute between-subject covariances and are zero because of the assumption of intersubject independence, and (2) the covariance framework is assumed to be the same for all subjects.

A large number of standard covariance formats can be applied to Σ matrices in repeated measures mixed models. The flexibility (and unwieldy character) of the analyses will increase with the number of estimated parameters. The simplest covariance structure is the *independent covariance* or *diagonal model* in which the off-diagonal elements of each within-subject correlation matrix, representing the correlations between each pair of times, are zero. This is the default structure for Σ in both *nlme* and *lme4*. Slightly more complex is *compound symmetry* in which the correlation of errors is assumed to be constant regardless of the temporal proximity of repeated measurements. Still more complex is *first-order autoregressive (AR1)* covariance structure. In this case, the correlation between measurements adjacent in time (e.g., times 1 and 2, or times 4 and 5) is ρ, the correlation between errors for observations separated by a single time interval is ρ^2, the correlation between errors for observations separated by two intervals is ρ^3, and so on. AR1 often provides effective representations of serial autocorrelation and is one of the few structures that allow noninteger time measurements (Pinhiero and Bates 2000). Another frequently used autocorrelation structure is *autoregressive moving average* or *(ARIMA)*. In this case, the covariance between errors is a function of both autoregressive and moving average processes from time series analysis. The order of the autoregressive process is given by the parameter p, while the number of prior time units used in the moving average calculations is given by the parameter q. For AR1 processes, $p = 1$ and $q = 0$. The most complex covariance structure is *unstructured covariance* or *general covariance* in which the within-subject errors for each pair of time intervals can have unique correlations. Because the number of estimated parameters increases quadratically with the number of within-subject observations, this structure will generally lead to overparameterized models (Pinhiero and Bates 2000). The covariance frameworks described above are demonstrated in a brief vignette contained in *asbio* that can be accessed by typing: `vignette(ranef.cov)`.

EXAMPLE 10.19

Freund et al. (1986) described a longitudinal study of exercise therapies. In the experiment, 37 patients were randomly assigned to one of two weightlifting programs. In the first program, repetitions with weights were increased as subjects became stronger. In the second program, the number of repetitions was fixed, but weights were increased as subjects became stronger. An index measuring strength was recorded for each subject at day 0, 2, 4, 6, 8, 10, and 12.

We first create a term to contain random effects of the subjects associated with day.

```
data(exercise.repeated)
strength.g <- groupedData(strength ~ day|ID, data = exercise.repeated,
order.groups = F)
```

The `groupedData` function allows straightforward formulation and depiction of complex nested and repeated measures designs. For instance, try `plot(strength.g)`.

Here is our reference model:

```
str.ref <- lme(strength ~ TRT * day, random = ~day, data = strength.g)
```

We can compare the reference model (which uses an independent covariance structure) to models with AR1, AR2, and ARIMA($p = 2$, $q = 2$) covariance structures.

```
str.AR1 <- update(str.ref, correlation = corAR1())
str.AR2 <- update(str.ref, correlation = corARMA(p = 2, q = 0))
str.ARIMA <- update(str.ref, correlation = corARMA(p = 2, q = 2))
```

The AR1 model has significantly higher log-likelihood than the original reference model, and is more parsimonious (has smaller information criteria values) than the AR2 and ARIMA autocorrelation models.

```
anova(str.ref, str.AR1, str.AR2, str.ARIMA)
          Model df      AIC      BIC    logLik   Test  L.Ratio p-value
str.ref       1  8 834.5395 862.2162 -409.2698
str.AR1       2  9 827.1674 858.3037 -404.5837 1 vs 2 9.372089  0.0022
str.AR2       3 10 829.0988 863.6946 -404.5494 2 vs 3 0.068680  0.7933
str.ARIMA     4 12 832.8736 874.3887 -404.4368 3 vs 4 0.225131  0.8935
```

★ **10.13 ANCOVA**

ANCOVA provides an extension to methods discussed in Section 9.15 for testing the equality of regression parameters (e.g., slopes and intercepts) associated with two treatments by allowing consideration of more than one factor, each with ≥2 factor levels. ANCOVA augments ANOVA by adding a quantitative predictor called the *concomitant variable* to a model with one or more factors. The method often increases power for distinguishing factor-level effects by accounting for variability in the response due to the concomitant variable that would otherwise be lumped with model error. Thus, we can think of ANCOVA as an ANOVA performed on the residuals from a regression of the response on the concomitant variable. ANCOVA also equates to a comparison of regression Y-intercepts that are equivalent to the grand mean plus the factor-level effects. This is illustrated in Figure 10.12. We note that the variability around the ANCOVA regression lines is smaller than the variability of observations with respect to their factor-level means. This results in increased statistical power for detecting factor-level effects.

The ANCOVA framework is extremely flexible. For instance, the relationship between the concomitant variable and Y need not be linear, and multiple concomitant variables and multiple random and/or fixed factors can both be used. For comprehensive information, see Milliken and Johnson (2002).

For a single-factor ANCOVA with fixed-factor levels, the model can be given as

$$Y_{ij} = \mu + \alpha_i + \beta_1(X_{ij} - \bar{X}) + \varepsilon_{ij},\tag{10.42}$$

where Y_{ij} is the jth observation from the ith level in A, μ is a constant, α_i is the ith treatment effect from A ($\alpha_i = \mu_i - \mu$), β_1 is the slope coefficient describing the linear change in

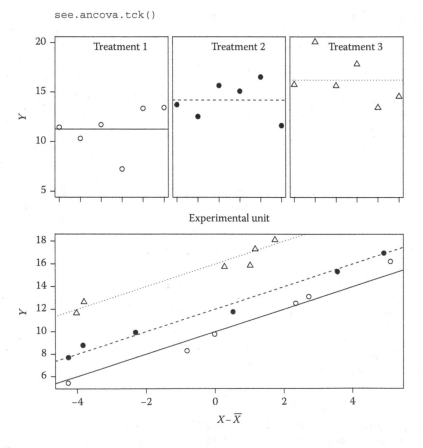

FIGURE 10.12
Mechanics of ANCOVA. Decreasing model error variance compared to factor-level population variance (due to an association between Y and the concomitant variable) results in increased explanatory power for the categorical explanatory variable.

Y as a function of the concomitant variable X, $\bar{X}$ is the sample mean of the concomitant observations and ε_{ij}s are random errors that are independent and follow $N(0, \sigma^2)$.

We center X so that comparisons of treatment Y-intercepts on the ANCOVA lines are not meaningless extrapolations. Examining Equation 10.42, we see that if the effect of the concomitant variable is zero then the ANCOVA model becomes a one-way ANOVA, and if the effect of the factor is zero, the model becomes a simple linear regression. That is

$$Y_{ij} = \mu + \alpha_i + 0 + \varepsilon_{ij},$$

$$Y_{ij} = \mu + 0 + \beta_1(X_{ij} - \bar{X}) + \varepsilon_{ij}.$$

In addition to the general assumptions of ANOVA (Section 10.7), the ANCOVA concomitant variable should not be confounded with the categorical predictors in the model. Consider a study to compare scores for biometry students taking online and conventional classroom courses using study time as a concomitant variable. Further, assume that one teaching method always requires more study time than the other (Figure 10.13). An ANCOVA distinguishes treatments by finding the difference in the elevations of the treatment lines. In this case the elevations appear similar. However, because of confounding, a comparison of the courses requires an extrapolation of the treatment lines.

Equation 10.42 is only valid if the slopes for the lines corresponding to the treatments are equal. If this is not true, then a model with less generality must be used:

$$Y_{ij} = \mu + \alpha_i + \beta_{1i}(X_{ij} - \bar{X}_i) + \varepsilon_{ij}, \tag{10.43}$$

where β_{1i} is the slope coefficient corresponding to the ith factor level, and $\bar{X}_i$ is the mean of the subset of concomitant values corresponding to the ith factor level. We can test for the

Fig.10.13() ## see Ch. 10 code

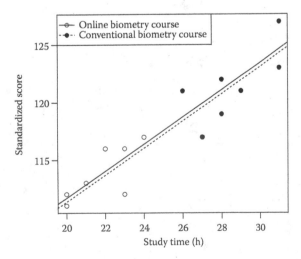

FIGURE 10.13
A scatterplot of the response on the concomitant variable can be used to assess the independence of the concomitant variable and the treatments. Nonindependence will often cause observations for the treatments to be clustered along the X axis.

presence/absence of constant slopes by running a nonadditive model, and checking for significance of the interaction of the concomitant variable and the factor. Large *P*-values provide little evidence against the null hypothesis of equal slopes.

The key ANCOVA inferences are identical to ANOVA; namely, do treatment effects exist, and if so what are they? Fixed treatment effects result in the same hypotheses as one-way fixed effect ANOVA, that is

$$H_0: \alpha_1 = \alpha_2 = \cdots = \alpha_a = 0,$$

versus

$$H_A: \text{not all } \alpha_i = 0.$$

Of lesser interest is the relationship of the concomitant variable and *Y*. Here our hypotheses are identical to simple linear regression:

$$H_0: \beta_1 = 0,$$

versus

$$H_A: \beta_1 \neq 0.$$

EXAMPLE 10.20

Partridge and Farqaur (1981) studied the effect of the number of mating partners on the longevity of male fruit flies (*Drosophila melanogaster*) by randomly assigning individual male fruit flies to one of five treatments: (1) one virgin female per day, (2) eight virgin females per day, (3) a control with one newly inseminated female per day, (4) a control with eight newly inseminated females per day, and (5) a control with no added females. Male "compliance" in the two experimental groups was verified by examining females in these groups for fertile eggs. As a concomitant variable, thorax length (a known predictor of male longevity), was also measured.

```
data(fly.sex)
```

First, we center the concomitant variable

```
thorax1 <- fly.sex$thorax - mean(fly.sex$thorax)
```

To provide linear model effect size measures that are differences from the estimated grand mean, we use sum contrasts. Longevity was log-transformed to compensate for its exponential scale.

```
contrasts(fly.sex$treatment) <- contr.sum(5)
fly.lm <- lm(log(longevity) ~ treatment + thorax1, data = fly.sex)
summary(fly.lm)
Coefficients:
            Estimate Std. Error t value Pr(>|t|)
(Intercept)  3.99678    0.01723 232.016  < 2e-16  ***
treatment1   0.16523    0.03463   4.771 5.25e-06 ***
treatment2   0.08123    0.03462   2.346 0.020626  *
treatment3   0.13326    0.03447   3.866 0.000181 ***
treatment4  -0.04268    0.03466  -1.231 0.220577
thorax1      2.74895    0.22795  12.060  < 2e-16  ***
```

The value 3.997 is the mean of the factor-level means. This is the estimated average male life span among all factor levels in log units. The value 2.75 is the common slope coefficient of the ANCOVA regression line. Factor-level means are the effect sizes *minus* the estimated grand mean. Consequently, the estimated grand mean *plus* the effect sizes will give the ANCOVA Y-intercepts. Note that the effect size for the last treatment is not given in the output, but is $\overline{Y}_5 - \hat{\mu} = 3.60 - 3.997 = -0.395$. A plot of the linear model can now be made (Figure 10.14).

As a final step, we partition the sums of squares from the linear model using ANOVA. We use type II sums of squares to address the presence of the quantitative concomitant variable (see Section 10.13).

```
library(car)
Anova(fly.lm)
Anova Table (Type II tests)
Response: log(longevity)
          Sum Sq  Df  F value    Pr(>F)
treatment 4.1499   4    27.97  2.231e-16 ***
thorax1   5.3946   1   145.43  < 2.2e-16 ***
Residuals 4.4141 119
```

A nonsignificant test for interactions indicates that a single slope model is adequate.

```
Anova(update(fly.lm, ~. + treatment:thorax1))
                  Sum Sq  Df   F value Pr(>F)
treatment:thorax1 0.2273   4    1.5611 0.1894
```

We conclude that the number of sex partners positively affects longevity in fruit flies after accounting for thorax length.

If the interaction was significant, we could have obtained the separate slopes and intercepts for the treatment lines by specifying a cell means model with no intercept, and without the covariate as a main effect. In the output below, the coefficients for the

```
with(fly.sex, plot(thorax1, log(longevity), pch = as.numeric(treatment)))
y.int <- c(0.165, 0.0812, 0.1333, -0.0427, -0.395) + 3.997
for(i in 1:5) abline(y.int[i], 2.75, lty = [i])
```

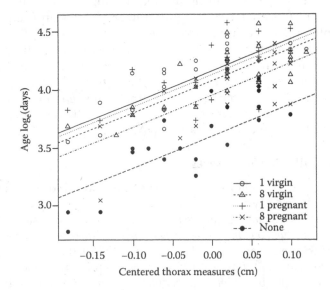

FIGURE 10.14
Depiction of the fly longevity ANCOVA analysis.

main effects are the *Y*-intercepts for their respective treatments, and the interactions between treatment and thorax are the slopes.

```
as.matrix(update(fly.lm, ~. -1 - thorax1 +  treatment:thorax1)
$"coefficients")
                       [,1]
treatment1        4.159335
treatment2        4.083816
treatment3        4.132915
treatment4        3.951554
treatment5        3.680838
treatment1:thorax1 2.574833
treatment2:thorax1 2.362532
treatment3:thorax1 2.128307
treatment4:thorax1 2.901558
treatment5:thorax1 3.755438
```

Factor-level comparisons will often follow an omnibus ANCOVA with significant treatment effects and valid additivity assumptions. Sum contrasts (unadjusted for familywise inference) were a part of the lm output in Exercise 10.17. However, we may also be interested in pairwise factor-level comparisons. The use of model MSE in conjunction with Scheffé or Bonferroni procedures is appropriate. However, Tukey's method is not appropriate for ANCOVA (Kutner et al. 2005). *Post hoc* procedures will be complicated by significant interactions between the factor and the concomitant variable because treatment differences will then vary with the value of the concomitant variable (see Milliken and Johnson 2002).

★   ## 10.14 Unbalanced Designs

Whether by intention or not, a researcher may wind up with an unequal number of replicates within treatments, that is, an unbalanced design. In one-way ANOVAs, lack of balance only constitutes a problem in significance testing when *P*-values are close to the significance level. If results are far from significant, or highly significant, then one can be confident in the inferential conclusions from the test (Kutner et al. 2005).

On the other hand, unbalanced multifactor ANOVAs will greatly complicate analyses. This is because faulty models result from conventional type I sums of squares when designs have at least one explanatory variable that is either quantitative, or categorical *and* unbalanced. For instance, we learned in Section 9.5.3 that such models will give different results depending on the order that predictors are specified. Type `see.lmu.tck()` for a graphical demonstration in the context of ANOVA.

10.14.1 Type II and III Sums of Squares

To address this problem, two alternatives have been proposed for type I sums of squares (denoted for brevity here as Type I SS).

Type II (or partial) sums of squares compare a reference model to a nested model without the variable of interest, and *without the term's higher-order relatives* (i.e., interactions with other predictors). Type II SS are always equivalent to type I SS for the last predictor specified in the model. *Type III (or marginal) sums of squares* compare a reference model to a nested model without a variable of interest but, *with the term's higher-order relatives*. Type II SS will be identical to type III SS if no interactions are specified in a model. If an ANOVA design is balanced *or* if the design has only one factor, then Type I = Type II = Type III sum of squares.

There is no consensus as to the relative superiority of type II or type III sums of squares. Unlike hypotheses for type I and II SS, type III SS hypotheses are not dependent on the number of observations in cells (Littell et al. 2006, p. 187). In addition, Milliken and Johnson (2009, pp. 217–222) recommend the use of type III SS because, given no empty cells, this method will have the same hypotheses, concerning the marginal means, that one would expect given a balanced layout. In contrast, type II SS may result in uninformative hypotheses concerning the marginal means. On the other hand, type III sum of squares have been critiqued for violating the *principle of marginality* (Fox 2002, Langsrud 2003). To illustrate, in type III SS, we consider the effect of factor A by comparing a reference model containing A and its higher-order relatives, to a nested model without A, but containing interactions with A. Thus, the nested model still contains A via its interactions. Fox (2002) also noted disapprovingly that type III sums of squares will vary depending on the type of contrast being used. Specifically, sum contrasts are necessary to obtain conventional type III SS. With respect to commercial software Type III SS is the default method for unbalanced linear models in SAS®, SPSS®, and Systat®, and the only option in MINITAB®. Type II SS is the default method in the **R** function Anova.

Conceptually, type II and III sums of squares are calculated in the same way. We find the sum of squares for a factor A by subtracting the SSE from a nested model that excludes A from the SSE of a reference model containing A:

$$SS_A = SSE_N - SSE_R. \tag{10.44}$$

The difference lies in what defines the nested model. For type II SS, interactions with A are left out of the nested model. For type III SS, higher-level interactions with A are retained in the nested model.

Table 10.13 summarizes the calculation of type I, II, and III SS sums of squares for a two-way ANOVA. Note that the interaction sum of squares ($SS_{A:B}$) and residual sum of squares

TABLE 10.13

Comparison of Sums of Squares for a Two-Way ANOVA

	Computation	Conditionality
Type I (Sequential)		
A	$SS_A = SSE(\mu) - SSE(A)$	$SS(A \mid \mu)$
B	$SS_B = SSE(A) - SSE(A + B)$	$SS(B \mid A)$
$A{:}B$	$SS_{A:B} = SSE(A + B) - SSE(A^* B)$	$SS(A{:}B \mid A, B)$
Error	$SSE(A^* B)$	
Type II (Partial)		
A	$SS_A = SSE(B) - SSE(A + B)$	$SS(A \mid B)$
B	$SS_B = SSE(A) - SSE(A + B)$	$SS(B \mid A)$
$A{:}B$	$SS_{A:B} = SSE(A + B) - SSE(A^*B)$	$SS(A{:}B \mid A, B)$
Error	$SSE(A^* B)$	
Type III (Marginal)		
A	$SS_A = SSE(B + A{:}B) - SSE(A^*B)$	$SS(A \mid B, A{:}B)$
B	$SS_B = SSE(A + A{:}B) - SSE(A^*B)$	$SS(B \mid A, A{:}B)$
$A{:}B$	$SS_{A:B} = SSE(A + B) - SSE(A^*B)$	$SS(A{:}B \mid A, B)$
Error	$SSE(A^*B)$	

Note: The terms in parentheses following SSE denote factors of the linear model. For instance, $SSE(A)$ indicates the sum of squares error for a model with factor A only. $SSE(\mu)$ indicates the sum of squares error for an intercept-only model. In correspondence to model specification in `lm`, the interaction of A and B is denoted as $A{:}B$, and $A^*B = A + B + A{:}B$. For type I SS, we assume that factor A is specified in the linear model before factor B.

TABLE 10.14

Sums of Squares Comparisons for Balanced, Unbalanced,
and Missing Cell Designs

	Balanced	Unbalanced	Empty Cells
A	I = II = III		
B	I = II = III	I = II	I = II
A:B	I = II = III	I = II = III	I = II = III

Note: We assume that factor *A* is specified in the linear model before
factor *B*, and that empty cells occurred in *A*.

(SSE) are calculated in the same way for all three types of sums of squares. Table 10.14 provides comparisons of sums of squares for balanced, unbalanced, and empty-celled designs.

EXAMPLE 10.21

Neter et al. (1996) described an experiment that measured the confounding effects of gender and bone development on the results from growth hormone therapy for prepubescent children. The response was the difference in subject height before and after hormone application (after–before). There were two factors: subject gender and bone development class. The latter had three levels indicating the severity of growth impediment before therapy: 1 = severely depressed, 2 = moderately depressed, and 3 = mildly depressed. At the start of the experiment, three children were assigned to each of the six treatment combinations. However, four of the children were unable to complete the experiment, resulting in an unbalanced design. The data are in the dataframe bone.

For notational simplicity, we designate gender as factor *A* and bone development as factor *B*. To calculate type II and III sums of squares, we create a design matrix whose elements have the value –1, 0, and 1.[*] The matrix will require $ab - 1 = (2 \times 3) - 1 = 5$ columns. Specifically, we will need $a - 1 = 1$ column for *A*, $b - 1 = 2$ columns for *B*, and $(a - 1)(b - 1) = 2$ columns for the interaction. We use the following rules for designating elements in the variables:

$$A = \begin{cases} 1 & \text{if factor level M} \\ -1 & \text{if factor level F} \end{cases} \quad B_1 = \begin{cases} 1 & \text{if factor level 1} \\ -1 & \text{if factor level 3} \\ 0 & \text{otherwise} \end{cases}$$

$$B_2 = \begin{cases} 1 & \text{if factor level 2} \\ -1 & \text{if factor level 3} \\ 0 & \text{otherwise} \end{cases}$$

$$AB_1 = A \times B_1$$

$$AB_2 = A \times B_2$$

---

[*] We cannot compute type III SS directly in R using lm to create reference and nested models. This is because lm will ignore the interaction *A:B* if both *A* and *B* are not in the nested model.

In **R** we have:

```
data(bone)
A.B <- data.frame(A = c(1, 1, 1, 1, 1, 1, 1, -1, -1, -1, -1, -1, -1, -1),
B1 = c(1, 1, 1, 0, 0, -1, -1, 1, 0, 0, 0, -1, -1, -1), B2 = c(0, 0, 0, 1,
1, -1, -1, 0, 1, 1, 1, -1, -1, -1))

X <- with(A.B, data.frame(cbind(A.B, AB1 = A * B1, AB2 = A * B2)))
X
    A B1 B2 AB1 AB2
1   1  1  0   1   0
2   1  1  0   1   0
3   1  1  0   1   0
4   1  0  1   0   1
5   1  0  1   0   1
6   1 -1 -1  -1  -1
7   1 -1 -1  -1  -1
8  -1  1  0  -1   0
9  -1  0  1   0  -1
10 -1  0  1   0  -1
11 -1  0  1   0  -1
12 -1 -1 -1   1   1
13 -1 -1 -1   1   1
14 -1 -1 -1   1   1

Y <- bone$growth


#-------------- Models required for type II SS-----------------#
## B removed, A in model
B.rem.A.in <- lm(Y ~ X[,1])
## A removed, B in model
A.rem.B.in <- lm(Y ~ X[,2] + X[,3])
## A:B removed, A and B in model
AB.rem.A.B.in <-lm(Y ~ X[,1] + X[,2] + X[,3])


#-------------- Models required for type III SS-----------------#
## B removed, A and A:B in model
B.rem.A.AB.in <- lm(Y ~ X[,1] + X[,4] + X[,5])
## A removed, B and A:B in model
A.rem.B.AB.in <-lm(Y ~ X[,2] + X[,3] + X[,4] + X[,5])


#----------Model required for both type II and III SS-----------#
full <-lm(Y ~ as.matrix(X))


Type.I.SS <- anova(lm(growth ~ gender * devel, data = bone))$"Sum Sq"


Type.II.SS <-t(data.frame(
SSA = anova(A.rem.B.in)$"Sum Sq"[3] - anova(AB.rem.A.B.in)$"Sum Sq"[4],
SSB = anova(B.rem.A.in)$"Sum Sq"[2] - anova(AB.rem.A.B.in)$"Sum Sq"[4],
SSAB = anova(AB.rem.A.B.in)$"Sum Sq"[4] - anova(full)$"Sum Sq"[2],
SSE = anova(full)$"Sum Sq"[2], row.names = "Type.II.SS"))


Type.III.SS <-t(data.frame(
SSA = anova(A.rem.B.AB.in)$"Sum Sq"[5] - anova(full)$"Sum Sq"[2],
SSB = anova(B.rem.A.AB.in)$"Sum Sq"[4] - anova(full)$"Sum Sq"[2],
SSAB = anova(AB.rem.A.B.in)$"Sum Sq"[4] - anova(full)$"Sum Sq"[2],
SSE = anova(full)$"Sum Sq"[2], row.names = "Type.III.SS"))
```

Here is a summary:

```
          Type.I.SS  Type.II.SS  Type.III.SS
SSA   0.002857143  0.09257143   0.12000000
SSB   4.396000000  4.39600000   4.18971429
SSAB  0.075428571  0.07542857   0.07542857
SSE   1.300000000  1.30000000   1.30000000
```

The type II SS results agree with those from Anova.

```
library(car)
Anova(lm(growth ~ gender * devel, data = bone), type="II")
Anova Table (Type II tests)
            Sum Sq Df F value  Pr(>F)
gender      0.0926  1  0.5697 0.472022
bone        4.3960  2 13.5262 0.002713 **
gender:bone 0.0754  2  0.2321 0.798034
Residuals   1.3000  8
```

To get identical type III SS to those calculated above, we must use sum contrasts.

```
options(contrasts = c('contr.sum','contr.poly'))
Anova(lm(growth ~ gender * devel, data = bone), type="III")
Anova Table (Type III tests)
            Sum Sq Df F value    Pr(>F)
(Intercept) 34.680  1 213.4154 4.729e-07 ***
gender       0.120  1   0.7385 0.415160
bone         4.190  2  12.8914 0.003145 **
gender:bone  0.075  2   0.2321 0.798034
Residuals    1.300  8
```

In this case, results from type II and III SS do not differ greatly from each other. Specifically, there is a significant effect for bone, but not for gender or the interaction of gender and bone. As a result, the choice of sums of squares will not affect the conclusions from the study.

★ 10.15 Robust ANOVA

Heteroscedasticity and nonnormality can often be effectively handled with conventional transformations. Outliers, however, generally cannot. As a result, a myriad of robust methods, resistant to outliers, nonnormality, and/or heteroscedasticity, have been created to address the many forms of ANOVA.

The applicability of conventional general linear model ANOVAs should be thoroughly assessed before using robust methods. There are three reasons for this. First, robust tests will be less powerful than general linear model procedures when normal assumptions are met (although robust methods will often have greater power if these assumptions are violated). Second, many multifactor (but not all) robust analyses require balanced designs. Third, robust methods do not exist for many complex ANOVA designs (e.g., split-split plot, strip-split plot, and fractional factorial). However, Wilcox (2005) and Maronna et al. (2006) describe many robust ANOVA analogs, including nested, split-plot, and repeated measures designs.

There are three general approaches that have led to methods that are robust to violations of at least some of the assumptions of ANOVA. These are the so-called permutation, rank-based permutation, and robust estimator procedures introduced in Chapter 6. Succinct descriptions of a wide variety of robust procedures follow a brief consideration of these three frameworks.

10.15.1 Permutation Tests

Several modern biometry texts (e.g., Quinn and Keough 2002, Gotelli and Ellison 2004) recommend purely permutational procedures over rank-based methods because rank-transformation results in a loss of information for continuous variables. Permutation methods do not require assumptions of normality or independence (Manly 1997). However, they do assume homoscedasticity (Boik 1987). In addition, theoretical problems exist for permutation testing of interactions (Edgington 1995). Finally, purely permutational tests result in inferences that are theoretically limited to the sample (Chapter 6, but see Manly 1997 and Gonzales and Manly 1998).

10.15.2 Rank-Based Permutation Tests

Rank-based permutation procedures perform implicit permutational analyses on rank transformed data. These methods will be extremely robust to outliers and do not require normality. However, they generally assume a similar distributional form for the factor-level populations being compared. In addition, the procedures are generally made more complex by the presence of ties. Problems with interactions have also traditionally plagued rank-based multifactor permutation tests (Quinn and Keough 2002), although this issue has been recently addressed with the introduction of procedures for factorial designs (Wilcox 2005).

10.15.3 Robust Estimator Tests

Robust estimators can be implemented to create ANOVA-type analyses that are extremely robust to outliers. For instance, Wilcox (2005) describes and provides **R**-code for bootstrap tests using Huber *M*-estimates and other *M*-estimators. Trimmed means and Winsorized means can also be used in this capacity.

10.15.4 One-Way ANOVA

10.15.4.1 Permutation Tests

Repeated permutation of a vector of observations (or residuals) across all factor-level combinations can be used to obtain an empirical distribution of test statistics to compare to an original observed test statistic (see Chapter 6). The function `oneway_test` from package *coin* provides a robust permutational alternative to a *t*-test or a one-way ANOVA. The test statistic is simply the largest factor-level sum of responses, and is asymptotically normal with a known conditional expectation and variance (Hothorn et al. 2006). As a result, the test statistic is reported as a standardized *z*-score. Permutation *P*-values can be obtained for `oneway_test` and many other nonparametric hypothesis testing algorithms in *coin* by specifying `distribution = approximate (B = perm)` among the function arguments, where `perm` defines the number of response vector permutations.

10.15.4.2 Rank-Based Permutation Tests

The most frequently used rank-based permutation alternative to a one-way ANOVA is the *Kruskal–Wallis test*. The method addresses the null hypothesis of unshifted populations, based on a comparison of standardized factor-level mean ranks (for details, see Hollander and Wolfe 1999). The function `kruskal_test` in the package *coin* can be used to run the procedure. The function can provide exact *P*-values based on the Kruskal–Wallis rank-based distribution. By default, however, it uses an asymptotic χ^2 null distribution for hypothesis testing. The function `kw.pairw` from *asbio* provides all possible pairwise tests of factor-level mean ranks in association with a Kruskal–Wallis test, adjusted for familywise type I error.

A problem with the Kruskal–Wallis test is that, while it does not assume normality for groups, it assumes that the groups have the same distributional form. As a solution, Brunner et al. (1997) proposed the *Brunner–Dette–Munk* test, which along with being robust to non-normality is also robust to heteroscedasticity, and is not made more complex by the presence of ties. The function `BDM.test` in *asbio* runs the analysis. See Wilcox (2005) for details.

10.15.4.3 Robust Estimator Tests

Well-known robust adaptations to one-way ANOVA include Welch's *t*-test procedure extended to >2 factor levels (see Section 6.2.7). While the method assumes underlying normality for factor-level populations, it is resistant to heteroscedasticity. Welch's extended procedure can be run using the function `oneway.test`.

A variant on ANOVA can be used which replaces sample means and standard errors estimates with trimmed estimates. The method assumes normality for trimmed factor-level populations. Wilcox (2005) described a trimmed method using Satterthwaite degrees of freedom for the test statistics that is resistant to both outliers and heteroscedasticity. The function `trim.test` in *asbio* performs the procedure. Robust linear models incorporating *M*, *S*, and *MM* estimators can be called using the functions `rlm` and `lqs` from *MASS* (see Section 9.18).

As noted in Section 10.7, traditional random effect models provide poor control over type I error when assumptions of normality and homoscedasticity are violated, or in the presence of outliers. Jeyaratnam and Othman (1985) described a heteroscedastic random effects model that can be extended to trimmed means estimators (Wilcox 1994b). Computational and theoretical details are given in Wilcox (2005, p. 303). The test can be run using the function `trim.ranef.test` from *asbio*.

10.15.5 Multiway ANOVA

10.15.5.1 Permutation Tests

Reliable purely permutational procedures exist for two-way designs that assume additivity. For instance, the function `oneway_test` from *coin* allows blocking for permutation tests. Nonadditive approaches, however, are poorly developed. The procedures recommended by Manly (1997) are implemented in the function `perm.fact.test` in *asbio*, which allows permutation interaction tests for up to three factors.

10.15.5.2 Rank-Based Permutation Tests

A number of rank-based permutational methods exist for additive two-way designs. The most frequently used is the *Friedman's Rank F*-test variant for randomized complete block designs (blocked designs with no replication). See Hollander and Wolfe (1999) for details. The function `friedman_test` from *coin* runs the procedure. The function `pairw.fried` from *asbio* runs all possible pairwise comparisons associated with Friedman's test, adjusted for familywise type I error. While less known and less frequently used, the *Agresti–Pendergrast test* (Agresti and Pendergrast 1986) is more powerful than Friedman's test, given normality, and remains powerful in heavier tailed distributions (Wilcox 2005).

The test can be run using the function AP.test from *asbio*. Randomized block designs with replication are often analyzed with the *Mack–Skillings test* (Mack and Skillings 1980); see Hollander and Wolfe (1999) for details. The function MS.test from *asbio* provides this analysis. Factorial designs have traditionally been a problem for rank-based permutational procedures because of the potential importance of interactions. Wilcox (2005) presented a two-way extension of the Brunner–Dette–Munk test that allowed testing for main effects and interactions. As with the one-way BDM test, the two-way test is powerful and allows heteroscedasticity. The function BDM.2way from *asbio* runs the analysis.

10.15.5.3 Robust Estimator Tests

As with one-way layouts, robust linear models incorporating M, S, and MM estimators can be called using the functions rlm and lqs from *MASS* (see Section 9.18).

> **EXAMPLE 10.22**
>
> Hollander and Wolfe (1999) compared young of year lengths at four sites for American gizzard shad, *Dorosoma cepedianum*, a fish of the herring family. Diagnostics indicated potential problems with both underlying nonnormality and heteroscedasticity, and prompted use of a robust procedure.
>   Here is the Brunner–Dette–Munk test result:
>
> ```
> data(shad)
> with(shad, BDM.test(length, site))
>
> One way Brunner-Dette-Munk test
>
>         df1       df2       F*    P(F > F*)
> X 2.866912 33.65639 16.98268 9.14567e-07
> ```
>
> Here are linear model coefficients resulting from M-estimation:
>
> ```
> summary(rlm(length ~ site, data = shad))
> Coefficients:
>               Value    Std. Error t value
> (Intercept)  40.7928    1.6549    24.6503
> siteII        4.8725    2.3403     2.0820
> siteIII     -12.6451    2.3403    -5.4032
> siteIV      -12.9928    2.3403    -5.5517
> ```

10.16 Bayesian Approaches to ANOVA

The Bayesian procedures for regression described in Section 9.23 can be extended to other general linear models and even, with adaptation, generalized linear models (see Gelman et al. 2003). Recall that the results of these analyses are posterior distributions for model parameters. In the case of ANOVA, these will be distributions for the true intercept term, effect sizes, and the error variance.

   The function bayes.lm allows Bayesian implementation of simple general linear models, including one-way ANOVA, with standard noninformative priors. These priors will be uniform on α_i and log σ (the treatment effects and error standard deviation) resulting in prior likelihoods of the model parameters (i.e., σ^2 and α_i), conditional on X, to be proportional to $1/\sigma^2$. The results from this approach will closely approximate those from a classic general linear model, but are often useful for predictive inference and model checking (see Gelman et al. 2003, p. 355). Computational details for the function are given in Section 9.23.

EXAMPLE 10.23

Here we reanalyze the baby walk experiment (Example 10.2) using a Bayesian approach with the *asbio* function bayes.lm. While any design matrix can be specified in the model, we will use sum contrasts.

```
data(baby.walk)
contrasts(baby.walk$treatment) <- contr.sum(4)
X <- with(baby.walk, model.matrix(~treatment))
bayes.lm(baby.walk$date, X)
```

```
Bayesian linear model with standard uniform priors
              2.5%       Median      97.5%
sigma.sq  0.9183187   1.6114957   3.2694454
mu        10.6554047 11.2027854  11.7907802
alpha1    -2.0517435 -1.0727190  -0.1427693
alpha2     0.1412789  1.1515719   2.1146605
alpha3    -1.5277810 -0.5466335   0.4593158
```

As expected, the credible intervals for the model are very similar (though slightly wider) than classical ANOVA model confidence intervals.

```
confint(lm(date ~ treatment , data = baby.walk))
                2.5 %      97.5 %
(Intercept) 10.6481520 11.7685146
treatment1  -2.0237344 -0.1429323
treatment2   0.1424348  2.1408985
treatment3  -1.5575652  0.4408985
```

10.17 Summary

- ANOVA is an estimation and hypothesis-testing procedure used when the response variable is quantitative and predictors (factors) are categorical. With fixed factors, inferences are made concerning the factor-level populations or the true effect sizes, while with random factors (random samples of levels from a population of related treatments), inferences concern the variance of the population of factor-level means.

- ANOVAs often serve as a precursor to more detailed comparisons of factor levels. Planned factor-level comparisons can be defined using a contrast matrix of linear constraints. Multiple factor-level comparisons, however, raise concerns about familywise type I error: the probability that one test from a family of tests will reject a true null hypothesis. To address this concern, a large number of methods have been developed.

- ANOVA is a type of general linear model. That is, it can be expressed in the form: $Y = \beta X + \varepsilon$, where ε defines a vector of random normal deviates. The design matrix X can take on a large number of forms, but as long as the contrast matrix specifies $a - 1$ unique comparisons (where a is the number of unique treatments), it will not affect fitted values from the linear model or the results (*P*-values) from the omnibus ANOVA.

- Assumptions for ANOVA follow the form of all general linear models. Specifically, we assume that outcomes from ε across all treatments are independent and identically normal. Random factors require additional assumptions of normality and independence for the random effects and independence among the random effects and marginal errors.

- The general ANOVA model has been modified for the analysis of a number of common experimental designs, and vice versa. Factorial designs require the consideration of interaction effects before assessing the main effects. Randomized complete block designs have no replication within block levels, requiring the assumption of additivity (no interactions) among blocks and treatments within blocks. Random blocks require the additional assumption of sphericity. Nested designs frequently contain random factors at lower hierarchical strata. Split-plot designs generally have random blocks that serve as the level of replication for whole-plot treatments, which in turn are the level of replication for split plots. Repeated measures designs will have repeated measurements on experimental units, for example, subjects, which are generally random effects.

- Temporal autocorrelation will cause asphericity in randomized block and split-plot repeated measures layouts, preventing analysis with classical mixed models. As a solution, one can use an analysis of contrasts, which decreases the degrees of freedom (and power) with asphericity, or modern mixed-model approaches, which allow user-specified autocorrelative covariance structures.

- ANCOVA is a hybrid of regression and ANOVA. It contains at least one categorical explanatory variable and at least one quantitative (concomitant) explanatory variable. The concomitant variable can be used to reduce model error and to increase model power in detecting factor-level differences.

- Lack of balance will result in untrustworthy multifactor analyses when using (conventional) type I sums of squares. As a solution, one can use type II sums of squares (which, for a factor *A*, create a nested model without *A* or its higher-order relatives) or type III sums of squares (which create a nested model without *A*, but with interactions that include *A*). Given balance, type I = type II = type III SS.

- Robust ANOVA can be used to address models containing outliers. New developments allow testing of interaction effects in models, which are resistant to outliers, nonnormality, and heteroscedasticity.

- Bayesian methods allow the use of prior information to inductively define distributions of ANOVA parameters. A naïve model with standard uniform priors is presented in Section 10.16.

EXERCISES

1. What is the scope of inference for the study described in Exercise 10.2? Is it a completely randomized design? To what population can we infer the results?

2. Type `see.anova.tck()` or click on the *ANOVA mechanics* link in the Chapter 10 pull-down menu.

   a. Explain what is happening with treatment effect sizes, $E(MSTR)$ and $E(MSE)$ as you change the positions of μ_1, μ_2, μ_3, the value of σ, and the sample size.

b. Set up a situation in which the null hypothesis $\mu_1 = \mu_2 = \mu_3$ is true. Sample from the populations a large number of times (>30) repeatedly by clicking on the refresh button. Is it still possible to reject H_0? Why? What is this called?

c. Does $E(MSTR)$ grow larger as you increase σ? Why?

d. Does MSE appear to be unbiased for $E(MSE)$? Why?

3. Define the following:

a. Familywise type I error

b. False discovery rate

c. Planned comparison

d. Post hoc comparison

e. Data snooping

f. Orthogonal contrast

4. Summarize the current views of experts concerning the need to address familywise type I error.

5. As one in a series of studies examining the relationship of caloric intake and longevity, Weindruch et al. (1986) compared life expectancy of field mice given different diets. To accomplish this, the authors randomly assigned 244 mice to one of four diet treatments. These were

- N/N85: Mice were fed normally both before and after weaning (the slash distinguishes pre- and postweaning). After weaning, the diet consisted of 85 kcal/week, a conventional total for mice rearing.

- N/R50: Mice were fed normally before weaning, but their diet was restricted to 50 kcal/week after weaning.

- R/R50: Mice were restricted to 50 kcal/week before and after weaning.

- N/R40: Mice were fed normally before weaning, but were given a severely restricted diet of 40 kcal/week after weaning.

The data are contained in the object `life.exp`.

a. What are the correct hypotheses?

b. Calculate the SSE and SS_A by hand (without using `lm`, `anova`, or `aov`).

c. Create the ANOVA table by hand, calculate F^*, and calculate the P-value. What do you conclude?

d. Verify your results by running the ANOVA using `lm` and `anova`. Attach the results.

e. Conduct and summarize diagnostic procedures.

f. Run all possible pairwise tests using Tukey's procedure with the function `pairw.anova`. Summarize the conclusions of this procedure.

g. Make a barplot summarizing the results of your pairwise comparisons using the functions `asbio::plot.pairw` and `bplot`. Indicate significant differences among treatments using letters.

6. Using data from Question 5, derive the vector of parameter estimates with matrix algebra (using **R** to help). Find these for both treatment and sum contrast design matrices.

7. A horticulturalist is investigating the phosphorus content of leaves from three different varieties of apple trees (1, 2, and 3). Random samples of five leaves from each of the three varieties were analyzed for P content.

```
Q7 <- data.frame(P=c(0.35, 0.4, 0.58, 0.5, 0.47, 0.65, 0.7, 0.9,
0.84, 0.79, 0.6, 0.8, 0.75, 0.73, 0.66), variety=factor(rep(1:3,
each=5)))
```

Create linear models (using lm) with treatment, sum, and helmert contrasts. Summarize your results. In particular, what hypotheses do the *P*-values correspond to?

8. Support the following statement: ANOVA and regression are fundamentally related because they are both linear models.

9. Type see.random() or use the book pull-down menu to eAccess the *random effect* GUI in Chapter 10. Describe, in detail, what is happening in the three graphs.

★ 10. Tippett (1950, p. 106) counted the number of yarn breakages per loom with respect to two factors: type of wool (A or B) and tension (L, M, or H). The dataset is located in the **R** base library under the name warpbreaks.

   a. What kind of a design is this?

   b. Correctly state the hypotheses.

   c. Run the ANOVA using lm and anova.

   d. Check for interaction effects using interaction.plot and ANOVA results. Why is checking for interactions important?

   e. Is it OK to examine the main effects? Why?

★ 11. For the warpbreaks dataset, treat wool type as a random factor.

   a. Specify and run the correct model in lmer.

   b. Find the REML variance component for wool type, and express it as a percentage.

   c. Conduct likelihood ratio tests to ascertain importance of the wool and tension factors.

★ 12. A researcher studied the effect of three experimental diets with varying fat contents on the total lipid level of plasma. Fifteen male subjects who were within 20% of their ideal body weight were put into five groups according to their age (i.e., three subjects wound up in each class). Within each age group, one of three experimental diets were randomly assigned to the three subjects. Data on reduction in lipid level (g L^{-1}) were recorded after the subjects were on the diets for 3 months each.

```
Q12 <- data.frame(lipid.loss=c(.73,.67,.15,.86,.75,.21,.94,.81,.26,
1.4, 1.32,.75, 1.62, 1.41,.78), age=factor(c(rep("15 to 24",
3),rep("20 to 34", 3),rep("35 to 44", 3),rep("45 to 54", 3),rep("55
to 64", 3))), fat.in.diet=factor(c(rep(c("low", "med", "hi"), 5))))
```

   a. What sort of experimental design do we have here? Does age help us account for potential issues in the design? How?

   b. State your hypotheses.

    c. Decide on the proper ANOVA model. Provide **R** code and output and interpret the results correctly.

    d. Conduct the appropriate diagnostics.

    e. Was the experimental design effective? Why?

★   13. We want to know if average AP biology test scores are related to the location of high schools in the state of Idaho. Six high schools are examined; two each, in three towns. At each school, data are gathered for the students of two instructors. Note that instructors are nested within school, which in turn is nested within town. That is, instructor 1 at school 1 is not the same person as instructor 1 at school 2.

```
Scores <- data.frame(APscores = c(4.6, 4.8, 4, 3.4, 3.4, 2.1, 4.1,
3.9, 4, 4.1, 3.8, 1.2, 4.5, 4.9, 4.3, 3.5, 3.6, 3.2, 3.9, 4.3, 3.8,
4.2, 4.4, 2.4), school = factor(rep(c(1,2), each = 6)), town =
c("Burley", "Pocatello", "Boise"), instructor = factor(rep(c(rep(1,
3), rep(2, 3)), 2)), class=factor(rep(c(1,2), each=12)))
```

    a. Treat town as a fixed factor, and treat instructor and school as random factors. Perform a variance components analysis using `lmer` output and describe your results.

    b. Perform an appropriate hypothesis test for the town fixed factor using nested and reference models, using `lmer` to help.

★   14. Calculate the variance components of the rat analysis in Example 10.14 by hand (using only `lm` and `anova`). Express these as percentages. Is this approach valid? Why?

★   15. Using the ANOVA results from Fisher's potato yield experiment (Example 10.9), calculate the method of moments variance estimates assuming that fertilizer is fixed but variety is random. Is there an issue here? Why, and what can we do about it?

★   16. Hoshmand (2006) described a split-plot design to test grain yield of corn (in bushels per acre) with respect to corn hybrids (split plots) and nitrogen, in lbs per acre (whole plots). The experiment was replicated at two blocks. The data are in the dataframe `corn`.

    a. Analyze the data correctly (with block in hybrid as a random effect) using `aov`.

    b. Verify that the degrees of freedom obtained in (a) are correct by calculating them by hand.

    c. Reanalyze the data, and repeat the hypothesis tests using `lmer`.

★   17. Crawley (2007) presented a dataset describing the effect of grazing on seed production in the invasive plant scarlet gilia (*Ipomopsis aggregata*). Forty plants were allocated to two treatments: grazed and ungrazed. Grazed plants were exposed to rabbits during the first 2 weeks of stem elongation. They were then protected from subsequent grazing by the erection of a fence and allowed to continue growth. Rabbit access was prevented entirely for the ungrazed treatment. Because initial plant size may influence subsequent fruit production, the diameter of the top of the rootstock was measured (in mm) before the experiment began. At the end of the experiment, fruit production (dry weight in mg) was recorded for each of the 40 plants. The data are in the dataframe `ipomopsis`.

    a. State your hypotheses.

b. Does a symbolic scatterplot indicate clustering; that is, is the concomitant variable confounded with the treatment?

c. Center the concomitant variable.

d. Run the ANCOVA in `lm`, and interpret the results: What is the slope of the concomitant regression line? What are the Y-intercepts of the grazed and ungrazed treatments?

e. Run a type III SS ANOVA using `Anova` from library *car* on the linear model from (d) and interpret the output.

f. Create a symbolic scatterplot that includes regression lines and multiple symbols and line types for distinguishing treatments.

18. Gurevitch (1986) presented data describing change in CO_2 concentration over time for airstreams passing over Wright's stonecrop (*Sedum wrightii*) plants. Concentrations were assumed to vary with plant respiration/photosynthetic processes. Two treatments were used. For "dry" plants, water was withheld for several weeks, while "wet" plants were well watered. CO_2 exchange was measured as [change in CO_2 concentration (g/mg)]/plant fresh mass (g). Thus, units were mg^{-1}. Positive values indicate net CO_2 uptake (due to photosynthesis) while negative values indicate net CO_2 output (due to respiration). Data are in `sedum.ts`.

a. Use `pairs` plots to compare the intervals with respect to CO_2 exchange. Is there any indication of temporal autocorrelation?

b. Analyze the data using an analysis of contrasts. Describe your results. What is the meaning of the Greenhouse–Geisser epsilon value?

c. Reanalyze as a mixed model using `lme`, and assess the optimality of using AR1 and ARIMA ($p = 1$, $q = 2$) autocorrelative covariance structures.

19. The questions below concern type I, II, and III sums of squares.

a. Briefly distinguish, in your own words, type I, II, and III sums of squares.

b. For the dataset below, calculate type II and III sums of squares by hand (using anova but not Anova).

```
Q18 <- data.frame(Y=c(2, 4, 2, 6, 7, 8, 2, 2, 1, 1, 1, 3, 3, 4, 5),
A=factor(c(rep(1, 7), rep(2,8))), B=factor(c(1, 1, 1, 2, 2, 2, 1,
1, 1, 2, 2, 2, 2, 1, 1)))
```

20. Reanalyze the random effects soil potassium experiment in Example 10.7 using a robust approach.

21. Reanalyze the leaf phosphorus data in Question 7 using a Bayesian approach.

11

Tabular Analyses

Whenever you can, count.

Sir Thomas Galton

11.1 Introduction

In Chapter 9, we dealt with regression analyses in which the predictor and response variables were both quantitative. In Chapter 10, we were introduced to ANOVA analyses in which the predictors were categorical and the response variable was quantitative. In this chapter, I describe models in which both the response and predictors are categorical (although response/predictor labels are not necessary or useful in many cases).

EXAMPLE 11.1

Drug abusers generally consider cocaine to be a relatively safe addiction compared to heroin. Bozarth and Wise (1985) assessed the legitimacy of this view by randomly assigning Long–Evans rats (an outbred strain of *Rattus norvegicus*) free access to one of the drugs. One group of rats ($n = 12$) had unlimited access to cocaine hydrochloride (1 mg/kg per infusion), while the other group ($n = 11$) had unlimited access to heroin hydrochloride (100 mg/kg per infusion). The rats self-administered the drugs by pressing a lever that activated a motor-driven syringe pump. Their data are shown in Table 11.1. Along with higher mortality rates (shown in the table), the cocaine group exhibited more behavioral disturbances, a greater degree of addiction, and poorer health, countering the supposition that cocaine is a relatively safe drug.

11.1.1 How to Read This Chapter

Sections 11.1 through 11.7 cover relatively simple topics including one-way formats (Section 11.3), confidence intervals for binomial proportions (Section 11.4), two-way tables (Section 11.6), ordinal variables (Section 11.7), and effect size and power (Section 11.8). Thus, these materials are appropriate for coverage in introductory classes.

The later sections, however, will be more appropriate for advanced students. A list of topics includes three-way tables (Section 11.9) and GLM variants on categorical data analysis, including conceptually difficult log-linear models (Section 11.10).

TABLE 11.1

Mortality Rates after 30 Days for Rats with Unlimited Access to Cocaine
or Heroin

	Dead	Alive	Total
Cocaine group	11	1	12
Heroin group	4	7	11
Total	15	8	23

Source: Adapted from Bozarth, M. A., and Wise, R. A. 1985. *Journal of the American Medical Association* 254: 81–83.

11.2 Probability Distributions for Tabular Analyses

The normal distribution and pdfs derived from the normal distribution (including the *t*- and *F*-distributions) are of primary importance for inferences concerning continuous response variables (essentially all those discussed so far). In contrast, the important distributions for categorical response variables are binomial and Poisson pdfs (described in Chapter 3) and the multinomial distribution.

The *multinomial distribution* describes the probability for outcomes given independent and identical trials with more than two categories. Let *c* denote the number of outcome categories and let the probabilities for each category be the set $\{\pi_1, \pi_2, ..., \pi_c\}$, where $\Sigma_{i=1}^{c} \pi_i = 1$. Then the probability that y_1 successes occur in category 1, y_2 successes occur in category 2, and so on, is given by

$$f(y_1, y_2, ..., y_c) = \frac{n!}{y_1! y_2! ... y_c!} \pi_1^{y_1} \pi_2^{y_2} ... \pi_c^{y_c}, \qquad (11.1)$$

where $\Sigma_{i=1}^{c} y_i = n$. Just as the Bernoulli distribution is a special case of the binomial distribution, the binomial pdf is a special case of the multinomial distribution in which $c = 2$ categories (Figure 11.1). Recall that if $Y \sim \text{BIN}(n, \pi)$, then $E(Y) = n\pi$ and $\text{Var}(Y) = \pi(1 - \pi)n$. Accordingly, if $Y_1, Y_2, ..., Y_c$ follow a multinomial distribution, we denote this as $Y_1, Y_2, ..., Y_c \sim \text{MNOM}(n, \pi_1, \pi_2, ..., \pi_c)$. In this case, $E(Y) = n\pi_i$, while $\text{Var}(Y) = \pi_i(1 - \pi_i)n$.

EXAMPLE 11.2

Haroldson (2010) reported that the proportions of grizzly bear (*Ursus arctos*) litters in the greater yellowstone ecosystem (GYE) consisting of one, two, three, and four cubs were approximately 0.242, 0.521, 0.232, and 0.005, respectively. Given this, what is the probability of producing four grizzly cubs from two sows?

Assuming that both sows reproduce, there are two ways that two sows can produce four cubs.
A = One sow has one cub and the other sow has three cubs.
B = Each sow has two cubs.

The probability for both outcomes can be calculated using the multinomial pdf. We have

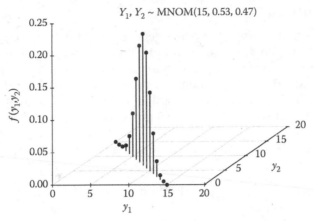

```
see.mnom.tck()
```

$Y_1, Y_2 \sim \text{MNOM}(15, 0.53, 0.47)$

FIGURE 11.1

Multinomial (binomial) distribution. For a particular count total, n, y_1 results in $y_2 = n - y_1$. Thus, $\pi_1 + \pi_2 = 1$. The distribution encompasses the Bernoulli and binomial pdfs, since the former is a subset of the latter.

$$P(A) = f(1,0,1,0) = \frac{2!}{1! \times 0! \times 1! \times 0!} 0.242^1 0.521^0 0.232^1 0.005^0 = 2(0.056) = 0.112,$$

$$P(B) = f(0,2,0,0) = \frac{2!}{0! \times 2! \times 0! \times 0!} 0.15^0 0.32^2 0.43^0 0.1^0 = 0.102.$$

The probability of the union of these disjoint outcomes is

$$P(A \cup B) = P(A) + P(B) = 0.214.$$

The ML and OLS estimator of the binomial parameter π is the proportion of successes given n-observed trials. That is

$$\hat{\pi} = \frac{y}{n}. \tag{11.2}$$

Accordingly, the estimator for the ith multinomial proportion is

$$\hat{\pi}_i = \frac{y_i}{n}. \tag{11.3}$$

The ML estimator of the standard error for the sampling distribution of $\hat{\pi}_i$ is

$$\hat{\sigma}_{\hat{\pi}} = \sqrt{\frac{\hat{\pi}(1 - \hat{\pi})}{n}}, \tag{11.4}$$

while the ML estimator of the standard error for the multinomial sampling distribution of $\hat{\pi}_i$ is

$$\hat{\sigma}_{\hat{\pi}_i} = \sqrt{\frac{\hat{\pi}_i(1 - \hat{\pi}_i)}{n}}. \tag{11.5}$$

11.3 One-Way Formats

There are four hypothesis testing approaches generally used with a one-way format (a tabular layout consisting of a single row or column). These are: the exact binomial, score, Wald, and likelihood ratio tests. These methods can be used to address three types of null hypotheses: (1) the comparison of a single binomial proportion to a null (expected) value, (2) the simultaneous comparison of a set of multinomial proportions to a corresponding set of null values, and (3) the comparison of two binomial proportions to each other.

1. To compare a single binomial proportion, π, to a null value, π_0, we have the following two-tailed hypotheses:

$$H_0: \pi = \pi_0,$$

versus

$$H_A: \pi \neq \pi_0.$$

Lower- and upper-tailed alternative hypotheses are also possible. These are

$$H_A: \pi < \pi_0,$$

and

$$H_A: \pi > \pi_0.$$

2. For a null hypothesis test in which a set of multinomial proportions, $\{\pi_1, \pi_2, \ldots, \pi_c\}$, are compared to a corresponding set of null values, we have the following hypotheses:

$$H_0: \text{All } \pi_i = \pi_{i0},$$

versus

$$H_A: \text{At least one } \pi_i \text{ differs from its hypothesized value.}$$

3. For tests comparing two binomial proportions π_1 and π_2, we have the following two-tailed hypotheses:

$$H_0: \pi_1 - \pi_2 = D_0,$$

versus

$$H_A: \pi_1 - \pi_2 \neq D_0,$$

where D_0 is a specified null difference. As with earlier approaches (e.g., t-tests), we generally let $D_0 = 0$.

We specify a lower-tailed test that assumes that π_1 is smaller than π_2 with the alternative

$$H_A: \pi_1 - \pi_2 < D_0,$$

and an upper-tailed alternative that assumes that π_1 is larger than π_2 with the alternative

$$H_A: \pi_1 - \pi_2 > D_0.$$

Note that tests of the second type include tests of the third type. The methods using the 2nd type of test however, exclude the testing of one-tailed hypotheses because of their distributions under H_0 (see Sections 11.3.1.2 and 11.3.3.2).

The exact binomial test is used for the first type of hypothesis format, usually in the context of small sample sizes. This test is described in Section 11.3.4. The score and likelihood ratio tests can be used to address hypothesis formats 1, 2, and 3. The Wald test can be used in formats 1 and 3. These methods are described below. For all four tests, in the context of all three hypothesis formats, we assume

1. Observations are from random binomial or multinomial samples, that is, observations are classified into categories independently.

2. Conditions are sufficient to allow asymptotic convergence to null distributions (see Section 11.3.4).

11.3.1 Score Test

11.3.1.1 Tests for $\pi = \pi_0$

The *score test* capitalizes on the asymptotic normality of the sampling distribution of $\hat{\pi}$. Specifically, as $n \to \infty$, then

$$\hat{\pi} \sim N\left(\pi, \frac{\pi(1-\pi)}{n}\right). \tag{11.6}$$

To test H_0: $\pi = \pi_0$, we use the test statistic

$$z^* = \frac{\hat{\pi} - \pi_0}{\sqrt{\dfrac{\pi_0(1-\pi_0)}{n}}}. \tag{11.7}$$

Note that in a departure from previous test statistics, we divide the deviation of the observed and expected values by the *null standard error* (the standard error under H_0) instead of the sample standard error. The test statistic z^* will be asymptotically standard normal if H_0 is true. Thus, P-values for two-tailed hypothesis tests are calculated as $2P(Z \geq |z^*|)$, while lower- and upper-tailed P-values are obtained by finding $P(Z \leq z^*)$ and $P(Z \geq z^*)$, respectively.

Several important extensions to Equation 11.7 can be made. First, by multiplying by n/n (i.e., by 1), we have the score test statistic expressed in counts:

$$\frac{\hat{\pi} - \pi_0}{\sqrt{\dfrac{\pi_0(1-\pi_0)}{n}}} \frac{n}{n} = \frac{n\hat{\pi} - n\pi_0}{\sqrt{n\pi_0(1-\pi_0)}} = \frac{y - n\pi_0}{\sqrt{n\pi_0(1-\pi_0)}}, \tag{11.8}$$

where y is the number of observed successes. Because the standard normal distribution squared is equivalent to a χ^2 distribution with one degree of freedom, the test statistic

$$X^2 = \frac{(y - n\pi_0)^2}{n\pi_0(1-\pi_0)} = \frac{(y - E)^2}{E - \pi_0 E}, \tag{11.9}$$

has an asymptotic χ_1^2 distribution under H_0. The term $n\pi_0$ gives the expected number of successes under H_0 and is denoted as E

$$E = n\pi_0. \tag{11.10}$$

Because the deviations between the observed and expected value under H_0 are squared, only a two-tailed alternative hypothesis $\pi \neq \pi_0$ is possible using this approach.[*] *P*-values are obtained by finding $P(\chi_1^2 \geq X^2)$.

11.3.1.2 Tests for All $\pi_i = \pi_{i0}$

Consider a binomial format with defined successes and failures. Let y_1 be the observed number of successes and y_2 be the number of failures; then Equation 11.9 can be expressed as

$$X^2 = \frac{(y - n\pi_0)^2}{n\pi_0(1-\pi_0)} = \frac{(y_1 - n\pi_0)^2}{n\pi_0} + \frac{(y_2 - n(1-\pi_0))^2}{n(1-\pi_0)} = \frac{(y_1 - E_1)^2}{E_1} + \frac{(y_2 - E_2)^2}{E_2}.$$

An extension to a multinomial framework can be made accordingly, that allows testing of the null hypothesis that all $\pi_i = \pi_{i0}$. The resulting test statistic is

$$X^2 = \sum_{i=1}^{c} \frac{(y_i - E_i)^2}{E_i}. \tag{11.11}$$

Here, X^2 will asymptotically follow a χ_{c-1}^2 distribution under H_0. *P*-values are obtained by finding $P(\chi_{c-1}^2 \geq X^2)$. This procedure is generally called a χ^2 *goodness of fit test*. Its derivation is due to Karl Pearson (Chapter 1).

11.3.1.3 Tests for $\pi_1 = \pi_2 = \cdots = \pi_c$ (Includes Two-Proportion Test: $\pi_1 = \pi_2$)

By setting the value of the parameter π_{i0} equal to $1/c$ for all c categories in Equation 11.11, we can use the χ^2 goodness of fit test to test the hypothesis H_0: $\pi_1 = \pi_2 = \cdots = \pi_c$. In this case, expected counts in each cell will be equal and will be computed as n/c. The test statistic and its distribution under H_0 will be identical to those described above.

> **EXAMPLE 11.3**
>
> In the GYE in 2007, 18 grizzly litters out 50 had three cubs. Is this different from the long-term proportions reported in Example 11.2?
> We have
>
> $$H_0: \pi = 0.232$$
>
> versus
>
> $$H_A: \pi \neq 0.232.$$

---

[*] Note that the expected value of the test statistic under null is zero because $E - O = 0$ if H_0 is true. Because the mean of a chi-squared distribution will always equal its degrees of freedom (Chapter 3), this means that the entire null distribution can serve as evidence against H_0. This is unlike the majority of test statistics whose value under H_0 will be equivalent to the expectation (close to the center) of the null distribution.

$$z* = \frac{0.36 - 0.232}{\sqrt{\dfrac{0.232(1 - 0.232)}{50}}} = 2.14225,$$

$$2P(Z \geq 2.14225) = 0.03201$$

Reanalysis using Equations 11.9 through 11.11 provides identical results.

$$X^2 = \frac{(y - n\pi_0)^2}{n\pi_0(1 - \pi_0)} = \frac{(18 - 11.6)^2}{11.6 - 2.6912} = 4.5977.$$

$$X^2 = \frac{(y_1 - n\pi_0)^2}{n\pi_0} + \frac{(y_2 - n(1 - \pi_0))^2}{n(1 - \pi_0)}$$

$$= \frac{(18 - 11.6)^2}{50(0.232)} + \frac{(32 - 50(1 - 0.232))^2}{50(1 - 0.232)} = 3.53104 + 1.066667 = 4.5977$$

$$P(\chi_1^2 \geq 4.5977) = 0.03201$$

We reject H_0 at $\alpha = 0.05$ and conclude that the proportion of three cub litters is different from 0.232. In **R**, we can use the functions `prop.test` or `chisq.test` to run these analyses.

```
prop.test(18, 50, 0.232, correct = FALSE)
X-squared = 4.5977, df = 1, p-value = 0.03201
chisq.test(c(18,32),p=c(0.232,0.768), correct = FALSE)
X-squared = 4.5977, df = 1, p-value = 0.03201
```

The `correct` argument in the functions above refers to *Yates' continuity correction* (Yates 1934), which is given as

$$X_{Yates}^2 = \sum_{i=1}^{n} \frac{(|y_i - E_i| - 0.5)^2}{E_i}. \tag{11.12}$$

The procedure was developed to help account for discrepancies in the binomial and multinomial distributions, which are discrete, and the χ^2 null distribution, which is continuous. The approximation of X^2 to the null distribution can be particularly poor when n is small. To address this, Yates' correction makes the χ^2 test statistic smaller and more conservative with respect to type I error. The correction, however, is frequently too conservative, even for sample sizes ($n \approx 20$), leading to type II error (Maxwell 1976, Sokal and Rohlf 2012). Given the availability of tabular methods for exact small sample inference (see Section 11.3.4), the procedure has debatable value and is not considered in an authoritative survey by Agresti (2012).[*]

11.3.2 Wald Test

11.3.2.1 Tests for $\pi = \pi_0$

The Wald test statistic (introduced in Section 9.20.5) is the result of a null value subtracted from an ML parameter estimate, all divided by the ML estimator's standard error. Because of the asymptotic normality of all ML estimators, the one-sample Wald test statistic

---

[*] Despite this, the default for the argument `correct` is TRUE for both `prop.test` and `chisq.test`.

$$z^* = \frac{\hat{\pi} - \pi_0}{\sqrt{\frac{\hat{\pi}(1 - \hat{\pi})}{n}}},$$

(11.13)

will be asymptotically standard normal under H_0. As before, the two-tailed P-value is $2P(Z \geq |z^*|)$, while lower- and upper-tailed P-values are $P(Z \leq z^*)$ and $P(Z \geq z^*)$, respectively.

EXAMPLE 11.4

Reanalysis of Example 11.3 reveals

$$z^* = \frac{0.36 - 0.232}{\sqrt{\frac{0.36(1 - 0.36)}{50}}} = 1.88562.$$

Here, $2P(Z \geq 1.88562) = 0.05934645$, resulting in a reversal of the significance decision in Example 11.3.

11.3.2.2 Tests for $\pi_1 = \pi_2$ (Two-Proportion Test)

To compare two proportions, π_1 and π_2, we use the estimators $\hat{\pi}_1 = \frac{x_1}{n_1}$ and $\hat{\pi}_2 = \frac{x_2}{n_2}$. Because the asymptotic joint sampling distribution of $\hat{\pi}_1 - \hat{\pi}_2$ is:

$$N\left(\pi_1 - \pi_2, \left(\frac{\pi_1(1 - \pi_1)}{n_1} + \frac{\pi_2(1 - \pi_2)}{n_2} \right) \right),$$

(11.14)

the two-sample Wald test statistic is:

$$z^* = \frac{(\hat{\pi}_1 - \hat{\pi}_2) - D_0}{\sqrt{\frac{\hat{\pi}_1(1 - \hat{\pi}_1)}{n_1} + \frac{\hat{\pi}_2(1 - \hat{\pi}_2)}{n_2}}}.$$

(11.15)

The test statistic will be asymptotically standard normal under H_0. Thus, the two-tailed P-value is $2P(Z \geq |z^*|)$, while lower- and upper-tailed P-values are given as $P(Z \leq z^*)$ and $P(Z \geq z^*)$, respectively.

EXAMPLE 11.5

Over the span of 5 years, 22,071 physicians served as subjects in a study to determine the effect of aspirin on the incidence of myocardial infarction (MI) (heart attacks). Subjects were randomly assigned to either an aspirin or a control group. Those in the aspirin group took an aspirin each day for 5 years, while those in the placebo group took a placebo over the same time span. The study was *double blind*. That is, neither the investigators nor the participant knew which type of pill they were receiving. The design addressed potential investigator biases and allowed the generation of causal evidence (Chapter 7). The incidence of MIs was examined at the end of the experiment. Of the 11,034 physicians assigned to the placebo group, 189 suffered MI, while of the 11,037 physicians in the aspirin group, 104 suffered MI (Steering Committee of PHSRG 1989).

The primary question was whether aspirin decreased the probability of MI. Thus, we have

$$H_0: \pi_A \geq \pi_P,$$

versus

$$H_A: \pi_A < \pi_P.$$

The probability estimates are

$$\hat{\pi}_A = 104/11{,}037 = 0.00942.$$

$$\hat{\pi}_P = 189/11{,}034 = 0.01703.$$

From the lessons learned in Chapter 6, we are careful to specify the treatments in the test statistic in the same order that they were given in the hypotheses.

$$z^* = \frac{(0.00942 - 0.01703) - 0}{\sqrt{\dfrac{0.00942(1 - 0.00942)}{11{,}037} + \dfrac{0.01703(1 - 0.01703)}{11{,}034}}} = -5.00403.$$

The *P*-value is

$$P(Z \leq -5.00403) = 0.000000281.$$

We reject H_0 at $\alpha = 0.05$, ignoring the effect of the enormous sample size on power, and conclude that aspirin reduces the incidence of MI for the population under study.

11.3.3 Likelihood Ratio Test

11.3.3.1 Tests for $\pi = \pi_0$

For the null hypothesis $\pi = \pi_0$, the appropriate likelihood ratio test statistic (see Sections 9.20.5 and 10.5.2) is 2 times the log of the binomial likelihood of the maximum likelihood estimate, divided by the binomial likelihood of π_0. That is

$$G^2 = 2\log\left[\frac{\mathcal{L}(\hat{\pi} \mid \text{data})}{\mathcal{L}(\pi_0 \mid \text{data})}\right] = 2[\ell(\hat{\pi} \mid \text{data}) - \ell(\pi_0 \mid \text{data})]. \tag{11.16}$$

If H_0 is true, then G^2 will be asymptotically χ^2 distributed with one degree of freedom. The *P*-value for the two-tailed alternative test (the only test possible) is given as $P(\chi_1^2 \geq G^2)$.

EXAMPLE 11.6

To reanalyze the sow data in Example 11.3, we first determine the log likelihood of the maximum likelihood estimate under H_0, given the data:

```
log(dbinom(18, size = 50, 18/50))
[1] -2.146547
log(dbinom(18, size = 50, 0.232))
[1] -4.220858
```

We have

$$G^2 = 2[-2.146547 + 4.220858] = 2.074311, \quad \text{which results in}$$

$$P(\chi_1^2 \geq 2.074311) = 0.1498.$$

As with Example 11.4, this result suggests a reversal of the significance decision made in Example 11.3.

11.3.3.2 Tests for All $\pi_i = \pi_{i0}$ (Includes Two-Proportion Test)

For a goodness of fit test that all $\pi_i = \pi_{i0}$, the likelihood ratio test statistic will be

$$G^2 = 2\sum_{i=1}^{c} y_i \log\left(\frac{y_i}{E_i}\right). \tag{11.17}$$

In this case, G^2 will have an asymptotic χ_{c-1}^2 distribution under H_0. P-values are $P(\chi_{c-1}^2 \geq G^2)$.

11.3.4 Comparisons of Methods and Requirements for Valid Inference

All three testing methods discussed in Section 11.3 (score, Wald, and likelihood) provide approximate P-values because they rely on asymptotic null distributions. As a result, Agresti (2012) recommends reporting P-values to not more than the third decimal place.

11.3.4.1 Tests for $\pi = \pi_0$

For small-to-moderate sample sizes, the Wald test is the least reliable for testing the null hypothesis $\pi = \pi_0$ (Agresti 2012). If results from the three tests differ markedly (as they do in Examples 11.3, 11.4, and 11.6), this generally indicates that the sampling distribution of $\hat{\pi}$ may be far from normal. In this case, or if $n\hat{\pi} < 5$ and $n(1 - \hat{\pi}) < 5$, *small sample inference* should be used by calling on the binomial pdf directly. This procedure is called an *exact binomial test*.

> **EXAMPLE 11.7 EXACT BINOMIAL TEST**
>
> In Example 11.3, we observed 18 "successes" out of 50 trials and used $\pi_0 = 0.232$. To calculate an upper-tailed P-value using an exact binomial test, we would find $P(Y \geq 18)$, where $Y \sim \text{BIN}(50, 0.232)$.
>
> ```
> pbinom(17, size = 50, p = 0.232, lower.tail = FALSE)
> [1] 0.02809654
> ```
>
> Note that to get $P(Y \geq 18)$, I had to specify $P(Y > 17)$ in pbinom. This procedure is provided in **R** by the function binom.test.[*]
>
> ```
> binom.test(18, 50, 0.232, alternative = "greater")
> number of successes = 18, number of trials = 50, p-value = 0.0281
> ```

---

[*] Note that a two-tailed test is problematic because of the discrete character of the binomial distribution and its asymmetry at $\pi = 0.232$. To obtain two-tailed P-values, the function binom.test identifies the expected value under H_0, that is, $E = n\pi_0$. Then, if y is greater than E (as in the case above), the function creates a sequence of integers starting at zero and ending at E, rounded down to the nearest integer. It then identifies densities of this sequence under the null binomial distribution that are less than the likelihood of maximum likelihood estimate for π multiplied by the error constant $1 + 1 \times 10^{-7}$. The sum of this subset is defined as v. The two-tailed P-value is now given as $P(Y \leq v) + P(Y \geq y)$.

The exact binomial test will be more conservative (with respect to type I error), and less powerful, than the score, Wald, and likelihood ratio tests (Agresti 2012).

11.3.4.2 Tests for All $\pi_i = \pi_{i0}$

The approximation to the null distribution for χ^2 and likelihood ratio goodness of fit test statistics will depend on both the number of categories and the number of trials. Agresti (2012) argued that no single rule can be used to designate an adequate sample size for all possible situations. Many texts, however, adhere to the suggestions of Cochran (1954), who recommended that no E_i should be less than 1 and that not more than 20% of E_is should be less than 5.

11.3.4.3 Tests for $\pi_1 = \pi_2$ (Two-Proportion Test)

Most experts recommend that the Wald test for the difference of two binomial proportions should be used only if $n_1\hat{\pi}_1 > 5$, $n_1(1 - \hat{\pi}_1) > 5$, $n_2\hat{\pi}_2 > 5$, and $n_2(1 - \hat{\pi}_2) > 5$. When using the χ^2 goodness of fit test for this null hypothesis, E_is should not be less than 5.

11.4 Confidence Intervals for π

Interval estimation for the binomial parameter π is problematic. Biologists conventionally depend on the asymptotic normal sampling distribution of $\hat{\pi}$ (see the Wald method below). However, the coverage properties for this approach will be poor when n is not large—and generally, and one does not know what large enough is—or when π is close to 0 or 1 (Vollset 1993, Brown et al. 2002). Many alternative approaches to the Wald method have been developed, both asymptotic and exact (Vollset 1993, Agresti and Coull 1998, Reiczigel 2003, Agresti 2012). Several popular examples, including analogs to hypothesis testing procedures discussed in Section 11.3, are described below.

11.4.1 Wald Method

Given the large sample normality of ML estimators (Chapter 4), an asymptotic $(1 - \alpha)100\%$ confidence interval for the parameter π can be found using

$$\hat{\pi} \pm z_{1-(\alpha/2)}\hat{\sigma}_{\hat{\pi}}, \tag{11.18}$$

where $\hat{\pi}$ is the ML estimator for π, $\hat{\sigma}_{\hat{\pi}}$ is the ML estimator for the true standard error of $\hat{\pi}$, and $z_{1-(\alpha/2)}$ is the outcome from the Z-quantile function at the probability $1 - (\alpha/2)$.

An asymptotic $(1 - \alpha)100\%$ Wald confidence interval for the true difference of two binomial proportions can be obtained using

$$\hat{\pi}_1 - \hat{\pi}_2 \pm z_{1-(\alpha/2)}\hat{\sigma}_{\hat{\pi}}, \tag{11.19}$$

where $\hat{\sigma}_{\hat{\pi}} = \sqrt{\dfrac{\hat{\pi}_1(1-\hat{\pi}_1)}{n_1} + \dfrac{\hat{\pi}_2(1-\hat{\pi}_2)}{n_2}}.$

11.4.2 Score Method

Agresti and Coull (1998) reported that Wald intervals are often far from nominal, even when sample sizes are large. They suggested that a more reliable interval estimate is obtained by inverting the score test statistic in Equation 11.7 and varying values for π_0 while holding $\hat{\pi}$ constant. Under this approach, a $(1 - \alpha)100\%$ score confidence interval for π will consist of values of π_0 that result in a failure to reject H_0 at α. Bounds can be obtained analytically by finding the roots of a quadratic expansion of the binomial log-likelihood function (see Exercise 2 at the end of this chapter).

11.4.3 Agresti–Coull Method

A simple approximation to the score confidence interval method can be obtained by adding the number two to the number of successes and failures (Wilson 1927, Agresti and Coull 1998). The resulting estimators are

$$\hat{\pi}_{AC} = \frac{y+2}{n+4}, \tag{11.20}$$

$$\hat{\sigma}_{\hat{\pi}AC} = \sqrt{\frac{\hat{\pi}(1-\hat{\pi})}{n+4}}. \tag{11.21}$$

A $(1 - \alpha)100\%$ confidence interval for the parameter π can now be found using

$$\hat{\pi}_{AC} \pm z_{1-(\alpha/2)}\hat{\sigma}_{\hat{\pi}AC}.$$

11.4.4 Clopper–Pearson Exact Method

With the *Clopper–Pearson method* (Clopper and Pearson 1934), the lower and upper bounds of an exact $(1 - \alpha)100\%$ confidence interval for π are obtained directly from the beta distribution (Chapter 3)

$$C_L = \text{BETA}(\alpha/2,\ y,\ n-y+1),$$
$$C_U = \text{BETA}(1-\alpha/2,\ y+1,\ n-y), \tag{11.22}$$

where $\text{BETA}(\alpha/2,\ y,\ n-y+1)$ denotes the beta quantile function, with shape parameters y and $n-y+1$, evaluated at probability $\alpha/2$.

11.4.5 Likelihood Ratio Method

Confidence intervals can also be obtained by inverting the likelihood ratio test statistic and finding values for which we fail to reject a null hypothesis of equal likelihood for the maximum likelihood estimate and varying values of π_0. The $(1 - \alpha)100\%$ confidence bounds for π occur in the support of the binomial log-likelihood function where the difference between the maximized log likelihood function $\chi_1^2(1-\alpha)/2$ intersects the function. The expression

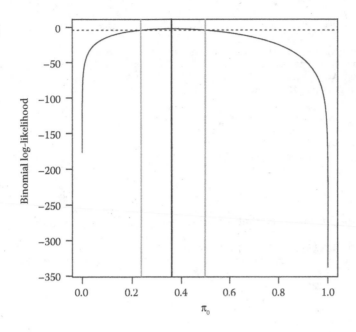

FIGURE 11.2
Illustration of the likelihood ratio confidence bounds for π for Example 11.7. The horizontal dashed line indicates the binomial log-likelihood value in which $\ell(\hat{\pi} \mid \text{data}) - \ell(\pi_0 \mid \text{data}) = \chi_1^2(1 - \alpha)/2$. The points where the log-likelihood function crosses this line are the interval bounds (gray line). Note that the same approach is used to find the maximum likelihood estimates in Box–Cox and Box–Tidwell optimal power transformation procedures (cf. Figure 9.13).

$\chi_1^2(1 - \alpha)/2$ represents the chi-squared quantile function with one degree of freedom evaluated at probability $1 - \alpha$, all divided by 2 (see Figure 11.2).

EXAMPLE 11.7

For the grizzly bear litter size data in Example 11.3, we have $y = 18$ and $n = 50$. Using the Agresti–Coull method, we have

$$\hat{\pi} = \frac{18 + 2}{50 + 4} = \frac{20}{54} = 0.37037,$$

$$\hat{\sigma}_{\hat{\pi}} = \sqrt{\frac{0.37037(1 - 37{,}037)}{54}} = 0.0657.$$

Thus, the 95% confidence interval for π is

$$0.37037 \pm 1.96 \times 0.067 = \{0.24157, 0.49916\}.$$

The function `ci.p` from *asbio* calculates Agresti–Coull intervals by default.

```
ci.p(c(rep(1, 18),rep(0,32)))
95% Confidence Interval for Binomial Parameter pi
(method = Agresti-Coull)
  Estimate      2.5%       97.5%
0.3703704 0.2415715 0.4991692
```

The 95% score interval is

```
ci.p(c(rep(1, 18),rep(0,32)), method = "score")
95% Confidence Interval for Binomial Parameter pi (method = score)
 Estimate      2.5%      97.5%
0.3600000 0.2413875 0.4985898
```

The 95% Clopper–Pearson interval is

```
ci.p(c(rep(1, 18),rep(0,32)), method = "exact")
95% Confidence Interval for Binomial Parameter pi
(method = Clopper-Pearson)
   Estimate      2.5%      97.5%
0.3600000 0.2291571 0.5080686
```

Finally, the 95% likelihood ratio confidence interval is

```
ci.p(c(rep(1, 18),rep(0,32)), method = "LR")
95% Confidence Interval for Binomial Parameter pi (method = likelihood
ratio)
 Estimate  2.5%    97.5%
0.36000 0.23638 0.49763
```

11.4.6 Comparison of Methods and Requirements for Valid Inference

The Wald, Agresti–Coull, score, and likelihood ratio methods rely on asymptotic distributions. Thus, these procedures should only be used when $n\hat{\pi} > 5$ and $n(1-\hat{\pi}) > 5$ (Ott and Longnecker 2004). The Wald procedure, in particular, may produce extremely misleading results given small sample sizes. The score and likelihood ratio methods generally provide superior coverage across possible configurations of n and π. These methods, however, may also produce suspect results given small sample sizes (Agresti 2012). A Clopper–Pearson interval is bounded by the exact nominal level, but its actual coverage may be well below this depending on n and π. That is, it may often have overly conservative confidence intervals that are too wide (Agresti and Coull 1998).[*] For large sample sizes, the five methods will give similar results.

11.4.7 Inference for the Ratio of Two Binomial Proportions

The ratios of two independent proportions are frequently used in biological analyses. Important examples include relative risk (i.e., π_1/π_2) introduced in Chapter 2. Another example is the *selection ratio*, denoted here as ω, which divides the proportional use of the ith resource by its proportional availability. For this measure, a ratio greater than one indicates that the resource is selected for, while ratios below one indicate that the resource is avoided.

Inferences for the true selection ratio, including confidence intervals, are inherently problematic because the ratio numerator and denominator will each be composed of sets of multinomial probabilities (describing proportional use and availability). As a result, values for "avoided" resources will be distributed in a narrow range [0, 1), while values for resources that are "selected" are in the range (1, ∞), causing the sampling distribution of

---

[*] Note, however, that this is the approach used for obtaining confidence intervals in the functions `binom.test` and `prop.test`.

ratio estimates ($\hat{\omega}_i s$) to be potentially strongly right skewed, except in the case that $\omega = 1$. When considering a single resource, the multinomial problem is reduced to a ratio of binomial proportions, although individual ratios should not be considered orthogonal. Two approximate parametric methods have been developed that are appropriate for obtaining confidence intervals for the true relative risk or selection ratio (Katz et al. 1978, Koopman 1984, Aho and Bowyer 2015).

The simpler method involves log transformation of the ratio, resulting in a sampling distribution for $\hat{\omega}_i$ that is approximately normal with mean $\ln(\omega_i)$ and variance estimator $\hat{\sigma}^2_{\hat{\omega}_i} = (1/x_i) - (1/m_i) + (1/y_i) - (1/n_i)$, where x_i/m_i is observed proportion of times the ith resource is used and y_i/n_i is the observed proportional availability of the ith resource. The bounds to an approximate $(1 - \alpha)100\%$ confidence interval for ω_i in the original ratio units are

$$C_L = \hat{\omega}_i \exp(-z_{1-(\alpha/2)} \hat{\sigma}_{\hat{\omega}_i}),$$

$$C_U = \hat{\omega}_i \exp(z_{1-(\alpha/2)} \hat{\sigma}_{\hat{\omega}_i}),$$

(11.23)

where $z_{1-\alpha/2}$ is the standard normal quantile function evaluated at probability $1 - \alpha$.

The second method requires varying a null value, ω_0, and the use and availability of proportion estimates dependent on ω_0, in a χ^2 test statistic for the null hypothesis $\omega_i = \omega_0$. An approximate $(1 - \alpha)100\%$ confidence interval for ω_i can be found by identifying values ω_0 that do not allow rejection of H_0 at α. The χ^2 method generally produces intervals that are closer to nominal (Koopman 1984). The function ci.prat in *asbio* implements both approaches with exception handling for extreme values, and corrections for simultaneous inference given multiple resources or risk ratios.

11.4.8 Inference for Odds Ratios and Relative Risk

11.4.8.1 Odds Ratio

Recall that the odds of an outcome are simply the probability of the outcome divided by its complement (Chapter 2), that is

$$\text{Odds} = \Omega = \pi/(1 - \pi).$$

Odds greater than 1 indicate that probability for a success (given as π) is greater than the probability of a failure, while odds less than 1 indicate that a failure is more likely than a success. Odds of 2 indicate that a success is twice as likely as a failure, and so on. Another useful measure of association is the *odds ratio*, which is simply the ratio of two odds. That is

$$\theta_{1,2} = \frac{\Omega_1}{\Omega_2}.$$

(11.24)

If the odds ratio is greater than 1, this indicates that the odds of success for the numerator are greater than the odds of success for the denominator. The estimator for $\theta_{1,2}$ is

$$\hat{\theta}_{1,2} = \frac{\hat{\pi}_1/(1-\hat{\pi}_1)}{\hat{\pi}_2/(1-\hat{\pi}_2)} = \frac{(y_1/n_1)/(1-y_1/n_1)}{(y_2/n_2)/(1-y_2/n_2)}.$$

(11.25)

As with selection ratios, the sampling distribution of $\hat{\theta}_{1,2}$ is in the range $[0, \infty)$ and is strongly skewed. Given log transformation, however, the sampling distribution is approximately normal with mean $\ln(\theta_{1,2})$, and variance estimator $\hat{\sigma}_{\hat{\theta}}^2 = 2(1/n_1 + 1/n_2)$ (Agresti 2012). As a result, the bounds to an approximate $(1 - \alpha)100\%$ confidence interval for $\theta_{1,2}$ are given by

$$
\begin{aligned}
C_L &= \hat{\theta}_{1,2} \times \exp(-z_{1-(\alpha/2)}\hat{\sigma}_{\hat{\theta}}), \\
C_U &= \hat{\theta}_{1,2} \times \exp(z_{1-(\alpha/2)}\hat{\sigma}_{\hat{\theta}}),
\end{aligned}
\tag{11.26}
$$

where $z_{1-\alpha/2}$ is the standard normal quantile function at probability $1 - \alpha$.

11.4.8.2 Relative Risk

Recall from Chapter 2 that relative risk is simply the ratio of two probabilities. That is

$$
RR_{1,2} = \frac{\pi_1}{\pi_2}.
\tag{11.27}
$$

This is estimated with

$$
\widehat{RR}_{1,2} = \frac{\hat{\pi}_1}{\hat{\pi}_2}
\tag{11.28}
$$

Relative risk is used to contrast the relative probabilities for "success" for different groups or treatments. For instance, if the proportion of lung cancer diagnoses in a group of nonsmokers was 0.62, while the proportion of diagnoses in a smoker group was 0.5, then the relative risk of lung cancer is 1.24 times higher for smokers than nonsmokers.

Both the odds ratio and relative risk are often used to quantify results from epidemiological research. Their correct usage, however, will depend on the experimental design being used. Recall from Chapter 7 that in a prospective study, a researcher identifies exposed and control groups and follows them over time, while in a retrospective study, a researcher looks into the past to compare groups. Relative risk can be calculated to aid in the interpretation of prospective studies, for instance, the aspirin experiment described in Example 11.5. However, calculation of relative risk for retrospective studies will be inappropriate because the totals for the marginal table will be fixed. In this case, only the odds ratio should be used (Motulsky 1995).

EXAMPLE 11.8

Cat scratch disease causes inflammation of the lymph nodes, and, as the name suggests, usually follows a cat scratch. Zangwill et al. (1993) examined the possibility that the presence of fleas in cats may be increasing the risk of disease for their owners. The investigators contacted primary care providers in Connecticut to identify 56 cat owners who developed cat scratch fever. They then selected a random sample of 56 cat owners who did not have the disease. For both groups, owners with cats with fleas were identified. The investigators found that 34 of the 56 diseased subjects had cats with fleas, while 4 of the 56 control owners had cats with fleas. The proportion of cats with fleas that produced disease symptoms is $32/(32 + 4) = 0.89$, while the proportion of flea-free cats that produced disease symptoms is $24/(24 + 52) = 0.32$. These proportions, however, depend entirely on the number of controls that were used. For instance, the use of 112 controls instead of 56, given the same proportion of fleaed cats, would result in the proportions $32/(32 + 8) = 0.76$ and $24/(24 + 104) = 0.19$, and a very different relative risk estimate.

11.5 Contingency Tables

Two-way and higher tabular data are conventionally organized into contingency tables or more complex arrays with variables constituting array dimensions, for example, rows and columns. Table 11.2 summarizes the counts of vegetation quadrats with or without dead coolibah trees (*Eucalyptus coolibah*) at three landscape positions: dunes, lakeshore, and intermediate, in Southern Australia (Roberts 1993, Quinn and Keough 2002).

A two-way contingency table will have r rows and c columns, with cell locations indexed using i and j. Thus, $i = \{1, 2, ..., r\}$ and $j = \{1, 2, ..., c\}$. We would say that Table 11.2 has dimensions 3×2.

A contingency table can be broken down into three probabilistic components. The *joint distribution* describes the proportion of counts at the intersection of the rows and columns. This probability is denoted by π_{ij} and is estimated by dividing the count in the ijth cell by the total count. That is

$$\hat{\pi}_{ij} = \frac{y_{ij}}{n}, \tag{11.29}$$

where y_{ij} is the count in the ijth cell. In Table 11.2, the joint probabilities are 15/57, 4/57, 0/57, 13/57, 8/57, and 17/57.

Marginal distributions are proportions obtained by dividing row or column sums by the total count, n. The population ith (row) and jth (column) marginal probabilities are denoted by π_{i+} and π_{+j}. We estimate these values analogously using sample counts. For instance

$$\hat{\pi}_{i+} = \frac{n_{i+}}{n}, \tag{11.30}$$

TABLE 11.2

Observed Counts of Quadrats Containing Dead Coolibah Trees, Cross-Classified with Landscape Position

Environment	Dead Coolibahs		Total
	Present	Absent	
Lakeshore	15	13	28
Intermediate	4	8	12
Dune	0	17	17
Total	19	38	57

Source: Adapted from Roberts, J. 1993. *Australian Journal of Ecology* 18: 345–350.

where n_{i+} is the count total in the ith row. There will be a marginal distribution for each variable in the table. The sample marginal distribution of landscape types in Table 11.2 is $\hat{\pi}_{1+} = 28/57$, $\hat{\pi}_{2+} = 12/57$, and $\hat{\pi}_{3+} = 17/57$.

It is often possible to distinguish explanatory and response variables in two-way contingency tables. In this case, Y categories are conventionally given as table columns, and X categories are listed in rows. Given this framework, it is useful to determine the probability distribution of Y at each level of X. These are called *conditional distributions*. In Table 11.2, it is reasonable to see the presence/absence of dead coolibahs as a response variable and presence/absence of landscape type as an explanatory variable. Computation of the conditional distributions requires the equation for conditional probability given in Chapter 2 (Equation 2.14). For example, the conditional distribution of dead coolibah presence/absence, given a lakeshore landscape is

$$\hat{\pi}(\text{Dead coolibah}^- \mid \text{lake}) = \frac{\hat{\pi}(\text{dead coolibah}^- \cap \text{lake})}{\hat{\pi}(\text{lake})} = \frac{15/57}{28/57} = \frac{15}{28} = 0.5357,$$

$$\hat{\pi}(\text{Dead coolibah}^+ \mid \text{lake}) = \frac{\hat{\pi}(\text{dead coolibah}^+ \cap \text{lake})}{\hat{\pi}(\text{lake})} = \frac{13/57}{28/57} = \frac{13}{28} = 0.4643.$$

Two other conditional probability distributions are possible based on the other two landscape types.

11.6 Two-Way Tables

In two-way and higher tables, we generally want to make inferences concerning the independence among variables representing table rows and columns. Owing to the independence rule of multiplication (Chapters 2 and 8), the null and alternative hypotheses can then be expressed as

$$H_0: \pi_{ij} = \pi_{i+}\pi_{+j}, \quad \text{for all } i \text{ and } j,$$

versus

$$H_A: \pi_{ij} \neq \pi_{i+}\pi_{+j}.$$

Three different experimental designs for tabular analyses were discussed in Chapter 7.

The most commonly used is a model I framework wherein experimental units are classified as they are acquired. Thus, the total number of trials, n, will be fixed, but row and column totals will be allowed to vary freely. The coolibah data in Table 11.2 are the result of a model I design. Neither the totals for dead coolibah's presence/absence nor the totals for landscape types were fixed by the investigator.

In a model II design, totals for the categories of one variable (usually the explanatory variable) are fixed, while totals for other categorical variables are allowed to vary. Thus, P-values should be computed by finding the proportion of all possible two-way tables, with fixed totals in one dimension, in which departures from independence are greater than or equal to the observed dependence. The aspirin experiment described in Example 11.5 is a model II design because the number of subjects in the aspirin and placebo groups was fixed, while MI results were allowed to vary. In a model II design, a test of independence provides a test of the null hypothesis that the fixed (explanatory variable) populations are equal with respect to the second (response) variable. This is called a *test of homogeneity* (Ott and Longnecker 2004).

Model III designs fix both row and column totals, and are extremely rare in biological investigations (Quinn and Keough 2002). Of interest, the original type III design was used by R. A. Fisher to test the hypothesis that a subject could distinguish cups of tea in which milk had been added first, from those in which milk had been added last (the evidence indicated that she could).

Three general hypothesis testing approaches are possible for two-way tests: the χ^2, and likelihood ratio procedures, introduced in Section 11.3, and Fisher's exact test.

Likelihood ratio tests were originally designed to address model I designs, Fisher's exact test (see below) was developed to analyze model III designs, and the χ^2 score test—the conventional approach for analysis—was not developed for any particular design framework (Sokal and Rohlf 2012). Despite this, the three methods generally provide similar results, regardless of design, and are frequently used interchangeably.

As with one-way formats, two-way procedures assume

1. Observations are independent. That is, observations are classified into categories independently.
2. Conditions suffice to allow asymptotic convergence to null distributions (see Section 11.6.4).

11.6.1 Chi-Squared (Score) Test

Equation 11.11 can be extended to a two-way format (Pearson 1900) using

$$X^2 = \sum_{i,j} \frac{(y_{ij} - E_{ij})^2}{E_{ij}}. \tag{11.31}$$

The expected count in the ith row and jth column is the total table count multiplied by the joint probability under H_0 (under independence). That is

$$E_{ij} = n\hat{\pi}_{i+}\hat{\pi}_{+j}. \tag{11.32}$$

Under H_0, X^2 will have an asymptotic χ^2 distribution with $(r-1)(c-1)$ degrees of freedom. P-values are given as $P(\chi^2_{(r-1)(c-1)} \geq X^2)$.

11.6.2 Likelihood Ratio Test

An alternative to the χ^2 test is provided by extending Equation 11.17 to

$$G^2 = 2\sum_{i=1}^{c} y_{ij} \log\left(\frac{y_{ij}}{E_{ij}}\right). \tag{11.33}$$

For large samples, G^2 will have an approximate χ^2 distribution under H_0 with $(r-1)(c-1)$ degrees of freedom. P-values are calculated by finding $P(\chi^2_{(r-1)(c-1)} \geq G^2)$.

11.6.3 Fisher's Exact Test

Fisher's exact test was created specifically for 2×2 tables resulting from a model III design. Under H_0, and given fixed row and column totals, the count in the ijth cell will determine the counts of the other three cells. The noncentral hypergeometric distribution

$$f(y_{11}) = \frac{\dbinom{n_{1+}}{y_{11}}\dbinom{n_{2+}}{n_{+1}-y_{11}}}{\dbinom{n}{n_{+1}}}, \tag{11.34}$$

expresses the probability for the count configuration of all four cells, under H_0, using only one cell, commonly y_{11} (Agresti 2012). This framework gives the probability for the *observed* configuration under H_0. However, the P-value will be the proportion of all possible two-way tables, with fixed row and column totals, containing counts equal to or more extreme than the observed y_{11} under H_0. Along with two-tailed tests, upper- and lower-tailed tests are also possible using Fisher's exact test. These are stated with respect to the odds ratio.

In a 2×2 table, the sample odds ratio can be found using joint probabilities. For instance

$$\hat{\theta}_{1,2} = \frac{\hat{\pi}_{11}/\hat{\pi}_{12}}{\hat{\pi}_{21}/\hat{\pi}_{22}} = \frac{\hat{\pi}_{11}\hat{\pi}_{22}}{\hat{\pi}_{12}\hat{\pi}_{11}} = \frac{(y_{11}/n)(y_{22}/n)}{(y_{12}/n)(y_{21}/n)} = \frac{y_{11}y_{22}}{y_{12}y_{21}}. \tag{11.35}$$

When the rows and columns are independent, then $\theta_{1,2} = 1$. This indicates that true row counts do not vary with true column counts. When $\theta_{1,2} > 1$, the odds of success are higher in row one than row two, while if $\theta_{1,2} < 1$, the odds of success are higher in row two than row one. This measure will be unchanged if the contingency table is transposed (rows become columns and columns become rows). It will be inverted to $1/\theta$ if categories in one variable are inverted; however, this will not affect two-tailed inferences. For instance, if the 95% confidence bounds for θ are (2, 3), then inverting the rows results in the bounds (1/3, 1/2). In both cases, 1 is not contained in the interval.

The two-tailed hypotheses for Fisher's exact test can be specified as

$$H_0: \theta = 1,$$

versus

$$H_A: \theta \neq 1.$$

Lower- and upper-tailed alternatives are specified as

$$H_A: \theta < 1,$$

$$H_A: \theta > 1.$$

The calculation of two-tailed P-values in Fisher's exact test is hampered by the general asymmetry and discrete character of the hypergeometric distribution. A complex series of exception-handling steps are used by the **R** function `fisher.test`. Fisher's exact tests for two-way tables larger than 2×2 are also possible. However, this requires a multivariate hypergeometric null distribution and prevents exact tests (Agresti 2012). Larger tables will also have more than one odds ratio because such tables will require partitioning into 2×2 subsections.

EXAMPLE 11.9

Temeles and Kress (2003) studied purple-throated caribs (*Eulampis jugularis*), a type of hummingbird that lives in the forest preserves of the Caribbean island of Saint Lucia. Males and females of this species have slightly different bill shapes, potentially affecting the flowers they visit for food. The researchers observed a population of 70 birds: 49 females and 21 males. They noted that of the female birds, 20 visited the flowers of red palulu (*Heliconia bihai*) while none of the males did (Table 11.3).

We will examine all three hypothesis testing methods discussed in Sections 11.6.1 through 11.6.3 in this example. The null hypothesis will be that the true number of *H. bihai* visits is independent of *E. jugularis'* sex.

Chi-Squared (Score) Test

The expected cell counts are

$$E_{11} = 70(20/70)(49/70) = 14, \quad E_{12} = 70(20/70)(21/70) = 6,$$

$$E_{21} = 70(50/70)(49/70) = 35, \quad E_{22} = 70(50/70)(21/70) = 15.$$

Thus, the χ^2 test statistic is

$$X^2 = (20 - 14)^2/14 + (0 - 6)^2/6 + (29 - 35)^2/35 + (21 - 15)^2/15 = 12.$$

Since the degrees of freedom are $(2 - 1)(2 - 1) = 1$, the *P*-value is

$$P(\chi_1^2 \geq 12) = 0.000532.$$

Using **R**, we have

```
jugularis <- matrix(c(20, 29, 0, 21), nrow = 2)
chisq.test(jugularis, correct = F)
     Pearson's Chi-squared test
```

TABLE 11.3

Data Describing Visitation of the Hummingbird *E. jugularis* to the Plant *H. bihai*

	Sex		
H. bihai Visits	♂	♀	**Total**
Yes	20	0	20
No	29	21	50
Totals	49	21	70

Source: Adapted from Temeles, E. J., and Kress, W. J. 2003. *Science* 300: 630–633.

```
data: jugularis
X-squared = 12, df = 1, p-value = 0.000532
```

Likelihood Ratio Test

We have

$$G^2 = 2\{[20 \times \ln(20/14)] + 0 + [29 \times \ln(29/35)] + [21 \times \ln(21/15)]\} = 17.4918.$$

Thus, the *P*-value is

$$P(\chi_1^2 \geq 17.4918) = 0.000029.$$

Using **R**, we have

```
g.test(jugularis)
Contingency table likelihood ratio test
 G.statistic        p-value
      17.4918 2.885492e-05
```

Fisher's Exact Test

The sample odds ratio is $(20 \times 21)/(0 \times 29) = \infty$. While undefined, this ratio obviously supports the conclusion that a female carib will visit *H. bihai* with much higher odds than a male carib. Using Fisher's exact test, the probability that $y_{11} = 20$ under H_0 is

$$\frac{\binom{20}{20}\binom{50}{29}}{\binom{70}{49}} = 0.000175.$$

Computation of an actual *P*-value involves the addition of this probability to the probabilities associated with all possible values of y_{11} that are more extreme than 20 under H_0. Fortunately, the function `fisher.test` runs the analysis for us.

```
fisher.test(jugularis)
p-value = 0.0002934
alternative hypothesis: true odds ratio is not equal to 1
95 percent confidence interval:
 3.055615        Inf
odds ratio
       Inf
```

P-values from all three tests are small. Thus, we reject H_0 and conclude that *H. bihai* visits are not independent of sex.

11.6.4 Comparison of Methods and Requirements for Valid Inference

As with one-way formats, the score and likelihood ratio tests provide approximate *P*-values because the test statistic sampling distribution will only approximate the null distribution given large sample sizes. The X^2 statistic will converge to $\chi^2_{(r-1)(c-1)}$ more quickly than G^2, and is therefore more trustworthy for moderate sample sizes. For large expected frequencies, the tests will have the same χ^2 distribution and will give very similar *P*-values (Agresti 2012).

Most experts recommend that when using G^2 and X^2 tests, no E_i should be less than 1 and not more than 20% of the E_is should be less than 5 (Cochran 1954). However, null approximations for G^2 tests can be poor when any E_i is less than about 5 (this likely explains the divergence of X^2 and G^2 results in Example 11.9) and X^2 approximations can be quite good when some E_is are as small as 1, if the number of rows and columns is large (Agresti 2012).

When the expected values under H_0 approach 5, it is safer to use Fisher's exact test, because its test statistic follows an exact null distribution instead of an asymptotic approximation. The test, however, is conservative with respect to type I error (the actual α may be much smaller than the nominal α). As a result, Agresti (2012) recommended using a *mid* *P*-value approach. Under this framework, the *P*-value is calculated by adding one-half the probability of the observed y_{ij} configuration, under H_0, to the full densities of the more extreme y_{ij}s under H_0. Small E_is can also be addressed using randomization tests. The function chisq.test allows specification of permutation *P*-values when E_is are small, but table dimensions are too large for exact tests.

11.7 Ordinal Variables

Ordinal variables are often identified as categorical variables with ordered classes.

Both variables being ordinal in a two way table, allows the usage of Pearson's correlation to test independence (Chapter 8). To use this approach, one simply calculates the correlation of the ordinal "levels" given the counts of their intersections. For a test of row and columns' independence, an appropriate test statistic is

$$M^2 = (n-1)r^2. \tag{11.36}$$

For large values of n, M^2 will have an approximate χ_1^2 distribution under H_0. Two-tailed *P*-values are calculated as $P(\chi_1^2 \geq M^2)$. When a true association exists, the M^2 test will have more power than the X^2 or G^2 test, which treat variables as categorical. In addition, M has an asymptotic approximate standard normal distribution, allowing one-tailed tests. Kendall's τ or Spearman's ρs (Chapter 8) can also be used directly as correlation measures and in independence tests and in cases with two ordinal variables.

EXAMPLE 11.10

Consider a cross-classification of locations by subjective scales of site windiness ($1 < 2 < 3 < 4$) and soil moisture ($1 < 2 < 3 < 4 < 5$).

```
wind.lvl <- 1:4; moist.lvl <- 1:5
```

We wish to test the null hypothesis that the true counts for the wind and soil moisture indices are independent.

The function expand.grid generates all possible level combinations of wind and moist.

```
cats <- expand.grid(wind.lvl, moist.lvl)
```

Assume we have the following counts:

```
count <- c(2, 3, 6, 10, 1, 3, 6, 8, 6, 5, 2, 6, 6, 3, 4, 1, 9, 8, 0, 0)
site <- data.frame(count = count, wind = cats[,1], moist = cats[,2])
```

The function xtabs can be used to create the contingency table.

```
xtabs(count ~ wind + moist, data = site)
    moist
wind  1  2  3  4  5
   1  6  2  1  1  1
   2  1  6  2  3  0
   3  2  2 10  4  4
   4  1  1  2  7  5
```

Here is Kendall's τ:

```
cor(rep(site$wind, count), rep(site$moist, count), method = "k")
[1] -0.4407509
```

There is a negative association between the rows and columns. Presumably, the soil moisture is decreased by windiness because the boundary layer (still layer of air over the ground) is eroded, resulting in increased evaporation. The M^2 test statistic and P-value are

```
r <- cor(rep(site$wind, count), rep(site$moist, count))
M2 <- r^2 *(sum(count) - 1)
pchisq(M2, 1, lower.tail = F)
[1] 9.989148e-05
```

Thus, we reject H_0 and conclude that soil moisture and site windiness are not independent.

In a $2 \times c$ table where the response variable is ordinal and the explanatory variable is dichotomous, the M^2 test will be identical to the Wilcoxon test under its normal approximation (Section 6.6). As with the Wilcoxon test, the M^2 test is inappropriate when the explanatory variable has more than two categories. In an $r \times 2$ table, with an ordinal explanatory variable and a dichotomous response variable, the M^2 statistic can be used to detect trends in the response successes across levels of X. This is called the *Cochran–Armitage* trend test. Here, significant P-values indicate that the population linear trend among proportions is nonzero (Agresti 2012).

11.8 Power, Sample Size, and Effect Size

The observed effect sizes in tabular models can be measured in a number of ways including proportions of counts in table cells, odds ratios, and the difference in proportions in two-proportion tests.

Cohen (1988) proposed that the true effect size for a test of two proportions could be defined as

$$h = 2\arcsin(\sqrt{\pi_1}) - 2\arcsin(\sqrt{\pi_2}). \tag{11.37}$$

Cohen (1992) suggested that the values $h = 0.2, 0.5$, and 0.8 could be used as lower bounds for small, medium, and large effect sizes (but see Section 9.9).

Analogously, Cohen (1988) recommended that the true effect size for χ^2 contingency table tests be measured as

$$w = \sqrt{\sum_{i=1}^{c} \frac{(\pi_{0i} - \pi_{1i})^2}{\pi_{0i}}},$$ (11.38)

where π_{0i} is the probability in the ith table cell under H_0 and π_{1i} is the expected probability under the alternative. Cohen (1992) proposed that the values $w = 0.1, 0.3$, and 0.5 could be used as the lower bounds for small, medium, and large effect sizes (see Section 9.9).

The functions `pwr.2p.test` and `pwr.2p2n.test` from library *pwr* can be used to perform power analyses for balanced and unbalanced two-proportion tests using the definition of effect size given in Equation 11.37. The function `pwr.chisq.test` performs χ^2 test power analyses using the effect size definition in Equation 11.38. Many other approaches have been developed to examine effect size and power for particular tabular analyses. Sokal and Rohlf (2012, pp. 801–801) provide a review in a biological context.

EXAMPLE 11.11

For a χ^2 test with five categories, we assume that $w = 0.3$. What is the sample size required to reject H_0: $\pi_1 = \pi_2 = \pi_3 = \pi_4 = \pi_5 = 1/5$ with probability 0.8, given $\alpha = 0.05$?
  We have

```
library(pwr)
pwr.chisq.test(w=0.3, df=4, sig.level=0.05, power=0.8)
          w=0.3
          N=132.6143
          df=4
  sig.level=0.05
      power=0.8
```

132 total observations are required.

11.9 Three-Way Tables

Three-way (and higher) contingency analyses are possible. The three variables are generally denoted as X, Y, and Z, with Y assigned to the response variable, if this variable exists. Extending earlier conventions, the number of levels in X, Y, and Z are denoted as r, c, and l (l indicates layer). Levels in the additional Z dimension will be indexed with k. Thus, $k = \{1, 2, \ldots, l\}$. For a three-way table, Pearson's χ^2 statistic can now be adapted to

$$X^2 = \sum_{i,j,k} \frac{(y_{ijk} - E_{ijk})^2}{E_{ijk}},$$ (11.39)

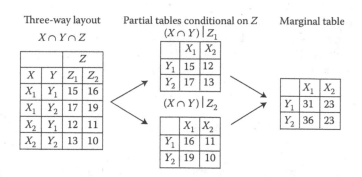

FIGURE 11.3

Example of partial and marginal tables for a dataset with three categorical variables X, Y, and Z. Partial conditional tables in the middle of the diagram were created by examining the intersection of X and Y categories while holding Z constant at Z_1 and Z_2. Other partial tables are possible by holding X or Y constant. A marginal table is created by summing the partial tables.

where y_{ijk} is the observed count (number of joint successes) at the ijkth intersection of the three-dimensional array and E_{ijk} is the corresponding ijkth expected value under H_0. Under independence, we have

$$E_{ijk} = \frac{n_{i++} \times n_{+j+} \times n_{++k}}{n^2}. \tag{11.40}$$

If H_0 is true, then X^2 will have an asymptotic χ^2 distribution with $rcl - r - c - l + 2$ degrees of freedom. P-values are obtained by finding $P(\chi^2_{rcl-r-c-l+2} \geq X^2)$.

Partial tables in three-way layouts result from the removal of two-dimensional cross sections from the three-dimensional array. For instance, X and Y could be cross-classified at each level of Z producing partial tables. Associations within partial tables are called *conditional associations* because they demonstrate the conditional relationship of X and Y when fixing Z at some level. Two-way tables created by summing the partial tables are called *marginal tables*. In this case, the third level is ignored instead of being held constant (Figure 11.3).

For the same data, the conditional associations in an array can be very different from the marginal associations, resulting in Simpson's paradox (see Example 11.12 below).

As with one-way and two-way formats, three-way inferential procedures assume

1. Observations are independent. That is, observations are classified into categories independently.

2. Conditions are sufficient to allow asymptotic convergence to the null distributions (see Section 11.9.1).

EXAMPLE 11.12

Appleton et al. (1996) reported on a study from the Whickham district of England to quantify the association of smoking, age, and death. Researchers interviewed 1314 women in the early 1970s with respect to their smoking habits. Twenty years later, the women were relocated and classified with respect to three factors: survival at the time of the follow-up {yes or no}, whether they smoked at the time of the original interview {yes or no}, and age (in years) at the time of the original interview {1 = 18–24, 2 = 35–64, and 3 = >65}. The data are shown in Table 11.4 and are contained in the dataframe whickham.

The function chisq.test currently handles only one- and two-dimensional arrays. However, the function xtabs has no upper limit for the number of array dimensions. It requires that we specify the cell count to be a linear function of the table dimensions. For instance

```
summary(xtabs(count ~ smoke + survival + age, data = whickham))
Test for independence of all factors:
    Chisq = 702.8, df = 7, p-value = 1.678e-147
```

We strongly reject the hypothesis of dimensional independence and conclude that at least one variable is dependent on at least one other variable. The partial tables of smoking and survival, when *holding age constant*, are

```
x <- print(xtabs(count ~ smoke + survival + age, data = whickham))
, , age = 1
    survival
smoke   N    Y
    N   6    5
    Y 213  174
, , age = 2
    survival
smoke   N    Y
    N  59   92
    Y 261  262
, , age = 3
    survival
smoke   N    Y
    N 165   42
    Y  28    7
```

The conditional odds ratios for the partial tables for ages 1 = (18–24), 2 = (35–64), and 3 = (>65) are $\frac{6 \times 174}{213 \times 5} = 0.98$, $\frac{59 \times 262}{261 \times 92} = 0.64$, and $\frac{165 \times 7}{28 \times 42} = 0.98$.

The sample odds of a nonsmoker dying are slightly lower than the odds of a smoker dying for age classes 1 and 3 and are dramatically lower for age class 2. For this category, smokers are predicted to be $1/0.64 = 1.5$ times more likely to die. However, the marginal table provides a different story (Table 11.4).

TABLE 11.4

Cross-Classification of Women from Whickham, United Kingdom, with Respect to Survivorship, Age, and Smoking

Age	Smoke	Survival		Total
		Yes	No	
1	No	6	5	11
	Yes	213	174	387
2	No	59	92	151
	Yes	261	262	524
3	Yes	165	42	207
	No	28	7	35
Total		732	582	1314

Source: Adapted from Appleton, D. R., French, J. M., and Vanderpump, M. P. J. 1996. *The American Statistician* 50(4): 340–341.

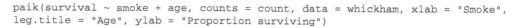

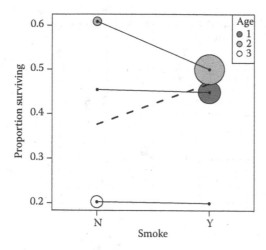

FIGURE 11.4

Paik diagram of the Whickham dataset that cross-classified age, smoking, and death. Discrepancies in the marginal association (dashed line) and conditional associations (solid lines) are evident as a result of Simpson's paradox.

```
x[, , 1]+x[, , 2]+x[, , 3]
      survival
smoke   N    Y
     N 230  139
     Y 502  443
```

In this case, the sample odds ratio is $\dfrac{230 \times 443}{502 \times 129} = 1.49$.

Here, the odds of a nonsmoker dying are much greater than the odds of a smoker dying. This discrepancy between the marginal and conditional odds is called Simpson's paradox. It occurs, in this case, because the marginal table *ignores* the confounding variable age, while the partial tables *condition* on age.

This is clearly demonstrated with a *Paik diagram* (Paik 1985; Figure 11.4). The dashed line in the figure connects the marginal proportions of the surviving subjects in the smoker and nonsmoker classes. Given this information, it appears that smokers survive at a higher rate than nonsmokers (the slope of the line is positive). The centers of the circles show the conditional probabilities of survival for the age and smoker classes. From this perspective, the nonsmokers' survival appears (the slope of the lines are negative). The proportion of both nonsmokers from the youngest age class and smokers in the oldest age class are very small, resulting in mere dots in the diagram. The discrepancy in the conditional and marginal associations occurs because the older subjects, who were primarily nonsmokers, were more at risk from dying of old age, while the young and middle-aged subjects, who were mostly smokers, were less susceptible to death from old age.

11.9.1 Comparison of Methods and Requirements for Valid Inference

As with earlier methods, the test statistic sampling distribution will only effectively approximate the null distribution for large sample sizes. Thus, the minimal sample sizes of Cochran (1954) given in Sections 11.4.6 and 11.6.4 also hold here.

The *Cochran–Mantel–Haenszel* (CMH) test provides an alternative analytical framework for $2 \times 2 \times l$ tables. The procedure tests the null hypothesis that the conditional odds ratios for all partial tables corresponding to levels in Z equal one. As in Fisher's exact test, the outcome y_{11} will determine the outcome of the other three cells in partial tables under H_0. This results in a noncentral hypergeometric null distribution. The test works best when the associations between X and Y are similar for each partial table (of course, this is not true for the conditional odds ratios in the previous example). If H_0 is true, the CMH test statistic (see Agresti 2012) will asymptotically follow a χ^2 distribution with one degree of freedom. The function `mantelhaen.test` provides the procedure.

★ 11.10 Generalized Linear Models

To minimize confusion and frustration, the introduction of GLMs in Section 9.20 should be reviewed here before proceeding further.

GLMs have been previously described as a unifying framework that allows response variables with an "unconventional" form, for example, binary or count outcomes to be expressed as a linear function of predictors. Because both the response and explanatory variables in GLMs can be categorical, these models allow the analysis of tabular data. GLMs are particularly useful for analyzing high-dimensional tabular formats with complex associations.

11.10.1 Binomial GLMs

Binomial GLMs with a canonical logit link function were examined in Chapter 9 as an analytical approach for regression models with dichotomous response variables. The approach can be readily extended to models with categorical predictors.

As with the design matrices of ANOVAs, tabular logit models require $c - 1$ indicator variables for specification of a one-way layout (see Section 10.4). Two-way layouts with no interactions require $(r - 1)(c - 1)$ indicator variables, and so on. Consider a model with a single predictor X with $c = 3$ levels. The logit model can be specified as

$$\text{logit}(\pi_i) = \beta_0 + \beta_1 X_{1i} + \beta_2 X_{2i} \tag{11.41}$$

where β_0 is a constant, and β_1 and β_2 are effects corresponding to the dummy variables X_1 and X_2, respectively. In Equation 11.41, we assume the data are in a nontabular format. That is, $i \in \{1, 2, \ldots, n\}$. We test the null hypothesis of row and column independence using a likelihood ratio test statistic that compares the residual deviance of reference (R) and nested (N) models.

$$G^2 = \hat{D}(N) - \hat{D}(R) = 2\ln\left[\frac{\hat{\mathcal{L}}(R)}{\hat{\mathcal{L}}(N)}\right] = 2[\hat{\ell}(R) - \hat{\ell}(N)]. \tag{11.42}$$

Given large sample sizes, G^2 will be approximately χ^2 distributed under H_0. The degrees of freedom will be the difference in the number of parameters estimated in the reference and nested models.

TABLE 11.5

Snoring and Heart Disease Counts

Snoring	Heart Disease		Proportion Yes
	Yes	No	
Never	23	1356	0.017
Occasional	35	603	0.055
Nearly every night	21	192	0.099
Every night	30	224	0.118
Total	109	2375	0.289

Source: Adapted from Norton, P. G., and Dunn, E. V. 1985. *British Medical Journal* 291: 630–632.

EXAMPLE 11.13

Norton and Dunn (1985) compiled data from four family-practice clinics in Toronto to quantify the association between snoring and heart disease for 2484 subjects (Table 11.5). The study allowed only correlative inference because confounded variables were not addressed by randomly assigning treatments (snoring levels were assigned by the subject's spouses). As a result both snoring and heart disease were confounded with the subject's age. Nonetheless, the size of the study allows trust in the precision of associative inferences.

```
data(snore)
snore.glm <- glm(disease ~ snoring, data = snore, family = "binomial")
```

We require the estimation of four parameters for the fitted model (the intercept and three indicator variable effects corresponding to the four levels in snoring). The nested null model consists of only an intercept and requires the estimation of only one parameter. Hence, the χ^2 distribution under H_0 will have $4 - 1 = 3$ degrees of freedom. We test the null hypothesis of independence by finding the difference in the residual and null deviance, and using this as the likelihood ratio test statistic. The difference is 65.904. Thus, the P-value is $P(\chi_3^2 \geq 65.904) = 3.2 \times 10^{-14}$.

```
library(car)
Anova(snore.glm)
Analysis of Deviance Table (Type II tests)

Response: disease
        LR Chisq Df Pr(>Chisq)
snoring   65.904  3  3.213e-14 ***
```

Although the sample size has undoubtedly resulted in overwhelming power, we reject H_0 and conclude that heart disease and snoring are associated.

11.10.2 Log-Linear Models

In Section 9.20, we learned that a Poisson log-linear model is a GLM that assumes Poisson errors and uses a canonical log link function. This approach can be used to effectively model cell counts in contingency tables.

11.10.2.1 Two-Way Tables

In a two-way table with variables X and Y, the expected count under independence in the ijth cell is given by $E_{ij} = n\pi_{i+}\pi_{+j}$. Taking the log of both sides of this equation yields an additive model that serves as the basis for the linear model

$$\log(E_{ij}) = \lambda + \lambda_i^X + \lambda_j^Y, \tag{11.43}$$

where λ is a constant, λ_i^X is the effect of the ith level in X, and λ_i^Y is the effect of the jth level in Y. Thus, the superscripts are merely labels. They are not exponents.

Equation 11.43 is called the *log-linear model of independence*. This is because the hypothesis of independence under H_0 is equivalent to the hypothesis that the model is true. Given the default treatment contrasts used by glm (Section 9.20, Chapter 10), $\hat{\lambda}$ will be given as the fitted value of the first alphanumeric level in the first factor (in log units). Estimates for other parameters (e.g., λ_i^Y) define the effect of the remaining variables on the expected log counts in terms of deviations from $\hat{\lambda}$. For tabular analyses, the corresponding saturated model can be expressed as

$$\log(E_{ij}) = \lambda + \lambda_i^X + \lambda_j^Y + \lambda_{ij}^{XY}. \tag{11.44}$$

Equation 11.44 satisfies situations in which X and Y are *not* independent. The additional terms λ_{ij}^{XY} reflect deviation from independence. Under independence, all λ_{ij}^{XY}s will equal zero and Equation 11.43 will equal Equation 11.44.

The parsimony of log-linear models can be gauged using information-theoretic criteria (Section 9.17). A null hypothesis test of model adequacy is provided by a likelihood ratio test. In this framework, the residual deviance is the test statistic, G^2. The null and alternative hypotheses are

$$H_0: \text{The model is correct,}$$

versus

$$H_A: \text{The model is incorrect.}$$

Note that large P-values here provide *more* support for the model.

Under H_0, G^2 will be approximately χ^2 distributed, given large sample sizes, with the degrees of freedom taken from the residual degrees of freedom, that is, $(r-1)(c-1)$. The importance of parameters within the optimal model can be explored using the magnitude of parameter coefficients and likelihood ratio tests that compare reference and nested models using Equation 11.42.

11.10.2.2 Three-Way Tables

The analysis of three-way tables should involve the consideration of specific patterns of independence and association. The three-way log-linear model that specifies no interactions

$$\log(E_{ijk}) = \lambda + \lambda_i^X + \lambda_j^Y + \lambda_k^Z, \tag{11.45}$$

is called the *mutual independence model*. It assumes that each pair of variables is independent both conditionally and marginally. The model is simple. Unfortunately, when variables for a table are chosen carefully, it will seldom be appropriate (Agresti 2012).

More complex, but often still inadequate are *partial independence models*. For instance, in the model

$$\log(E_{ijk}) = \lambda + \lambda_i^X + \lambda_j^Y + \lambda_k^Z + \lambda_k^{YZ}, \tag{11.46}$$

X is partially independent of Y and Z because it is independent of the interaction of Y and Z. Conversely, Y and Z are specified to be dependent. There will be three possible partial independence models, one for each interaction. Log-linear models are generally notated by listing the highest-order term(s). Thus, Equation 11.45 can be summarized as (X, Y, Z) while Equation 11.46 can be denoted as (YZ).

A possible *conditional independence model* is (XZ, YX)

$$\log(E_{ijk}) = \lambda + \lambda_i^X + \lambda_j^Y + \lambda_k^Z + \lambda_{ik}^{XZ} + \lambda_{jk}^{YZ}. \tag{11.47}$$

This model considers the interaction of the other variables with Z, but does not contain the XY interaction parameter. Thus, the model specifies the conditional independence of X and Y, while controlling for Z. Again, there will be three conditional independence models. One for each deleted interaction parameter.

The *homogenous association model* (XY, XZ, YX) has the form

$$\log(E_{ijk}) = \lambda + \lambda_i^X + \lambda_j^Y + \lambda_k^Z + \lambda_{ij}^{XY} + \lambda_{ik}^{XZ} + \lambda_{jk}^{YZ}. \tag{11.48}$$

This model allows all three variables to have conditional associations, but assumes that the variables have *joint independence*, that is, there is no three-way association. Under its constraints, the conditional odds between any two variables will be the same at each level of the third variable.

The three-way saturated model (XYZ) that will fit the data perfectly, assumes complete dependence, and has no degrees of freedom, is

$$\log(E_{ijk}) = \lambda + \lambda_i^X + \lambda_j^Y + \lambda_k^Z + \lambda_{ij}^{XY} + \lambda_{ik}^{XZ} + \lambda_{jk}^{YZ} + \lambda_{ijk}^{XYZ}. \tag{11.49}$$

Under the null hypothesis of model adequacy and large sample sizes, the residual deviance for the models above will follow χ^2 distributions. Degrees of freedom will vary with the fitted model (Table 11.6).

Of primary interest are drop in deviance tests comparing reference and nested models that (generally) differ with respect to a single variable. This allows testing for particular types of independence. For instance, to test the null hypothesis of the conditional independence of X and Y when holding Z constant, we would designate the homogenous association model (XY, XZ, YZ) as the reference model and let (XZ, YZ) be the nested model. Under H_0, the likelihood ratio test statistic (difference in residual deviance) will follow a χ^2 distribution with the degrees of freedom taken from the difference in the degrees of freedom in the reference and nested models.

TABLE 11.6

Degrees of Freedom and Independence Frameworks for Types of Three-Way Models

Log-Linear Model	Residual *df*	Independence
$X + Y + Z$	$rcl - r - c - l + 2$	Mutual
$X + Y + Z + X{:}Y$	$(l - 1)(rc - 1)$	Partial
$X + Y + Z + X{:}Z$	$(c - 1)(rl - 1)$	Partial
$X + Y + Z + Y{:}Z$	$(r - 1)(rl - 1)$	Partial
$X + Y + Z + X{:}Y + X{:}Z$	$r(l - 1)(c - 1)$	Conditional
$X + Y + Z + X{:}Y + Y{:}Z$	$c(r - 1)(l - 1)$	Conditional
$X + Y + Z + Y{:}Z + X{:}Z$	$l(r - 1)(c - 1)$	Conditional
$X + Y + Z + X{:}Y + Y{:}Z + X{:}Z$	$(r - 1)(c - 1)(l - 1)$	Joint
$X + Y + Z + X{:}Y + Y{:}Z + X{:}Z + X{:}Y{:}Z$	0	None

EXAMPLE 11.14

To test the "predation-sensitive food" hypothesis, which predicts that both food and predation limit prey populations, Sinclair and Arcese (1995) examined wildebeest (*Connochaetes taurinus*) carcasses in the Serengeti. The degree of malnutrition in animals was measured by marrow content since marrow will contain the last fat reserves in ungulates. Carcasses were cross-classified with respect to three categorical variables: sex {♀, ♂}, cause of death {predation, nonpredation}, and marrow type {SWF = solid white fatty, indicating healthy animals, OG = opaque gelatinous, indicating malnourishment, and TG = translucent gelatinous, indicating severe malnourishment}. Their data are shown in Table 11.7 and are contained in the dataframe wildebeest.

We first establish a set of candidate models.[*]

```
data(wildebeest)
SPM = glm(count ~ (sex + predation + marrow)^3, data = wildebeest,
family = "poisson")
SP.SM.PM = update(SPM, ~. - sex : predation : marrow)
SM.PM = update(SP.SM.PM, ~. - sex : predation)
SM.SP = update(SP.SM.PM, ~. - predation : marrow)
PM.SP = update(SP.SM.PM, ~. - sex : marrow)
PM = update(PM.SP, ~. - sex : predation)
SP = update(SM.SP, ~. - sex : marrow)
SM = update(SM.PM, ~. - predation : marrow)
S.P.M = update(SM, ~. - sex : marrow)
```

Model comparisons are given in Table 11.8. The *P*-values correspond to tests for the null hypotheses of model adequacy. From this perspective, all the models are somewhat poor (*P*-values are small). Unfortunately, the saturated model (which provides no inferential simplification and has an incalculable goodness of fit *P*-value) appears to be the best choice here. This is true because (1) the (SPM) model had the lowest *AIC* of all models and (2) the model adequacy test for the homogenous association model (SP.SM.PM) also provides a test for joint independence. The *P*-value for this test was significant at $\alpha = 0.05$ resulting in a simultaneous rejection of the null hypothesis of model adequacy for the homogenous association model and support for the joint dependence model.

Despite this, for pedagogic purposes, we will explore the second best (SM, SP) model in greater detail. Recall that this model holds sex constant while considering marrow

---

[*] See Quinn and Keough (2002) for additional perspectives on a log-linear analysis of this dataset.

TABLE 11.7

Characteristics of Wildebeest Carcasses Taken from the Serengeti from 1977 to 1991

| Cause of Death | Sex | Marrow | | | Total |
		SWF	OG	TG	
Predation	♂	26	32	8	66
	♀	14	43	10	67
Nonpredation	♂	6	26	16	48
	♀	7	12	26	45
Total		53	113	60	226

Source: Adapted from Sinclair, A. R. E., and Arcese, P. 1995. *Ecology* 76(3): 882–891.

and predation. Thus, (SM, SP) specifies that the marrow and predation categories are independent for both sexes. Below are partial tables that examine marrow and predation when holding sex constant. Fitted values for the (SM, SP) model are in the table cells.

```
print(xtabs(fitted(SM.PM) ~ marrow + predation + sex, data = wildebeest))
, , sex = F
     predation
marrow          N         P
  OG  19.504425 38.495575
  SWF  7.849057 24.150943
  TG  16.800000  7.200000
, , sex = M
     predation
marrow          N         P
  OG  18.495575 36.504425
  SWF  5.150943 15.849057
  TG  25.200000 10.800000
```

Levels of predation are predicted to be higher for both males and females in the OG and SWF marrow groups. In parallel, levels of nonpredation death are predicted to be higher for both sexes in the TG marrow group. Thus, values vary dramatically with respect to the predation and marrow categories, and the patterns of conditional association are consistent for both sexes. A drop in deviance test for the null hypothesis of conditional independence of predation and marrow of a given sex is given by

TABLE 11.8

Comparison of Wildebeest Models from Least Complex to Most Complex (Saturated)

Model	G^2	df	P-Value	ΔAIC	Akaike Weights
(S,P,M)	42.76	7	<0.001	28.76	0.00
(PM)	13.24	5	0.0212	5.16	0.05
(SM)	37.98	5	<0.001	27.98	0.00
(SP)	42.68	6	<0.001	30.68	0.00
(SM,SP)	37.9	4	<0.001	29.90	0.00
(SM,PM)	8.47	3	0.0373	2.46	0.18
(PM,SP)	13.16	4	0.0105	5.16	0.05
(SP,SM,PM)	7.19	2	0.0275	3.19	0.12
(SPM)	0	0	–	0.00	0.61

Note: S, sex; P, predation/no predation; M, marrow.

```
anova(SM.SD, SD.SM.DM, test = "LRT")
Analysis of Deviance Table
   Resid. Df Resid. Dev Df Deviance    Pr(>Chi)
1         4    37.8980
2         2     7.1883  2  30.7097 2.1452e-07 ***
```

We would reject H_0 (although this decision is precluded by the significant three-way interaction) and conclude that marrow and predation are associated when holding sex constant.

11.10.2.3 Four-Way Tables

Four-way models constitute natural extensions of three-way models. However, their interpretation is as difficult as the interpretation of four-way interactions in ANOVA (Sokal and Rohlf 2012). See Agresti (2002) for additional information including worked examples.

11.10.3 Comparison of Methods and Requirements for Valid Inference

As with binomial regression, the likelihood ratio test statistic for a binomial GLM with categorical predictors will have an approximate χ^2 distribution under H_0 given large sample sizes. Although a single rule for defining adequate sample size does not exist, Agresti (2012) suggested that fitted cell counts should be not less than 1 and that not more than 20% of fitted counts should be less than 5. The appropriateness of log-linear models can be assessed with GLM diagnostic plots (Section 9.20.7), checks for over-dispersion (Section 9.20.7), and likelihood ratio tests for log-linear model adequacy as described above. The validity of logit models can also be checked with GLM diagnostic plots, goodness of fit tests such as the Hosmer Lemeshow test (see Section 9.20), and ROC curves (Section 9.20.7).

Unlike logit models, a log-linear approach combines experimental unit responses within a category. Thus, the number of log-linear responses will be the number of cells in the contingency table and the response variable outcomes will be counts from the categorical intersections. Because of this, causal relationships cannot be specified among table dimensions in log-linear models. It follows that log-linear models should only be used with tables comprised of response variables. If a table includes both response and explanatory variables, it will be better to use a logit approach for meaningful interpretations of causality. Log-linear models with categorical predictors will have equivalent logit models and logits can be created from log-linear models with a dichotomous response variable. Important extensions to GLMs, including logit analysis of paired data, random and mixed effect models, and ordinal variables are discussed in Agresti (2012). Bayesian approaches to GLMs including categorical log-linear models are discussed by Gelman et al. (2003).

11.11 Summary

- Tabular analyses are used when both the response and predictors variables are categorical, although the designation of explanatory variables is frequently undesirable or necessary. Of great importance in these analyses are the binomial distribution, and its extension, the multinomial distribution. As noted in Chapter 3, the binomial distribution describes the probability for a particular number of successes given a set number of independent identical trials and the probability of an individual success, π. The multinomial distribution describes the probability for c distinct possible outcomes, whose intersection must be the product of their marginal probabilities for success (due to independence) and whose union must be 1.

- Inferential tests for π may involve three possible null hypotheses: $\pi = \pi_0$, all $\pi_i = \pi_{i0}$, or $\pi_1 = \pi_2$ (tests of the second type encompass tests of the third type). These hypotheses can be addressed with three possible methods: the score test (which can be extended to a χ^2 test), the Wald test, and the likelihood ratio test. Confidence interval inferences for π are problematic because binomial outcomes are discrete and absolute interval limits (0, 1) exist. Four methods are described that provide alternatives to the conventional (and unreliable) Wald approach.

- Data from two- and three-way tabular layouts are typically presented in two-dimensional contingency tables. These can be summarized with respect to joint probabilities (the number of counts divided by the total number of counts), marginal probabilities (the total for a row or column divided by the total number of counts), and conditional probabilities obtained after subsetting a variable or variables from a table with respect to levels in another variable.

- Inferences for multiway tables include hypothesis tests for the independence of array dimensions. If rows and columns in a 2×2 table are independent, then their true odds ratio will equal one. Hypothesis testing in two-way tabular layouts is usually accomplished with one of the three approaches: the score test (i.e., Pearson's χ^2 test), the likelihood ratio test, or Fisher's exact test. The latter is usually used in the context of small sample sizes because it has an exact null distribution for any sample size. χ^2 and likelihood ratio approaches can be naturally extended to higher-dimensional formats, although they require the consideration of conditional and marginal counts and probabilities.

- The unifying format of generalized linear models can be applied to tabular analysis. Logit analyses, which assume binomial errors and use a canonical logit link function, require the examination of individual successes and failures. Log-linear models use counts from tables directly, assume Poisson errors, and use a canonical log link function. Logit models are more useful when both response and explanatory variables occur in a table. Log-linear models allow a user to identify complex forms of dependence among a set of response variables.

EXERCISES

1. Examine and contrast the binomial distributions depicted in `see.bin.tck()` and `see.mnom.tck()`. Do they represent the same thing? Why?

2. In 2001, it was estimated that 56,200 Americans would be diagnosed with non-Hodgkin's lymphoma and that 26,300 of these would die from it (Cerhan et al. 2002).

   Estimate the probability of death given the diagnosis of Hodgkin's lymphoma and calculate a 99% confidence interval for π by hand (without `ci.p`) using the Agresti–Coull and score procedures. For the score method, confidence bounds can be obtained by finding the roots of $ax^2 + bx + c$ using the quadratic equation. This requires letting $a = 1 + \dfrac{z^2_{(1-\alpha/2)}}{n}$, $b = -2\hat{\pi} - \dfrac{z^2_{(1-\alpha/2)}}{n}$, and $c = \hat{\pi}^2$. Verify your results using the function ci.p.

3. Hegazy (1990) found that a random sample of individuals from endangered Mediterranean shrub *Cleome droserifolia* produced approximately 10,000 seeds, but that only 63 of these seeds germinated. Calculate the approximate likelihood ratio confidence interval for the true probability of germination by hand. Confirm your result with ci.p.

4. Counts of GYE grizzly bear litter sizes with 1, 2, 3, and 4 cubs spanning the period 1973 through 2008 can be found in the dataset bear. From 1967 to 1971, the proportions of grizzly bear litter sizes with 1, 2, 3, and 4 cubs in Glacier National Park (not a part of the GYE) were 0.33, 0.53, 0.14, and 0.0, respectively (Martinka 1974). Were the GYE proportions different from the Glacier proportions?

5. Branson (2005) reported that the Reveal™ G2 serum rapid HIV test had a sensitivity of 0.998 and a specificity of 0.991. Interpret the meaning of these values.

6. In the late 1980s, The U. S. veterans administration studied the effect of the antiviral drug AZT on AIDS symptoms for 338 veterans who were just beginning to express symptoms of the disease. AZT treatment was withheld on a random component of veterans until helper T cells showed even greater depletion, while the remaining subjects received the drug immediately. The subjects were also classified by race. The data are contained in the dataframe aids.

a. Create a 2×2 table from the data.

b. Calculate and interpret the odds ratio.

c. Run the Pearson χ^2 test and the likelihood ratio test for independence by hand (using **R** to help).

7. Are the analyses used in Example 11.9 valid? Why or why not?

★ 8. Agresti (2012) discussed tabular analyses concerning death penalty sentences given for homicide convictions in Florida from 1976 to 1987 (Radelet and Pierce 1991). He examined three categorical variables: the defendant's race {Black, White}, the victims' race {Black, White}, and the death penalty verdict {Yes, No}. The data are contained in the dataframe death.penalty.

a. Create two partial tables conditioning by victims' race, that is, cross-classify by the defendant's race and verdict while holding the victims' race constant.

b. Run two-way analyses on the partial tables using Pearson's χ^2 test and the likelihood ratio test.

c. Calculate and interpret the odds ratios for each table.

d. Sum the partial tables to create a single 2×2 marginal table. This should be a defendant race by the verdict two-way table.

e. Redo (b) and (c) using the marginal table from (d). Compare your results. Create a Paik diagram using the function paik to help you summarize your findings.

f. Click on *Simpson's paradox* tab in Chapter 11 pulldown menu for additional insights on this problem. A complete summary of the output can be obtained by typing vignette("simpson"). Change your answers (without plagiarizing) as necessary.

★   9. Schoener (1968) examined the resource partitioning of *Anolis* lizards on the
          Caribbean island of South Bimini. He cross-classified lizard counts in habitats
          (branches in trees) with respect to three variables: lizard species {*A. sagrei, A. disti-*
          chus}, branch height {high and low}, and branch size {small and large}. His data are
          contained in the dataframe anolis.

      a. Conduct a log-linear analysis of the data. This analysis will include model
         specification, model comparison, model selection, optimal model checking,
         and optimal model interpretation including tests of independence.

      b. Conduct a logit analysis (using *A. sagrei* presence as a success) using the same
         steps.

      c. Contrast the analysis in (a) and (b). In retrospect, which is more appropriate?

Appendix: A Mathematical Reference

A.1 Algebra Rules Frequently Used in Mathematical Statistics

A.1.1 General Rules for Arithmetic Operations

$$a(b+c) = ab + ac \qquad \frac{a}{b} + \frac{c}{d} = \frac{ad+cb}{bd}$$

$$\frac{a+c}{b} = \frac{a}{b} + \frac{c}{b} \qquad \frac{a}{b} \Big/ \frac{c}{d} = \frac{a}{b} \times \frac{d}{c} = \frac{ad}{bc}$$

A.1.2 Special Products

$$(a+b)^2 = a^2 + 2ab + b^2 \qquad\qquad (a-b)^2 = a^2 - 2ab + b^2$$

$$(a+b)^3 = a^3 + 3a^2b + 3ab^2 + b^3 \qquad (a-b)^3 + a^3 - 3a^2b + 3ab^2 - b^3$$

A.1.3 Factoring

$$a^2 - b^2 = (a+b)(a-b) \qquad\qquad a^3 + b^3 = (a+b)(a^2 - ab + b^2)$$

$$a^3 - b^3 = (a-b)(a^2 + ab + b^2)$$

A.1.4 Exponents

$$a^0 = 1 \qquad\qquad a^1 = a \qquad\qquad a^m a^n = a^{m+n}$$

$$(a^m)^n = a^{mn} \qquad (ab)^n = a^n b^n \qquad \left(\frac{a}{b}\right)^n = \frac{a^n}{b^n}, \quad b \neq 0$$

$$\frac{a^m}{a^n} = a^{m-n} \qquad \frac{1}{a^n} = a^{-n} \qquad \frac{1}{a} = a^{-1}$$

$$\sqrt[n]{a} = a^{1/n} \qquad -\sqrt[n]{a} = a^{-1/n} = \frac{1}{a^n}$$

A.1.5 Radicals

$$\sqrt[n]{ab} = \sqrt[n]{a}\sqrt[n]{b} \qquad \sqrt[n]{\frac{a}{b}} = \frac{\sqrt[n]{a}}{\sqrt[n]{b}} \qquad \sqrt[n]{a^m} = (a^m)^{1/n} = a^{m/n}$$

$$\sqrt[n]{\sqrt[n]{a}} = ((a)^{1/mn})^{1/n} = a^{1/mn} = \sqrt[mn]{a}$$

A.1.6 Logarithms

$$\log_b b = 1 \qquad\qquad \log_b 1 = 0$$

$$\log_b(MN) = \log_b(M) + \log_b(N) \qquad \log_b\left(\frac{M}{N}\right) = \log_b(M) - \log_b(N)$$

$$\log_b(N^y) = y\log_b(N) \qquad \log_b\left(\frac{1}{N}\right) = \log_b N^{-1} = -\log_b N$$

$$\ln = \log_e \qquad \ln(e) = 1 \qquad e^{\ln} = 1$$

A.1.7 Quadratic Formula

If $ax^2 + bx + c = 0$ and $a \neq 0$, then

$$x = \frac{-b \pm \sqrt{b^2 - 4ac}}{2a}.$$

A.1.8 Factorial Rules

$$n! = n(n-1)(n-2)\ldots[n-(n+1)] \qquad 0! = 1 \qquad (n+1)! = (n+1)n!$$

If n and j are integers with $n \geq j \geq 0$, then we define

$$\binom{n}{j} = \frac{n!}{(n-j)!j!}.$$

A.1.9 Additive Series Rules

Where a and b are variables and c is a constant

$$\sum_{i=1}^{n}(a_i + b_i) = \sum_{i=1}^{n}a_i + \sum_{i=1}^{n}b_i \qquad \sum_{i=1}^{n}(a_i - b_i) = \sum_{i=1}^{n}a_i - \sum_{i=1}^{n}b_i$$

$$\sum_{i=1}^{n}ca_i = c\sum_{i=1}^{n}a_i \qquad\qquad \sum_{i=1}^{n}c = nc$$

$$\sum_{i=1}^{n}i = \frac{n(n+1)}{2} \qquad\qquad \sum_{i=1}^{n}r^{i-1} = \frac{1-r^n}{1-r}, \quad r \neq 0, r \neq 1$$

Let a_n be an arithmetic sequence with a common difference within the sequence, then

$$\sum_{i=1}^{n} a_i = \frac{n}{2}(a_1 + a_n)$$

$$\sum_{i=1}^{n} a_i = \sum_{i=1}^{n} a_1 r^{i-1} = a_1 \frac{1 - r^n}{1 - r}, \quad r \neq 0, r \neq 1.$$

A.1.10 Infinite Geometric Series

If $|r| < 1$, then

$$\sum_{i=0}^{\infty} r^i = \frac{1}{1 - r}$$

$$\sum_{i=1}^{\infty} r^i = \frac{r}{1 - r}$$

$$\sum_{i=1}^{\infty} ar^{i-1} = \frac{a}{1 - r}$$

A.1.11 Product Series Rules

Where c is a constant

$$\prod_{i=1}^{n} c = c^n$$

$$\prod_{i=1}^{n} e^{x_i} = e^{\sum x_i}$$

A.2 Calculus Rules Frequently Used in Mathematical Statistics

A.2.1 Differentiation

Where c is a constant, $f(x)$ and $g(x)$ represent differentiable functions involving x, and $f'(x)$ is the first derivative of $f(x)$.

$$\frac{d}{dx} f(x) = \text{change in the function } f(x) \text{ with respect to the variable } x.$$

$$\frac{d}{dx}c = 0 \qquad\qquad \frac{d}{dx}cf(x) = cf'(x).$$

$$\frac{d}{dx}[f(x) + g(x)] = f'(x) + g'(x) \qquad\qquad \frac{d}{dx}[f(x) - g(x)] = f'(x) - g'(x)$$

$$\frac{d}{dx}e^x = e^x \qquad\qquad \frac{d}{dx}\ln(x) = \frac{1}{x}, \quad \text{for } x > 0$$

$$\frac{d}{dx}c^x = c^x \ln(c) \qquad\qquad \frac{d}{dx}\log_b(x) = \frac{1}{x\ln(b)}, \quad \text{for } x > 0$$

A.2.2 Power Rule

$$\frac{d}{dx}x^c = cx^{c-1}.$$

Example: $\dfrac{d}{dx}x^5 = 5x^4.$

A.2.3 Chain Rule

$$\frac{d}{dx}[f(g(x))] = f'(g(x))g'(x).$$

Example: $\dfrac{d}{dx}[(1+x)^{-1}] = -1(1+x)^{-2} \cdot 1 = -\dfrac{1}{(1+x)^2}.$

A.2.4 Product Rule

$$\frac{d}{dx}[f(x)g(x)] = f'(x)g(x) + g'(x)f(x).$$

Example 1: $\dfrac{d}{dx}(xe^x) = 1(e^x) + x(e^x) = e^x(1 + x).$

Example 2:
$$\frac{d}{dx}\left(\frac{e^x}{1+x}\right) = \frac{d}{dx}[e^x(1+x)^{-1}]$$

$$= e^x(1+x)^{-1} + e^x\left[-\frac{1}{(1+x)^2}\right]$$

$$= \frac{e^x(1+x)}{(1+x)^2} - \frac{e^x}{(1+x)^2} = \frac{xe^x}{(1+x)^2}$$

A.2.5 L'Hôpital's Rule

If we have an indeterminate form near a, for example, 0/0 or ∞/∞, then

$$\lim_{x \to a} \frac{f(x)}{g(x)} = \lim_{x \to a} \frac{f'(x)}{g'(x)}$$

A.2.6 Antiderivatives

A function F is called an antiderivative of f for an interval I if $F'(x) = f(x)$ for all x in I. In the table below, c is a constant.

Function $f(x)$	Antiderivative $F(x)$
$cf(x)$	$cF(x)$
$f(x) + g(x)$	$F(x) + G(x)$
$x^n, n \neq -1$	$\dfrac{x^{n+1}}{n+1}$
$1/x$	$\ln(x)$
e^x	$\dfrac{e^x}{x'} = e^x$

A.2.7 Integration

$$\int f(x)dx = F(x) + c, \quad \text{where } c \text{ is an unknown constant.}$$

A.2.8 Fundamental Theorem of Calculus

If f is continuous on $[a, b]$ and $a \geq x \geq b$, then the area beneath a curve described by $f(x)$ that is between $x=a$ and $x = b$ is described by

$$\int_b^a f(x)dx = F(a) - F(b).$$

Example: Integrate $\int_4^5 3x^2 dx$.

$$\int_4^5 3x^2 dx = x^3 \Big|_4^5 = 5^3 - 4^3 = 125 - 64 = 61.$$

This is the area under the curve described by $3x^2$ between $x = 5$ and $x = 4$.

A.2.9 Integration by u Substitution

This approach corresponds to the chain rule for derivatives.

$$\int f(g(x))g'(x)dx = \int f(u)du.$$

Example: Integrate $\int 2x\sqrt{1+x^2}\,dx$.

We will let $u = 1 + x^2$. Then, we have $du = 2x\,dx$ and $dx = du/2x$. Substituting this back into our integral, we have

$$\int 2x\sqrt{u}\,\frac{du}{2x} = \int \frac{2x\sqrt{u}}{2x}\,du = \int \sqrt{u}\,du = \frac{u^{3/2}}{3/2} + c = \frac{2u^{3/2}}{3} + c = \frac{2(1+x^2)^{3/2}}{3} + c$$

A.3 Matrix Algebra

A.3.1 Terms

- *A matrix* is an array with row and column dimensions. The dimensionality of a matrix is defined by number of rows $\times$ number of columns. Matrices are denoted with capitalized bold-faced letters; that is, matrix A is denoted as **A**.

- *Vectors* are matrices composed of a single row or column. They are denoted with bold lowercase letters, for example, **a**.

- *Scalars* are single-number variables. They are denoted as lowercase italicized letters, for example, s.

- *Square matrices* will have the same number of rows and columns.

- *Symmetric matrices* will have mirrored upper and lower triangles. For instance, the matrix

$$\begin{bmatrix} 2 & 0 & 5 \\ 0 & -1 & 4 \\ 5 & 4 & 3 \end{bmatrix}$$

  is symmetric.

- *Diagonal matrices* consist of zeros except for their diagonal elements. For instance, $\begin{bmatrix} 5 & 0 \\ 0 & -3 \end{bmatrix}$.

- *Identity matrices* are square diagonal matrices with ones on the diagonal. For instance, $\begin{bmatrix} 1 & 0 \\ 0 & 1 \end{bmatrix}$ is an identity matrix.

A.3.2 Addition and Subtraction

Adding or subtracting a scalar to/from a matrix is a simple proposition. One simply adds or subtracts the scalar from every element in the matrix. The resulting array will have the same dimensions as the matrix in the arithmetic. That is

If s is a scalar and $\mathbf{A} = \begin{bmatrix} a & c \\ b & d \end{bmatrix}$, then $s \pm \mathbf{A} = \begin{bmatrix} (s \pm a) & (s \pm c) \\ (s \pm b) & (s \pm d) \end{bmatrix}$.

When adding or subtracting two matrices, we require that they have the same dimensions. Given this, the parallel elements in the matrices are simply added or subtracted. The resulting array will be a matrix with the same dimensions as the matrices being added or subtracted.

That is, if $\mathbf{A} = \begin{bmatrix} a & c \\ b & d \end{bmatrix}$ and $\mathbf{B} = \begin{bmatrix} e & g \\ f & h \end{bmatrix}$, then $\mathbf{A} - \mathbf{B} = \begin{bmatrix} (a \pm e) & (c \pm g) \\ (b \pm f) & (d \pm h) \end{bmatrix}$.

A.3.3 Transposition

In a transposed matrix (e.g., $\mathbf{A}'$), the rows become the columns and the columns become the rows. That is, if $\mathbf{A} = \begin{bmatrix} a & c \\ b & d \end{bmatrix}$, then $\mathbf{A}' = \begin{bmatrix} a & b \\ c & d \end{bmatrix}$. Transposition in **R** is accomplished using the function t. For instance

```
A <- matrix(2, 2, data = c(2, 1, 3, 4))
A
      [,1] [,2]
[1,]    2    3
[2,]    1    4
t(A)
      [,1] [,2]
[1,]    2    1
[2]     3    4
```

A.3.4 Multiplication

To multiply a matrix by a scalar, one simply finds the product of the scalar and each element in the matrix. The resulting array will have the same dimensions as the matrix in multiplication.

Thus, if $\mathbf{A} = \begin{bmatrix} a & c \\ b & d \end{bmatrix}$ and s is a scalar, then $s \cdot \mathbf{A} = \begin{bmatrix} sa & sc \\ sb & sd \end{bmatrix}$.

When multiplying two matrices, the number of columns in the first matrix has to be equal to the number of rows in the second matrix. If this is not true, then the solution to the multiplication does not exist.

The product **AB** is calculated by finding the cross-products of rows of **A** with the columns of **B** and then summing the cross-products.

That is, if $\mathbf{A} = \begin{bmatrix} a & c \\ b & d \end{bmatrix}$ and $\mathbf{b} = \begin{bmatrix} e \\ f \end{bmatrix}$ then $\mathbf{Ab} = \begin{bmatrix} ae + cf \\ be + df \end{bmatrix}$.

Similarly, if $\mathbf{A} = \begin{bmatrix} a & c \\ b & d \end{bmatrix}$ and $\mathbf{B} = \begin{bmatrix} e & g \\ f & h \end{bmatrix}$, then $\mathbf{AB} = \begin{bmatrix} (ae + cf) & (ag + ch) \\ (be + df) & (bg + dh) \end{bmatrix}$.

For instance, if $\mathbf{A} = \begin{bmatrix} 1 & 3 \\ 2 & 4 \end{bmatrix}$ and $\mathbf{B} = \begin{bmatrix} 2 & -2 \\ 1 & 0 \end{bmatrix}$, then $\mathbf{AB} = \begin{bmatrix} (1 \times 2) + (3 \times 1) & (1 \times -2) + (3 \times 0) \\ (2 \times 2) + (4 \times 1) & (-2 \times 2) + (4 \times 0) \end{bmatrix} = \begin{bmatrix} 5 & -2 \\ 8 & -4 \end{bmatrix}.$

We can perform this multiplication in **R** using the following commands:

```
A <- matrix(2, 2, data = c(1, 2, 3, 4))
B <- matrix(2, 2, data = c(2, 1, -2, 0))
AB <- A%*%B
AB
      [,1] [,2]
[1,]    5   -2
[2,]    8   -4
```

A.3.5 Inverse and Determinant

In a square matrix, a *determinant* allows a planar interpretation of the matrix components. For instance, a 2×2 matrix with determinant -2, when applied to a plane, will transform the region into one with twice the area and reverse its orientation.

Let $\mathbf{A} = \begin{bmatrix} a & c \\ b & d \end{bmatrix}$, then the *determinant* of the matrix is

$$D = \det(\mathbf{A}) = ad - bc.$$

The *inverse* of a matrix $\mathbf{A}$, denoted by $\mathbf{A}^{-1}$ is the matrix equivalent to scalar division. It can be found by rearranging the elements of the matrix and dividing by the determinant of the matrix. Specifically

$$\mathbf{A}^{-1} = \begin{bmatrix} d/D & -c/D \\ -b/D & a/D \end{bmatrix}.$$

Determinants and inverses are much more difficult to calculate for larger matrices. Using **R**, we can calculate the determinant by using the det function:

```
A<- matrix(data = c(1, 3, 2, 4), ncol = 2, nrow = 2, byrow = T)
det(A)
[1] -2
```

and find $\mathbf{A}^{-1}$ by using the solve command:

```
A.inv <- solve(A)
      [,1] [,2]
[1,]   -2  1.5
[2,]    1 -0.5
```

A.3.5.1 Eigenanalysis

In eigenanalysis, we find the roots of the equations resulting from

$$\mathrm{Det}(\mathbf{S} - \lambda\mathbf{I}) = 0.$$

where λ is an unknown scalar, S is a symmetric matrix, and I is an identity matrix with the same dimensions as S.

A.3.5.2 *Orthogonality*

An orthogonal matrix allows a large number of statistical linear algebra operations including QR decomposition. Among many other important characteristics, if the columns of a matrix A are orthogonal, then $A^{-1} = A'$.

A.3.5.2 *Positive Definite Matrices*

A positive definite matrix is analogous to a positive real scalar. Among many other important characteristics, if a matrix, A, is positive definite, then all of its eigenvalues will be positive.

A.4 Set Theory and Probability

$$0 \le P(A) \le 1 \qquad P(A) - 1 = P(A').$$

A.4.1 Unions

$$P(A \cup B) = P(A) + P(B) - P(A \cap B).$$

$$P(A \cup B \cup C) = P(A) + P(B) + P(C) - P(A \cap B) - P(A \cap C) - P(B \cap C) + P(A \cap B \cap C).$$

A.4.2 Intersections

$$A - B = A \cap B'.$$

$$P(A \cap B) = P(B)P(A|B) = P(A)P(B|A).$$

A.4.3 Conditional Probability

$$P(A|B) = \frac{P(A) \cap P(B)}{P(B)}, \quad P(B) \ne 0.$$

A.4.4 Bayes' Theorem (General)

$$f(\theta|x) = \frac{f(x|\theta)f(\theta)}{f(x)}.$$

A.4.5 Independence

- If $P(A|B) = P(A)$ or $(B|A) = P(B)$, then A and B are independent.
- If $P(A \cap B) = P(A)P(B)$, then A and B are independent.
- If A and B are independent, then A and B' are independent, A' and B are independent, and A' and B' are independent.

A.4.6 Set notation

The notation (a, b) indicates that the set includes all possible numbers between a and b, conversely. "[" and "]" are inclusive brackets.

A.5 Statistical Expectations

Where c is a constant (this includes parameters), X and Y are random variables $E(X) = \mu$, and Var $(X) = \sigma^2$.

$$\text{Var}(X) = E(X^2) - E(X)^2 = \sigma^2$$

$$E(X^2) = \text{Var}(X) + E(X)^2 \qquad E(X)^2 = E(X^2) - \text{Var}(X)$$

$$E(X + Y) = E(X) + E(Y) \qquad E(cX) = cE(X)$$

$$\text{Var}(cX) = c^2\text{Var}(X) \qquad E(c) = c$$

$$E\sum_{i=1}^{n}(X_i) = \sum_{i=1}^{n}E(X_i) = n\mu \qquad E\prod_{i=1}^{n}(X_i) = \prod_{i=1}^{n}E(X_i) = \mu^n$$

Given: $E(X) = \mu_X$ and $E(Y) = \mu_Y$

$$\begin{aligned}
\text{Cov}(X, Y) &= E[(X - \mu_X)(Y - \mu_Y)] \\
&= E(XY - X\mu_Y - Y\mu_X + \mu_X\mu_Y) \\
&= E(XY) - \mu_Y E(X) - \mu_X E(Y) + \mu_X\mu_Y \\
&= E(XY) - 2\mu_Y\mu_X + \mu_X\mu_Y \\
&= E(XY) - \mu_X\mu_Y.
\end{aligned}$$

If X and Y are independent, then $\text{Cov}(X,Y) = 0$.
Let $X \sim (\mu_X, \sigma_X^2)$ and $Y \sim (\mu_Y, \sigma_Y^2)$.

If $Z = X + Y$

$$\mu_Z = \mu_X + \mu_Y.$$

$$\sigma_Z^2 = \sigma_X^2 + \sigma_Y^2 + 2\,\text{Cov}(X,Y)$$

If $Z = X - Y$

$$\mu_Z = \mu_X - \mu_Y.$$

$$\sigma_Z^2 = \sigma_X^2 + \sigma_Y^2 - 2\,\text{Cov}(X,Y)$$

A.6 Exponential Family

The form for all distributions that are members of the exponential family can be written as

$$f(y) = \exp\left\{\frac{y\theta - b(\theta)}{\phi} + c(y,\phi)\right\},$$

where $b(.)$ and $c(.)$ are known functions. We call θ the *canonical parameter* of the distribution, and ϕ the *dispersion parameter*.

In this framework, $b(.)$ will only be a function of θ (it will not include ϕ and y), and $c(.)$ will only be a function of y and ϕ (it will not include θ). The function $c(y,\phi)$ serves as a normalizing term to make the function sum or integrate to 1 along its support (Hardin and Hilbe 2007).

The first derivative of $b(\theta)$ with respect to θ, $b'(\theta)$ will be the mean of the particular exponential distribution. The second derivative of $b(\theta)$, $b''(\theta)$, is called the *variance function*, and is denoted as $V(\theta)$. The variance of the distribution of Y will be the product of the variance function and ϕ. That is, $\text{Var}(Y) = V(\theta)\phi$.

A.6.1 Normal Distribution

We can demonstrate that the normal distribution is a member of the exponential family by utilizing the principle $e^{\ln(x)} = x$. That is,

$$f(y) = \frac{1}{\sqrt{2\pi\sigma^2}}\,e^{-\frac{(y-\mu)^2}{2\sigma^2}} = \exp\left[\ln\left\{\frac{1}{\sqrt{2\pi\sigma^2}}\,e^{-\frac{(y-\mu)^2}{2\sigma^2}}\right\}\right].$$

We take the natural log of the pdf first.

$$\ln\left\{\frac{1}{\sqrt{2\pi\sigma^2}}\,e^{-\frac{(y-\mu)^2}{2\sigma^2}}\right\} = -\frac{1}{2}\ln(2\pi\sigma^2) - \frac{(y-\mu)^2}{2\sigma^2}.$$

Then, we exponentiate the logged pdf.

$$f(y) = \exp\left\{-\frac{1}{2}\ln(2\pi\sigma^2) - \frac{(y-\mu)^2}{2\sigma^2}\right\} = \exp\left\{\frac{-(y^2 - 2y\mu + \mu^2)}{2\sigma^2} - \frac{1}{2}\ln(2\pi\sigma^2)\right\}$$

$$= \exp\left\{\frac{2y\mu - \mu^2}{2\sigma^2} - \frac{y^2}{2\sigma^2} - \frac{1}{2}\ln(2\pi\sigma^2)\right\} = \exp\left\{\frac{y\mu - \mu^2/2}{\sigma^2} - \frac{y^2}{2\sigma^2} - \frac{1}{2}\ln(2\pi\sigma^2)\right\}.$$

We can see that the normal pdf fits the form of the exponential family.

The canonical parameter, θ, equals μ. In addition, $b(\theta) = \mu^2/2$, $\phi = \sigma^2$, and $c(y, \phi) = -(y^2/2\sigma^2) - (1/2)\ln(2\pi\sigma^2)$.

Thus

$$E(Y) = b'(\theta) = \frac{d}{d\theta}\frac{\theta^2}{2} = \theta = \mu.$$

$$V(\theta) = b''(\theta) = \frac{d}{d\theta}\theta = 1, \quad \text{and} \quad \text{Var}(Y) = V(\theta)\sigma^2 = 1 \times \sigma^2 = \sigma^2.$$

As a link function, the canonical parameter μ is conventionally given the value 1. As a result, we say that general linear models have an identity link.

A.6.2 Binomial Distribution

$$f(y) = \binom{n}{y}\pi^y(1-\pi)^{n-y} \qquad\qquad f(y) = \binom{n}{y}\pi^y(1-\pi)^{-y}(1-\pi)^n$$

$$f(y) = \binom{n}{y}\pi^y\frac{1}{(1-\pi)^y}(1-\pi)^n \qquad f(y) = \binom{n}{y}\left(\frac{\pi}{1-\pi}\right)^y(1-\pi)^n.$$

Taking the exponential of the log, we have

$$f(y) = \exp\left\{\ln\left[\binom{n}{y}\left(\frac{\pi}{1-\pi}\right)^y(1-\pi)^n\right]\right\} = \exp\left\{\ln\binom{n}{y} + y\ln\left(\frac{\pi}{1-\pi}\right) + n\ln(1-\pi)\right\}$$

$$= \exp\left\{\frac{y\ln(\pi/(1-\pi)) - [-n\ln(1-\pi)]}{1} + \ln\binom{n}{y}\right\}.$$

In this case, the canonical parameter is $\theta = \ln\left(\frac{\pi}{1-\pi}\right)$. Recall that this is the logit function (and the logistic inverse cdf). In addition, we have $b(\theta) = -n\ln(1-\pi)$, $\phi = 1$, and $c(y, \phi) = \ln\binom{n}{y}$.

Manipulating the logit function, we have

$$e^\theta = \frac{\pi}{1-\pi}, \quad e^\theta - \pi e^\theta = \pi, \quad e^\theta = \pi + \pi e^\theta, \quad e^\theta = \pi(1+e^\theta), \quad \pi = \frac{e^\theta}{1+e^\theta}$$

and

$$1 - \pi = 1 - \frac{e^\theta}{1+e^\theta} = \frac{1}{1+e^\theta}.$$

Thus, if $\theta = \ln\left(\dfrac{\pi}{1-\pi}\right)$, then $\pi = \dfrac{e^\theta}{1+e^\theta}$ and $1 - \pi = \dfrac{1}{1+e^\theta}$.

As a result, $b(\theta) = -n\ln(1-\pi) = -n\ln\left(\dfrac{1}{1+e^\theta}\right) = -n\ln(1+e^\theta)^{-1} = n\ln(1+e^\theta)$.

Using this information, we can now find the mean and variance function

$$E(Y) = b'(\theta) = \frac{d}{d\theta} n\ln(1+e^\theta) = n\frac{e^\theta}{1+e^\theta} = \pi n.$$

$$V(\theta) = b''(\theta) = \frac{d}{d\theta} ne^\theta(1+e^\theta)^{-1} = n\{e^\theta(1+e^\theta)^{-1} + e^\theta[-e^\theta(1+e^\theta)^{-2}]\}$$

$$= n\left[\pi - \frac{e^{2\theta}}{(1+e^\theta)^2}\right] = n\left[\pi - \frac{e^\theta e^\theta}{(1+e^\theta)^2}\right] = n\pi - n\pi^2 = n\pi(1-\pi),$$

$$\text{Var}(Y) = \phi V(\theta) = n\pi(1-\pi).$$

A.6.3 Poisson Distribution

Taking the exponential of the log, we have

$$f(y) = \exp\left\{\ln\left[\frac{\lambda^y e^{-\lambda}}{y!}\right]\right\} = \exp\{[\ln(\lambda^y) + \ln(e^{-\lambda})] - \ln(y!)\} = \exp\{[y\ln(\lambda) - \lambda] - \ln(y!)\}$$

$$= \exp\left\{\frac{y\ln(\lambda) - \lambda}{1} - \ln(y!)\right\}.$$

In this case, the canonical parameter $\theta = \log(\lambda)$, $b(\theta) = \lambda$, $\phi = 1$, and $c(y, \phi) = \ln(y!)$. Further, since $\theta = \log(\lambda)$, $e^\theta = \lambda$. Thus

$$E(Y) = b'(\theta) = \frac{d}{d\theta} e^\theta = e^\theta = \lambda.$$

$$V(\theta) = b''(\theta) = \frac{d}{d\theta} e^\theta = e^\theta = \lambda, \quad \text{and} \quad \text{Var}(Y) = V(\theta) \times \phi = \lambda \times 1 = \lambda.$$

Note that while a large number of important pdfs fit into the exponential family format—for example, normal, lognormal, chi-squared, exponential, gamma, geometric, Poisson, and the binomial family (assuming a fixed number of trials, n)—a number of others do not, including the t-, Weibull, and F-distributions.

References

Aaboe, A. 1974. Scientific astronomy in antiquity. *Philosophical Transactions of the Royal Society* 276(1257): 21–42.

Abdullah, M. B. 1990. On a robust correlation coefficient. *The Statistician* 39: 455–460.

Achinstein, P. 2004. *Science Rules: A Historical Introduction to Scientific Methods.* Johns Hopkins University Press, Baltimore, MD.

Agresti, A. 2002. *Categorical Data Analysis*, 2nd ed. Wiley, New York.

Agresti, A. 2012. *An Introduction to Categorical Data Analysis*, 3rd ed. Wiley, New York.

Agresti, A. and Coull, B. A. 1998. Approximate is better than "exact" for interval estimation of binomial proportions. *The American Statistician* 52: 119–126.

Agresti, A. and Pendergrast, J. 1986. Comparing mean ranks for repeated measures data. *Communications in Statistics—Theory and Methods* 15: 1417–1433.

Aho, K. 2006. *Alpine Ecology and Subalpine Cliff Ecology in the Northern Rocky Mountains.* Doctoral dissertation, Montana State University, 458 pp.

Aho, K. 2012. Management of introduced mountain goats in Yellowstone National Park (vegetation analysis along a mountain goat gradient). PMIS# 105289. Report prepared for USDA National Park Service, 150 pp.

Aho, K. 2013. *Asbio*; a collection of statistical tools for biologists. R package version 1.0. http://cran.r-project.org/web/packages/asbio/index.html

Aho, K. and Bowyer, R.T. 2015. Confidence intervals for ratios of proportions: Implications for selection ratios. *Methods in Ecology and Evolution* 6: 121–132.

Aho, K. and Weaver, T. 2010. Ecology of alpine nodes: Environments, communities, and ecosystem evolution (Mount Washburn; Yellowstone National Park). *Arctic, Antarctic, and Alpine Research* 40(2): 139–151.

Aho K., Derryberry, D., and Peterson, T. 2014. Model selection for ecologists: The worldviews of AIC and BIC. *Ecology* 95(3): 631–636.

Aho, K., Weaver, T., and Regle, S. 2011. Identification and siting of native vegetation types on disturbed land: Demonstration of statistical methods. *Journal of Applied Vegetation Science* 14(2): 277–290.

Aho, K., Huntly, N., Moen, J., and Oksanen, T. 1998. Pikas (*Ochotona princeps*: Lagomorpha) as allogenic engineers in an alpine ecosystem. *Oecologia* 114(3): 405–409.

Akaike, H. 1973. Information theory and an extension of the maximum likelihood principle. In Petrov, B. N., and Caski, S., eds., *Proceedings of the Second International Symposium on Information Theory*, Akademiai Kaido, Budapest, 267–281.

Albert, J. 2007. Bayesian computation with R. Springer, New York, NY.

Allard, R. W. 1966. *Principles of Plant Breeding.* John Wiley & Sons, New York.

Altman, D. G. 1991. *Practical Statistics for Medical Research.* Chapman-Hall, London.

Altman, D. G. 1994. The scandal of poor medical research. *British Medical Journal* 308: 283–284.

Altman, D. G. and Bland, J. M. 2005. Standard deviations and standard errors. *British Medical Journal* 31(7521): 903.

Andersen, P.K., Borgan, O., Gill, R.D., and Keiding, N. 1993. *Statistical Models Based on Counting Processes.* Springer-Verlag, New York, NY.

Andrew, N. L. and Underwood, A. J. 1993. Density dependent foraging in the sea urchin *Centrostephanus rogersii* on shallow subtidal reefs in New South Wales, Australia. *Marine Ecology Progress Series* 99: 89–98.

Anscombe, F. J. 1973. Graphs in statistical analysis. *American Statistician* 27(1): 17–21.

Antelman, G. 1997. *Elementary Bayesian Statistics.* In Madansky, A., and McCullough, R., eds., Edward Elgar, Cheltenham, UK.

Appleton, D. R., French, J. M., and Vanderpump, M. P. J. 1996. Ignoring a covariate: An example of Simpson's paradox. *The American Statistician* 50(4): 340–341.

Artz, C. P., McFarland, J. B., and Banett, W. O. 1977. Clinical evaluation of gastric freezing for the peptic ulcer. *Annals of Surgery* 159(5): 758–764.

Austin, J. H. 1978. *Chase, Chance and Creativity, the Lucky Art of Novelty*. Columbia University Press, New York, NY.

Ayala, F. J. et al. 2008. *Science, Evolution, and Creationism*. The National Academies Press, Washington, DC.

Bacastow, R. and Keeling, C. D. 1974. Atmospheric carbon dioxide and radiocarbon in the natural carbon cycle II: Changes for AD 1700 to 2070 as deduced from a geochemical model. In Woodwell, G. M., and Pecan, E. V., eds., *Carbon and the Biosphere*. BHNL/CONF 702510. Springfield, VA.

Bain, L. J. and Engelhardt, M. 1992. *Introduction to Probability and Mathematical Statistics*. Duxbury Press, Belmont, CA, USA.

Baio, J. 2012. Prevalence of autism spectrum disorders—Autism and developmental disabilities monitoring network, 14 sites, United States, 2008. *Morbidity and Mortality Weekly Report Surveillance Summaries*, Centers for Disease Control and Prevention, 1–19.

Baker, M. C. and Mewaldt, L. R. 1978. Song dialects as barriers to dispersal in white-crowned sparrows, *Zonotrichia leucophrys nuttalli*. *Evolution* 32(4): 712–722.

Barbour, M. G. 1973. Desert dogma reexamined: Root/shoot productivity and plant spacing. *American Midland Naturalist* 89: 41–57.

Barnard, G. 1951. The theory of information. *Journal of the Royal Statistical Society B* II: 115–149.

Batternik, M. and Wijffles, G. 1983. Een verglijkend vegetatiekundig onderzoek naar de typologies en invloeden van het beheer van 1973 tot 1982 in de Duinweilanden op Terschelling. Report of the Agricultural University, Department of Vegetation Science, Plant Ecology and Weed Science, Wageningen.

Bauer, D. F. 1972. Constructing confidence sets using rank statistics. *Journal of the American Statistical Association* 67: 687–690.

Beals, E. W. 1968. Spatial pattern of shrubs on a desert plain in Ethiopia. *Ecology* 49: 744–746.

Begon, M., Harper, J., and Townsend, C. R. 1996. *Ecology, Individuals, Populations and Communities*, 3rd ed. Blackwell Science, Malden, MA.

Benjamini, Y. and Hochberg, Y. 1995. Controlling the false discovery rate: A practical and powerful approach to multiple testing. *Journal of the Royal Statistical Society Series B* 57: 289–300.

Benson, M. A., Schmalzer, K. M., and Frank, D. W. 2009. A sensitive fluorescence-based assay for the detection of ExoU mediated PLA2 activity. *Clinica Chimica Acta* 411(3–4): 190–197.

Berry, D. A. and Stangl, D. K. 1996. Bayesian methods in health related research. In Berry, D. A., and Stangl, D. K., eds., *Bayesian Biostatistics*. Marcel-Dekker, New York.

Best, D. J. and Roberts, D. E. 1975. Algorithm AS 89: The upper tail probabilities of Spearman's rho. *Applied Statistics* 24: 377–379.

Bialek, W. and Botstein, D. 2004. Introductory science and mathematics education for 21st-century biologists. *Science* 303(5659): 788–790.

Bliss, C. I. and Fisher, R. A. 1953. Fitting the negative binomial distribution to biological data. *Biometrics* 9: 176–200.

Bloch, D. A. and Gastwirth, J. L. 1968. On a simple estimate of the reciprocal of the density function. *The Annals of Mathematical Statistics* 39: 1083–1085.

Boik, R. 1987. The Fisher–Pittman test permutation test, a non-robust alternative to normal theory *F*-tests when variances are heterogeneous. *British Journal of Mathematical and Statistical Psychology* 40: 26–42.

Box, G. E. P. 1976. Science and statistics. *Journal of the American Statistical Association* 71: 791–799.

Box, G. E. P. and Cox, D. R. 1964. An analysis of transformations. *Journal of the Royal Statistical Society B* 26: 211–246.

Box, G. E. P. and Jenkins, G. M. 1970. *Time Series Analysis: Forecasting and Control*. Hoden-Day, London.

Box, G. E. P. and Tiao, G. C. 1973. Bayesian inference in statistical analysis. Wiley, New York.

Box, G. E. P. and Tidwell, P. W. 1962. Transformation of the independent variables. *Technometrics* 4: 531–550.

Box, M. J. 1971. Bias in non-linear estimation. *Journal of the Royal Statistical Society B* 33: 171–201.

Boyd, R. N. 1983. On the current status of the issue of scientific realism. *Erkenntnis* 19: 45–90.

Boyd, R., Gasper, P., and Trout, J. D. 1991. *The Philosophy of Science*. MIT Press, Cambridge.

Bozarth, M. A. and Wise, R. A. 1985. Toxicity associated with long-term intravenous heroin and cocaine self-administration in the rat. *Journal of the American Medical Association* 254: 81–83.

Bozdogan, H. 1987. Model selection and Akaike's information criterion (AIC): The general theory and its analytical extensions. *Psychometrika* 52(3): 345–370.

Bradford, M. M. 1976. A rapid and sensitive method for the quantitation of microgram quantities of protein utilizing the principle of protein-dye binding. *Analytical Biochemistry* 72: 248–254.

Branson, B. M. 2005. Rapid HIV testing: 2005 update. Presentation by the Center for Disease Control. http://www.cdc.gov/hiv/topics/testing/resources/slidesets/pdf/USCA_Branson.pdf. (Accessed 6/1/2013).

Brenner, D. E. et al. 1993. The relationship between cigarette smoking and Alzheimer's disease in a population based case control study. *Neurology* 43: 293–300.

Brockman, H. J. 1996. Satellite male groups in horseshoe crabs, *Limulus polyphemus*. *Ethology* 102(1): 1–21.

Brown, L. D., Cai, T. T., and Dasgupta, A. 2002. Confidence intervals for a binomial proportion and asymptotic expansions. *The Annals of Statistics* 30(1): 160–201.

Brown, W. J. et al. 1996. Women's health in Australia: Establishment of the Australian longitudinal study on women's health. *Journal of Women's Health* 5: 467–472.

Brunner, E., Dette, H., and Munk, A. 1997. Box-type approximations in nonparametric factorial designs. *Journal of the American Statistical Association* 92: 1494–1502.

Burke da Silva, K., Mahan, C., and da Silva, J. 2002. The trill of the chase: Eastern chipmunks call to warn kin. *Journal of Mammalogy* 83: 546–552.

Burnham, K. P. and Anderson, D. R. 2002. *Model Selection and Multimodel Inference, A Practical Information-Theoretic Approach*, 2nd ed. Springer, New York.

Byrd, R. S. and Joad, J. P. 2006. Urban asthma. *Current Opinion in Pulmonary Medicine* 12(1): 68–74.

Canty, A. and Ripley, B. 2012. Boot: Bootstrap R (S-Plus) functions. R package version 1.3–4.

Carmer, S. and Swanson, M. 1973. An evaluation of ten multiple pairwise procedures by Monte-Carlo methods. *Journal of the American Statistical Association* 92: 392–415.

Carnap, R. 1963. *Logical Foundations of Probability*, 2nd ed. University of Chicago Press, Chicago.

Carpenter, J. and Bithell, J. 2000. Bootstrap confidence intervals: When, which, what? A practical guide for medical statisticians. *Statistics in Medicine* 19(9): 1141–1164.

Casella G. and Berger, R. L. 1987. Reconciling Bayesian and frequentist evidence in the one-sided testing problem. *Journal of the American Statistical Association* 82: 106–111.

Cech, T. R. 1991. Overturning the dogma: Catalytic DNA. In Pfenninger, K. H., and Shubik, H., eds., *The Origins of Creativity*. Oxford University Press. 5–19.

Centers for Disease Control and Prevention, National Center for Health Statistics. Underlying cause of death 1999–2009. http://wonder.cdc.gov/ucd-icd10.html. (Accessed 6/19/2012).

Cerhan, J. R., Janney, C. A., Vachon, C. M., Habermann, T. M., Kay, N. E., Potter, J. D., Sellers, T. A., and Folsom, A. R. 2002. Anthropometric characteristics, physical activity, and risk of non-Hodgkin's lymphoma subtypes and B-cell chronic lymphocytic leukemia: A prospective study. *American Journal of Epidemiology* 156(6): 507–516.

Chalmers, A. F. 1999. *What Is This Thing Called Science?* Hackett Publishing Co., Indianapolis, IN.

Chatterjee, S., and Firat, A. 2007. Generating data with identical statistics but dissimilar graphics: A follow up to the Anscombe dataset. *American Statistician* 61(3): 248–254.

Chen, J., Shiyomi, M., Yamamura, Y., and Hori, Y. 2006. Distribution model and spatial variation of cover in grassland vegetation. *Grassland Science* 52(4): 167–173.

Chok, N. S. 2010. Pearson's versus Spearman's and Kendall's correlation coefficients for continuous data. Masters thesis, University of Pittsburgh.

Clark, J. S., Silman, M., Kern, R., Macklin, E., and Lambers, J. H. R. 1999. Seed dispersal near and far: Patterns across temperate and tropical forests. *Ecology* 80(5): 1475–1494.

Clarke, M. R. B. 1980. The reduced major axis of a bivariate sample. *Biometrika* 67: 441–446.

Clarke, R. T. 2002. Estimating trends in data from the Weibull and a generalized extreme value distribution. *Water Resources Research* 38(6): 25-1–25-10.

Cleveland, W. S. 1979. Robust locally weighted regression and smoothing scatterplots. *Journal of the American Statistical Association* 74: 829–836.

Cleveland, W. S., Grosse, E., and Shyu, M. J. 1992. A package of C and Fortran routines for fitting local regression models. http://www.netlib.org/a/ (Accessed 9/19/2012).

Clopper, C. and Pearson, S. 1934. The use of confidence or fiducial limits illustrated in the case of the binomial. *Biometrika* 26: 404–413.

Cochran, W. 1954. Some methods for strengthening the common χ^2 test. *Biometrics* 10: 417–451.

Cochran, W. G. and Cox, G. M. 1957. *Experimental Designs*, 2nd ed. John Wiley and Sons, New York.

Cohen, J. 1988. *Statistical Power Analysis for the Behavioral Sciences*. Lawrence Erlbaum Associates.

Cohen, J. 1992. A power primer. *Psychological Bulletin* 112: 155–159.

Cohen, J. 1994. The earth is round ($p < 0.05$). *American Psychologist* 49: 997–1003.

Condit, R., Pitman, N., Leigh, E. G., Chave, J., Terborgh, J., Foster, R. B., Nunez, P., Aguilar, S., Valencia, R., Villa, G., Muller-Landau, H. C., Losos, E., and Hubbell, S. P. 2002. Beta-diversity in tropical forest trees. *Science* 295: 666–669.

Conover, W. J., Johnson, M. E., and Johnson, M. M. 1981. A comparative study of tests for homogeneity of variances, with applications to the outer continental shelf bidding data. *Technometrics* 23: 351–361.

Cook, R. D. 1998. *Regression Graphics: Ideas for Studying Regressions through Graphics*. Wiley, New York.

Coombs, W. T., Algina, J., and Oltman, D. O. 1996. Univariate and multivariate hypothesis tests selected to control type I error rate when variances are not necessarily equal. *Review of Educational Research* 66: 137–179.

Cooper, D. A., Gatell, J. M., Kroon, S., Clumeck, N., Millard, J., Goebel, F.-D., Bruun, J. et al. 1993. Zidovudine in persons with asymptomatic HIV infection and CD4+ counts greater than 400 per cubic millimeter. *New England Journal of Medicine* 329: 297–303.

Cornfield, J., Haenszel, W., Hammond, E., Lilienfeld, A., Shimkin, M., and Wynder, E. 1959. Smoking and lung cancer: Recent evidence and a discussion of some questions. *Journal of the National Cancer Institute* 22: 173–203.

Coste, J., Fermanian, J., and Venot, A. 1995. Methodological and statistical problems in the construction of composite measurement scales: A survey of six medical and epidemiological journals. *Statistics in Medicine* 14: 331–345.

Craig, B. A., Black, M. A., and Doerge, R. W. 2003. Gene expression data: The technology and statistical analysis. *Journal of Agricultural, Biological, and Environmental Statistics* 8: 1–28.

Crainiceanu, C. M., and Ruppert, D. 2004. Likelihood ratio tests in linear mixed models with one variance component. *Journal of the Royal Statistical Society: Series B (Statistical Methodology)* 66: 165–185.

Crawley, M. J. 2007. *The R book*. Wiley, New York.

Crowley, P. H. 1992. Resampling methods for computation-intensive data analysis in ecology and evolution. *Annual Review of Ecology and Systematics* 23: 405–447.

Dass, S. C. and Lee, J. 2004. A note on the consistency of Bayes factors for testing point null versus non-parametric alternatives. *Journal of Statistical Planning and Inference* 119(1): 143–152.

Davidian, M. and Carroll, R. J. 1987. Variance function estimation. *Journal of the American Statistical Association* 82: 1079–1091.

Davis, D. L., Webster, P., Stainthorpe, H., Chilton, J., Jones, L., and Doi, R. 2007. Declines in sex ratio at birth and fetal deaths in Japan, and in U.S. whites but not African Americans. *Environmental Health Perspectives* 115(6): 941–946.

Davis, T. 2007. *2007 Mountain Goat Surveys. Memo to Glenn Plumb, Branch Chief, Yellowstone Center for Resources*. Yellowstone National Park, Mammoth, Wyoming.

Davison, A. C. and Hinkley, D. V. 1997. *Bootstrap Methods and Their Application*. Cambridge University Press, Cambridge, UK.

Dawkins, R. 1995. *River Out of Eden*. Basic Books, New York.

Day, R. W., and Quinn, G. P. 1989. A comparison of treatments after an analysis of variance in ecology. *Ecological Monographs* 59: 433–463.

de Angelis, M., Bertoldi, A., Cacciapuoti, L., Giorgini, A., Lamporesi, G., Prevedelli, M., Saccorotti, G., Sorrentino, F., and Tino, G. M. 2009. Precision gravimetry with atomic sensors. *Measurement Science and Technology* 20(2): 022001.

DeMuth, J. 2006. *Basic Statistics and Pharmaceutical Statistical Applications*. Chapman and Hall, Boca Raton, FL.

Dennis, B. 1996. Discussion: Should ecologists become Bayesians? Ecological Applications 6: 1095–1103.

Denny, M. and Gaines, S. 2002. Chance in Biology, Using Probability to Explore Nature. Princeton University Press, Princeton, NJ.

Dickersin, K. 1990. The existence of publication bias and risk factors for its occurrence. *Journal of the American Medical Association* 263(10): 1385–1359.

Dixon, P. M. 1993. The bootstrap and the jackknife: Describing the precision of ecological indices. In Scheiner, S., and Gurevitch, J., eds., *Design and Analysis of Ecological Experiments*. Chapman and Hall, New York, 210–318.

Doak, D. F., Bigger, D., Harding, E. K., Marvier, M. A., O'Malley, R. E., and Thompson, D. 1998. The statistical inevitability of stability–diversity relationships in community ecology. *The American Naturalist* 151: 264–276.

Dobson, A. J. 2001. *An Introduction to Generalized Linear Models*, 2nd ed. Chapman and Hall, London.

Dobzhansky, T. 1950. Evolution in the tropics. *American Scientist* 38: 209–221.

Doornik, J. A. and Hansen, H. 2008. An Omnibus test for univariate and multivariate normality. *Oxford Bulletin of Economics and Statistics* 70: 927–939.

Driscoll, D. A. and Roberts, J. D. 1997. Impact of fuel-reduction burning on the frog *Geocrinia lutea* in southwest Western Australia. *Australian Journal of Ecology* 22: 334–339.

Duncan, G. T. and Layard, M. W. 1973. A Monte Carlo study of asymptotically robust tests for correlation. *Biometrika* 60: 551–558.

Edgington, E. S. 1995. *Randomization Tests*. Marcel Dekker, New York, NY.

Edwards, A. W. F. 1972. *Likelihood, Expanded Edition*. Johns Hopkins University Press, Baltimore, MD.

Edwards, D. 1996. Comment: The first data analysis should be journalistic. *Ecological Applications* 6: 1090–1094.

Edwards, W., Lindman, H., and Savage, L. J. 1963. Bayesian statistical inference for psychological inference. *Psychological Review* 70(3): 193–242.

Efron, B. 1978. Regression and ANOVA with zero-one data: Measures of residual variation. *Journal of the American Statistical Association* 73: 113–121.

Efron, B. 1979. Bootstrap methods: Another look at the jackknife. *The Annals of Statistics* 7(1): 1–26.

Efron, B. 1981. Nonparametric estimates of standard error: The jackknife, the bootstrap and other methods. *Biometrika* 68: 589–599.

Efron, B. 1982. *The Jackknife, the Bootstrap, and Other Resampling Methods*. Society for Industrial and Applied Mathematics, CBMS-NSF, Monograph 38, Philadelphia.

Efron, B. 1986. Why isn't everyone a Bayesian (a discussion). *American Statistician* 40: 1–11.

Efron, B. and Stein, C. 1981. The jackknife estimate of variance. *Annals of Statistics* 9(3): 586–596.

Efron, B. and Tibshirani, R. J. 1993. *An Introduction to Bootstrap*. Chapman and Hall, New York.

Einstein, A. 1931. *Living Philosophies*. Simon and Schuster, New York.

Everitt, B. 2005. *An R and S-plus Companion to Multivariate Analysis*. Springer, New York.

Feinstein, A. R. 1997. Beyond statistics: What is really important in medicine? *Cleveland Clinical Journal of Medicine* 64: 127–128.

Finney, D. J. 1995. Statistical science and effective scientific communication. *Journal of Applied Statistics* 22(2): 293–308.

Fisher, F. E. 1973. *Fundamental Statistical Concepts*. Canfield Press, San Francisco.

Fisher, N. I. and Switzer, P. 2001. Graphical assessment of dependence: Is a picture worth 100 tests? *The American Statistician* 55: 233–239.

Fisher, R. A. 1915. Frequency distribution of the values of the correlation coefficient in samples of an indefinitely large population. *Biometrika* 10(4): 507–521.

Fisher, R. A. 1921. On the "probable error" of a coefficient of correlation deduced from a small sample. *Metron* 1: 3–32.

Fisher, R. A. 1922. On the mathematical foundations of theoretical statistics. *Philosophical Transactions of the Royal Society of London. Series A, Containing Papers of a Mathematical or Physical Character* 222(594–604): 309–368.

Fisher, R. A. 1925. *Statistical Methods for Research Workers*, 1st ed. Oliver and Boyd, Edinburgh.

Fisher, R. A. 1926. The arrangement of field experiments. *Journal of the Ministry of Agriculture of Great Britain* 33: 504.

Fisher, R. A. 1936. Has Mendel's work been rediscovered? *Annals of Science* 1: 115–137.

Fisher, R. A. 1947. The analysis of covariance method for the relation between a part and the whole. *Biometrics* 3: 65–68.

Fisher, R. A. 1949. *Design of Experiments*, 5th ed. Oliver and Boyd, Edinburgh.

Fisher, R. A. 1954. *Statistical Methods for Research Workers*, 1954 edition. Oliver and Boyd, Edinburgh.

Fisher, R. A. 1956. *Statistical Methods and Scientific Inference*. Oliver and Boyd, Edinburgh.

Fisher, R. A. 1958. Lung cancer and cigarettes. *Nature* 182: 108.

Fitzmaurice, G. M., Laird, N. M., and Waird, J. H. 2004. *Applied Longitudinal Analysis*. Wiley, New York.

Flesch, E. P. and Garrott, R. A. 2011. Population trends of bighorn sheep and mountain goats in the Greater Yellowstone Area. (Unpublished report). http://www.gyamountainungulateproject. com/annual_report_2011/PopulationTrends.pdf (Accessed 6/6/2012).

Fox, J. 2002. *An R and S-Plus Companion to Applied Regression*. Sage Publications, Thousand Oaks, CA.

Freeman, S. 2010. *Biological Science*, 4th ed. Prentice-Hall, Upper Saddle River, NJ.

Freund, R. J., Littell, R. C., and Spector, P. C. 1986. *SAS System for Linear Models*. SAS Press, Cary, NC.

Galilei, G. 1989. *Two New Sciences, Including Centres of Gravity and Force of Percussion, Translated with a New Introduction by Drake, S*, 2nd ed. Wall and Thompson, Toronto.

Garret, K. A. and Bowden, R. L. 2002. An Allee effect reduces the invasive potential of *Tilletia indica*. *Phytopathology* 92: 1152–1159.

Garrott, R., Rotella, J., O'Reilly, M., DeVoe, J., Butler, C., Flesch, E., and Sawaya, M. 2011. The Greater Yellowstone Area mountain ungulate project 2011 annual report. http://www.gyamountainun gulateproject.com/. (Accessed 1/28/2013).

Gaukroger, S. 2001. Objectivity, history of. In Smelser, N. J., and Baltes, B. B., eds., *Encyclopedia of the Social and Behavioral Sciences*. Elsevier Science, Oxford, UK.

Gause, C. 2012. Desiccation tolerance in a desert moss *(Grimmia alpestris)*: Implications of duration of dry period for quantum yield and photosynthetic recovery upon rehydration. Unpublished.

Gelman, A. 2009. Bayes, Jeffreys' prior distributions and the philosophy of statistics. *Statistical Science* 24(2): 176–178.

Gelman, A., Carlin, J. B., Stern, H. S., and Rubin, D. B. 2003. *Bayesian Data Analysis*, 2nd ed. Chapman and Hall/CRC Press, Boca Raton, FL.

Gelman, A. and Rubin, D. B. 1992. Inference from iterative simulation using multiple sequences (with discussion). *Statistical Science* 7: 457–511.

Gelman, A. and Rubin, D. B. 1996. Markov chain Monte Carlo methods in biostatistics. *Statistical Methods in Medical Research* 5(4): 339–355.

Geman, S. and Geman, D. 1984. Stochastic relaxation, Gibbs distributions, and the Bayesian restoration of images. *IEEE Transactions on Pattern Analysis and Machine Intelligence* 6(6): 721–741.

Gibbs, H. L. and Grant, P. R. 1987. Oscillating selection on Darwin's finches. *Nature* 327: 511–513.

Gigerenzer, G. and Gaissmaier, W. 2011. Heuristic decision making. *Annual Review of Psychology* 62: 451–482.

Gill, J. 1999. The insignificance of null hypothesis significance testing. *Political Research Quarterly* 52(3): 647–674.

Gillies, D. 2000. *Philosophical Theories of Probability*. Routledge, London.

Gillies, D. 2002. Causality, propensity, and Bayesian networks. *Synthese* 132(1–2): 63–88.

Glass, G. V., Peckham, P. D., and Sanders, J. R. 1972. Consequences of failure to meet assumptions of underlying fixed effects analysis of variance and covariance. *Review of Educational Research* 42: 237–288.

Gleason, H. A. 1922. On the relation between species and area. *Ecology* 3: 158–162.

Goldberg, K. M. and Ingelwicz, B. 1992. Bivariate extensions of the boxplot. *Technometrics* 34: 307–320.

Gonzales, L. and Manly, B. F. J. 1998. Analysis of variance by randomization with small data sets. *Envirometrics* 9: 53–65.

Gosline, J. M., Denny, M. W., and Demont, M. E. 1984. Spider silk as rubber. *Nature* 309: 551–552.

Gotelli, N. J. and Ellison, A. M. 2004. *A Primer of Ecological Statistics*. Sinauer, Sunderland, MA.

Gough, K. F. and Kerley, G. I. H. 2006. Demography and population dynamics in the elephants *Loxodonta africana* of Addo Elephant National Park, South Africa: Is there evidence of density dependent regulation? *Oryx* 40: 434–441.

Gould, S. J. 1981. *The Mismeasure of Man*. W. W. Norton, New York.

Goutis, C. and Casella, G. 1995. Frequentist post-data inference. *International Statistical Review* 63: 325–344.

Gower, B. 1997. *Scientific Method, an Historical and Philosophical Introduction*. Routledge, London.

Graven, H. D., Guilderson, T. P., and Keeling, R. F. 2012. Observations of radiocarbon in CO_2 at La Jolla, California, USA 1992–2007: Analysis of the long-term trend. *Journal of Geophysical Research* 117: D02302: 1–14.

Graves, A. B. et al. 1991. Alcohol and tobacco consumption as risk factors for Alzheimer's disease: A collaborative re-analysis of case-control studies. *International Journal of Epidemiology* 20(2): 548–557.

Graybill, F. A. 1976. *Theory and Application of the Linear Model*. Duxbury Press, North Scituate, MA.

Greig-Smith, P. 1983. *Quantitative Plant Ecology*. University of California Press, Berkeley, CA.

Greig-Smith, P. and Chadwick, M. J. 1965. Data on pattern within plant communities. *Journal of Ecology* 53: 465–474.

Guisan, A. and Zimmermann, N. E. 2000. Predictive habitat distribution models in ecology. *Ecological Modelling* 135: 147–186.

Gurevitch, J. and Chester, S. T. Jr. 1986. Analysis of repeated measures experiments. *Ecology* 67(1): 251–255.

Gurland, J. and Tripathi, R. C. 1971. A simple approximation for the unbiased estimation of the standard deviation. *The American Statistician* 25: 30–32.

Hacking, I. 1965. *The Logic of Scientific Inference*. Cambridge University Press, Cambridge, UK.

Hald, A. 1998. *A History of Mathematical Statistics*. Wiley, New York.

Hall, P. 1992. *The Bootstrap and Edgeworth Expansion*. Springer-Verlag, New York.

Hall, P. and Wang, Q. Y. 2004. Exact convergence rate and leading term in central limit theorem for student's *t* statistic. *Annals of Probability* 32(2): 1419–1437.

Hall, R. and Welsch, A. H. 1985. Limit theorems for the median deviation. *Annals of the Institute of Statistical Mathematics* 37(A): 27–36.

Hancock, G. R. and Klockars, A. J. 1996. The quest for α: Developments in multiple comparison procedures in the quarter century since Gaines (1971). *Review of Educational Research* 66: 269–306.

Hand, D. J. 1994. Deconstructing statistical questions. *Journal of the Royal Statistical Society, Series A* 157: 317–356.

Hand, D. J. and Crowder, M. J. 1996. *Practical Longitudinal Data Analysis*, Chapman and Hall, London, UK.

Hansen, L. P. 2002. Method of moments. In Smelser, N. J., and Bates, B. B., eds., *International Encyclopedia of the Social and Behavior Sciences*. Pergamon, Oxford.

Hansen, S. G., Ford, J. C., Lweis, M. S., Ventura, A. B., Hughes C. M., Coyne-Johnson, L., Whizin, N. et al. 2011. Profound early control of highly pathogenic SIV by an effector memory T-cell vaccine. *Nature* 473: 523–527.

Harder, L. D. and Thomson, J. D. 1989. Evolutionary options for maximizing pollen dispersal in animal pollinated plants. *American Naturalist* 133: 323–344.

Hardin, J. W. and Hilbe, J. M. 2007. *Generalized Linear Models and Extensions*. Stata Press, College Station, TX.

Haroldson, M. A. 2010. Assessing trend and estimating population size from counts of unduplicated females. In Schwartz, C. C., Haroldson, M. A., and West, K., eds., *Yellowstone Grizzly Bear*

Investigations: Annual Report of the Interagency Grizzly Bear Study Team. U.S. Geological Survey, Bozeman, MT, 10–15.

Harris, H. 2013. *How Fire and Debris Flows Affect Headwater Linkages to Downstream and Riparian Ecosystems*. Masters thesis, Idaho State University, Pocatello, Idaho.

Hartung, J. B. 1976. Was the formation of a 20-km diameter impact crater on the moon observed on June 18, 1178? *Meteoritics* 11(3): 187.

Hawking, S. 1988. *A Brief History of Time*. Bantam Books, New York, NY.

Hays, W. L. 1994. *Statistics*, 5th ed. Harcourt-Brace, Forth Worth, TX.

Hedges, L. V. 1981. Distribution theory for Glass's estimator of effect size and related estimators. *Journal of Educational Statistics* 6(2): 107–128.

Heffner, R. A., Butler, M. J. IV, and Reilly, C. K. 1996. Pseudoreplication revisited. *Ecology* 77(8): 2558–2562.

Hegazy, A. K. 1990. Population ecology and the implications for conservation of *Cleome droserifolia*, a threatened xerophyte. *Journal of Arid Environments* 19: 269–282.

Heimpel, G. E., Rosenheim, J. A., and Mangel, M. 1997. Predation on adult *Aphytis* parasitoids in the field. *Oecologia* 110: 346–352.

Heisenberg, W. 1927. Über den anschaulichen Inhalt der quantentheoretischen Kinematik und Mechanik. *Zeitschrift für Physik* 43: 172–198.

Heltshe, J. F. and Forrester, N. E. 1983. Estimating species richness using the jackknife procedure. *Biometrics* 39: 1–12.

Hempel, C. G. 1966. *Philosophy of Natural Science*. Prentice-Hall, Engelwood Cliffs, NJ.

Hempel, C. G. 1990. Empiricist criteria of cognitive significance. In Boyd, R. et al., eds. *The Philosophy of Science*. MIT Press, Cambridge.

Hettmansperger, T. P. 1984. *Statistical Inference Based on Ranks*. Wiley, New York.

Hill, R. 2004. Multiple sudden infant deaths—Coincidence or beyond coincidence? *Pediatric and Perinatal Epidemiology* 18: 320–326.

Hill, R. W. and Dixon, W. J. 1982. Robustness in real life: A study of clinical laboratory data. *Biometrics* 38: 377–396.

Hillborn, R. and Mangel, M. 1997. *The Ecological Detective: Confronting Models with Data*. Princeton University Press, Princeton, NJ.

Hines, A. H., Whitlatch, R. B., Thrush, S. F., Hewitt, J. E., Cummings, V. J., Dayton, P. K., and Legendre, P. 1997. Nonlinear foraging response of a large marine predator to benthic prey: Eagle ray pits and bivalves in a New Zealand sandflat. *Journal of Experimental Marine Biology and Ecology* 216: 191–210.

Hoenig, J. M. and Heisey, D. M. 2001. The abuse of power: The pervasive fallacy of power calculations for data analysis. *The American Statistician* 55(1): 1–6.

Hofstadter, D. R. 1979. *Gödel, Escher, Bach: An Eternal Golden Braid*. Basic Books, New York, NY.

Hollander, M. and Wolfe, D. A. 1999. *Nonparametric Statistical Methods*. John Wiley & Sons, New York.

Holm, S. 1979. A simple sequentially rejective multiple test procedure. *Scandinavian Journal of Statistics* 6: 65–70.

Horgan, G.W., Elston, D.A., Franklin, M. F., Glasbey, C. A., Hunter, E. A., Talbot, M., Kempton, R. A., McNichol, J. W., and Wright, F. 1999. Teaching statistics to biological research scientists. *The Statistician* 48(3): 393–400.

Hoshmand, A. R. 2006. *Design of Experiments for Agriculture and the Natural Sciences*, 2nd ed. CRC Press, Boca Raton, FL.

Hosmer, D. W. and Lemeshow, S. 2002. *Applied Logistic Regression*, 2nd ed. Wiley, New York.

Hosmer, D. W., Hosmer, T., le Cessie, S., and Lemeshow, S. 1997. A comparison of goodness-of-fit tests for the logistic regression model. *Statistics in Medicine* 16: 965–980.

Hothorn, T., Hornik, K. van de Wiel, M. A., and Zeileis, A. 2006. A lego system for conditional inference. *The American Statistician* 60(3): 257–263.

Hougaard, P. 1985. The appropriateness of the asymptotic distribution in a non-linear regression in relation to curvature. *Journal of the Royal Statistical Society B* 47: 103–114.

Howell, D. C. 2010. *Statistical Methods for Psychology*, 7th ed. Thomson Wadsworth, Belmont, CA.

Howson, C. 2001. Evidence and confirmation. In Newton-Smith, W. H., ed., *A Companion to the Philosophy of Science*. Wiley, New York, 108–117.

Hubbell, S. P. 2001. *The Unified Neutral Theory of Biodiversity and Biogeography*. Princeton University Press, Princeton and Oxford, UK.

Huber, P. J. 2004. *Robust Statistics*. Wiley, New York.

Huberty, C. J. 1993. Historical origins of statistical testing practices: The treatment of Pearson versus Neyman–Pearson views in textbooks. *Journal of Experimental Education* 61: 317–333.

Hughes, K. D., Weselogh, D. V., and Braune, B. M. 1998. The ratio of DDE to PCB concentrations in Great Lakes herring gull eggs and its use in interpreting contaminants data. *Journal of Great Lakes Research* 24(1): 12–31.

Humane Society United States (HSUS). 2013. Pet overpopulation estimates; http://www.humanesociety. org/issues/pet_overpopulation/facts/pet_ownership_statistics.html (Accessed 8/8/2013).

Hume, D. 1996. *Enquiry of Human Understanding*. In Selby-Bigge, ed., Oxford University Press, New York, NY.

Hurlbert, S. H. 1984. Pseudoreplication and the design of ecological field experiments. *Ecological Monographs* 54(2): 187–211.

Hurvich, C. M. and Tsai, C-L. 1989. Regression and time series model selection in small samples. *Biometrika* 76(2): 297–307.

Jacobsen, J. S., Lorber, S. H., Schaff, B. E., and Jones, C. A. 2002. Variation in soil fertility test results from selected northern Great Plains laboratories. *Communications in Soil Science and Plant Analysis* 33(3&4): 303–319.

James, F. C. and McCulloch, C. E. 1985. Data analysis and the design of experiments in ornithology. In Lovie, A. D., ed., *Current Ornithology*, Vol. 2. Plenum Press, New York, 1–63.

Janssen, B. J. et al. 2008. Global gene expression analysis of apple fruit development from the floral bud to ripe fruit. *BMC Plant Biology* 8: 1–29.

Jaynes, E. T. 1976. Confidence intervals versus Bayesian intervals. In Harper, W. L., and Hooker, C. A., eds., *Foundations of Probability Theory, Statistical Inference, and Statistical Theories of Science*. Dordrecht, Boston, MA.

Jeffreys, H. 1935. Some tests of significance treated by the theory of probability. *Proceedings of the Cambridge Philosophical Society* 31: 203–222.

Jefferys, H. 1961. *Theory of Probability*, 3rd ed. Clarendon Press, Oxford.

Jeyaratnam, S. and Othman, A. B. 1985. Test of hypothesis in a one way random effects model with unequal variances. *Journal of Statistical Computation and Simulation* 21: 51–57.

Johnson, D. H. 1999. The insignificance of statistical significance testing. *Journal of Wildlife Management* 63: 763–772.

Johnson, D. S., Barry, R. P., and Bowyer, R. T. 2004. Estimating timing of life-history events with coarse data. *Journal of Mammalogy* 85(5): 932–939.

Johnson, N. L., Kotz, S., and Balakrishnan, N. 1995. *Continuous Univariate Distributions*, Vol. 2. Wiley, New York.

Jolicoeur, P. 1973. Imaginary confidence limits of the slope of the major axis of a bivariate normal distribution: A sampling experiment. *Journal of the American Statistical Association* 68: 866–871.

Jolicoeur, P. and Mosimann, J. E. 1968. Intervalles do confidance pout la pent de l'axe majuer d'une distribution normal bidimensionelle. *Biom-Praxim* 9: 121–140.

Kaeding, L. R., Boltz, G. D., and Carty, D. G. 1995. Lake trout discovered in Yellowstone Lake. In Varley, J. D., and Schullery, P., eds., *The Yellowstone Lake Crisis: Confronting a Lake Trout Invasion*, a report to the director of the National Park Service. http://www.nps.gov/yell/planyourvisit/ upload/laketrout2.pdf. (Accessed 6/6/2012).

Kandel, R. 1991. *Our Changing Climate*. McGraw-Hill, New York.

Kaplan, E. L. and Meier, P. 1958. Nonparametric estimation from incomplete observations. *Journal of the American Statistical Association* 53: 457–481.

Kärki, T., Maltamo, M., and Eerikäinen, K. 2000. Diameter distribution, stem volume and stem quality models for grey alder (*Alnus incana*) in eastern Finland. *New Forests* 20: 65–86.

Kass, R. E. and Raftery, A. E. 1993. Bayes factors and model uncertainty. Technical Report #254. Department of Statistics, University of Washington, Seattle, WA.

Kass, R. E. and Raftery, A. E. 1995. Bayes factors. *Journal of the American Statistical Association* 90: 773–795.

Katz, D., Baptista, J., Azen, S. P., and Pike, M. C. 1978. Obtaining confidence intervals for the risk ratio in cohort studies. *Biometrics* 34: 469–474.

Kendall, M. G. and Stuart, A. 1966. *The Advanced Theory of Statistics*, Vol. 3. Hafner Publishing Co., New York.

Kenney, J. F. and Keeping, E. S. 1951. *Mathematics of Statistics*, Pt. 2, 2nd ed. Van Nostrand, Princeton, NJ.

Kenward, M. G. and Roger, J. H. 1997. Small sample inference for fixed effects from restricted maximum likelihood. *Biometrics* 53: 983–997.

Keppel, G. 1991. *Design and Analysis: A Researcher's Handbook*, 3rd ed. Prentice-Hall, Englewood Cliffs, NJ.

Kermack, K. A. and Haldane, J. B. S. 1950. Organic correlation and allometry. *Biometrika* 37: 30–41.

Keynes, J. M. 1921. *A Treatise on Probability*. Macmillan, London.

Kirk, R. E. 1995. *Experimental Design*. Brooks/Cole, Pacific Grove, CA.

Koel, T. M., Bigelow, P. E., Doepke, P. D., Ertel, B. D., and Mahony, D. L. 2005. Nonnative lake trout result in Yellowstone cutthroat trout decline and impacts to bears and anglers. *Fisheries* 30(11): 10–19.

Kokko, H. 2007. *Modelling for Field Biologists and Other Interesting People*. Cambridge University Press, Cambridge.

Koopman, P. A. R. 1984. Confidence intervals for the ratio of two binomial proportions. *Biometrics* 40: 513–517.

Kramer, C. Y. 1956. Extension of multiple-range tests to group means with unequal numbers of replications. *Biometrics* 12: 307–310.

Krewski, D. and Rao, J. N. K. 1981. Inference from stratified samples: Properties of the linearization, jackknife and balanced repeated replication methods. *Annals of Statistics* 9(5): 1010–1019.

Krochmal, A. R. and Sparks, D. W. 2007. Timing of birth and estimation of age of juvenile *Myotis septentrionalis* and *Myotis lucifugus* in West-Central Indiana. *Journal of Mammalogy* 88(3): 649–656.

Kuhn, T. S. 1963. *The Structure of Scientific Revolutions*. University of Chicago Press, Chicago, IL.

Kuhn, T. S. 1970. *The Structure of Scientific Revolutions*, 2nd ed. University of Chicago Press, Chicago, IL.

Kullback, S. 1959. *Information Theory and Statistics*. John Wiley & Sons, New York, NY.

Kullback, S. and Leibler, R. A. 1951. On information and sufficiency. *The Annals of Mathematical Statistics* 22(1): 79–86.

Kutner, M. H., Nachtsheim, C. J., Neter, J., and Li, W. 2005. *Applied Linear Statistical Models*, 5th ed. McGraw-Hill, Boston.

Lakatos, I. 1978. *The Methodology of Scientific Research*. Cambridge University Press, New York.

Lambert, C. M. S., Wielgus, R. B., Robinson, H. S., Katnik, D. D., Cruickshank, H. S., Clarke, R., and Almack, J. 2006. Cougar population dynamics and viability in the Pacific Northwest. *Journal of Wildlife Management* 70: 246–254.

Langsrud, Ø. 2003. ANOVA for unbalanced data: Use type II instead of type III sums of squares. *Statistics and Computing* 13: 163–167.

Larrick, J. 1993. Smoking and neurodegenerative diseases (letter). *Lancet* 342: 1238.

Laudan, L. 1977. *Progress and Its Problems: Towards a Theory of Scientific Growth*. University of California Press, Berkley, CA.

Lax, D. A. 1975. An interim report of a Monte Carlo study of robust estimators of width. Technical Report 93, Series 2. Department of Statistics, Princeton University, Princeton, NJ.

Lax, D. A. 1985. Robust estimators of scale: Finite sample performance in long-tailed symmetric distributions. *Journal of the American Statistical Association* 80: 736–741.

Lebreton, J. D., Gosselin, F., and Niel, C. 2007. Extinction and viability of populations: Paradigms and concepts of extinction models. *Ecoscience* 14(4): 472–481.

Legendre, P. 2008. *Model II Regression Users Guide, R Edition*. http://cran.r-project.org/web/packages/lmodel2/vignettes/mod2user.pdf. (Accessed 5/8/2013).

Legendre, P. and Legendre, L. 1998. *Numerical Ecology*, 2nd English ed. Elsevier, Amsterdam, The Netherlands.

Letourneau, D. K. and Dyer, L. A. 1998. Experimental test in lowland tropical forest shows top-down effects through four trophic level. *Ecology* 79: 1667–1687.

Levene, H. 1960. Robust tests for equality of variances. In Olkin, I., and Hotelling, H. et al. eds., *Contributions to Probability and Statistics: Essays in Honor of Harold Hotelling*. Stanford University Press Stanford, CA, 278–292.

Limpert, E., Stahel, W. A., and Abbt, M. 2001. Log-normal distributions across the sciences: Keys and clues. *Bioscience* 51(5): 341–352.

Lindén, A. and Mäntyniemi, S. 2011. Using the negative binomial distribution to model overdispersion in ecological count data. *Ecology* 92: 1414–1421.

Lindley, D. V. and Phillips, L. D. 1976. Inference for a Bernoulli process (a Bayesian view). *American Statistician* 30: 112–119.

Lindsay, J. K. 1997. *Applying Generalized Linear Models*. Springer, New York.

Littell, R. C., Stroup, W. W., and Fruend, R. J. 2002. *SAS for Linear Models*. Wiley, New York.

Littell, R. C., Milliken, G. A., Stroup, W. W., Wolfinger, R. D., and Schabenberger, O. 2006. *SAS for Mixed Models*, 2nd ed. SAS Press, Cary, NC.

Little, R. J. A. and Rubin, D. B. 1987. *Statistical Analysis with Missing Data*. Wiley, New York.

Lloyd, G. E. R. 1979. *Magic, Reason, and the Development of Science*. Cambridge University Press, Cambridge.

Lloyd, S. J., Garlid, K. D., Reba, R. C., and Seeds, A. E. 1969. Permeability of different layers of the human placenta to isotopic water. *Journal of Applied Physiology* 26: 247–276.

Lohr, S. L. 1999. *Sampling: Design and Analysis*. Duxbury Press, Pacific Grove, USA.

Lomolino, M. V., Brown, J. H., and Davis, R. 1989. Island biogeography of montane forest mammals in the American Southwest. *Ecology* 70: 180–194.

Lunneborg, C. E. 2000. *Data Analysis by Resampling: Concepts and Applications*. Duxbury Press, Pacific Grove, CA.

Lynch, H. J. and Fagan, W. F. 2009. Survivorship curves and their impact on the estimation of maximum population growth rates. *Ecology* 90: 1116–1124.

Macarthur, R. H. and Wilson, E. O. 1963. An equilibrium theory of insular zoogeography. *Evolution* 17: 373–387.

Mack, G. A. and Skillings, J. H. 1980. A Friedman-type rank test for main effects in a two-factor ANOVA. *Journal of the American Statistical Association* 75: 947–951.

Madansky, A. 1959. The fitting of straight lines when both variables are subject to error. *Journal of the American Statistical Association* 54: 173–205.

Madsen, V., Balsby, T. J. S., Dabelsteen, T., and Osorno, J. L. 2004. Bimodal signaling of a sexually selected trait: Gular pouch drumming in the magnificent frigatebird. *Condor* 106: 156–160.

Magurran, A. 1988. *Ecological Diversity and Its Measurement*. Princeton University Press, Princeton, NJ.

Manly, B. F. J. 2007. *Randomization and Monte Carlo Methods in Biology*, 3rd ed. Chapman and Hall, London.

Mann, H. B. and Whitney, D. R. 1947. On a test of whether one of two random variables is stochastically larger than the other. *Annals of Mathematical Statistics* 18: 50–60.

Mann, M. E. and Kump, L. R. 2009. *Dire Predictions: Understanding Global Warming*. DK Press, London.

Mapstone, B. D. 1995. Scalable decision rules for environmental impact studies: Effect size, type I and type II errors. *Ecological Applications* 5: 401–410.

Maronna, R., Martin, D., and Yohai, V. 2006. *Robust Statistics*. Wiley, New York.

Martin, H. 2009. Comparison of two types of forest disturbance using multitemporal Landsat TM/ETM + imagery and field vegetation data. *Remote Sensing of the Environment* 113(4): 835–845.

Martinka, C. J. 1974. Population characteristics of grizzly bears in Glacier National Park, Montana. *Journal of Mammalogy* 55(1): 21–29.

Maxwell, E. A. 1976. Analysis of contingency tables and further reasons for not using Yates' correction in 2 × 2 tables. *Canadian Journal of Statistics* 4: 277–290.

Maxwell, S. E. and Delaney, H. D. 1990. *Designing Experiments and Analyzing Data: A Model Comparison Perspective*. Wadsworth Publishing, Belmont, CA.

Mayer, R. E. 1999. Designing instruction for constructivist learning. In Reigeluth, C. M., ed. *Instructional Design Theories and Models*. Erlbaum, Lawrence.

Mayo, D. G. 1996. *Error and the Growth of Experimental Knowledge*. University of Chicago Press, Chicago, IL.

McArdle, B. H. 1988. The structural relation; regression in biology. *Canadian Journal of Zoology* 66: 2329–2339.

McCarthy, M. A. 1997. Competition and dispersal from multiple nests. Ecology 78(3): 873–883.

McCune, B. and Grace, J. B. 2002. *Analysis of Ecological Communities*. MjM Software Design, Gelenden Beach, OR.

McDonald, J. H. 2009. *Handbook of Biological Statistics*, 2nd ed. Sparky House Publishing, Baltimore, MD.

McDonald, J. H. and Kreitman, M. 1991. Adaptive protein evolution at the *Adh* locus in *Drosophila*. *Nature* 351: 652–654.

McDonald, M. E. and Paul, J. F. 2010. Timing of increased autistic disorder cumulative incidence. *Environmental Science and Technology* 44: 2112–2118.

McDowell, M. A., Fryar, C. D., Hirsch, R., and Ogden, C. L. 2005. Anthropometric reference data for children and adults: US population, 1999–2002. Advance data from vital health and statistics. 361. http://www.cdc.gov/nchs/data/ad/ad361.pdf. (Accessed 5/28/2013).

McGill, R., Tukey, J. W., and Larsen, W. A. 1978. Variations of box plots. *The American Statistician* 32: 12–16.

McKean, J. W. and Schrader, R. M. 1984. A comparison of methods for studentizing the sample median. *Communications in Statistics: Simulation and Computation* 13: 751–773.

McNamara, J. M., Green, R. F., and Olsson, O. 2006. Bayes' theorem and its applications in animal behaviour. *Oikos* 112: 243–251.

McQuarrie, A. D. R. and Tsai, C.-L. 1998. *Regression and Time Series Model Selection*. World Scientific Publishing Company, Singapore.

Mehta, T., Tanik, M., and Allison, D. B. 2004. Towards sound epistemological foundations of statistical methods for high-dimensional biology. *Nature Genetics* 36(9): 943–947.

Melo, A. S., and Froehlich, C. G. 2001. Evaluation of methods for estimating macro in vertebrate species richness using individual stones in tropical streams. *Freshwater Biology* 46(6): 711–721.

Menard, S. 2000. Coefficients of determination for multiple logistic regression analysis. *The American Statistician* 54: 17–24.

Menotti-Raymond, M. and O'Brien, S. J. 1993. Dating the genetic bottleneck of the African cheetah. *Proceedings of the National Academy of Science* 90: 3172–3176.

Mentis, M. T. 1988. Hypothetico-deductive and inductive approaches in ecology. *Functional Ecology* 12: 5–14.

Merganic, J. and Sterba, H. 2006. Characterization of diameter distribution using the Weibull function: Method of moments. *European Journal of Forestry Research* 125: 427–439.

Miao, L. L. 1977. Gastric freezing: An example of the evaluation of medical theory by randomized clinical trials. In Bunker, J. P., Barnes, B. A., and Mosteller, F., eds., *Costs, Risks and Benefits of Surgery*. Oxford University Press, New York, NY, 198–211.

Miller, G. L. and Carroll, B. W. 1989. Modeling vertebrate dispersal distances: Alternatives to the geometric distribution. *Ecology* 70: 977–986.

Miller, R. G. 1974. The jackknife—A review. *Biometrika* 61: 1–16.

Miller, R. G. 1981. *Simultaneous Statistical Inference*. Springer-Verlag, New York, NY.

Miller, R. I. and Wiegert, R. G. 1989. Documenting completeness, species–area relations, and the species–abundance distribution of a regional flora. Ecology 70(1): 16–22.

Milliken, G. A. and Johnson, D. E. 2002. *Analysis of Messy Data: Analysis of Covariance*. CRC Press, Boca Raton, FL.

Milliken, G. A. and Johnson, D. E. 2009. *Analysis of Messy Data: Vol. I. Designed Experiments*, 2nd ed. CRC Press, Boca Raton, FL.

Monteith, K. L, Schmitz, L. E., Jenks, J. A., Delger, J. A., and Bowyer, R. T. 2009. Growth of male white-tailed deer: Consequences of maternal effects. *Journal of Mammalogy* 90(3): 651–660.

Moore, G. S. and McCabe, B. P. 2004. *Introduction to the Practice of Statistics*. Freeman, New York, NY.

Motulsky, H. 1995. *Intuitive Biostatistics*. Oxford University Press, New York, NY.

Mueller-Dombois, D. and Ellenberg, H. 1974. *Aims and Methods of Vegetation Ecology*. John Wiley and Sons, New York.

Müller, H., Chiow, S. W., Herrmann, S., Chu, S., and Chung, K.-Y. 2008. Atom–interferometry tests of the isotropy of post-Newtonian gravity. *Physical Review Letters* 100: 031101.

Müller, J. and Hothorn, T. 2004. Maximally selected two-sample statistics as a new tool for the identification and assessment of habitat factors with an application to breeding-bird communities in oak forests. *European Journal of Forest Research* 123: 219–238.

Murakami, H., Ogawara, H., Morita, K., Saitoh, T., Matsushima, T., Tamura, J., Sawamura, M. et al. 1997. Serum beta-2-microglobulin in patients with multiple myeloma treated with alpha interferon. *Journal of Medicine* 28(5–6): 311–318.

Nadaraya, E. A. 1964. On estimating regression. *Theory of Probability and Its Applications* 9(1): 141–142.

Nair, R. T. et al. 2008. Biochemical and hematologic alterations following percutaneous cryoablation of liver tumors: Experience in 48 procedures. *Radiology* 248(1): 303–311.

National Library of Medicine. http://www.nlm.nih.gov/about/index.html. (Accessed 6/15/2012).

Nefzger, M. D. and Drasgow, J. 1957. The needless assumption of normality in Pearson's *r*. *The American Psychologist* 12: 623–625.

Nelder, J. 1999. From statistics to statistical science. *The Statistician* 48(2): 257–269.

Nelder, J. and Wedderburn, R. 1972. Generalized linear models. Journal of the Royal Statistical Society, Series A (General) 135: 370–384.

Neter, J., Kutner, M. H., Nachtsheim, C. J., and Wasserman, W. 1996. *Applied Linear Statistical Models*. McGraw-Hill, Boston, MA.

Neyman, J. and Pearson, E. 1928. On the use and interpretation of certain test criteria for purposes of statistical inference: Part I. *Biometrika* 20A: 175–240.

Neyman, J. and Pearson, E. 1933. On the problem of the most efficient tests of statistical hypotheses. *Philosophical Transactions of the Royal Society of London, Series A* 231: 289–337.

Niselman, A.V., Ben, M.G., and Rubio, M. C. 1998. Robust methods in bioequivalence assay; preliminary results. *European Journal of Drug Metabolism and Pharmacokinetics* 23(2): 148–152.

Norton, P. G. and Dunn, E. V. 1985. Snoring as a risk factor for disease: An epidemiological survey. *British Medical Journal* 291: 630–632.

Oakes, M. 1986. *Statistical Inference: A Commentary for the Social and Behavioral Sciences*. Wiley, Chichester, UK.

Okasha, S. 2002. *Philosophy of Science (A Very Short Introduction)*. Oxford University Press, New York, NY.

Ott, R. L. and Longnecker, M. T. 2004. *A First Course in Statistical Methods*. Thompson Brooks/Cole, Belmont, CA.

Owen, D. B. 1974. *On the History of Statistics and Probability*. Marcel Dekker, New York.

Paik, M. 1985. A graphical representation of a three-way contingency table: Simpson's paradox and correlation. *American Statistician* 39: 53–54.

Palmer, M.W. 1990. The estimation of species richness by extrapolation. *Ecology* 71: 1195–1198.

Palmer, M. W. 1991. Estimating species richness: The second-order jackknife reconsidered. *Ecology* 72: 1512–1513.

Partridge, L. and Farquhar, M. 1981. Sexual activity and the lifespan of male fruitflies. *Nature* 294: 580–581.

Pastor, J. 2008. *Mathematical Ecology*. Wiley-Blackwell, Chichester, UK.

Patterson, H. D. and Thompson, R. 1971. Recovery of inter-block information when block sizes are unequal. *Biometrika* 58: 545–554.

Pearl, J. 2000. *Causality: Models of Reasoning and Inference*. Cambridge University Press, Cambridge, UK.

Pearson, K. 1900. On the criterion that a given system of deviations from the probable in the case of a correlated system of variables is such that it can be reasonably supposed to have arisen from random sampling. *Philosophical Magazine Series* 550(302): 157–175.

Pearson, K. 1901. On lines and planes of closest fit to systems of points in space. *Philosophical Magazine Series* 6(2): 559–572.

Pease, C. M. and Mattson, D. J. 1999. Demography of the Yellowstone grizzly bears. *Ecology* 80: 957–975.

Pedersen, O. 1993. *Early Physics and Astronomy: A Historical Introduction*. Cambridge University Press, Cambridge, UK.

Peng, C. Y. J., Lee, K. L., and Ingersoll, G. M. 2002. An introduction to logistic regression analysis and reporting. *Journal of Educational Research* 96(1): 3–14.

Pereira, H. M. and Daily, G. C. 2006. Modeling biodiversity dynamics in countryside landscapes. *Ecology* 87: 1877–1885.

Peterson, I. 1993. *Newton's Clock: Chaos in the Solar System*. Freeman, New York.

Petit, L. I. 1986. Diagnostics in Bayesian model choice. *The Statistician* 35: 183–190.

Phillips, D. L. and MacMahon, J. A. 1981. Competition and spacing patterns in desert shrubs. Journal of Ecology 69(1): 97–115.

Piattelli-Palmarini, M. 1994. *Inevitable Illusions: How Mistakes of Reason Rule Our Minds*. John Wiley, Chichester, UK.

Pinheiro, J. and Bates, D. 2000. *Mixed Effect Models in S and S-Plus*. Springer-Verlag, New York.

Pitman, E. J. G. 1948. *Notes on Nonparametric Statistical Inference*. Columbia University Press, New York, NY.

Plackett, R.L. 1950. Some theorems in least squares. *Biometrika* 37: 149–157.

Pollock, K. H., Winterstein, S. R., and Curtis, P. D. 1989. Survival analysis in telemetry studies: The staggered entry design. *Journal of Wildlife Management* 53(1): 7–1.

Popper, K. 1959. *The Logic of Scientific Discovery*. Routledge, London, UK.

Popper, K. 1983. *Realism and the Aim of Science. Postscript to the Logic of Scientific Discovery*. Rowman and Littlefield, Totowa, NJ.

Port, S., Demer, L., Jennrich, R., Walter, D., and Garfinkel, A. 2000. Systolic blood pressure and mortality. *The Lancet* 355: 175–179.

Potvin, C. and Roff, D. A. 1993. Distribution-free and robust statistical methods. In Gurevitch, J., and Scheiner, S., eds., *Design and Analysis of Ecological Experiments*. Chapman and Hall, New York, 46–68.

Preston, F. W. 1948. The commonness and rarity of species. *Ecology* 29: 254–283.

Preston, F. W. 1962. The canonical distribution of commonness and rarity. *Ecology* 43: 185–215.

Price, M. 2010. Malagasy spiders spin the world's toughest biological material. *Science Now*. (Accessed 1/28/2013).

Price, R. M. and Bonett, D. G. 2001. Estimating the variance of the sample median. *Journal of Statistical Computation and Simulation* 68: 295–305.

Prosser, J. I. 2010. Replicate or lie. *Environmental Microbiology* 12(7): 1806–1810.

Psilos, S. 1999. *Scientific Realism, How Science Attacks Truth*. Routledge, London, UK.

Pyke, C. R., Condit, R., Aguilar, S., and Lao, S. 2001. Floristic composition across a climatic gradient in a neotropical lowland forest. *Journal of Vegetation Science* 12: 553–566.

Quinn, G. P. and Keough, M. J. 2002. *Experimental Design and Data Analysis for Biologists*. Cambridge University Press, Cambridge, UK.

R Development Core Team. 2012. R: A language and environment for statistical computing. *R Foundation for Statistical Computing*, Vienna, Austria. ISBN 3-900051-07-0, http://www.R-project.org. (Accessed 5/28/2013).

Radelet, M. L. and Pierce, G. L. 1991. Choosing those who will die: Race and the death penalty in Florida. *Florida Law Review* 43(1): 1–34.

Raj, G. G. 1994. Effective period for control of the brown spiny field mouse (*Mus plantythrix*) in dry land crops. *Proceedings of the Sixteenth Vertebrate Pest Conference*, February 28–March 3, Santa Clara, CA.

Rakison, D. H. 2009. Does women's greater fear of snakes and spiders originate in infancy? *Evolution and Human Behavior* 30(6): 438–444.

Ramsey, P. H. 1993. Multiple comparisons of independent means. In Edwards, L., ed., *Applied Analysis of Variance in the Behavioral Sciences*. Marcel Dekker, New York, 25–61.

Ramsey, F. L. and Schafer, D. W. 1997. *The Statistical Sleuth: A Course in Methods of Data Analysis.* Duxbury Press, Belmont, CA.

Rao, P. M. et al. 1998. Effect of computed tomography of the appendix on treatment of patients and use of hospital resources. *The New England Journal of Medicine* 338(3): 141–147.

Ratbi, I., Legendre, M., Niel, F., Martin, F., Soufir, J., Izard, V., Costes, B., Costa, C., Goossens, C. M., and Girodon, E. 2007. Detection of cystic fibrosis transmembrane conductance regulator (CFTR) gene rearrangements enriches the mutation spectrum in congenital bilateral absence of the vas deferens and impacts on genetic counseling. *Human Reproduction* 22(5): 1285–1291. http://humrep.oxfordjournals.org/cgi/content/full/22/5/1285. (Accessed 5/28/2013).

Reece, J. D., Taylor, M. R., Simon, E. J., and Dickey, J. L. 2011. *Campbell Biology: Concepts and Connections*, 7th ed. Pearson, San Francisco, CA.

Reiczigel, J. 2003. Confidence intervals for the binomial parameter: Some new considerations. *Statistics in Medicine* 22: 611–621.

Rényi, A. 1961. On measures of entropy and information. In Neyman, J., ed., *Proceedings of the Fourth Berkeley Symposium of Mathematical Statistics and Probability*, University of California Press, Berkeley, 547–561.

Ricklefs, R. E. and Miller, G. L. 2000. *Ecology*. Freeman, San Francisco.

Ripley, B. D. 1996. *Pattern Recognition and Neural Networks*. Cambridge University Press, Cambridge.

Robert, C. P. 2007. *The Bayesian Choice*, 2nd ed. Springer, New York, NY.

Roberts, J. 1993. Regeneration and growth of coolibah (*Eucalyptus coolibah* ssp. *arida*) a riparian tree, in the Cooper Creek region of South Australia. *Australian Journal of Ecology* 18: 345–350.

Robertson, C. 1991. Computationally intensive statistics. In Lovie, P., and Lovie, A. D., eds., *New Developments for Statistics for Psychology and Social Sciences*, Vol. 2. BPS and Routledge, London, 49–90.

Robinson, C. T., Minshall, G. W., and Rushforth, S. R. 1994. The effects of the 1988 wildfires on diatom assemblages in the streams of Yellowstone National Park. Technical Report NPS/NRYELL/NRTR-93/XX.

Rosenberg, D. K., Overton, W. S., and Anthony, R. G. 1995. Estimation of animal abundance when capture probabilities are low and heterogeneous. *Journal of Wildlife Management* 59(2): 252–261.

Rousseeuw, P.J. and Yohai, V. J. 1984. Robust regression by means of S-estimators. In Franke, J., Hardle, W., and Martin, R. D., eds., *Robust and Nonlinear Time Series. Lectures Notes in Statistics*, Vol. 26. Springer-Verlag, New York, NY, 256–272.

Royall, R. M. 1997. *Statistical Evidence, A Likelihood Paradigm*. Chapman and Hall, London.

Royston, P. 1982. An extension of Shapiro and Wilk's W test for normality to large samples. *Applied Statistics* 31: 115–124.

Ruppert, D., Sheather, S. J., and Wand, M. P. 1995. An effective bandwidth selector for local least squares regression. *Journal of the American Statistical Association* 90: 1257–1270.

Ruskey, F. and Weston, M. 2005. *A Survey of Venn Diagrams*. http://www.combinatorics.org/files/Surveys/ds5/VennEJC.html. (Accessed 1/28/2013).

Russell, B. 1912. *The Problems of Philosophy*. Williams and Norgate, London.

Russell, B. 1948. *Human Knowledge: Its Scope and Limits*. George Allen and Unwin, London.

Ruzycki, J. R., Beauchamp, D. A., and Yule, D. L. 2000. Effects of introduced lake trout on native cutthroat trout in Yellowstone Lake. *Ecological Applications* 13(1): 23–37.

Sainani, K. 2011. What biomedical computing can learn from its mistakes. *Biomedical Computation Review Fall* 2011: 12–19.

Saint-Germain, M., Drapeau, P., and Buddle, C. 2007. Occurrence patterns of aspen-feeding woodborers (Coleoptera: Cerambycidae) along the wood decay gradient: Active selection for specific host types or neutral mechanisms? *Ecological Entomology* 32: 712–721.

Sakamoto, Y., Ishiguro, M., and Kitagawa, G. 1986. *Akaike Information Criterion Statistics*. KTK Scientific Publishers, Tokyo, Japan.

Salib, E. and Hillier, V. 1997. A case-control study of smoking and Alzheimer's disease. *International Journal of Geriatric Psychiatry* 12(3): 295–300.

Salmon, W. C. 1963. *Logic*. Prentice-Hall, Englewood Cliffs, NJ.

Salmon, W. C. 1967. *The Foundations of Scientific Inference*. University of Pittsburgh Press, Pittsburgh, PA.

Salsburg, D. 2001. *The Lady Tasting Tea: How Statistics Revolutionized Science in the Twentieth Century*. Henry Holt and Co., New York.

Salsburg, D. S. 1985. The religion of statistics as practiced in medical journals. *American Statistician* 39: 220–223.

Santner, T. J. and Duffy, D. E. 1989. *The Statistical Analysis of Discrete Data*. Springer-Verlag, New York.

Schabenberger, O. and Gotway, C. A. 2005. *Statistical Methods for Spatial Data Analysis*. Chapman and Hall, Boca Raton, FL.

Schmidt, G. H. and Garbutt, D. J. 1985. Species abundance data from fouling communities conform to the gamma distribution. *Marine Ecology—Progress Series* 23: 287–290.

Schoenberg, I. J. 1964. Spline functions and the problem of graduation. *Proceedings of the National Academy of Science* 52: 947–950.

Schoener, T. W. 1968. The *Anolis* lizards of Bimini: Resource partitioning in a complex fauna. *Ecology* 49(4): 704–726.

Schwarz, G. 1978. Estimating the dimension of a model. *Annals of Statistics* 6: 461–464.

Seal, A. N., Pratley, J. E., Haig, T., and Lewin, L. G. 2004. Screening rice varieties for allelopathic potential against arrowhead (*Sagittaria montevidensis*), an aquatic weed infesting Australian Riverina rice crops. *Australian Journal of Agricultural Research* 55: 673–680.

Searle, S. R. 1971. *Linear Models*. Wiley, New York.

Searle, S. R., Casella, G., and McCullough, C. E. 1992. *Variance Components*. Wiley, New York.

Shaffer, J. P. 1995. Multiple hypothesis testing. *Annual Review of Psychology* 46: 561–576.

Shapin, S. and Schaffer, S. 1985. *Leviathan and the Air Pump: Hobbes, Boyle and the Experimental Life, Including a Translation of Thomas Hobbes, Dialogus Physicus De Natura Aeris*. Princeton University Press, Princeton, NJ.

Sharma, M. A. and Singh, J. B. 2010. Use of probability distribution in rainfall analysis. *New York Science Journal* 3(9): 40–49.

Shaw, D. J. and Dobson, A. P. 1995. Patterns of macroparasite abundance and aggregation in wildlife populations: A quantitative review. *Parasitology* 111(Suppl): S111–S127.

Sheather, S. J. and Jones, M. C. 1991. A reliable data-based bandwidth selection method for kernel density estimation. *Journal of the Royal Statistical Society, Series B* 53: 683–690.

Sheather, S. J. and McKean, J. W. 1987. A comparison of testing and confidence interval methods for the median. *Statistics and Probability Letters* 6: 31–36.

Shibata, R. 1976. Selection of the order of an autoregressive model by Akaike's information criterion. *Biometrika* 63: 117–126.

Shoemaker, L. H. and Hettmansperger, T. P. 1982. Robust estimates and tests for the one- and two-sample scale models. *Biometrika* 69: 47–53.

Shrader-Frechette, K. S. and McCoy, E. D. 1992. Statistics, costs and rationality in ecological inference. *Trends in Ecology and Evolution* 7: 96–99.

Shumway, R. H. and Stoffer, D. S. 2000. *Time Series Analysis and Its Applications*. Springer, New York.

Siegel, S. 1956. *Nonparametric Statistics for the Behavioral Sciences*. McGraw-Hill, New York.

Silvapulle, M. J. 2001. Tests against qualitative interaction: Exact critical values and robust tests. *Biometrics* 57: 1157–1165.

Simpson, E. H. 1949. Measurement of diversity. *Nature* 163: 688.

Simpson, E. H. 1951. The interpretation of interaction in contingency tables. *Journal of the Royal Statistical Society Series B* 13: 238–241.

Sinclair, A. R. E. and Arcese, P. 1995. Population consequences of predation-sensitive foraging: The Serengeti wildebeest. *Ecology* 76(3): 882–891.

Siniff, D. B. and Skoog, R. O. 1964. Aerial censusing of caribou using stratified random sampling. *Journal of Wildlife Management* 28: 391–401.

Smith, E. P. and van Belle, G. 1984. Nonparametric estimation species richness. *Biometrics* 40: 119–129.

Smith, J. N. M., Keller, L. F., Marr, A. B., and Arcese, P. 2006. *Conservation and Biology of Small Populations: The Song Sparrows of Mandarte Island*. Oxford University Press, New York.

Smithson, M. and Verkuilen, J. 2006. A better lemon squeezer? Maximum-likelihood regression with beta-distributed dependent variables. *Psychological Methods* 11(1): 54–71.

Snedecor, G. W. and Cochran, W. G. 1989. *Statistical Methods*, 8th ed. Iowa State College Press, Ames, Iowa.

Sober, E. 1999. *Philosophy of Biology*. Westview Press, Boulder, CA.

Sokal, R. R. and Rohlf, F. J. 2012. *Biometry*, 4th ed. W. H. Freeman and Co., New York.

Spiegel, M. R. and Stephens, L. J. 1998. *Theory and Problems of Statistics*, 3rd ed. McGraw-Hill, New York.

Stanford School of Medicine. 2009. http://bloodcenter.stanford.edu/about_blood/blood_types.html. (Accessed 1/28/2013).

Steering Committee of the Physicians' Health Study Research Group. 1989. Final report on the aspirin component of the ongoing physicians health study. *New England Journal of Medicine* 321: 129–139.

Stent, G. S. 1972. Prematurity and uniqueness in scientific discovery. *Scientific American* 227: 84–93.

Stewart, J. S. 2003. *Calculus, Early Transcendentals*, 5th ed. Brooks/Cole, Belmont, CA.

Strawson, P. F. 1952. *Introduction to Logical Theory*. Methuen and Co., London.

Student. 1908. The probable error of the mean. *Biometrika* 6: 1–25.

Suddath, R. L. et al. 1990. Anatomical abnormalities in the brains of monozygotic twins discordant for schizophrenia. *New England Journal of Medicine* 322(12): 789–793.

Suess, H. and Urey, H. 1956. Abundances of the elements. *Reviews of Modern Physics* 28: 53–74.

Sullivan, J. and Joyce, P. 2005. Model selection and phylogenetics. *Annual Review Ecology Evolution and Systematics* 36: 445–466.

Tauber, A. I. 2009. *Science and the Quest for Meaning*. Baylor University Press, Waco, TX.

Taylor, J. R. 1997. *An Introduction to Error Analysis: The Study of Uncertainties in Physical Measurements*. University Science Books, Sausalito, CA.

Temeles, E. J. and Kress, W. J. 2003. Adaptation in plant–hummingbird association. *Science* 300: 630–633.

Tempelman, R. J. and Gianola, D. 1999. Genetic analysis of fertility in dairy cattle using negative binomial mixed models. *Journal of Dairy Science* 82(8): 1834–1847.

Templeton, A. R., Brazeal, H., and Neuwald, J. L. 2011. The transition from isolated patches to a metapopulation in the eastern collared lizard in response to prescribed fires. *Ecology* 92: 1736–1747.

Ter Braak, C. F. J. 1992. Permutation versus bootstrap significance tests in multiple regression and ANOVA. In Jöckel, K. J., ed., *Bootstrapping and Related Techniques*. Springer-Verlag, Berlin.

Thiel-Egenter C. et al. 2009. Effects of species traits on the genetic diversity of high-mountain plants: A multi-species study across the Alps and the Carpathians. *Global Ecology and Biogeography* 18(1): 78–87.

Thomas, M. A. and Klaper, R. D. 2012. Psychoactive pharmaceuticals induce fish gene expression profiles associated with human idiopathic autism. *PLoS One* 7(6): e32917.

Thompson, B. 1998. Statistical significance and effect size reporting: Portrait of a possible future. *Research in the Schools* 5(2): 33–38.

Thornton, A. and Currey, D. 1991. *To Save an Elephant. The Undercover Investigation Into the Illegal Ivory Trade*. Doubleday, London, UK.

Tilman, D. 1996. Biodiversity: Population versus ecosystem stability. *Ecology* 80: 350–363.

Tilman, D., Lehman, C. L., and Bristow, C. E. 1998. Notes and comments: Diversity–stability relationships: Statistical inevitability or ecological consequence? *The American Naturalist* 151(3): 277–282.

Tippett, L. H. C. 1950. *Technological Applications of Statistics*. John Wiley & Sons, New York, NY.

Tippett, L. H. C. 1952. *The Methods of Statistics*, 4th ed. John Wiley & Sons, New York, NY.

Tobler, W. 1970. A computer movie simulating urban growth in the Detroit region. *Economic Geography* 46(2): 234–240.

Todd, C. D. and Keough, M. J. 1994. Larval settlement in hard substratum epifaunal assemblages: A manipulative field study of the effects of substratum filming and the presence of incumbents. *Journal of Experimental Marine Biology and Ecology* 181: 159–187.

Tukey, J. W. 1949. One degree of freedom test for non-additivity. *Biometrics* 5: 232–242.

Tukey, J. W. 1953. *The Problem of Multiple Comparisons.* Unpublished manuscript.

Tukey, J. W. 1977. *Exploratory Data Analysis.* Addison-Wesley, Reading, MA.

Underwood, A. J. 1990. Experiments in ecology and management: Their logic, functions, and interpretations. *Australian Journal of Ecology* 14: 365–389.

Underwood, A. J. 1997. *Experiments in Ecology: Their Logical Design and Interpretation Using Analysis of Variance.* Cambridge University Press, Cambridge, UK.

Underwood, A. J. 1999. Publication of so-called negative results in marine ecology. *Marine Ecology Progress Series* 191: 307–309.

USFWS-DOI. 2007. Grizzly bears; Yellowstone District population; notice of final petition finding; final rule. *Federal Register* 72(60): 14866–14936. http://www.fws.gov/mountain-prairie/species/mammals/grizzly/FR_Final_YGB_rule_03292007.pdf

Vallbona, C., Hazlewood, C. F., and Jurida, G. 1997. Response of pain to static magnetic fields in postpolio patients, a double blind pilot study. *Archives of Physical Medicine and Rehabilitation* 78: 1200–1203.

Van Dongen, S. 2006. Prior specification in Bayesian statistics: Three cautionary tales. *Journal of Theoretical Biology* 242(1): 90–100.

Van Duijn, C. M. and Hoffman, A. 1991. Relation between nicotine intake and Alzheimer's disease. *British Medical Journal* 302: 1491–1494.

Väre, H., Ohtonen, R., and Oksanen, J. 1995. Effects of reindeer grazing on understorey vegetation in dry *Pinus sylvestris* forests. *Journal of Vegetation Science* 6: 523–530.

Varona, L. and Sorensen, D. 2009. A genetic analysis of mortality in pigs. *Genetics* 184(1): 277–284.

Vaux, D. L. 2012. Know when your numbers are significant. *Nature* 492: 180–181.

Venables, W. N. and Ripley, B. D. 2002. *Modern Applied Statistics with S.*, 4th ed. Springer, New York.

Verbeke, G. and Molenberghs, G. 2000. *Linear Mixed Models for Longitudinal Data.* Springer-Verlag, Berlin.

Vinson, J. P., David, B., Jaffe, D. B., O'Neill, K., Karlsson, E. K., Stange-Thomann, N., Anderson, S. et al. 2005. Assembly of polymorphic genomes: Algorithms and application to *Ciona savignyi. Genome Research* 15: 1127–1135.

Vollset, S. E. 1993. Confidence intervals for a binomial proportion. *Statistics in Medicine* 12: 809–824.

von Storch, H., and Zwiers, F. W. 2001. *Statistical Analysis in Climate Research.* Cambridge University Press, Cambridge, UK.

Walter, H. and Leith, H. 1967. *Klimadiagramm Weltatlas.* In Fischer, G., ed., Jena. Stuttgart, GER.

Wand, M. P. 2012. KernSmooth: Functions for kernel smoothing for Wand and Jones, 1995. R package version 2.23–8. http://CRAN.R-project.org/package=KernSmooth.

Wand, M. P. and Jones, M. C. 1995. *Kernel Smoothing.* Chapman and Hall, London.

Wangensteen, O. H., Peter, E. T., Bernstein, A. I., Walder, A. I., Sosin, H., and Madsen, A. J. 1962. Can physiological gastrectomy be achieved by gastric freezing? *Annals of Surgery* 156: 579.

Wasserman, L. 2004. *All of Statistics: A Concise Course in Statistical Inference.* Springer, New York, NY.

Watson, G. S. 1964. Smooth regression analysis. *Sankhyā: The Indian Journal of Statistics Series A* 26(4): 359–372.

Wedderburn, R. W. M. 1974. Quasi-likelihood functions, generalized linear models and the Gauss–Newton method. *Biometrika* 61: 439–447.

Weindruch, R., Walford, R. L., Fligiel, S., and Guthrie, D. 1986. The retardation of aging in mice by dietary restriction: Longevity, cancer, immunity and lifetime energy intake. *The Journal of Nutrition* 116(4): 641–654.

Welch, B. L. 1947. The generalization of student's problem when several different population variances are involved. *Biometrika* 34: 28–35.

Welch, B. L. 1951. On the comparison of several mean values: An alternative approach. *Biometrika* 38: 330–336.

West, B. T. and Galecki, A. T. 2011. An overview of current software procedures for fitting linear mixed models. *The American Statistician* 65: 274–282.

West, B. T., Welch, K. B., and Galecki, A. T. 2008. *Linear Mixed Models, A Practical Guide Using Software.* Chapman and Hall, Boca Raton, FL.

White, P. J. 2003. *Mountain Goat Surveys. Memo to John Varley, Branch Chief, Yellowstone Center for Resources*. Yellowstone National Park, Mammoth, Wyoming.

Whitlock, M. and Schluter, D. 2009. *The Analysis of Biological Data*. Roberts and Co., Greenwood Village, CO.

Wilcox, R. R. 1986. Improved simultaneous confidence intervals for linear contrasts and regression parameters. *Communications in Statistics—Simulations and Computation* 15: 917–932.

Wilcox, R. R. 1994a. The percentage bend correlation coefficient. *Psychometrika* 59: 601–616.

Wilcox, R. R. 1994b. A one way random effects model for trimmed means. *Psychometrika* 59: 289–306.

Wilcox, R. R. 1997. Tests of independence and zero correlations among p random variables. *Biometrical Journal* 39: 183–193.

Wilcox, R. R. 2005. *Introduction to Robust Estimation and Hypothesis Testing*, 2nd ed. Elsevier, Burlington, MA.

Wilcox, R. R. and Muska, J. 2001. Inferences about correlations when there is heteroscedasticity. *British Journal of Mathematical and Statistical Psychology* 54: 39–47.

Wilcoxon, F. 1945. Individual comparisons by ranking methods. *Biometrics Bulletin* 1: 80–83.

Williams, D. A. 1987. Generalized linear model diagnostics using the deviance and single case deletions. *Applied Statistics* 36: 181–191.

Wilk, M. B. 1955. The randomization analysis of a generalized randomized block design. *Biometrika* 42: 70–79.

Wilson, E. B. 1927. Probable inference, the law of succession, and statistical inference. *Journal of the American Statistical Association* 22: 209–212.

Winer, B. J., Brown, D. R., and Michels, K. M. 1991. *Statistical Principles in Experimental Design*, 3rd ed. McGraw-Hill, New York.

Wood, S. N. 2006. *Generalized Additive Models, An Introduction with R*. Chapman and Hall, London.

Wood, S. N. 2011. Fast stable restricted maximum likelihood and marginal likelihood estimation of semiparametric generalized linear models. *Journal of the Royal Statistical Society B* 73(1): 3–36.

Wright, P. J., Bonser, R., and Chukwu, U. O. 2000. The size–distance relationship in the wood ant *Formica rufa*. *Ecological Entomology* 25(2): 226–233.

Wright, S. 1938. Size of population and breeding structure in relation to evolution. *Science* 87(2263): 430–431.

Wu, P. -C. 2002. Central limit theorem and comparing means, trimmed means, one step M-estimators and modified one step M-estimators under non-normality. Doctoral dissertation, University of Southern California.

Xia, J. Q. and Lo, J. Y. 2008. Dedicated breast computed tomography: Volume image denoising via a partial-diffusion equation based technique. *Medical Physics* 35(5): 1950–1958.

Xie, Y. 2012. Animation: A gallery of animations in statistics and utilities to create animations. *R*-package version 2.1. http://cran.r-project.org/web/packages/animation/index.html.

Yates, F. 1934. Contingency table involving small numbers and the χ^2 test. *Journal of the Royal Statistical Society* (Supplement)1: 217–235.

Ye, J. 1998. On measuring and correcting the effects of data mining and model selection. *Journal of the American Statistical Association* 93: 120–131.

Yoccoz, N. G. 1991. Use, overuse, and misuse of significance tests in evolutionary biology and ecology. *Bulletin of the Ecological Society of America* 72(2): 106–111.

Yohai, V., Stahel, W. A., and Zamar, R. H. 1991. A procedure for robust estimation and inference in linear regression. In Stahel, W. A., and Weisberg, S. W., eds., *Directions in Robust Statistics and Diagnostics, Part II*. Springer, New York.

Ypma, T. 1995. Historical development of the Newton–Raphson method. *SIAM Review* 37: 531–551.

Zador, S. G. and Piatt, J. F. 2007. Simulating the effects of predation and egg-harvest at a gull colony. In Piatt, J. F. and Gende, S. M., eds., *Proceedings of the Fourth Glacier Bay Science Symposium*, Juneau, Alaska, October 26–28, 2004. US Geological Survey Scientific Investigations Report 2007–5047, 188–192.

Zangwill, K. M., Hamilton, D. H., Perkins, B. A., Regnery, R. L., Plikaytis, B. D., Hadler, J. L., Cartter, M. L., and Wenger, J. D. 1993. Cat scratch disease in Connecticut. Epidemiology, risk factors, and evaluation of a new diagnostic test. *New England Journal of Medicine* 329(1): 8–13.

Zar, J. H. 1999. *Biostatistical Analysis*, 4th ed. Prentice-Hall, Upper Saddle River, NJ.

Zelazo, P. R., Zelazo, N. A., and Kolb, S. 1972. Walking in the newborn. *Science* 176: 314–315.

Index

A

Accuracy, 5
Addition and Subtraction, 546–547
Additivity, 282; *see also* ANOVA
    designs
  additive model, 533
  additive series rules, 542–543
  exercises, 291
Affirming the consequent, 13; *see also*
    Logic
  exercises, 19
Agresti–Coull method, 514, 538
  binomial parameter confidence intervals,
    514, 538
  exercise, 538
Agresti–Pendergrast test, 494
AIC, *see* Akaike information criterion
    (AIC)
AICc, *see* Second-order "corrected" Akaike
    information criterion (AICc)
Akaike information criterion (AIC), 372; *see also*
    Model selection
  exercises, 414, 417
  second-order "corrected", 374
Algebra rules
  additive series rules, 542–543
  arithmetic operations rules, 541
  exponents, 541
  factorial rules, 542
  factoring, 541
  infinite geometric series, 543
  logarithms, 542
  product series rules, 543
  quadratic formula, 542
  radicals, 542
  special products, 541
All possible subsets, 377–378; *see also* Model
    selection
Analysis of contrasts, 479
  example, 480–481
  Greenhouse–Geisser epsilon, 480
  Huynh–Felt epsilon, 480
  in longitudinal designs, 481
  MANOVA, 479, 480
Analysis of deviance, *see* Drop in deviance
    test

Analysis of variance (ANOVA), 78, 421,
    496–497; *see also* ANOVA designs;
    Factor; Factor-level; Fixed effects;
    Multiway ANOVA; One-way ANOVA;
    Random effects model; Randomized
    block design; Regression; Repeated
    measures design; Robust ANOVA;
    Split-plot design; Two-way factorial
    design; Unbalanced designs
  analyses, 503
  applications of, 421
  assumptions, 454, 456
  Bayesian approaches to, 495–496
  constant μ, 423
  diagnostic plots, 455
  effect size, 453
  error terms independence, 454–456
  error terms normality, 456
  error variance constancy, 456–457
  example, 421–422
  exercises, 497–501
  factor-level sample mean, 423
  as general linear model, 442–445
  linear regression, 331–333
  mechanics of, 426
  moments variance components estimators,
    472
  multivariate, 479
  nested design, 471–474
  omega squared, 453
  outliers, 457
  random block effects, 469–470
  sample size, 423, 453, 454
  table for balanced two-factor split-plot
    design, 476
  terminology, 276–277
ANCOVA, 483, 497; *see also* Analysis of variance
    (ANOVA); Generalized linear models
    (GLMs)
  to assess independence of concomitant
    variable and treatments, 485
  concomitant variable, 483, 485
  example, 486–488
  exercise, 500–501
  framework, 484
  inferences, 486

ANCOVA (*Continued*)
  mechanics of, 484
  model for single-factor, 484
ANOVA, *see* Analysis of variance (ANOVA)
ANOVA designs, 276, 288; *see also* Experimental design
  additivity, 282
  ANOVA terminology, 276–277
  compendium of, 278
  compound symmetry, 285
  demonstration of, 279
  exercises, 289, 291
  factorial design, 280
  factors, 276
  fixed and random effects, 277–278
  longitudinal design, 284
  matched pairs, 284
  nested design, 282–283
  one-way ANOVA, 278–280
  randomized block design, 281–282
  repeated measures design, 284
  sphericity, 284–286
  split-plot design, 283–284
anova function, 332, 340–341, 392, 428, 480
Anscombe's quartet, 301, 302
Antiderivatives, 545
aov function, 475
AR1 covariance, *see* First-order autoregressive covariance (AR1 covariance)
ARIMA, *see* Autoregressive moving average (ARIMA)
Aristotle, 6
Arithmetic mean, 109–110; *see also* Location measures
  example, 153
  exercises, 144
  sampling distribution of, 151
Arithmetic operations rules, 541
Asphericity, 284; *see also* ANOVA designs
auc function, 394
Autoregressive moving average (ARIMA), 482
Avicenna, 6
Azidovudine (AZT), 34
AZT, *see* Azidovudine (AZT)

B

Bacon, Francis, 7
Bar charts, 107
Bartlett's test, 457; *see also* Fligner–Killeen test
Basic bootstrap method, 175; *see also* Bootstrapping
Bates, Douglas, 449

Bayesian applications, 137, 182; *see also* Bayes rule; Confidence intervals; Resampling distributions; Sampling distributions
  advanced applications, 188–192
  Bayes rule, 137
  conjugacy, 138
  direct simulation, 183–184
  example, 139–142, 182–183
  exercises, 146, 147, 196
  Gibbs sampling, 189
  indirect simulation, 184–185
  informative priors, 138
  Markov chain, 185
  MCMC algorithm, 187–188
  MCMC convergence, 188
  Metropolis–Hastings algorithm, 189–190
  noninformative priors, 138
  priors, 138
  simple applications, 185–187
  transition probabilities, 185
Bayesian approaches, 7, 413
  to ANOVA, 495–496
  exercises, 419, 501
  to regression, 411–412
Bayesian hierarchical models, 39
Bayesian hypothesis tests, 237; *see also* Null hypothesis
  disadvantages, 240
  example, 238–239
  exercises, 244, 245
  posterior odds ratio, 238
  posterior probability, 237
Bayesian information criterion (BIC), 374; *see also* Model selection
  exercises, 417
Bayesian statistical methods, 26; *see also* Probability
bayes.lm function, 412, 495
Bayes rule, 38, 137; *see also* Bayesian applications; Probability
  application, 39
  Bayesian hierarchical models, 39
  and cystic fibrosis diagnosis, 40–42
  data distribution, 38
  examples, 39
  exercises, 43, 46
  total probability, 38
Bayes, Thomas, 7, 8, 38, 549
BCa, *see* Bias corrected and accelerated bootstrap (BCa)
BDM.2way function, 495
BDM.test function, 494
Benjamini–Hochberg method, 441

Bernoulli, Daniel, 7, 61
Bernoulli distribution, 55, 504; *see also*
    Probability density functions (pdf)
Bernoulli, Jakob, 7, 55
Bernoulli random variable, 388
Best linear unbiased estimators (BLUEs), 325
Beta distribution, 81; *see also* Continuous pdfs
  errors, 386
  exercises, 99
  parameters, 82
  pdf, 81
Between-subject effect, 478
Bias, 266
  exercises, 291
Bias corrected and accelerated bootstrap (BCa),
    176–177; *see also* Bootstrapping
BIC, *see* Bayesian information criterion (BIC)
bicubic function, 404
Binary trial, 55
Binomial coefficient, 37
Binomial distribution, 55, 552–553; *see also*
    Discrete pdfs; Probability density
    functions (pdf)
  examples, 57–59
  exercises, 96–97, 538
  parameters, 56
  parameter value effect on pdf and cdf, 57
Binomial errors, 386; *see also* Generalized linear
    models (GLMs)
Binomial GLMs, 387, 531–532; *see also*
    Generalized linear models (GLMs)
  Bernoulli random variable, 388
  expected probability of success, 388
  linear regression model, 387
  logistic mean function, 388
  logit link function, 389
Binomial parameter confidence intervals, 513
  Agresti–Coull method, 514, 538
  binomial proportions ratio inference,
    516–517
  Clopper–Pearson exact method, 514
  likelihood ratio method, 514–516, 539
  method comparison for valid inference, 516
  odds ratio inference, 517–518
  relative risk inference, 518–519
  score method, 514
  Wald method, 513
Binomial pdf, 504
Binomial proportions comparison, 506
binom.test function, 512, 516
Bins, 50
Bivariate normal distribution, 299; *see also*
    Pearson's correlation

  example, 299–300
  joint density, 299
  two-element vector, 299
Biweight midcorrelation, 315–316
Blocking factor, 474
BLUEs, *see* Best linear unbiased estimators
    .(BLUEs)
Bonferroni, 343
  confidence intervals, 343
  correction, 437, 440
  –Holm procedure, 441
  inequality, 31–32, 45; *see also* Boole's
    inequality; Probability
Bonferroni, Carlo Emilio, 31
Boole, George, 30
Boole's inequality, 30; *see also* Bonferroni's
    inequality; Probability
  exercises, 45
Bootstrap, 178–179
  confidence intervals, 174
  trap standard error, 173
Bootstrapping, 172; *see also* Confidence
    intervals; Resampling distributions
  approaches for, 172–173
  bias corrected and accelerated bootstrap,
    176–177
  bias estimation, 173
  exercises, 195
  methods, 175, 378–379
  normal approximation method, 174–175
  percentile method, 176
  sample, 173–174
  studentized bootstrap, 175–176
  worldview, 173
Bootstrap-t method, *see* Studentized
    bootstrap
Box–Cox procedure, 361; *see also* Optimal
    transformation
  exercises, 415
Box, David, 17
boxTidwell function, 361
Box–Tidwell procedure, 361; *see also* Optimal
    transformation
  exercises, 415
Boyd, Richard, 15
Brunner–Dette–Munk test, 494
Burn-in period, 185
bv.boxplot function, 304

C

$c(y, \phi)$ function, 551
Calculus, fundamental theorem of, 545

Calculus rules
  antiderivatives, 545
  chain rule, 544
  differentiation, 543–544
  integration, 545–546
  L'Hôpital's rule, 545
  power rule, 544
  product rule, 544
  theorem, 545
Canonical links, 386
Canonical parameter, 551
Carryover effects, 479
Categorical variables, 249; *see also* Variables
Cauchy distribution, 62
Causality, 248
  exercises, 289
cdf, *see* Cumulative distribution function (cdf)
Central limit theorem, 154–156
  exercises, 193
CFTR, *see* Cystic fibrosis transmembrane
    regulatory (CFTR)
Chain rule, 544
Chaotic processes, 15
Chebyshev inequality, 106; *see also* Parameters
  exercises, 144
Chebyshev, Pafnuty Lvovich, 106
chisq.test function, 524, 529
χ^2 distribution, *see* Chi-squared distribution
Chi-squared distribution, 74; *see also*
    Continuous pdfs
  degrees of freedom, 76
  example, 76
  exercises, 99
  pdf, 75
  goodness of fit test, 508
Chi-squared test in two-way tables, 521
ci.p function, 515
ci.prat function, 517
Clopper–Pearson exact method, 514
Clusters, 256
Cluster sampling, 256; *see also* Sampling design
  exercises, 289
CMH test, *see* Cochran–Mantel–Haenszel test
    (CMH test)
CMV, *see* Cytomegalovirus (CMV)
Cochran–Mantel–Haenszel test (CMH test), 531
Cochran, W.G., 17
Coefficient of determination, 346
Coefficient of partial determination, 347
Coefficient of variation, 121; *see also* Scale
    estimators
  advantages, 122
  example, 122

  exercises, 144
  population, 121
  sample, 121
Cognitive illusions, 102
Cohens *d*, 224
Combinations, 37; *see also* Combinatorial
    analysis; Probability
  exercises, 46
Combinatorial analysis, 35; *see also* Probability
  combinations, 37
  examples, 36
  exercises, 46
  multiplication principle, 35
  permutations, 36–37
Communalism, 5
  exercises, 18
Completely randomized design (CRD), 278
  exercises, 289, 291
Compound symmetry, 285; *see also* ANOVA
    designs
Computerized tomography (CT), 46
Concomitant variable, 483, 485
ConDis.matrix function, 313, 319
Conditional associations, 528
Conditional distributions, 520
Conditional independence model, 534
Conditional probability, 32, 549
Confidence bounds, 161
  for median, 169
Confidence intervals, 161; *see also*
    Bootstrapping; Credible intervals;
    Regression; Resampling distributions;
    Sampling distributions
  assumptions and requirements for, 172
  Bonferroni, 343
  calculation, 162
  dirty.dist, 171
  estimate precision, 165–167
  exact, 163
  example, 163–164, 166–167, 168, 170, 343–344,
    345–346
  exercises, 195, 415
  for β_k, 343–344
  and hypothesis testing, 209–210
  interpretation, 164–165
  logical estimator, 165
  margin of error, 171
  one-sided, 167–168
  pivotal quantity, 162–163
  for population mean, 161, 165
  for population median, 169–171
  for population variance, 168–169
  and sample size, 171

standard error of mean, 165
  for true fitted values and prediction
      intervals, 344–346
confint function, 344
Confounded variables, 252–253
  exercises, 288
Conjugate, 138
  example, 139–142
Contingency tables, 519–520; *see also* Partial
      tables; Two-way tables
  conditional distributions, 520
  joint distribution, 519
  marginal distributions, 519
Continuous pdfs, 51, 74; *see also* Chi-squared
      distribution; Continuous uniform
      distribution; Exponential distribution;
      F-distribution; Lognormal
      distribution; Probability density
      functions (pdf); Weibull distribution
  beta distribution, 81–82
  conceptualization of, 52
  exercises, 99
  expected value, 103
  gamma distribution, 82–83
  logistic distribution, 88–89
  similar to normal distributions, 75
  t-distribution, 75, 76–77
Continuous random variables, 49; *see also*
      Random variables
  cdf, 52
  exercises, 95–96
Continuous uniform distribution, 60; *see*
     also Probability density functions
     (pdf)
  cdf, 60
  example, 60–61, 104
  parameters, 60
Continuous variables, 250; *see also* Variables
contrast function, 432
Contrast matrix, 431
Contrasts, 429; *see also* Factor-level; Orthogonal
      comparisons
  argument, 324
  example, 430–431
  exercises, 498
  Helmert contrast, 433–435
  linear contrasts, 429
  lm contrast, 432
  matrix, 431
  pairwise comparisons, 430, 439
  reverse Helmert contrasts, 433
  sum contrast, 435
  treatment contrasts, 432–433, 443

Controls, 271–272
  exercises, 291
Cook's distance, 354–355; *see also* Linear
      regression
  computation, 459
  exercises, 415, 417
Cornfield, Jerome, 16, 17
Correction factor, 121
Correlation, 295, 318; *see also* Pearson's
      correlation
  analysis, 295
  comparisons of procedures, 316–318
  example, 295, 310–311
  exercises, 318–320
  Kendall's τ, 312
  percentage bend criterion, 315–316
  population, 296–297
  rank-based permutation approaches, 308
  robust, 308, 315
  sample biweight midvariance, 315–316
  Spearman's estimator, 308–311
  Winsorized correlation, 315
cor.test function, 306, 311, 315
Covariance structure, 482–483
Covariates, 272
  exercises, 291
Cox, Gertrude, 17
CRD, *see* Completely randomized design (CRD)
Credible intervals, 182, 193; *see also* Confidence
      intervals
Critical value, 201
  exercises, 241
CT, *see* Computerized tomography (CT)
cubic spline function, 407
Cumulative distribution function (cdf), 52; *see*
     also Empirical cdf (ecdf); Probability
     density functions (pdf)
  continuous uniform, 60
  exponential, 79
  extensions of, 53
  from goat sighting distribution, 54
  inverse, 53
  lower tail probability, 52
  normal, 62
  Weibull, 84
Cystic fibrosis transmembrane regulatory
     (CFTR), 40
Cytomegalovirus (CMV), 198

D

DA, *see* Definite appendicitis (DA)
Data argument, 324

Data
  distribution, 38
  dredging, 378; *see also* Model selection
DDT, *see* Dichlorodiphenyltrichloroethane
        (DDT)
Decision rule, 202
  exercises, 241
Deduction, 11; *see also* Logic
Definite appendicitis (DA), 47
Degrees of freedom, 74
de Laplace, Simon Pierre, 7
de Moivre, Abraham, 7, 61
Density, 50; *see also* Probability density
        functions (pdf)
  continuous, 50–51
  discrete, 50
  exercises, 98
Denying the consequent, 12; *see also* Logic
Descartes, René, 11
Determinant, 548
det function, 548
Deviance, 389–391; *see also* Generalized linear
        models (GLMs)
Deviations, 119
DH.test function, 304
Diagonal matrices, 546
Diagonal model, *see* Independent covariance
Dichlorodiphenyltrichloroethane (DDT), 211
Differentiation, 543–544
difftime function, 406
dirty.dist function, 171; *see also* Confidence
        intervals
Discrete pdfs, 51, 66; *see also* Binomial
        distribution; Geometric distribution;
        Hypergeometric distribution;
        Negative binomial distribution;
        Poisson distribution; Probability
        density functions (pdf)
  Bernoulli distribution, 55
  example, 104
Discrete random variables, 49; *see also*
        Probability mass functions (pmfs);
        Random variables
  exercises, 95
Discrete variables, 250; *see also* Variables
Disjoint, 28; *see also* Probability
Dispersion, 399
  parameter, 551
Distribution shape, 124; *see also* Statistics
  example, 126–127
  exercises, 145
  *i*th central sample moment, 126
  kurtosis, 124

  moment generating functions, 124
  parameters and estimators for, 124
  sample moments and MOM estimators, 126
  skewness, 124
  unbiased estimators, 127
Doornik–Hansen test, 304
dpik function, 405
Drop in deviance test, 391
Dummy coding, 443
Dunnett's method, 439, 441
Durbin–Watson test, 262

E

ecdf, *see* Empirical cdf (ecdf)
Edwards, A.W.F., 16, 17
Effective population size, 112
Effect size, 453, 526–527
  exercises, 497–498
eff.rbd function, 469
Eigenanalysis, 548–549
Einstein, Albert, 15
Elements, 23–24; *see also* Set theory
Empirical cdf (ecdf), 90–91; *see also* Cumulative
        distribution function (cdf)
Empirical distribution plots, 352
Empirical rule, 62
Empirical sciences, 4, 15
  exercises, 18
Empiricism, 6
Empodocles of Agrigentum, 6
Empty set, 24; *see also* Set theory
Equivocally appendicitis, *see* Probably
        appendicitis (PA)
Error matrix, 392
Error terms
  independence, 454–456
  normality, 456
Error variance constancy, 456–457
Estimators, 106, 143; *see also* Location
        measures; Maximum likelihood (ML);
        M-estimators; Method of moments
        (MOM); Ordinary least squares (OLS);
        Parameter; Scale estimators; Statistics
  approaches for deriving, 127
  breakdown point of, 113
  exercises, 143
  gauging estimator effectiveness, 108
  interval estimators, 108
  of location, 113
  point estimators, 108–109
  types of, 108
Euclidean distance, 404

Event, 24; *see also* Set theory
Exact binomial tests, 506, 507, 512–513
exactLRT function, 450
expand.grid function, 525
Expected value, 103; *see also* Parameter
    exercises, 143, 144
Experiment, 24; *see also* Set theory
Experimental design, 269, 287–288; *see also*
        ANOVA designs; Sampling design;
        Tabular designs
    appropriate covariates measurement, 272
    classification of, 273
    controls, 271–272
    exercises, 289, 290, 291
    approaches, 269
    linear models, 274
    manipulative vs. observational, 269
    multiple regression, 275–276
    prospective and retrospective studies,
        272–273
    randomized vs. nonrandomized, 270–271
    regression designs, 275
    simple linear regression, 275
Experimental units, 247
Explanatory variables, 248–249; *see also*
        Variables
    exercises, 288
Exponential distribution, 79; *see also*
        Continuous pdfs
    applications, 79–80
    cdf, 79
    example, 80–81, 85
    exercises, 99, 100
    pdf, 79
Exponential family, 138, 551
Exponents, 541
Extrapolation, 342
Extra sums of squares, *see* Type II sums of
    squares (Type II SS)

F

Factorial design, 280; *see also* ANOVA designs
    advantages, 280
    disadvantages, 280
    exercises, 291
Factorial rules, 542
Factoring, 541
Factor-level, 422–423; *see also* Contrasts;
        Simultaneous inference procedures
    comparisons, 442
    inferences for, 429
    multiple comparison issues, 436

    population effects, 423
    population variance, 424
    P-value adjustment methods, 441–442
Fallacious argument, 11; *see also* Logic
False discovery rate (FDR), 441
Falsifiability, 9–10
Familywise type I error rates, 436, 441
F-distribution, 75, 77, 158–160; *see also*
        Continuous pdfs; Test statistics
    applications, 78
    degrees of freedom, 77
    example, 78
    exercises, 99
FDR, *see* False discovery rate (FDR)
Fermat, Pierre de, 7
F function, 545
Finite population correction, 258–259; *see also*
        Sampling design
First-order autoregressive covariance (AR1
        covariance), 482
    exercise, 501
Fisher, Roland A., 16, 17
Fisher's exact test in two-way tables, 523
Fisher's least significant difference (LSD), 436
Fisher–Snedecor distribution, *see*
        F-distribution
fisher.test function, 523
fitted function, 342, 390
Fixed effects
    effect of random factor-level selection, 465
    intercept term, 452
    likelihood estimate based on REML, 450
    linear model, 446
    quadratic expressions of, 476
    standard errors for, 449
    variance–covariance components, 448
Fligner–Killeen test, 215, 457; *see also* Bartlett's
        test; Modified Levene test
Formula argument, 324
F-quantile function, 438
friedman_test function, 494
Friedman's Rank F-test variant, 494

G

Galilei, Galileo, 7
Galton, Francis, 16, 17
gam function, 409–410; *see also* Generalized
        additive models (GAMs)
Gamma distribution, 82–83; *see also* Continuous
        pdfs
    example, 85
    exercises, 100

GAMs, *see* Generalized additive models
(GAMs)
Gauging estimator, 108
Gauss, Carl Friedrich, 51, 61
Gauss–Newton method, 400
General covariance, *see* Unstructured
covariance
Generalized additive models (GAMs), 370, 408;
see also Smoothers
assumptions, 410
example, 410
exercises, 419
gam function, 409–410
relative parsimony of, 409
Generalized least squares (GLS), 452
Generalized linear models (GLMs), 17, 134, 274,
323, 386, 413, 531; *see also* Binomial
GLMs; GLM inferential methods;
Log-linear models; Poisson GLM;
Regression
ANOVA as, 442–445
assumptions, 350, 396
binomial errors, 386
binomial GLMs, 531–532
components, 386
deviance, 389–391
dispersion, 399
exercises, 418
fitted probabilities numerically 0 or 1, 400
glm function, 387
lm function, 324
methods comparison, 537
model estimation, 387
model goodness of fit, 397–398
with nonlinear transformations, 323–324
overdispersion, 399
Pearson residuals, 397
polynomial regression models, 323
predictor regression, 323
quasi-likelihood, 399
residuals, 396–397
Geometric distribution, 71; *see also* Discrete
pdfs
example, 72
pdf, 71
Geometric mean, 110–111; *see also* Location
measures
exercises, 144
Gibbs, Josiah Willard, 188
Gibbs sampler, 188
Gibbs sampling, 189
glm function, 387; *see also* Generalized linear
models (GLMs)

GLM inferential methods, 391; *see also*
Generalized linear models (GLMs)
concordance index, 393
error matrix, 392
information-theoretic criteria, 394–395
likelihood ratio test, 391–392
measures of explained variance, 394
package pROC, 394
ROC and AUC, 392–394
sensitivity, 392
specificity, 392
Wald test, 391
glm.nb function, 399
GLMs, *see* Generalized linear models (GLMs)
GLS, *see* Generalized least squares (GLS)
gls function, 452
Gosset, William Sealy, 16, 17
Greater Yellowstone Ecosystem (GYE), 259
Greenhouse–Geisser epsilon, 480
exercise, 501
Grizzly bear, 259
groupedData function, 483
GYE, *see* Greater Yellowstone Ecosystem
(GYE)

H

Harmonic mean, 111; *see also* Location
measures
example, 112
exercises, 144
population, 111
of population sizes, 112–113
sample, 112
Hastings, W.K., 189
Heisenberg uncertainty principle, 15
Helmert contrast, 433–435
Heteroscedasticity, 459
Histogram, 50
continuous pdf, 52
for simple random variable, 51
HLgof.test function, 398
Homogenous association model, 534
Hosmer Lemeshow goodness of fit test, 397
HSD, *see* Tukey's honest significant difference
(HSD)
Huber's estimator, 379
Hume, David, 7
Huygens, Christian, 7
Huynh–Felt epsilon, 480
Hypergeometric distribution, 69; *see also*
Discrete pdfs
application, 69

example, 70–71
exercises, 98
parameters, 69
pdf, 69
Hypothesis, 9; *see also* Scientific hypotheses
Hypothesis testing, 197, 240–241; *see also* Null
    hypothesis; Parametric frequentist
    null hypothesis testing; Permutation
    tests; Power
  confidence intervals and, 209–210
  exercises, 241–245
  groups, 197
  in multiple regression, 337–338
  in simple linear regression, 329–331
  type I and type II errors, 219–220

I

Idealized population, 112
Identity matrices, 546
Inclusion–exclusion principle, 30; *see also*
    Probability
Independence, 550
  model, 533–534
Independent covariance, 482
Inductive reasoning, 8, 10–11; *see also* Logic
Infinite geometric series, 543
Inflection point, 365
Infrared gas analyzer, 248
Integration, 545
Interquartile range (IQR), 109, 122, 123; *see also*
    Scale estimators
  exercises, 144
Intersect, 29; *see also* Probability
Intersections, 549
Interval estimators, 108; *see also* Estimators
Intraclass correlation coefficient, 448
Inverse, 548
  hyperbolic tangent, 306
IQR, *see* Interquartile range (IQR)

J

Jackknifing resampling, 179; *see also*
    Resampling distributions
  bias estimator, 180
  example, 180–181
  exercises, 195, 196
  first-order, 179
  jackknife estimate, 179
  pseudo.v, 181–182
  pth order, 180
Jensen's inequality, 110

joint.ci.bonf function, 343
Joint distribution, 519
Joint independence, 534

K

Kendall's rank correlation coefficient, 308
Kendall's τ, 312; *see also* Correlation
  assumptions for tests, 313
  example, 313–315
  exercises, 319
  sample correlation coefficient, 312
  test statistic, 313
Kernel-based approaches, 404–405
kernel function, 404
Kernel-smoothers, 404–405; *see also*
    Smoothers
  exercises, 419
Keynes, John Maynard, 7
KL-information, *see* Kullback–Leibler
    information (KL-information)
Knots, 405
Kolmogorov, A.N., 17
Kolmogorov–Smirnov test, 230; *see also*
    Parametric frequentist null
    hypothesis testing; Rank-based
    permutation tests
  exercises, 243
kruskal_test function, 493
Kruskal–Wallis test, 493
  problem with, 494
Kuhn, Thomas, 10
Kullback–Leibler information (KL-information),
    372
Kurtosis, 124; *see also* Distribution shape
  exercises, 145
kw.pairw function, 494

L

Lag, 262
Lakatos, Imre, 10
Large sample theory, 403; *see also* Nonlinear
    regression models
Laws of physics, 15
Least significant difference, 437
Leibniz, Gottfried Wilhelm, 7
L'Hôpital's rule, 545
Likelihood-based approaches, 240; *see also* Null
    hypothesis
  exercises, 244
Likelihood function, *see* Data distribution
Likelihood principle, 225

Likelihood ratios, 238, 391; *see also* GLM
　　inferential methods
　　binomial parameter confidence intervals,
　　　　514–516, 539
　　for confidence intervals, 514–516
　　exercise, 539
　　method, 514–516, 539
　　test for one-way format, 511–513
　　test in two-way tables, 521–522
　　tests in one-way formats, 506
Limiting theorems, 154
Linear regression, 413; *see also* Multicollinearity;
　　　　Regression
　　assumptions, 350
　　Cook's distance, 354–355
　　default diagnostic plots, 351
　　error variance constancy, 353
　　exercises, 414, 415, 416
　　Fligner–Killeen test for homoscedasticity,
　　　　353
　　general linear regression model, 350
　　hypotheses, 352
　　hypothesis testing procedures, 350
　　independence of error terms, 350–352
　　*i*th standardized residual, 352
　　model, 322–323
　　normality of error terms, 352
　　outliers, 354–355
　　relationship between X and Y is linear,
　　　　353–354
Linear transformations, 134, 267
　　and parameters, 135–136
　　and statistics, 136–137
Lineplots, 107
link function, 386
lm contrast, 432
Lme function, 450–451
lmer function, 451
lmer in *lmer4* function, 459
lm function, 324; *see also* Generalized linear
　　　　models (GLMs)
lmodel2 function, 384
lm.select function, 376
Locally weighted scatter plot smoother
　　　　(LOWESS), 404; *see also* Smoothers
　　exercises, 419
Location measures, 109, 143; *see also* Estimator;
　　　　M-estimators; Statistics
　　arithmetic mean, 109–110, 118
　　geometric mean, 110–111, 118
　　harmonic mean, 111–113, 118
　　location estimator selection, 118–119
　　median, 114–115

mode, 113
　　robust, 113
　　trimmed mean, 115
　　Winsorized mean, 115–116
Location parameter, 62
loess function, 407
Logarithms, 542
Logic, 10; *see also* Science
　　deduction, 11
　　exercises, 19
　　fallacious argument, 11
　　induction, 10–11
　　induction vs. deduction, 11–12
　　mathematical arguments, 11
　　modus tollens, 12–13
　　and null hypothesis testing, 12
　　reductio ad absurdum, 13–14
logistic cdf function, 388
Logistic distribution, 88; *see also* Continuous
　　　　pdfs
　　logit function, 89
　　pdf, 88
Logistic mean function, 388, 389; *see also*
　　　　Generalized linear models (GLMs)
logit function, 386, 388, 389, 552, 553
Logit link function, 389; *see also* Generalized
　　　　linear models (GLMs)
logLik function, 370
Log-linear models, 532; *see also* Poisson log-
　　　　linear model; Three-way tables
　　additive model, 533
　　exercise, 540
　　four-way tables, 537
　　of independence, 533
　　log-linear model of independence, 533
　　two-way tables, 533
log link function, 395
Lognormal distribution, 75, 86; *see also*
　　　　Continuous pdfs
　　application, 87
　　example, 86–87
　　exercises, 100
　　pdf, 86
　　Preston's lognormal distribution, 88
Longitudinal design, 284; *see also* ANOVA
　　　　designs
LOWESS, *see* Locally weighted scatter plot
　　　　smoother (LOWESS)
lqs function, 494, 495
LSD, *see* Fisher's least significant difference
　　　　(LSD)
Lurking variables, 251–252
　　exercises, 288

M

Mach, Ernst, 15
Mack–Skillings test, 495
MAD, *see* Median absolute deviation (MAD)
Major axis regression (MA regression), 381; *see also* Model II regression
    exercises, 418
    MA slope, 382
    OLS and random regression parameter estimation, 382
Mallows' C_p, 374–375; *see also* Model selection
Manipulative experiment, 269
MANOVA, *see* Multivariate analysis of variance (MANOVA)
mantelhaen.test function, 531
MA regression, *see* Major axis regression (MA regression)
Marginal distributions, 519
Marginality violation principle, 489
Marginal residual distribution, 459
Marginal sums of squares, 488, *see* Type III sums of squares
Marginal tables, 528
Margin of error, 171
Markov, Andrey, 185
Markov chain, 185
    burn-in period, 185
    exercises, 196
Markov chain Monte Carlo (MCMC), 150, 185
    algorithm, 187–188
    convergence, 188
    exercises, 196
Markov transition probabilities, 185
    exercises, 196
Matched pairs, 284; *see also* ANOVA designs
Mathematical arguments, 11; *see also* Logic
Matrix, 546
Matrix algebra, 546
    addition and subtraction, 546–547
    determinant, 548
    diagonal matrices, 546
    eigenanalysis, 548–549
    identity matrices, 546
    inverse, 548
    matrix, 546
    multiplication, 547–548
    orthogonality, 549
    positive definite matrices, 549
    scalars, 546
    square matrices, 546
    symmetric matrices, 546

    transposition, 547
    vectors, 546
Maximal model, 378
Maximum likelihood (ML), 103, 128, 143, 448; *see also* Estimator; Method of moments (MOM); Ordinary least squares (OLS)
    estimator, 131
    example, 131
    exercises, 145, 146
    likelihood vs. probability, 132–133
    MOM vs. OLS vs. ML estimation, 133–134
    normal likelihood function, 129
    normal log-likelihood function, 130
    sample size effect on log-likelihood functions, 132
Maximum likelihood estimation (MLE), 128
MCAR, *see* Missing completely at random (MCAR)
McCullagh, Peter, 17
MCMC, *see* Markov chain Monte Carlo (MCMC)
mean function, 553
Mean-preserving spread, 113
Mean squared error (MSE), 214, 326
Measurement error, 266
    exercises, 291
Median, 114–115; *see also* Location measures
    exercises, 144
Median absolute deviation (MAD), 122, 123; *see also* Scale estimators
    exercises, 144
Mendel, Gregor, 27
M-estimators, 113, 116, 143; *see also* Estimator; Location measures
    example, 118
    exercises, 144
    Newton–Raphson method, 117–118
    weighting function for Huber, 117
M estimators, 379
Method of moments (MOM), 126, 143; *see also* Estimator; Maximum likelihood (ML); Ordinary least squares (OLS)
    exercises, 145
    vs. OLS vs. ML estimation, 133–134
Metropolis algorithm, 189–190
    exercises, 196
Metropolis–Hastings algorithm, 189, 190
    exercises, 196
Metropolis, Nicholas Constantine, 189
MGF, *see* Moment generating functions (MGF)
Michaelis–Menten
    kinetics equation, 324
    model, 401

Midvariance, 315
Millimeters of mercury, 206
Mill, John Stuart, 10
Missing completely at random (MCAR), 266
Missing data, 467
Mixed effect models, 457; *see also* Random effects model
   diagnostic plots, 458
   heteroscedasticity, 459
   type I error, 45
ML, *see* Maximum likelihood (ML)
MLE, *see* Maximum likelihood estimation (MLE)
MM estimation, 379
Mode, 113; *see also* Location measures
   exercises, 144
Model II regression, 381; *see also* Regression
   assumptions, 385–386
   example, 384–385
   MA regression, 381, 382–383
   RMA regression, 384
   SMA regression, 383–384
Model selection, 371; *see also* Regression
   AIC, 372–374
   AIC_c, 374
   all possible subsets, 377–378
   BIC, 374
   data dredging, 378
   *i*th Akaike weight, 374
   KL-information, 372–373
   Mallows' C_p, 374–375
   model selection approaches, 372
   parsimonious model, 371
   PRESS, 375
   selection criteria comparison, 375–377
   stepwise regression, 377–378
Modern mixed-model approaches, 482–483; *see also* Analysis of contrasts
   advantages, 482
   between-subject effect, 482
   covariance structure, 482–483
   example, 483
   within-subject effect, 482
Modified Levene test, 457; *see also* Fligner–Killeen test
Modus tollens, 12; *see also* Logic
   exercises, 19
MOM, *see* Method of moments (MOM)
Moment correlation coefficient, 300–301
Moment generating functions (MGF), 124
   for continuous random variable, 125
   for discrete random variable, 125
   example, 125
   exercises, 145
Moments variance estimators, 463, 464
   exercise, 500
montane.island function, 412
MSE, *see* Mean squared error (MSE)
MS.test function, 495
Multicollinearity, 355; *see also* Linear regression
   consequences of, 356
   example, 355, 357–358
   partial residual plots, 356
   population sampling variance, 358
   VIFs, 358–359
Multinomial distribution, 504, 538
Multinomial proportions comparison, 506
Multiple regression, 275–276, 333; *see also* Experimental design; Regression
   ANOVA approach, 339–341
   combined effect of X on Y, 338
   example, 333, 336, 337–338, 339, 340–341
   exercises, 291, 417
   fitted values for response variable, 335
   hypothesis testing, 337
   nonlinearity in, 354
   parameter estimation, 333–336
   partial regression coefficients, 335
   QR decomposition, 334
   response surface, 333
   tests concerning regression parameters, 337–338
   test statistic, 337, 340
   vector of regression parameters, 334
Multiplication, 547–548
Multivariate analysis, 250
Multivariate analysis of variance (MANOVA), 479
Multiway ANOVA, 494; *see also* One-way ANOVA
   Agresti–Pendergrast test, 494
   Brunner–Dette–Munk test, 495
   example, 495
   Mack–Skillings test, 495
   permutation tests, 494
   rank-based permutation tests, 494
   robust estimator tests, 495
Mutual independence model, 533–534
Mutually exclusive events, *see* Disjoint

N

NA, *see* No appendicitis (NA)
Negative binomial distribution, 72; *see also* Binomial distribution; Discrete pdfs

applications, 73
example, 73
exercises, 97–98
gamma function, 74
for noninteger values, 73
parameters, 72
pdf, 72
Nelder, John, 17
Nested design, 282, 471–474; *see also* ANOVA
designs
advantages, 282–283
disadvantages, 283
exercises, 292, 500
Net primary productivity (NPP), 249
Newton, Isaac, 7
Newton–Raphson method, 117–118
Neyman, Jerzy, 17
nlme::plot.lme function, 458
nls function, 401; *see also* Nonlinear regression
models
No appendicitis (NA), 47
Noncentral t-distribution, 224
Nonempirical sciences, 4
exercises, 18
Nonlinear regression models, 400, 413; *see also*
Regression
advantages, 403
assumptions, 402–403
disadvantages of, 403
Gauss–Newton method, 400
large sample theory, 403
model examples, 401–402
nls function, 401
nonlinear least squares, 400
Nonlinear transformations, 268
Nonquantitative variables, *see* Categorical
variables
Normal approximation method, 174–175; *see also*
Bootstrapping
Normal distribution, 61, 504, 551–552; *see also*
Probability density functions (pdf);
Z-distribution
parameters, 62
pdf and cdf, 62
standard, 63–65
NPP, *see* Net primary productivity (NPP)
Null hypothesis, 197; *see also* Bayesian
hypothesis tests; Hypothesis testing;
Likelihood-based approaches; Null
hypothesis testing; Parametric
frequentist null hypothesis testing
alternatives to, 237
example, 198

exercises, 241
motivation, 198
notion of, 198
two-tailed, 205
Null hypothesis testing, 12, 200; *see also* Logic;
Parametric frequentist null hypothesis
testing
critical value, 201
decision rule, 202
five-step process, 200
hybrid schema, 202
Neyman and Pearson model, 201–202
nonsignificant results, 202–203
P-values, 200–201
significant results, 203
Null standard error, 507

O

Objectivity, 4
Odds ratio, 34; *see also* Probability
exercises, 46, 539
inference, 517–518
OLS, *see* Ordinary least squares (OLS)
Omega squared, 453
One-sample t-test, 208–209; *see also*
Parametric frequentist null
hypothesis testing
One-sample z-test, 206–208; *see also* Parametric
frequentist null hypothesis testing
oneway_test function, 493, 494
One-way ANOVA, 278423, 442–443, 493; *see
also* ANOVA designs; Multiway
ANOVA
advantages, 279
Brunner–Dette–Munk test, 494
disadvantages, 279–280
$E(MS_A)$, 447
$E(MSE)$, 424
factor-level population variance, 424
general form of, 425
interactions potential, 459
Kruskal–Wallis test, 493
lm revisited, 426–428
permutation tests, 493
rank-based permutation tests, 493–494
robust estimator tests, 494
test statistic, 424
variance of observations, 424
One-way formats, 506; *see also* Likelihood ratio;
Score test; Wald test
exact binomial tests, 506, 507, 512–513
oneway.test function, 457, 494

Optimal transformation, 361; *see also* Regression
  Box–Cox, 361
  Box–Tidwell, 361
  example, 361–363
  exercises, 415
Order
  effects, 479
  statistics, 123
Ordinal variables, 249, 525–526; *see also*
      Pearson's correlation; Variables
Ordinary least squares (OLS), 103, 127, 143; *see
      also* Estimator; Maximum likelihood
      (ML); Method of moments (MOM)
  estimator, 127
  example, 128
  exercises, 145
  MOM vs. OLS vs. ML estimation, 133–134
Orthogonal comparisons, 431–432
Orthogonality, 549
Outcome, 24; *see also* Set theory
Outliers, 265–266, 354–355; *see also* Linear
      regression
  exercises, 291
Overdispersion, 399

P

PA, *see* Probably appendicitis (PA)
Package pROC, 394; *see also* GLM inferential
      methods
p.adjust function, 441
Paired t-test, 211; *see also* Parametric frequentist
      null hypothesis testing
  example, 212–214
  paired designs, 211
  test statistic, 212
pairs function, 358
pairw.anova function, 439, 463
pairw.fried function, 494
Parameter, 54, 103, 143; *see also* Statistics
  Chebyshev inequality, 106
  estimation, 149
  examples, 104
  exercises, 143
  expected value, 103
  linear transformations, 135–136
  variance, 105
Parametric frequentist null hypothesis testing,
      197; *see also* Hypothesis testing; Null
      hypothesis; Paired t-test; Permutation
      tests; Pooled variance t-test; Rank-based
      permutation tests; Robust estimator
      tests; Welch's approximate t-test

  alternatives to, 227
  confidence intervals and hypothesis testing,
      209–210
  criticisms of, 225–227
  exercises, 244
  population means, 210
  significance testing, 199–200
  single population mean, 205
  tailed alternatives, 205
  tailed tests, 203–205
  t-test, 208–209
  z-test, 206–208
Partial correlation coefficient, 347
Partial independence models, 534
partial.R2 function, 347
Partial regression coefficients, 335
Partial residual plots, 356
Partial sums of squares, *see* Type II sums of
      squares
Partial tables, 528; *see also* Contingency tables;
      Three-way tables
  exercise, 539
Pascal, Blaise, 7
PCA, *see* Principal component analysis (PCA)
pdf, *see* Probability density functions (pdf)
Pearson, Egon, 17
Pearson, Karl, 16, 17
Pearson residuals, 397
Pearson's correlation, 296, 318; *see also*
      Correlation; Population—covariance
  Anscombe's quartet, 301, 302
  association and independence, 297
  bivariate normal distribution, 299–300
  confidence interval for ρ, 306
  confidence limits, 306
  effect size, 307
  estimation of ρ, 300
  example, 298, 304–306, 307–308
  hypothesis tests for ρ, 302–306
  least squares estimator, 302
  Pearson's product moment correlation
      coefficient, 300–301
  power, 307
  sample size, 307
Pearson's product moment correlation
      coefficient, 300–301
Penalized least squares, 406
Penalty term, 373
Percentage bend criterion, 315–316
Percentile method, 176; *see also* Bootstrapping
perm.fact.test function, 494
Permutations, 36–37; *see also* Combinatorial
      analysis; Probability

exercises, 46
tests, 493
Permutation tests, 227; *see also* Parametric
    frequentist null hypothesis testing;
    Rank-based permutation tests
  advantage of, 227
  example, 227–228
  exercises, 244
  problem with, 227
PEV, *see* Proportion of explained variation
    (PEV)
Pivotal quantity, 162–163
  exercises, 195
Planned comparisons, *see* Contrasts
Plato, 6
plot.lm() function, 350
pmfs, *see* Probability mass functions (pmfs)
Point estimators, 108–109; *see also* Estimators
Poisson distribution, 66, 553–554; *see also*
    Binomial distribution; Discrete
    pdfs
  application, 67
  example, 67–69
  pdf, 66
Poisson GLM, 386, 395; *see also* Generalized
    linear models (GLMs)
  example, 395–396
  exercises, 418
  mean function, 395
  Poisson loglinear model, 395
Poisson log-linear model, 532; *see also* Log-linear
    models
Poisson, Siméon-Denis, 7, 66
Polymerase chain reactions, 248
Polynomial regression, 365; *see also* Regression
  advantages, 365
  disadvantages, 366
  example, 366–367
  models, 323
  second-order, 365
Pooled variance t-test, 214; *see also* Parametric
    frequentist null hypothesis testing
  assumptions, 215
  example, 215–217
  Fligner–Killeen test, 215
  test statistic, 214
Popper, Karl, 7, 9, 15
Population, 103; *see also* Parameter; Statistics
  bottleneck, 112
  coefficient of variation, 121
  correlation, 296–297
  covariance, 296, 297; *see also* Pearson's
    correlation

idealized, 112
median, 114
mode, 113
idealized, 112
Positive definite matrices, 549
Posterior odds ratio, 239
Power, 220; *see also* Hypothesis testing
  basic form, 221
  Cohens *d*, 224
  effect size in t-tests, 224
  example, 221, 223
  exercises, 241, 243
  rule, 544
  sample adequacy, 223
  transformation, 359
  ways to increase, 223
Precision, 5
predict function, 345, 390
Prediction interval, 344
Prediction sum of squares (PRESS), 375; *see also*
    Model selection
Premises, 6
PRESS, *see* Prediction sum of squares
    (PRESS)
Principal component analysis (PCA), 359
Priors, 138; *see also* Bayesian applications
Probability, 3, 8, 21, 43; *see also* Bayes
    rule; Combinatorial analysis;
    Probability density functions
    (pdf); Random variables; Set
    theory; Statistics
  of blood types O and B in U.S., 29
  Bonferroni's inequality, 31–32
  Boole's inequality, 30
  classical, 26
  conditional, 32
  convergence in, 25
  degrees of belief, 26
  disjoint, 28–30
  distribution, 49
  exercises, 43–47
  free falling velocity equation, 22
  frequentist interpretation of, 24, 25
  inclusion–exclusion principle, 30
  independence, 31, 33–34
  intersect, 29
  odds, 34
  odds ratio and relative risk, 34–35
  philosophical conceptions of, 24
  Simpson's diversity index and, 32
  union, 28
  of union, 505
  Venn diagrams, 27

Probability density functions (pdf), 49,
        92; *see also* Continuous pdfs;
        Cumulative distribution function
        (cdf); Density; Discrete pdfs;
        Normal distribution; Random
        variables
    beta, 81
    chi-squared, 75
    distributions, 65
    empirical cdfs, 90–91
    exercises, 95–100
    exponential, 79
    geometric, 71
    from goat sighting distribution, 54
    hypergeometric, 69
    logistic, 88
    lognormal, 86
    negative binomial distribution, 72
    Poisson, 66
    reference tables, 91–92, 93, 94–95
    selection, 90
    tabular expression of discrete, 53
    Weibull, 84
Probability distributions for tabular analyses,
        504–505
    Bernoulli distribution, 504
    binomial pdf, 504
    example, 504–505
    multinomial distribution, 504, 504, 538
    normal distribution, 504
    probability of union, 505
Probability mass functions (pmfs), 49
Probability values (P-values), 199
Probably appendicitis (PA), 47
Product rule, 544
    series rules, 543
Proportion of explained variation (PEV),
        453
Prospective studies, 272
pseudo.v function, 181–182; *see also* Jackknifing
        resampling
    exercises, 196
Psuedoreplication, 263–265; *see also* Sampling
        design
    exercises, 291
P-values, *see* Probability values
        (P-values)
pwr.anova.test function, 453
pwr.chisq.test function, 527
pwr.f2.test function, 348
pwr.r.test function, 307
pwr.2p2n.test function, 527
pwr.2p.test function, 527

Q

QR decomposition, 334
qtukey function, 438
Quadratic formula, 542
quantile function, 53, 307, 343, 439, 517; *see also*
        Cumulative distribution function (cdf)
Quantitative random variables, 49; *see also*
        Random variables
Quantitative variable, 249–250; *see also* Variables
Quasi-likelihood, 399

R

Radicals, 542
Random block effects, 469–470
Random effects model, 445, 446, 457; *see also*
        Fixed effects—linear model; Mixed
        effect models; Two-way factorial
        design
    assumptions, 457–458
    Cook's distance computation, 459
    diagnostic plots, 458
    example, 446–447, 451–453
    heteroscedasticity, 459
    hypothesis testing, 449
    intraclass correlation coefficient, 448
    likelihood ratio test, 449–450
    lme functions, 450–451
    lmer functions, 451
    marginal residual distribution, 459
    R functions, 450–451
    type I error, 458
    variance components, 447–448
    variance–covariance matrix, 448
Random factor levels characteristics, 445
Randomization, 254–255
Randomized block design, 466
    ANOVA table for, 468
    example, 468–469
    missing data, 467
    random block effects, 469–470
    Tukey's test for additivity, 467
Randomized complete block design (RCBD),
        281; *see also* ANOVA designs
    advantages, 281
    disadvantages, 281–282
    exercises, 289, 291
Random sampling, 254–255
Random variables, 21; *see also* Probability;
        Probability distribution; Variables
    continuous, 49
    discrete, 49

example, 22–23
exercises, 95, 97
expected value of, 103
histogram for simple, 51
MGF, 125
models for, 21–23
quantitative, 49
Ranged major axis regression (RMA), 382; *see also* Model II regression
Rank-based permutation tests, 228, 493–494; *see also* Kolmogorov–Smirnov test; Parametric frequentist null hypothesis testing; Permutation tests; Wilcoxon rank sum test; Wilcoxon sign rank test
advantages, 228
disadvantages, 229
example, 229
exercises, 244
Rayleigh distribution, 84
r.bw function, 316
RCBD, *see* Randomized complete block design (RCBD)
r.dist function, 303
Realism, 4
exercises, 18
Receiver operating characteristic curve (ROC curve), 393; *see also* GLM inferential methods
Reduced major axis regression, *see* Standardized major axis regression (SMA)
Reductio ad absurdum, 13–14; *see also* Logic
exercises, 19
Regression, 321, 413; *see also* ANOVA; Confidence interval; Generalized linear models (GLMs); Linear regression; Model selection; Model II regression; Multiple regression; Nonlinear regression models; Optimal transformation; Polynomial regression; Simple linear regression; Smoothers; Weighted least squares (WLS)
adjusted R^2, 347
algebraic rearrangement of regression model, 342
analysis, 503, 321
Bayesian approaches, 411–412
bootstrapping methods, 378–379
coefficient of determination, 346, 347
confidence and prediction intervals, 343
effect size, 348
example, 321, 347, 348

exercises, 414–419
fitted and predicted values, 341–343
likelihood and general linear models, 369–371
linear regression model, 322–323
model slope comparison, 368–369
partial correlation coefficient, 347
power, 348
robust, 378
robust estimators, 379–380
sample size, 348
simple transformations, 360
transformation approaches, 359
Y-intercept, 364
Regression designs, 275; *see also* Experimental design
Regression sum of squares (SSR), 423
regsubsets function, 377
Reichenbach, Hans, 7
Relative risk, 34; *see also* Probability
REML, *see* Restricted maximum likelihood (REML)
Repeated measures ANOV, *see* Analysis of contrasts
Repeated measures design, 284; *see also* Analysis of contrasts; ANOVA designs; Modern mixed-model approaches
advantages, 285
between-subject effect, 478
carryover effects, 479
disadvantage, 285
order effects, 479
sphericity, 479
split plot in time, 478
within-subject effect, 478
Replication, 255
Resampling distributions, 172, 193; *see also* Bootstrapping; Confidence intervals; Jackknifing resampling; Sampling distributions
Research hypothesis, 253–254
exercises, 288
Residuals, 325
Residual sum of squares, *see* Sum of squares error (SSE)
Response variables, 248–249; *see also* Variables
exercises, 288
Restricted maximum likelihood (REML), 134, 448
Retrospective studies, 272
Reverse Helmert contrasts, 433
R functions, 450–451
Riechenbach, Hans, 8

Riemann, Bernhard, 51
Riemann sum, 51
rlm function, 379, 494, 495
RMA, *see* Ranged major axis regression (RMA)
Robust ANOVA, 492; *see also* Multiway ANOVA;
　　One-way ANOVA
　exercise, 501
　permutation tests, 493
　robust estimator tests, 493
Robust estimator tests, 237, 494, 495; *see also*
　　Parametric frequentist null hypothesis
　　testing
ROC curve, *see* Receiver operating
　　characteristic curve (ROC curve)
Rotary screw trap, 248
Rothamsted, 421
r.pb function, 315–316
R update function, 470

S

samp.dist function, 151; *see also* Sampling
　　distributions
Sample
　median, 114
　moments, 126
　size, 453, 454
　space, 24; *see also* Set theory
Sample variance, 119; *see also* Scale estimators
　deviations, 119
　example, 121
　proof, 119–120
　sample standard deviation, 121
　sum of squares, 119
Sampling design, 256, 287; *see also* Experimental
　　design
　adjustments for, 258, 259–261
　altering datasets, 269
　bias, 266
　cluster sampling, 256
　comparison of designs, 257
　demonstration of, 257
　exercises, 289, 291
　finite population correction, 258–259
　lack of independence in samples, 262
　measurement error and precision, 266
　missing data and nonresponse bias, 266–267
　outliers, 265–266
　psuedoreplication, 263–265
　randomized designs, 256
　random sampling, 256, 261
　sampling concerns, 265
　systematic sampling, 256

time series models, 262–263
　transforming data, 267–269
　true population estimation, 260
　true strata mean estimation, 260
　unbiased estimator for standard error, 260
　variance estimation, 260
Sampling distributions, 150, 193; *see also*
　　Confidence intervals; Resampling
　　distributions; Test statistics
　of arithmetic means, 151
　central limit theorem, 154–156
　example, 153, 156
　exercises, 193, 194
　of F-distribution, 158–160
　principles of, 152
　samp.dist function, 151
　of sample variance, 156
　standard deviation of, 151
　of t-distribution, 156–158
Sampling error, 14, 108, 266
　exercises, 291
Sampling units, *see* Experimental units
Satterthwaite adjusted t-test, 217
Scalars, 546
Scale estimators, 119; *see also* Coefficient of
　　variation; Estimator; Sample variance;
　　Statistics
　interquartile range, 123
　median absolute deviation, 123
　robust estimators, 122
　selection, 124
Scale–location plot, 353
Scale parameters, 62
Scatterplots, 107
Scheffé's procedure, 437–438, 440
Schlick, Moritz, 15
Schwarz criterion, *see* Bayesian information
　　criterion (BIC)
Science, 3, 18; *see also* Logic; Scientific
　　hypotheses; Scientific method;
　　Scientific principles; Statistics
　exercises, 18–19
　laws of physics, 15
　nature of, 3–4
　and statistics, 16–18
　subset, 4
　variability and uncertainty, 14
Scientific hypotheses, 9; *see also* Science
　exercises, 18
　falsifiability, 9–10
Scientific method, 5; *see also* Science
　experimentation, 8
　induction, 8

modern developments in, 7
probability, 8
steps, 6
terse history, 6
Scientific principles, 4; *see also* Science
accuracy and precision, 5
communalism, 5
exercises, 18
objectivity, 4
realism, 4
Scientism, 15
Score test, 506, 507; *see also* Likelihood ratio;
Wald test
for all $\pi i = \pi i0$, 508
$\chi 2$ goodness of fit test, 508
for $\pi = \pi 0$, 507–508
for $\pi 1 = \pi 2 =...= \pi c$, 508–509
null standard error, 507
Yates' continuity correction, 509
Second-order "corrected" Akaike information
criterion (AICc), 374; *see also* Model
selection
exercises, 417
see.nlm function, 401
see.r.dist.tck function, 303
see.smooth.tck function, 408
Selection ratio, 516
selfStart function, 402
Sensitivity, 392
S estimation, 379
Set notation, 550
Set theory, 23, 549; *see also* Probability
Bayes' theorem, 549
conditional probability, 549
elements, 23
empty set, 24
event, 24
experiment, 24
independence, 550
intersections, 549
outcome, 24
sample space, 24
set notation, 550
trial, 24
unions, 549
s function, 409
SIC criterion, *see* Bayesian information criterion
(BIC)
SIDS, *see* Sudden infant death syndrome (SIDS)
Significance testing, 199–200; *see also* Parametric
frequentist null hypothesis testing
exercises, 241
Simian immunodeficiency virus (SIV), 198

Simple linear regression, 275, 324; *see also*
Experimental design; Regression
ANOVA approach, 331–333
estimated regression model, 325
example, 327–329, 330–331, 332–333
exercises, 291, 414
hypothesis testing, 329–331
MSE, 326–327
OLS estimators, 325
parameter estimation, 325–329
population regression line, 349
SSE, 326
SSTO, 331
standard errors, 327
Simple random sampling, 256; *see also* Sampling
design
exercises, 289
Simple whole-plot design, 474
Simpson's paradox, 287
Simultaneous inference procedures, 436
Bonferroni correction, 437, 440
comparing, 439–441
Dunnett's method, 439, 441
familywise type I error rates, 436, 441
least significant difference, 437
P-value adjustment, 442
Scheffé's procedure, 437–438, 440
Tukey–Kramer method, 438, 442
Single binomial proportion comparison, 506
Single factor random effects model, 445
SIV, *see* Simian immunodeficiency virus (SIV)
Skewness, 124; *see also* Distribution shape
SMA, *see* Standardized major axis regression
(SMA)
Smoothers, 403, 413; *see also* Generalized
additive models (GAMs); Regression;
Spline smoother
advantages, 403
drawbacks, 403
example, 406–408
kernel-based approaches, 404–405
LOWESS, 404
Snedecor, George W., 17, 78
Snedecor's F-distribution, *see* F-distribution
Span, 404
Spatial autocorrelation, 262; *see also* Sampling
design
spearman_test function, 311
Spearman's estimator, 308; *see also* Correlation
Spearman's rank correlation coefficient, 308
exercises, 319
Special products, 541
Specificity, 392

Sphericity, 284–286, 479; *see also* ANOVA
          designs; Repeated measures design
     exercises, 291
Spline smoother, 405; *see also* Smoothers
     concern of, 406
     cubic spline, 405
     exercises, 419
     general linear model, 405
     penalized least squares, 406
Split-plot design, 283, 474; *see also* ANOVA
          designs
     advantages, 283
     ANOVA table for balanced two-factor, 476
     aov function, 475–476
     blocking factor, 474
     disadvantages, 283–284
     example, 476–477
     exercises, 292, 500
     simple whole-plot design, 474
Square matrices, 546
SSE, *see* Sum of squares error (SSE)
SSR, *see* Regression sum of squares (SSR)
SSTO, *see* Total sum of squares (SSTO)
SSTR, *see* Treatment sum of squares (SSTR)
Standard deviations, 62, 105
     of bootstrap statistics, 173
     of sampling distribution, 151
Standard error, 151; *see also* Sampling
          distributions; Standard deviation
Standardization, 267–268
Standardized major axis regression (SMA), 382;
          see also Model II regression
     confidence intervals, 384
     estimate for β_1, 382
     exercises, 418
     test statistic, 383
Statistical expectations, 550–551
Statistical hypothesis, 253
     exercises, 288
Statistics, 3, 16, 101; *see also* Distribution shape;
          Location measures; Parameter; Scale
          estimators; Science
     and biology, 16
     considerations, 106–108
     estimator types, 108
     exercises, 19
     gauging estimator effectiveness, 108
     goals of inferential, 247
     hagiography of statisticians, 17
     inferential procedures, 102
     linear transformations, 136–137
     nonlinear transformations, 268
     population, 103

sampling error, 108
Stemplots, 107
Stepwise regression, 377–378; *see also* Model
          selection
Stochastic processes, 15
Strata, 256
Stratified random sampling, 256; *see also*
          Sampling design
     exercises, 289
Studentized bootstrap, 175; *see also*
          Bootstrapping
Subjects, *see* Experimental units
Sudden infant death syndrome (SIDS), 33
Sum of squares, 119
Sum of squares error (SSE), 326, 423, 424
     exercise, 416, 501
Surber sampler, 248
Symmetric matrices, 546
Systematic sampling, 256
Systolic blood pressure, 206

T

Tabular analyses, 503, 537–538; *see also* Binomial
          parameter confidence intervals;
          Contingency tables; One-way formats;
          Generalized linear models (GLMs);
          Probability distributions for tabular
          analyses; Three-way tables; Two-way
          tables
     effect size, 526–527
     exercises, 538–540
     ordinal variables, 525–526
     response and predictors, 503
Tabular designs, 286; *see also* Experimental
          design
     advantage of, 286
     contingency table, 286
     disadvantage, 286–287
     exercises, 292
     Simpson's paradox, 287
     three, 287
t-distribution, 75, 76–77, 156–158; *see also*
          Continuous pdfs; Test statistics
     exercises, 99
Temporal autocorrelation, 262; *see also* Sampling
          design
Test of homogeneity, 521
Test statistics, 156, 199; *see also* Estimators;
          Sampling distributions
     calculation, 156
     example, 158, 160
     exercises, 241, 242

F-distribution, 158–160
t-distribution, 156–158
t function, 547
Thales of Miletus, 6
Theories, 9
Theory of island biogeography, 415
Three-way tables, 527, 533–537; *see also* Tabular analyses; Two-way tables
   conditional associations, 528
   degrees of freedom and independence frameworks for, 535
   example, 528–530, 535–537
   homogenous association model, 534
   independence model, 533–534
   joint independence, 534
   marginal tables, 528
   methods comparison, 530
   partial tables, 528, 539
Time series models, 262–263; *see also* Sampling design
Total sum of squares (SSTO), 331, 423
t-quantile function, 437
Translation, 267
Transposition, 547
Treatment contrasts, 432–433
Treatment sum of squares (SSTR), 423
Trial, 24; *see also* Set theory
Trimmed mean, 115; *see also* Location measures
   exercises, 144
trim.ranef.test function, 494
T-test, 208–209
Tukey–Kramer method, 438, 442
Tukey's honest significant difference (HSD), 436, 438
Tukey's test for additivity, 467
Two-way ANOVAs interactions potential, 459
Two-way factorial design, 459
   analytical operations for, 460, 461
   balanced, 462
   example, 462–463, 464–466
   exercise, 500
   expected mean squares for, 464
   interaction mean square, 463
   linear model, 461
   moments variance estimators, 463, 464
   random effects, 463–466
   effect of random factor-level selection on inferences, 465
Two-way tables, 520; *see also* Three-way tables
   Chi-squared test, 521
   example, 523–524
   Fisher's exact test, 523
   likelihood ratio test, 521–522

   method comparison for valid inference, 524–525
   test of homogeneity, 521
Type I error, 458
Type I SS, *see* Type I sums of squares (Type I SS)
Type I sums of squares (Type I SS), 339
Type II SS, *see* Type II sums of squares (Type II SS)
Type II sums of squares (Type II SS), 339, 488, 489
Type III sums of squares, 339, 488, 489

U

Unbalanced designs, 488
   example, 490–492
   sums of squares, 488–490
Union, 28, 549; *see also* Probability
Univariate analysis, 250
Universal set, *see* Sample space; Set theory
Unplanned comparisons, 429
Unstructured covariance, 487
update function, 347, 465
Usual splitplot design, 474

V

Variables, 21, 247, 287; *see also* Random variable
   categorical, 249
   confounding, 252–253
   discrete and continuous, 250
   exercises, 288
   explanatory and response, 248–249
   lurking, 251–252
   ordinal, 249
   quantitative, 249–250
   univariate and multivariate analysis, 250–251
Variance, 62, 105; *see also* Parameter
   components, 447–448
   –covariance matrix, 448
   exercises, 143
   function, 399, 551, 553
Variance inflation factors (VIFs), 358–359
   exercises, 417
vark function, 314
Vectors, 546
Velocity, 111
   of falling body, 21
Venn diagrams, 27
   exercises, 44
   for flower color outcomes, 28
   for herbivore problem, 29
Venn, John, 27
VIFs, *see* Variance inflation factors (VIFs)

W

Wald test, 391, 506, 509; *see also* GLM inferential methods; Likelihood ratio; Score test in one-way
  binomial parameter confidence intervals, 513
  exact binomial tests, 512–513
  for $\pi = \pi 0$, 509–510
  for $\pi 1 = \pi 2$, 510–511, 513
  reliability, 512
Weibull distribution, 75, 83; *see also* Continuous pdfs
  applications, 84–85
  cdf, 84
  example, 85
  exercises, 100
  pdf, 84
Weibull, Waloddi, 83
Weighted least squares (WLS), 321, 353; *see also* Regression
  example, 365
  exercises, 417
  maximum likelihood estimator, 364
  variance–covariance matrix for, 364
weighting function, 379
weights argument, 324
Welch's approximate t-test, 217; *see also* Parametric frequentist null hypothesis testing
  assumptions, 217
  example, 217–219
  test statistic, 217
Wilcoxon rank sum test, 233; *see also* Parametric frequentist null hypothesis testing; Rank-based permutation tests
  assumptions, 235
  example, 235–237
  exercises, 243
  population mean of W, 234
  test statistic calculation, 234

Wilcoxon sign rank test, 231; *see also* Parametric frequentist null hypothesis testing; Rank-based permutation tests
  alternative hypotheses, 231
  assumptions, 232
  example, 233
  lower and upper-tailed alternatives, 231
  population mean of V, 232
  population variance, 232
  test statistic calculation, 232
  two-tailed null hypothesis, 231
William of Occam, 371
win function, 315
Winsor, Charles P., 115
Winsorized correlation, 315; *see also* Correlation
  exercises, 320
Winsorized mean, 115–116; *see also* Location measures
  exercises, 144
Within-subject effect, 478
Wittgenstein, Ludwig, 15
WLS, *see* Weighted least squares (WLS)
Wright, Sewall, 112

X

xtabs function, 526, 529

Y

Yates' continuity correction, 509
Yihui Xie's package animation, 102

Z

Z-distribution, 63; *see also* Normal distribution
  example, 63–65
  exercises, 97
Z-score, 63
Z-test, 206–208